Dedication to Professor Agustín Costa-García: Screen-Printed Electrodes-Based (Bio)sensors: Development and New Challenges of the 21st Century

Dedication to Professor Agustín Costa-García: Screen-Printed Electrodes-Based (Bio)sensors: Development and New Challenges of the 21st Century

Editor

Elisa González-Romero

MDPI • Basel • Beijing • Wuhan • Barcelona • Belgrade • Manchester • Tokyo • Cluj • Tianjin

MDPI

Editor
Elisa González-Romero
Analytical and Food Chemistry
University of Vigo
Vigo
Spain

Editorial Office
MDPI
St. Alban-Anlage 66
4052 Basel, Switzerland

This is a reprint of articles from the Special Issue published online in the open access journal *Biosensors* (ISSN 2079-6374) (available at: www.mdpi.com/journal/biosensors/special_issues/SPE_biosensors).

For citation purposes, cite each article independently as indicated on the article page online and as indicated below:

LastName, A.A.; LastName, B.B.; LastName, C.C. Article Title. *Journal Name* **Year**, *Volume Number*, Page Range.

ISBN 978-3-0365-1322-5 (Hbk)
ISBN 978-3-0365-1321-8 (PDF)

Contents

About the Editor vii

Preface to "Dedication to Professor Agustín Costa-García: Screen-Printed Electrodes-Based (Bio)sensors: Development and New Challenges of the 21st Century" ix

Alberto Sánchez-Calvo, Maria Carmen Blanco-López and Agustín Costa-García
Paper-Based Working Electrodes Coated with Mercury or Bismuth Films for Heavy Metals Determination
Reprinted from: *Biosensors* **2020**, *10*, 52, doi:10.3390/bios10050052 1

Beatriz Pérez-Fernández, Agustín Costa-García and Alfredo de la Escosura- Muñiz
Electrochemical (Bio)Sensors for Pesticides Detection Using Screen-Printed Electrodes
Reprinted from: *Biosensors* **2020**, *10*, 32, doi:10.3390/bios10040032 15

Daniel Antuña-Jiménez, María Begoña González-García, David Hernández-Santos and Pablo Fanjul-Bolado
Screen-Printed Electrodes Modified with Metal Nanoparticles for Small Molecule Sensing
Reprinted from: *Biosensors* **2020**, *10*, 9, doi:10.3390/bios10020009 41

Ricarda Torre, Estefanía Costa-Rama, Henri P. A. Nouws and Cristina Delerue-Matos
Screen-Printed Electrode-Based Sensors for Food Spoilage Control: Bacteria and Biogenic Amines Detection
Reprinted from: *Biosensors* **2020**, *10*, 139, doi:10.3390/bios10100139 63

Paloma Yáñez-Sedeño, Susana Campuzano and José Manuel Pingarrón
Screen-Printed Electrodes: Promising Paper and Wearable Transducers for (Bio)Sensing
Reprinted from: *Biosensors* **2020**, *10*, 76, doi:10.3390/bios10070076 89

Estefanía Costa-Rama and María Teresa Fernández-Abedul
Paper-Based Screen-Printed Electrodes: A New Generation of Low-Cost Electroanalytical Platforms
Reprinted from: *Biosensors* **2021**, *11*, 51, doi:10.3390/bios11020051 119

Samuel J. Rowley-Neale, Dale A. C. Brownson, Graham Smith and Craig E. Banks
Graphene Oxide Bulk-Modified Screen-Printed Electrodes Provide Beneficial Electroanalytical Sensing Capabilities
Reprinted from: *Biosensors* **2020**, *10*, 27, doi:10.3390/bios10030027 143

Lasangi Dhanapala, Colleen E. Krause, Abby L. Jones and James F. Rusling
Printed Electrodes in Microfluidic Arrays for Cancer Biomarker Protein Detection
Reprinted from: *Biosensors* **2020**, *10*, 115, doi:10.3390/bios10090115 153

Ramón Lorenzo-Gómez, Daniel González-Robles, Rebeca Miranda-Castro, Noemí de-los-Santos-Álvarez and María Jesús Lobo-Castañón
On the Electrochemical Detection of Alpha-Fetoprotein Using Aptamers: DNA Isothermal Amplification Strategies to Improve the Performance of Weak Aptamers
Reprinted from: *Biosensors* **2020**, *10*, 46, doi:10.3390/bios10050046 181

About the Editor

Elisa González-Romero

Dr. Elisa González-Romero is a professor of analytical chemistry at the University of Vigo since 1991. She was an executive committee member of Real Sociedad de Química de España, RSQE-STGallega (2002–2020). She is the leader and founder of the research group "Electroanalysis and Biosensors". Her main research interests are in health and the environment. Her research applies analytical chemistry methodology, typically electro-analytical techniques, and biochemistry approaches to problem solving in food, clinical, and environmental analyses using biosensors and chemical sensors devices. She has co-authored more than 80 scientistic publications, 1 international patent, the design of several device, and several book chapters. She has presented more than 10 invited lectures in the most prestigious conferences on analytical nanotechnologies. She has also supervised five PhD students and several postdoctoral researchers.

Preface to "Dedication to Professor Agustín Costa-García: Screen-Printed Electrodes-Based (Bio)sensors: Development and New Challenges of the 21st Century"

Dedicated to Professor Agustín Costa-García (Meres, 1949-2019), an impressive chemist who knew how to look to the future and who contributed so much to science. Thanks Agustín!

In this book, the reader will find answers to the following questions and much more: What are the new trends in the design of (bio)sensor devices? Which are the key components? What are the critical points in the fabrication of these devices? What types of ink are useful for the fabrication of SPEs? What is the effect of material impurities on the analytical response? What is it that society demands?

Many thanks to all of the authors who participated in this project, making it an elegant and excellent book.

Elisa González-Romero
Editor

biosensors

MDPI

Article

Paper-Based Working Electrodes Coated with Mercury or Bismuth Films for Heavy Metals Determination

Alberto Sánchez-Calvo, Maria Carmen Blanco-López * and Agustín Costa-García †

Departamento de Química Física y Analítica, Facultad de Química, Universidad de Oviedo, 33006 Oviedo, Spain; albertosc14@hotmail.com (A.S.-C.); costa@uniovi.es (A.C.-G.)

* Correspondence: cblanco@uniovi.es

† In Memoriam to Prof. Agustín Costa-García.

Received: 27 March 2020; Accepted: 8 May 2020; Published: 13 May 2020

Abstract: Paper-based carbon working electrodes were modified with mercury or bismuth films for the determination of trace metals in aqueous solutions. Both modification procedures were optimized in terms of selectivity and sensitivity for the determination of different heavy metals, aiming their simultaneous determination. Cd (II), Pb (II) and In (III) could be quantified with both films. However, Cu (II) could not be determined with bismuth films. The modification with mercury films led to the most sensitive method, with linear ranges between 0.1 and 10 μg/mL and limits of detection of 0.4, 0.1, 0.04 and 0.2 μg/mL for Cd (II), Pb (II), In (III) and Cu (II), respectively. Nevertheless, the bismuth film was a more sustainable alternative to mercury. Tap-water samples were analyzed for the determination of metals by standard addition methodology with good accuracy, by using a low-cost and easily disposable paper-based electrochemical platform. This system demonstrated its usefulness for monitoring heavy metals in water.

Keywords: paper electrodes; bismuth films; mercury films; low-cost analysis; heavy metal determination

1. Introduction

Electrochemistry research is looking at new materials and surfaces for sustainable technological applications in microelectronics, energy or in the development of sensors. In this field, surface characteristics of the electrodes are important for electron transfer steps or non-faradaic interactions, like adsorption or ion-pair formation [1,2]. Mercury electrodes have been widely used for several decades because of their large cathodic window, reproducibility and low background [3,4]. However, mercury is a dangerous heavy metal because of its toxicity and bio-accumulation in many species. This has triggered the search for less-toxic alternatives [5–7]. Nowadays, carbon is one of the most advantageous materials for low-cost and flexible-design electrodes. Glassy carbon, carbon paste or screen-printed carbon electrodes (SPCE) are easy to fabricate and to modify for sensor development [8–10]. With the use of carbon electrodes, advantageous properties of mercury surfaces, such as interactions with thiol peptides or potential windows down −2.5 V (vs. Ag/AgCl), are restricted [2,11–14]. However, mercury can be electrochemically deposited on the conductive carbon surface, forming a thin film which can have the same applications as conventional mercury electrodes [15–17], but significantly reducing the required amount. Determination of trace metals by stripping voltammetry is one of the most developed applications on this surface because of the high affinity of mercury for metals [11,17–19]. The procedure followed for determining heavy metals with stripping techniques involves the preconcentration of metal species on a solid electrode surface, followed by a selective oxidation of each metal during an anodic potential sweep. The peak current, being proportional to the concentration of the heavy metal in the solution, is recorded.

Nowadays, bismuth films are being used as an alternative to mercury films, with the objective of using a more environmentally friendly platform [20–25]. Bismuth has very low toxicity and possesses suitable electrochemical properties, such as a wide range of operating negative potentials, low background currents and the ability to electrodeposit elements on its surface by formation of intermetallic compounds or fused alloys. It has also a high-quality stripping performance, and therefore is good for the replacement of mercury at the determination of heavy metals by stripping voltammetry [23,26]. Likewise, bismuth films can also be formed by different ways on carbon substrates, mainly by applying a negative potential in situ, (i.e., bismuth is in the sample solution together with the analytes) [27–32], or ex situ by previous deposition, in a solution different from this of the analyte [33–36]. Recently, heavy metals determination by using bismuth nanoparticles was documented [37,38].

Bulk bismuth is the semimetal with the smallest electron effective mass [39]. This confers advantages like more pronounced quantum confinement effects becoming a good option in the form of a nanostructure. There are several chemical ways to form nanoparticles [40], like pyrolysis of bismuth [41], adding reducing agents or photochemical activation [42]. Reductive methods like adding sodium borohydride to bulk bismuth allow to control nanoparticles in terms of size and shape [43]. However, bismuth films can be generated electrochemically, with the advantage of faster in situ modification, employing less reagents.

Bismuth and mercury films have been successfully formed on glassy carbon, carbon paste and screen-printed carbon electrodes (SPCEs) [11,31,44–49], for the determination of heavy metals such as Cd (II), Pb (II) or Cu (II), and they were also successfully applied to the determination of Tl (I) [50]. Recent trends in analytical chemistry encourage the development of simpler alternatives that could also be used in developing countries. Conventional glassy carbon and carbon paste electrodes are not very appropriate for decentralized analysis, and although screen-printed electrodes are an interesting alternative, very often they are developed on ceramic substrates. Searching for low-cost and easy-to-dispose materials is then of enormous interest.

The use of paper as substrate for sensors is considered a good option for electrochemical measurements [51–55]. It is composed of cellulose fibers which confer properties like roughness and hydrophilicity, becoming easy to handle, with the possibility to form three-dimensional sensors [40–43]. Paper can also be modified by the addition of (i) hydrophobic materials, like wax, to form barriers that allow customization of the electrochemical cell, (ii) appropriate inks to have a conductive surface or (iii) nanostructures to improve selectivity and/or sensitivity [56–60]. Research literature on the determination of heavy metals by using paper electrodes is scarce. Reports include colorimetric or electrochemical determinations [58,61,62]. Films (e.g., bismuth) generated on paper can also be modified with nanomaterials, to improve the sensitivity of the methodology [63].

In this work, we studied the ex situ deposition of bismuth films on a working paper-based electrode developed in previous works [59,64,65] and compared the performance with that obtained with more traditional mercury films.

These electrodes were tested with Cd (II), Pb (II), In (III) and Cu (II) aqueous solutions. The working paper electrode prepared was placed over the working electrode of a screen-printed carbon card, having also an auxiliary and reference electrode which can be reutilized without any interference between measurements. On this platform, the bismuth and mercury films were electrodeposited from the corresponding salts by applying a negative potential. After that, heavy metals previously preconcentrated were analyzed by anodic stripping voltammetry. With this method, the paper electrode used can be removed, allowing its use for other analysis without any contamination from the last measurement. Bismuth films were chosen as an environmentally friendly alternative to mercury. Both films were compared in terms of sensitivity and selectivity. The use of these kind of films allows the development of sensors for heavy metals, by using paper-based carbon electrodes and low volumes of reagents for the formation of films or the sample analysis. This results on great economical savings and lower toxic waste generated. The electrochemical detection allows the characterization of heavy metals by using only the previously formed film, instead of using different reagents for every heavy metal, as most colorimetric sensors do [66]. This also lowers the economic cost. Besides, the small size of these foldable paper electrodes makes them portable and suitable for easy waste treatment. Paper acts as storage, concentrating heavy metals in a porous solid substrate.

2. Materials and Methods

2.1. Reagents and Materials

Hg (II) acetate, sodium sulfate and Cu (II) nitrate trihydrate were purchased from Sigma-Aldrich (USA). Cd (II) standard solution (1000 µg/mL) was acquired from Merck KGaA (Germany) (https://www.merckgroup.com/en). Standard solutions of lead and bismuth for ICP were acquired from Fluka Analytical (https://www.lab-honeywell.com/products/brands/fluka/). In (III) chloride was acquired from Alfa Aesar (Germany) (https://www.alfa.com/es/). Ultrapure water (18.2 MΩ) was directly taken from a Millipore Direct-Q® 3 UV purification system from Millipore Ibérica (Spain) (https://www.tecnoaqua.es/empresas/millipore-iberica-sa). N,N-dimethylformamide anhydrous (DMF) was acquired from Sigma-Aldrich (https://www.sigmaaldrich.com).

A 0.1 M acetate buffer solution was prepared by mixing 0.1 M acetic acid with 0.5 M sodium sulphate as background electrolyte and adjusting the pH with sodium hydroxide, until pH 4.0. A 10^{-3} M bismuth solution was prepared in the acetate buffer solution, whereas a 10^{-3} M solution of mercury was prepared by solving the corresponding amount of mercury (II) acetate in 0.1 M HCl. All the solutions of heavy metals were prepared in 0.1 M acetate buffer pH 4, 0.5 M in sodium sulphate (employed as background electrolyte).

Carbon paste (ref. C10903P14) was acquired from Gwent group (United Kingdom) (http://www.gwent.org/). A wax printer (Xerox Colorqube 8570) was used to wax-print the paper. A thermostat model (Nabertherm d-2804) was used to melt the wax. An ultrasonic bath sonicator (Elmasonic P) was employed to homogenize carbon ink solutions. The spray adhesive 3M Spray Mount™ was acquired from a local store.

Screen-printed electrode cards were obtained from Metrohm.Dropsens (http://www.dropsens.com/) (SPCEs, ref. DRP-110, Spain). Working and auxiliary electrodes were made of carbon ink, and the pseudo-reference electrode was silver. They were connected to the potentiostat by a DSC connector (ref. DRP-DSC) from the same company. Electrochemical measurements were carried out with a potentiostat (Autolab, PGSTAT 10) controlled by the Autolab GPES software.

2.2. Fabrication of Paper-Based Electrodes

Paper-based working electrodes were prepared by a procedure previously developed in our group [64]. Hydrophobic wax patterns designed by Inkscape software were printed on chromatography paper Whatman Grade 1 as cellulose substrate. A temperature of 80 °C was applied on the paper until wax was melted, followed by cooling at room temperature.

The next step was the modification of the paper by the addition of 2 µL of a carbon ink suspension by drop casting on one of the sides ("bottom side"). The measuring solutions were added by the other side ("upper side"). The carbon ink suspensions used to modify the chromatography paper were made by dilution of commercial carbon paste in anhydrous N,N-dimethylformamide (DMF) to a final concentration of 40% (w/w). Homogenization was performed by sonication for 1 h.

The whole process of preparation of the paper platform is shown in Figure 1. Each cellulosic electrode was cut to obtain a circle with a diameter of 4 mm. The area was wide enough to cover the working electrode (WE) of a screen-printed platform, as well as the ceramic surface between the WE and the auxiliary (AE) and reference (RE) electrodes. The area of the cellulosic working electrode was covered with a protective polymeric circular film before spraying with an adhesive. Then, after removing the protective film, the working paper-based electrode was placed over the WE of a screen-printed carbon electrode card (SPCE). Following this procedure, the carbon ink in the paper and that of the WE of the SPCE were in contact. Moreover, there was no interference from the adhesive spray, which was only spread on the crown surrounding the protector. Thus, the WE of the SPCE acted just as a connection. Before performing the measurements, it was verified that there was full overlap, and also that the solution added on the surface of the paper WE did not wet the WE of the SPCE. By ensuring this, the SPCE could be reused, without any interference from previous measurements.

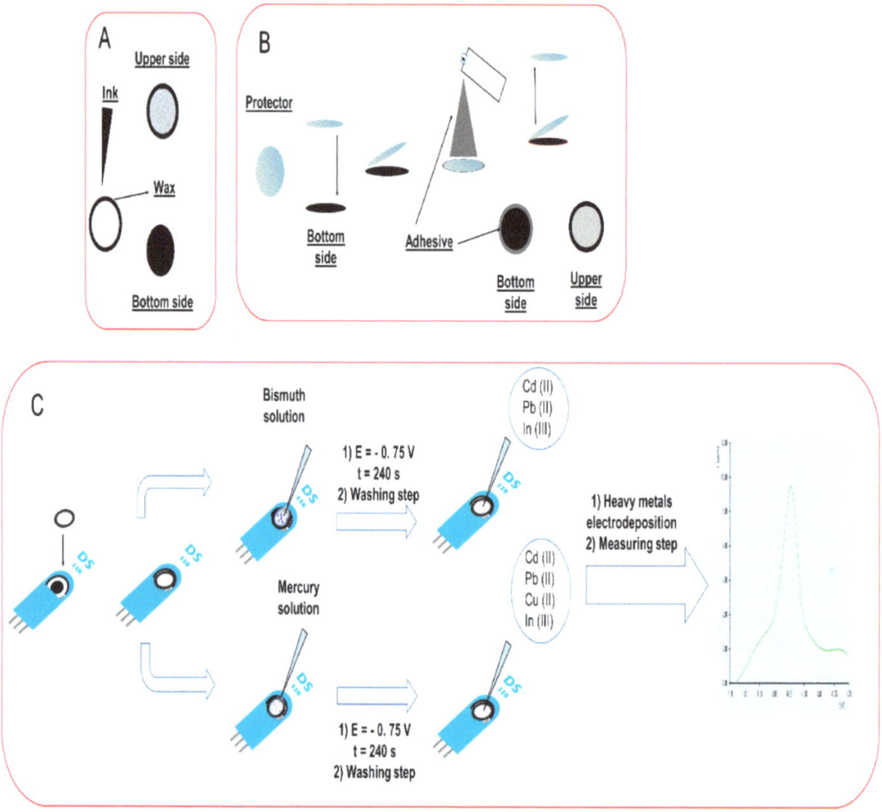

Figure 1. Schematic diagram showing the preparation of the paper-based electrodes. (**A**) Modification of cellulose substrate by addition of carbon ink on one side, having the platform with two visually different sides. Solution was added with a micropipette through the upper side. (**B**) Adhesive spraying on the protected working electrode. (**C**) Overlapping process of the bottom side of the paper electrode (the one with the ink) on the screen-printed card and modification with mercury (II) acetate or bismuth (III) standard solutions.

2.3. Modification of Paper Electrodes: Deposition of Mercury and Bismuth Films

Mercury and bismuth films were generated in the paper electrode by using standard solutions of 10^{-3} M Bi (III) in 0.1 M acetate buffer (with 0.5 M sodium sulphate as background electrolyte) pH 4, or 10^{-3} M Hg (II) in 0.1 M HCl. A potential of - 0.75 V was applied in both cases. Then, the paper electrode was washed with Milli-Q water and was ready for the measurements.

2.4. Electrochemical Procedures

Linear sweep voltammetry was used to characterize the redox processes of each heavy metal (Cu (II), In (III), Pb (II) and Cd (II)) and to optimize the formation of the film on the cellulose-based electrode.

A potential of −1.2 V was applied to preconcentrate the analytes before the measurement. Then, the reduced analyte was stripped away according to the next parameters: initial potential (E_i) = −1.2 V, final potential (E_f) = 0 V, step potential (E_s) = 2 mV and scan rate (v) = 50 mV/s.

All heavy metals' measurements were carried out in 0.1 M acetate buffer pH 4, 0.5 M in sodium sulphate, employed as background electrolyte.

Relative standard deviations (RSDs) were calculated for all analytes at concentrations located in the middle of the linear range of each analyte. Different paper electrodes were tested (n = 3), using different SPCEs for different analytes.

2.5. Water Analysis

Samples of tap water were collected in our laboratory for heavy metal determination. They were spiked with standards of Cd (II) and Pb (II) before their analysis. The solutions were analyzed by anodic stripping linear sweep voltammetry, but the parameters for the modification and measurement steps in water samples were different depending on the type of film used.

For the determination at paper-based electrodes with mercury films, water samples were spiked with different volumes of 10 µg/mL of Cd (II) and Pb (II). The spiked samples were diluted 1:10 in 0.1 M acetate buffer, pH 4, with 0.5 M sodium sulphate used as background electrolyte. Metals were preconcentrated on paper electrodes modified with mercury films, by applying a potential of −1.2 V for 600 s. Then, deposited heavy metals were stripped away by linear sweep voltammetry by applying the following conditions: E_i = −1.2 V, E_f = 0 V, E_s = 2 mV and v = 50 mV/s.

On the other hand, samples of tap water collected to be quantified by using paper-based electrodes modified with bismuth films were spiked with different volumes of a standard solution of 25 µg/mL of Zn (II), Cd (II) and Pb (II). The spiked tap-water sample was 1:10 diluted in 0.1 M acetate buffer, pH 4, with 0.5 M sodium sulphate. A preconcentration potential of -1.3 V was applied for 600 s according to a procedure previously published [36]. Afterward, deposited metals were stripped anodically by sweeping the potential linearly from −1.3 to −0.6 V at 50 mV/s, with 2 mV of step potential. Measurements were recorded in triplicate.

3. Results and Discussions

3.1. Electrochemical Characterization of Mercury/Bismuth Films at Paper Electrodes

The mercury or bismuth films electrochemically formed on the carbon ink deposited at the cellulose matrix of paper electrodes were characterized by linear sweep voltammetry (LSV). The optimization of the film formation method was done by measuring the intensity of the oxidation current of a Cd (II) solution deposited on the paper electrode after the film formation. Cd (II) solution was chosen because, among the heavy metals tested, this was the one with the most negative reduction potential. A negative potential of −1.2 V was used to preconcentrate this metal for different deposition times, varying from 60 to 300 s. Then, the reduced Cd (0) was stripped away by scanning the potential anodically. Regarding the film generation, the application of −0.75 V for 240 s on an aliquot of 40 µL of 10^{-3} M of bismuth or mercury standard solutions was the option that produced the most precise signals with highest intensities for both mercury and bismuth films. Higher deposition times did not increase the intensity of the oxidation currents. Therefore, this time was used for the film deposition at the analysis of the heavy metals in water samples.

3.2. Determination of Heavy Metals at Paper-Based Electrodes with Mercury/Bismuth Films

Standard solutions of heavy metals such as Cd (II), Pb (II), In (III) and Cu (II) were prepared in 0.1 M acetate buffer pH 4 with 0.5 M sodium sulphate. This salt was used to increase the conductivity of the solution. Firstly, all of them were electrochemically characterized by linear sweep voltammetry on an individual base. Standard solutions of each analyte were added on paper electrodes modified with mercury films, to perform individual measurements. A potential of −1.2 V was applied for the preconcentration step, and then the analyte was stripped away by applying an anodic potential scan from −1.2 to 0 V. Different preconcentration times were tested, varying from 240 to 720 s, with 600 s being the option with the highest peak current intensity for Cd (II), In (III), Pb (II) and Cu (II). Their oxidation peak potentials were −905, −800, −720 and −180 mV, respectively, on paper electrodes modified with mercury films.

In order to evaluate a more sustainable alternative with less toxicity, the same procedure was carried out on paper electrodes modified with bismuth films. Following a similar procedure, in this case, the oxidation peaks for Cd (II), In (III) and Pb (II) occurred at −1000, −920 and −780 mV. Cu (II) could not be detected on paper-based electrodes modified with bismuth films because its analytical signal overlapped with a bismuth oxidation peak that appeared at −300 mV. Cd (II), In (III) and Pb (II) were successfully detected in both films. For a 10 µg/mL concentration of Pb (II), oxidation intensity currents of 8 and 19 µA were obtained in paper electrodes modified with bismuth and mercury films, respectively. For In (III), this was also higher in paper electrodes modified with mercury films, obtaining values of 10 µA for a concentration of 2.5 µg/mL, and 2.5 µA for 2 µg/mL in bismuth films. However, in the case of Cd (II), the intensity of peak currents was nearly similar in both cases (19 and 16 µA, respectively, for a concentration of 10 µg/mL). However, the linear range is wider in paper electrodes modified with mercury films, and the limits of detection are lower.

Calibrations curves for the single determination of heavy metals (Cd (II), In (III), Pb (II) and Cu (II)) were carried out by anodic stripping linear sweep voltammetry by employing the same parameters (n = 3). Apart from LSV, different electrochemical techniques (differential pulse voltammetry (DPV) and square wave voltammetry (SWV)) were tested. LSV was the best option because of the good reproducibility obtained with all analytes tested. Thus, LSV was employed for the remaining work. Table 1 shows the equations for the calibration curves, along with analytical characteristics, such as the linear ranges and limits of detection obtained with paper electrodes modified with mercury or bismuth films. Similar slopes for Cd (II) and Pb (II) obtained in mercury films can be explained because they have similar diffusion coefficients in mercury (1.5×10^{-5} cm^2/s for Cd (II), 1.25×10^{-5} for Pb (II)), respectively [67].

Table 1. Analytical characteristics for Cd (II), Pb (II), In (III) and Cu (II) on paper-based electrodes modified with bismuth or mercury films.

		Bismuth Films	
	RSD (%)	**Linear Range (µg/mL)**	**Limit of Detection (µg/mL)**
Cd	4.3	2.5–10	1
Pb	14.6	1–10	0.7
In	9.6	1–4	0.6
		Mercury films	
	RSD (%)	Lineal range (µg/mL)	Limit of detection (µg/mL)
Cd	6.4	0.5–10	0.4
Pb	10.9	0.5–10	0.1
In	9.1	0.1–5	0.04
Cu	4.3	0.25–6.35	0.2

Therefore, mercury films showed higher slopes for all the analytes, except for Cd (II). However, Cd (II) in bismuth films is measured in a narrower linear range. In conclusion, paper electrodes with mercury films have higher sensitivity for all analytes tested.

The limits of detection (LOD) for Cd (II), Pb (II), In (III) and Cu (II) were calculated as the concentration corresponding to a signal equivalent to $3S_a/m$, where S_a is the standard deviation of the blank (intercept), and m is the slope of the calibration line of paper electrodes with mercury or bismuth films. The lowest LODs were achieved in all cases for the paper electrodes modified with mercury films. The relative standard deviation (RSD) corresponding to a concentration located in the middle of the linear range for each analyte is also reported. They were done by using the same screen-printed electrode card for all the paper electrodes. Nevertheless, the RSD did not change by using different SPCEs. Care should be taken in fixing and removing the paper electrodes, in order to avoid ink losses and therefore, connection damage. In this way, screen-printed cards can be reused, at least for ca. seven measurements.

Although for Cd (II) the methodology that employs paper electrodes modified with mercury films has a better limit of detection (LOD), the precision was better for mercury films. In the case of Pb (II), both sensitivity and reproducibility were better when mercury films were used. In (III) had the best LOD for both types of films, being lower for mercury films. The RSDs are acceptable. Cu (II) was only measured on paper electrodes modified with mercury films, obtaining the a wide linear range. Therefore, in general, mercury films gave better results with higher sensitivity and allowed the determination of Cu (II). However, bismuth films showed reliable results with the advantage of being a material less toxic than mercury.

Among all the heavy metals tested, Cd (II) and Pb (II) are the most toxic, having legal limits more restrictive in drinking water in comparison with In (III) and Cu (II). Legislation sets limits of 5 and 10 μg/L for Cd (II) and Pb (II), respectively, whilst Cu (II) has a limit of 2000 μg/L, making Cd (II) and Pb (II) the most important ones for quality analysis [68,69]. There is no established limit for In (III), an element whose toxicity and environmental effects are still not confirmed, but its use as a component in alloys and also in the semiconductor industry makes its determination useful for quality control purposes. With this procedure, Cd (II) and Pb (II) provided reliable results (using mercury and bismuth films) in terms of linear range, reproducibility or sensitivity, concluding that they can be determined in water samples different from drinking water because of the legislation limits. However, our research group has proved that it is possible to decrease 800-fold the LOD of the determination of a contaminant (diclofenac) in waters with this type of electrodes, by simply depositing several aliquots of the sample and allowing the preconcentration in the paper matrix before launching the measurement [70]. This opens the possibility to improve the sensitivity of the detections if required. In the case of In (III), the methodology also showed reliable results, resulting in an In (III) sensor in aqueous solutions for the semiconductor industry. Finally, in the case of Cu (II), mercury-film paper-based electrodes can be used as a low-cost sensor for drinking water.

On the other hand, in order to study the possibility of making multianalyte determination, with simultaneous measurements for all heavy metals studied, solutions containing different concentrations of Cd (II), In (III), Pb (II) and Cu (II) were prepared. Determinations were carried out by anodic stripping linear sweep voltammetry, employing the same parameters as before. It could be observed that the oxidation reactions for Cd and In for concentrations of 10 μg/mL in paper electrodes with mercury films overlapped, with potentials of −845 and −780 mV, respectively. In the case of bismuth films, another overlap between the anodic peaks of Bi and Cu (with −234 and −140 mV respective peak potentials) was observed. However, Pb (II) and Cd (II) could be identified with enough resolution. Linear sweep voltammograms recorded in solutions containing Cd (II), In (III), Pb (II) and Cu (II) for both films are shown in Figure 2.

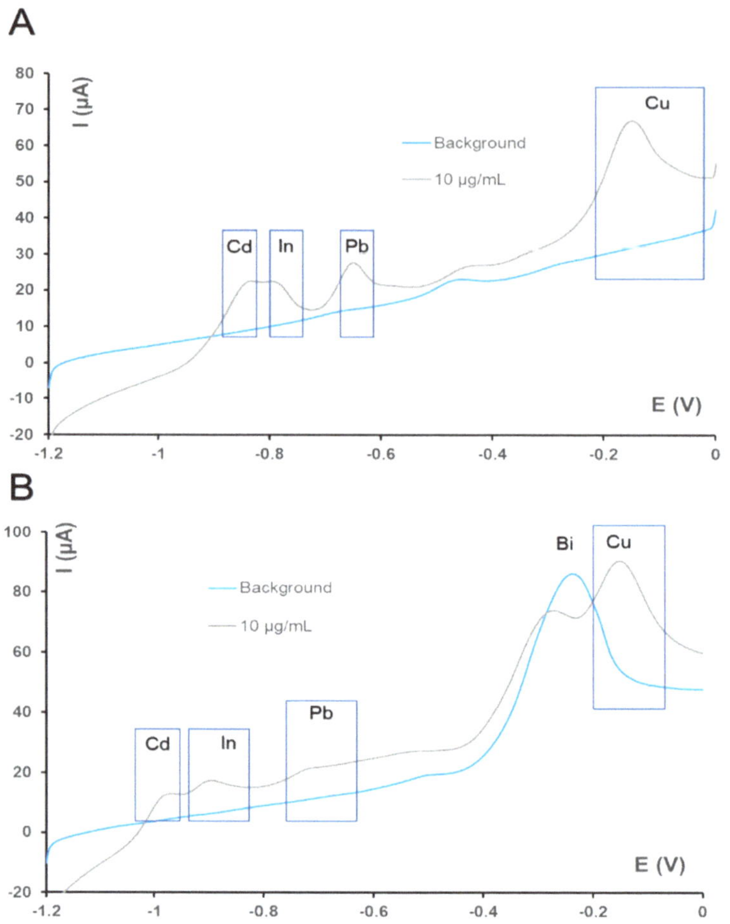

Figure 2. Linear sweep voltammograms for Pb (II), Cd (II), Cu (II) and In (III) (10 µg/mL) solutions recorded in paper-based electrodes modified with (**A**) mercury or (**B**) bismuth films. Preconcentration step: $E_d = -1.2$ V, $t_d = 600$ s). Stripping step: $E_i = -1.2$ V, $E_f = 0$ V, $E_s = 2$ mV, $v = 50$ mV/s).

3.3. Analysis of Water: Determination of Cd (II) and Pb (II)

Water samples obtained from our laboratory were collected as a substitute for sewage-water samples and spiked for Cd (II) and Pb (II) determination, using paper electrodes modified with mercury and bismuth films.

Samples were spiked with 10 µg/mL of Cd (II) and 10 µg/mL of Pb (II) when paper electrodes were modified with mercury films and with 25 µg/mL of each analyte for bismuth films.

For the determination on paper electrodes modified with mercury films, standard additions were made with solutions prepared with an initial aliquot of 100 µL of the spiked sample (10 µg/mL of Cd (II) and 10 µg/mL of Pb (II)). Then, different aliquots (100, 200, 300 and 400 µL) of a 10 µg/mL of Cd (II) and Pb (II) standard solution in 0.1 M acetate buffer (pH 4) with 0.5 M sodium sulphate as a background electrolyte were added. Finally, buffer solution was added until a final volume of 1 mL. Solutions were analyzed by the stripping voltammetry method developed. The quantification of both analytes was successfully carried out at mercury films. However, Pb (II) and Cd (II) quantification on paper electrodes modified with bismuth films was quite irreproducible. This could be due to the formation of intermetallic compounds (Pb/Cd) between the oxidation peaks for Pb (II) and Cd (II).

This could alter the intensity of the peaks analyzed yielding lower concentration in solution for both heavy metals. It has been described that the presence of Zn (II) in the solution could avoid the formation of those intermetallic compounds with Cd and Pb and eliminate this interference [36]. Then, for the determination of Cd (II) and Pb (II) on paper-based electrodes modified with bismuth films, Zn (II) was added to the solutions employed for the determination, following the standard addition methodology. Therefore, the determination procedure was modified according to the method previously published [36].

With this aim, samples of water were spiked with Zn (II), Cd (II) and Pb (II) standards, obtaining a spiked sample of 25 μg/mL of Zn (II), 25 μg/mL Cd (II) and 25 μg/mL Pb (II) in 0.1 M acetate buffer pH 4 with 0.5 M sodium sulphate. For the determination on paper electrodes modified with bismuth films, standard additions were carried out by taking 100 μL of the spiked sample. Then, different aliquots (100, 200, 300 and 400 μL) of a standard solution with 10 μg/mL of Cd (II) and 10 μg/mL of Pb (II) were added. Finally, buffer solution was added until a final volume of 1 mL. Solutions were analyzed by stripping voltammetry with the next parameters: preconcentration step at −1.3 V for 600 s, followed by anodic stripping linear sweep voltammetry and scanning the potential from −1.3 to 0 V at 50 mV/s with 2 mV of step potential. With this procedure, samples of water spiked with Cd (II) and Pb (II) were analyzed with a reproducibility similar to that obtained for mercury films.

Results obtained indicated a higher recovery of the spiked sample in analysis made on paper electrodes modified with mercury films in comparison with bismuth films. Pb (II) determination had a recovery of 98.8% on mercury films, whereas a recovery of 89.2% was obtained on bismuth films. Recoveries obtained for Cd (II) determination were of 99.2% and 86.3%, concluding the efficacy of using paper electrodes with mercury films. The results obtained for the sample with paper-based electrodes with mercury films developed in this study are similar to those obtained on an analysis of the sample by a method previously developed at SPCEs modified with bismuth films (103% Pb (II), 97.5% Cd (II)) [71]. In this case, the possibility of reusing the SPCEs with paper electrodes allows the development of a sensor with a lower cost.

3.4. Comparison of Methods

The analytical characteristics reported in the literature for trace metals' determination by using paper-based electrodes are shown in Table 2. Heavy metal determination on paper devices was carried out by different physicochemical principles, mainly spectrophotometric and electrochemical. However, electrochemical devices can be used for the quantification of metals with less reagent consumption: only those for the formation of the metallic films and electrolytes; therefore, it could be considered that electrochemical methods offer a more economic option. In addition, this methodology was tested with a higher variety of analytes obtaining characteristics like linear range, sensitivity and reproducibility, in comparison with devices based on a colorimetric determination (Table 2).

Electrochemical devices for heavy metal determination in paper devices are scarce. The LOD of our procedure might seem high compared with some electrochemical paper devices previously published, but with the method developed in this work, other analytes, such as Cu (II) and In (III), could be successfully analyzed, increasing the scope of application for this heavy metal sensor. The LOD for Pb (II) and Cd (II) make it suitable for monitoring the levels in contaminated water. In the case of Cu (II), the linear range obtained, together with the high reproducibility, is a promising low-cost alternative to colorimetric sensors. Our work has an additional advantage regarding waste management, since all heavy metals analyzed, as well as the films formed, stay in the working paper electrode. Therefore, the SPCE platform can be reused without any contamination, and the paper electrodes used can be easily disposed of due to their small size and easy portability.

Table 2. Analytical characteristics of other paper devices described in the literature.

Sensor	Detection Technique	Analyte	Linear Range (µg/mL)	LOD (µg/mL)
Solid phase extraction-µPADs	Colorimetric [72]	Cu (II)	0.02–500	0.02
Rotational paper-based device	Colorimetric [73]	Ni (II)	1.5–60	4.8
		Cu (II)	0.5–80	1.6
		Cr (VI)	0.5–10	0.18
Table style paper device	Colorimetric [66]	Ni (II)	0.3–5	0.3
		Cu (II)	0.6–3	0.6
		Cr (VI)	0.2–3	0.2
Three-dimensional microfluidic device	Colorimetric [58]	Cu (II)	5–20	0.29
		Cd (II)	0.05–0.4	0.19
		Ni (II)	5–20	0.33
		Cr (VI)	0.2–1	0.35
Double-sided conductive adhesive carbon tape with bismuth	Electrochemical (SWASV) [62]	Pb (II)	0.002–0.5	0.002
		Cd (II)	0.1–0.2	0.1
		Zn (II)	0.1–0.2	0.1
BDDPE-µPADs	Electrochemical (SWASV) [61]	Pb (II)	0.001–0.2	0.001
		Cd (II)	0.025–0.2	0.025
Graphite paper electrode with sulfonated polyaniline/antimony	Electrochemical (DPASV) [74]	Pb (II)	0.002–0.07	0.0002
		Cd (II)	0.002–0.07	0.00041
Electrochemical device with silver ink and office paper	Electrochemical (SWASV) [75]	Pb (II)	1.87–9.95	0.35

4. Conclusions

An analytical methodology for the study of heavy metals in paper electrodes modified with mercury or bismuth films was successfully developed. Films were freshly generated by applying a reduction potential in solutions containing mercury (II) or bismuth (III) ions, in order to develop a paper platform to quantify heavy metals. The paper working electrodes were overlapped on the working electrode of a screen-printed card. This working-to-working configuration provided the following: (i) a homogeneous connection for all the electrode's circular area; (ii) the possibility of reusing the screen-printed card, since the ink in the paper covers the pores and the ink in the ceramic card is not wetted; and (iii) auxiliary and reference electrodes required for film generation and further measurements. Cd (II), Pb (II) and In (III) could be detected in both films, whereas Cu (II) could only be detected in paper electrodes modified with mercury films. Furthermore, mercury films offered higher sensitivity for all heavy metals tested, in comparison with paper electrodes modified with bismuth films, with limits of detection of 0.4, 0.1, 0.04 and 0.2 µg/mL for Cd (II), Pb (II), In (III) and Cu (II). Cd (II) and Pb (II) could be determined in spiked water samples with recoveries of 99.2% and 98.8% for mercury films and 88.7% and 89.2% for bismuth films. Both types of paper-based electrodes modified with films could be considered as low-cost sensors to the determination of heavy metals in contaminated water. Analytical characteristics like linear range, sensitivity and reproducibility were calculated in a higher variety of metals than paper sensors based on colorimetric detection. Mercury films were a better option to determine heavy metals, and bismuth films were a more sustainable option. The small size of the platform makes it very manageable and easy to treat in terms of waste management.

Author Contributions: This work is part of A.S.-C. PhD thesis, under the supervision of A.C.-G. M.C.B.-L. contributed to the discussion of the results and the preparation of the manuscript. All authors have read and agreed to the published version of the manuscript.

Funding: This research was funded by Ministerio de Economía y Competitividad, grant number CTQ2014-58826-R, MAT2017-84959-C2-1-R and Principality of Asturias (Plan of Science, Technology and Innovation 2013–2017), grant number GRUPIN14-021, IDI/2018/000185.

Acknowledgments: This may include administrative and technical support, or donations in kind (e.g., materials used for experiments). Alberto Sánchez Calvo thanks the Ministry of Economy and Competitiveness for the award of an FPI Grant (BES-2015-072220). Support from the European Regional Development Fund (ERDF) is gratefully acknowledged.

Conflicts of Interest: The authors declare no conflict of interest.

References

1. Svancara, I.; Baldrianova, L.; Vlcek, M.; Metelka, R.; Vytras, K. A role of the plating regime in the deposition of bismuth films onto a carbon paste electrode. Microscopic study. *Electroanalysis* **2005**, *17*, 120–126. [CrossRef]
2. Vyskočil, V.; Barek, J. Mercury electrodes-possibilities and limitations in environmental electroanalysis. *Crit. Rev. Anal. Chem.* **2009**, *39*, 173–188. [CrossRef]
3. Domingos, R.F.; Huidobro, C.; Companys, E.; Galceran, J.; Puy, J.; Pinheiro, J.P. Comparison of AGNES (absence of gradients and Nernstian equilibrium stripping) and SSCP (scanned stripping chronopotentiometry) for trace metal speciation analysis. *J. Electroanal. Chem.* **2008**, *617*, 141–148. [CrossRef]
4. Aguilar, D.; Galceran, J.; Companys, E.; Puy, J.; Parat, C.; Authier, L.; Potin-Gautier, M. Non-purged voltammetry explored with AGNES. *Phys. Chem. Chem. Phys.* **2013**, *15*, 17510. [CrossRef] [PubMed]
5. Kumar, A.; Vij, R.; Gupta, M.; Sharma, S.; Singh, S. Risk assessment of exposure to radon concentration and heavy metal analysis in drinking water samples in some areas of Jammu & Kashmir, India. *J. Radioanal. Nucl. Chem.* **2015**, *304*, 1009–1016.
6. Teigen, S.W.; Andersen, R.A.; Daae, H.L.; Skaare, J.U. Heavy metal content in liver and kidneys of grey seals (Halichoerus grypus) in various life stages correlated with metallothionein levels: Some metal-binding characteristics of this protein. *Environ. Toxicol. Chem.* **1999**, *18*, 2364–2369. [CrossRef]
7. Malar, S.; Sahi, S.V.; Favas, P.J.C.; Venkatachalam, P. Assessment of mercury heavy metal toxicity-induced physiochemical and molecular changes in Sesbania grandiflora L. *Int. J. Environ. Sci. Technol.* **2014**, *12*, 3273–3282. [CrossRef]
8. Wang, J.; Tian, B. Mercury-free disposable lead sensors based on potentiometric stripping analysis at gold-coated screen-printed electrodes. *Anal. Chem.* **1993**, *65*, 1529–1532. [CrossRef]
9. Baldo, M.A.; Bragato, C.; Mazzocchin, G.A.; Daniele, S. Lead and copper deposition from dilute solutions onto carbon disc microelectrodes. Assessment of quantification procedures by anodic stripping voltammetry. *Electrochim. Acta* **1998**, *43*, 3413–3422. [CrossRef]
10. Navrátil, T.; Šebková, S.; Kopanica, M. Voltammetry of lead cations on a new type of silver composite electrode in the presence of other cations. *Anal. Bioanal. Chem.* **2004**, *379*, 294–301. [CrossRef]
11. Pérez-Ràfols, C.; Serrano, N.; Díaz-Cruz, J.M.; Ariño, C.; Esteban, M. Mercury films on commercial carbon screen-printed devices for the analysis of heavy metal ions: A critical evaluation. *Electroanalysis* **2015**, *27*, 1345–1349. [CrossRef]
12. Heyrovsky, M. Early polarographic studies on proteins. *Electroanalysis* **2004**, *16*, 1067–1073. [CrossRef]
13. Serrano, N.; Šestáková, I.; Díaz-Cruz, J.M.; Ariño, C. Adsorptive accumulation in constant current stripping chronopotentiometry as an alternative for the electrochemical study of metal complexation by thiol-containing peptides. *J. Electroanal. Chem.* **2006**, *591*, 105–117. [CrossRef]
14. Gusmão, R.; Cavanillas, S.; Ariño, C.; Díaz-Cruz, J.M.; Esteban, M. Circular dichroism and voltammetry, assisted by multivariate curve resolution, and mass spectrometry of the competitive metal binding by phytochelatin PC 5. *Anal. Chem.* **2010**, *82*, 9006–9013. [CrossRef]
15. Economou, A.; Fielden, P.R. Mercury film electrodes: Developments, trends and potentialities for electroanalysis. *Analyst* **2003**, *128*, 205–212. [CrossRef]
16. Metelka, R.; Vytřas, K.; Bobrowski, A. Effect of the modification of mercuric oxide on the properties of mercury films at HgO-modified carbon paste electrodes. *J. Solid State Electrochem.* **2000**, *4*, 348–352. [CrossRef]
17. Yosypchuk, B.; Fojta, M.; Barek, J. Preparation and properties of mercury film electrodes on solid amalgam surface. *Electroanalysis* **2010**, *22*, 1967–1973. [CrossRef]
18. Wong, D.K.Y.; Ewing, A.G. Anodic stripping voltammetry at mercury films deposited on ultrasmall carbon-ring electrodes. *Anal. Chem.* **1990**, *62*, 2697–2702. [CrossRef]
19. Fertonani, F.L.; Benedetti, A.V.; Servat, J.; Portillo, J.; Sanz, F. Electrodeposited thin mercury films on Pt ± Ir alloy electrodes. *Thin Solid Films* **1999**, *349*, 147–154. [CrossRef]

20. Wang, J.; Lu, J.; Hocevar, S.B.; Farias, P.A.M.; Ogorevc, B. Bismuth-coated carbon electrodes for anodic stripping voltammetry. *Anal. Chem.* **2000**, *72*, 3218–3222. [CrossRef]
21. Pauliukaitė, R.; Brett, C.M.A. Characterization and application of bismuth-film modified carbon film electrodes. *Electroanalysis* **2005**, *17*, 1354–1359. [CrossRef]
22. Vereecken, P.M.; Rodbell, K.; Ji, C.; Searson, P.C. Electrodeposition of bismuth thin films on n-GaAs (110). *Appl. Phys. Lett.* **2005**, *86*, 121916. [CrossRef]
23. Wang, J.; Lu, J.; Kirgöz, U.A.; Hocevar, S.B.; Ogorevc, B. Insights into the anodic stripping voltammetric behavior of bismuth film electrodes. *Anal. Chim. Acta* **2001**, *434*, 29–34. [CrossRef]
24. Hutton, E.A.; Ogorevc, B.; Hočevar, S.B.; Weldon, F.; Smyth, M.R.; Wang, J. An introduction to bismuth film electrode for use in cathodic electrochemical detection. *Electrochem. Commun.* **2001**, *3*, 707–711. [CrossRef]
25. Królicka, A.; Bobrowski, A. Bismuth film electrode for adsorptive stripping voltammetry—Electrochemical and microscopic study. *Electrochem. Commun.* **2004**, *6*, 99–104. [CrossRef]
26. Martín-Yerga, D.; Álvarez-Martos, I.; Blanco-López, M.C.; Henry, C.S.; Fernández-Abedul, M.T. Point-of-need simultaneous electrochemical detection of lead and cadmium using low-cost stencil-printed transparency electrodes. *Anal. Chim. Acta* **2017**, *981*, 24–33. [CrossRef]
27. Caldeira, A.; Gouveia-Caridade, C.; Pauliukaite, R.; Brett, C.M.A. Application of square wave anodic stripping voltammetry for determination of traces of Ti(I) at carbon electrode in situ modified with Bi films. *Electroanalysis* **2011**, *23*, 1301–1305. [CrossRef]
28. Hocevar, S.B.; Wang, J.; Deo, R.P.; Ogorevc, B. Potentiometric stripping analysis at bismuth-film electrode. *Electroanalysis* **2002**, *14*, 112–115. [CrossRef]
29. Hočevar, S.B.; Ogorevc, B.; Wang, J.; Pihlar, B. A study on operational parameters for advanced use of bismuth film electrode in anodic stripping voltammetry. *Electroanalysis* **2002**, *14*, 1707–1712. [CrossRef]
30. Baldrianova, L.; Svancara, I.; Economou, A.; Sotiropoulos, S. Anodic stripping voltammetry at in situ bismuth-plated carbon and gold microdisc electrodes in variable electrolyte content unstirred solutions. *Anal. Chim. Acta* **2006**, *580*, 24–31. [CrossRef]
31. Kefala, G.; Economou, A.; Voulgaropoulos, A.; Sofoniou, M. A study of bismuth-film electrodes for the detection of trace metals by anodic stripping voltammetry and their application to the determination of Pb and Zn in tapwater and human hair. *Talanta* **2003**, *61*, 603–610. [CrossRef]
32. Baldrianova, L.; Svancara, I.; Vlcek, M.; Economou, A.; Sotiropoulos, S. Effect of Bi(III) concentration on the stripping voltammetric response of in situ bismuth-coated carbon paste and gold electrodes. *Electrochim. Acta* **2006**, *52*, 481–490. [CrossRef]
33. Vladislavić, N.; Buzuk, M.; Brinić, S.; Buljac, M.; Bralić, M. Morphological characterization of ex situ prepared bismuth film electrodes and their application in electroanalytical determination of the biomolecules. *J. Solid State Electrochem.* **2016**, *20*, 2241–2250. [CrossRef]
34. Serrano, N.; Díaz-Cruz, J.M.; Ariño, C.; Esteban, M. Ex situ deposited bismuth film on screen-printed carbon electrode: A disposable device for stripping voltammetry of heavy metal ions. *Electroanalysis* **2010**, *22*, 1460–1467. [CrossRef]
35. Arribas, A.S.; Bermejo, E.; Chicharro, M.; Zapardiel, A. Voltammetric detection of the herbicide metamitron at a bismuth film electrode in nondeaerated solution. *Electroanalysis* **2006**, *18*, 2331–2336. [CrossRef]
36. Mandil, A.; Pauliukaite, R.; Amine, A.; Brett, C.M.A. Electrochemical characterization of and stripping voltammetry at screen printed electrodes modified with different brands of multiwall carbon nanotubes and bismuth films. *Anal. Lett.* **2012**, *45*, 395–407. [CrossRef]
37. Ping, J.F.; Wu, J.; Ying, Y.B. Determination of trace heavy metals in milk using an ionic liquid and bismuth oxide nanoparticles modified carbon paste electrode. *Chin. Sci. Bull.* **2012**, *57*, 1781–1787. [CrossRef]
38. Niu, P.; Fernández-Sánchez, C.; Gich, M.; Ayora, C.; Roig, A. Electroanalytical assessment of heavy metals in waters with bismuth nanoparticle-porous carbon paste electrodes. *Electrochim. Acta* **2015**, *165*, 155–161. [CrossRef]
39. Yang, H.; Li, J.; Lu, X.; Xi, G.; Yan, Y. Reliable synthesis of bismuth nanoparticles for heavy metal detection. *Mater. Res. Bull.* **2013**, *48*, 4718–4722. [CrossRef]
40. Cadevall, M.; Ros, J.; Merkoçi, A. Bismuth nanoparticles integration into heavy metal electrochemical stripping sensor. *Electrophoresis* **2015**, *36*, 1872–1879. [CrossRef]

41. Niu, P.; Fernández-Sánchez, C.; Gich, M.; Navarro-Hernández, C.; Fanjul-Bolado, P.; Roig, A. Screen-printed electrodes made of a bismuth nanoparticle porous carbon nanocomposite applied to the determination of heavy metal ions. *Microchim. Acta* **2016**, *183*, 617–623. [CrossRef]
42. Luz, A.; Feldmann, C. Reversible photochromic effect and electrochemical voltage driven by light-induced Bi0-formation. *J. Mater. Chem.* **2009**, *19*, 8107–8111. [CrossRef]
43. Cui, L.; Wu, J.; Ju, H. Synthesis of bismuth-nanoparticle-enriched nanoporous carbon on graphene for efficient electrochemical analysis of heavy-metal ions. *Chem. Eur. J.* **2015**, *21*, 11525–11530. [CrossRef] [PubMed]
44. Wang, J.; Lu, J. Bismuth film electrodes for adsorptive stripping voltammetry of trace nickel. *Electrochem. Commun.* **2000**, *2*, 390–393. [CrossRef]
45. Wang, J.; Anik Kirgöz, Ü.; Lu, J. Stripping voltammetry with the electrode material acting as a "built-in" internal standard. *Electrochem. Commun.* **2001**, *3*, 703–706. [CrossRef]
46. Krolicka, A.; Bobrowski, A.; Kalcher, K.; Mocak, J.; Svancara, I.; Vytras, K. Study on Catalytic Adsorptive Stripping Voltammetry of Trace Cobalt at Bismuth Film Electrodes. *Electroanalysis* **2003**, *15*, 1859–1863. [CrossRef]
47. Banks, C.E.; Kruusma, J.; Hyde, M.E.; Salimi, A.; Compton, R.G. Sonoelectroanalysis: Investigation of bismuth-film-modified glassy carbon electrodes. *Anal. Bioanal. Chem.* **2004**, *379*, 277–282. [CrossRef]
48. Wang, J.; Lu, J.; Hocevar, S.B.; Ogorevc, B. Bismuth-coated screen-printed electrodes for stripping voltammetric measurements of trace lead. *Electroanalysis* **2001**, *13*, 13–16. [CrossRef]
49. Honeychurch, K.C.; Hart, J.P. Screen-printed electrochemical sensors for monitoring metal pollutants. *TrAC Trends Anal. Chem.* **2003**, *22*, 456–469. [CrossRef]
50. Domańska, K.; Tyszczuk-Rotko, K. Integrated three-electrode screen-printed sensor modified with bismuth film for voltammetric determination of thallium(I) at the ultratrace level. *Anal. Chim. Acta* **2018**, *1036*, 16–25. [CrossRef]
51. Mettakoonpitak, J.; Boehle, K.; Nantaphol, S.; Teengam, P.; Adkins, J.A.; Srisa-Art, M.; Henry, C.S. Electrochemistry on paper-based analytical devices: A review. *Electroanalysis* **2016**, *28*, 1420–1436. [CrossRef]
52. Meredith, N.A.; Quinn, C.; Cate, D.M.; Reilly, T.H.; Volckens, J.; Henry, C.S. Paper-based analytical devices for environmental analysis. *Analyst* **2016**, *141*, 1874–1887. [CrossRef] [PubMed]
53. Amor-Gutiérrez, O.; Costa-Rama, E.; Fernández-Abedul, M.T. Sampling and multiplexing in lab-on-paper bioelectroanalytical devices for glucose determination. *Biosens. Bioelectron.* **2019**, *135*, 64–70. [CrossRef] [PubMed]
54. Wang, S.; Ge, L.; Yan, M.; Yu, J.; Song, X.; Ge, S.; Huang, J. 3D microfluidic origami electrochemiluminescence immunodevice for sensitive point-of-care testing of carcinoma antigen 125. *Sens. Actuators B Chem.* **2013**, *176*, 1–8. [CrossRef]
55. Zang, D.; Ge, L.; Yan, M.; Song, X.; Yu, J. Electrochemical immunoassay on a 3D microfluidic paper-based device. *Chem. Commun.* **2012**, *48*, 4683. [CrossRef]
56. Medina-Sánchez, M.; Cadevall, M.; Ros, J.; Merkoçi, A. Eco-friendly electrochemical lab-on-paper for heavy metal detection. *Anal. Bioanal. Chem.* **2015**, *407*, 8445–8449. [CrossRef]
57. Chen, G.H.; Chen, W.Y.; Yen, Y.C.; Wang, C.W.; Chang, H.T.; Chen, C.F. Detection of mercury(II) ions using colorimetric gold nanoparticles on paper-based analytical devices. *Anal. Chem.* **2014**, *86*, 6843–6849. [CrossRef]
58. Wang, H.; Li, Y.J.; Wei, J.F.; Xu, J.R.; Wang, Y.H.; Zheng, G.X. Paper-based three-dimensional microfluidic device for monitoring of heavy metals with a camera cell phone. *Anal. Bioanal. Chem.* **2014**, *406*, 2799–2807. [CrossRef]
59. Sánchez-Calvo, A.; Núñez-Bajo, E.; Fernández-Abedul, M.T.; Blanco-López, M.C.; Costa García, A. Optimization and characterization of nanostructured paper-based electrodes. *Electrochim. Acta* **2018**, *265*, 717–725. [CrossRef]
60. Nantaphol, S.; Kava, A.A.; Channon, R.B.; Kondo, T.; Siangproh, W.; Chailapakul, O.; Henry, C.S. Janus electrochemistry: Simultaneous electrochemical detection at multiple working conditions in a paper-based analytical device. *Anal. Chim. Acta* **2019**, *1056*, 88–95. [CrossRef]
61. Nantaphol, S.; Channon, R.B.; Kondo, T.; Siangproh, W.; Chailapakul, O.; Henry, C.S. Boron Doped Diamond Paste Electrodes for Microfluidic Paper-Based Analytical Devices. *Anal. Chem.* **2017**, *89*, 4100–4107. [CrossRef] [PubMed]

62. Feng, Q.M.; Zhang, Q.; Shi, C.G.; Xu, J.J.; Bao, N.; Gu, H.Y. Using nanostructured conductive carbon tape modified with bismuth as the disposable working electrode for stripping analysis in paper-based analytical devices. *Talanta* **2013**, *115*, 235–240. [CrossRef] [PubMed]
63. Figueredo, F.; Jesús González-Pabón, M.; Cortón, E. Low cost layer by layer construction of CNT/Chitosan flexible paper-based electrodes: A versatile electrochemical platform for point of care and point of need testing. *Electroanalysis* **2018**, *30*, 497–508. [CrossRef]
64. Nunez-Bajo, E.; Blanco-López, M.C.; Costa-García, A.; Fernández-Abedul, M.T. Electrogeneration of Gold Nanoparticles on Porous-Carbon Paper-Based Electrodes and Application to Inorganic Arsenic Analysis in White Wines by Chronoamperometric Stripping. *Anal. Chem.* **2017**, *89*, 6415–6423. [CrossRef] [PubMed]
65. Amor-Gutiérrez, O.; Costa Rama, E.; Costa-García, A.; Fernández-Abedul, M.T. Paper-based maskless enzymatic sensor for glucose determination combining ink and wire electrodes. *Biosens. Bioelectron.* **2017**, *93*, 40–45. [CrossRef]
66. Li, M.; Cao, R.; Nilghaz, A.; Guan, L.; Zhang, X.; Shen, W. "Periodic-Table-Style" Paper Device for Monitoring Heavy Metals in Water. *Anal. Chem.* **2015**, *87*, 2555–2559. [CrossRef]
67. Galus, Z. Diffusion coefficients of metals in mercury. *Pure Appl. Chem.* **1984**, *56*, 635–644. [CrossRef]
68. De Sanidad, G. Real Decreto 140/2003 de 7 de febrero, por el que se establecen los criterios sanitarios de la calidad del agua de consumo humano. Available online: https://www.lalosa.es/wp-content/uploads/2019/01/decret_legis_AGUA.pdf (accessed on 13 May 2020).
69. European Commission. Council Directive 98/83/EC of 3 november 1998 on the quality of water intended for human consumption. *Off. J. Eur. Commun.* **1998**, 1–23.
70. Costa-Rama, E.; Nouws, H.P.A.; Delerue-Matos, C.; Blanco-López, M.C.; Fernández-Abedul, M.T. Preconcentration and sensitive determination of the anti-inflammatory drug diclofenac on a paper-based electroanalytical platform. *Anal. Chim. Acta* **2019**, *1074*, 89–97. [CrossRef]
71. Chao, H.; Fu, L.; Li, Y.; Li, X.; Du, H. Sensitive stripping determination of cadmium (II) and lead (II) on disposable graphene modified screen-printed electrode. *Electroanalysis* **2013**, *25*, 2238–2243.
72. Quinn, C.W.; Cate, D.M.; Miller-Lionberg, D.D.; Reilly, T.; Volckens, J.; Henry, C.S. Solid-phase extraction coupled to a paper-based technique for trace copper detection in drinking water. *Environ. Sci. Technol.* **2018**, *52*, 3567–3573. [CrossRef] [PubMed]
73. Sun, X.; Li, B.; Qi, A.; Tian, C.; Han, J.; Shi, Y.; Lin, B.; Chen, L. Improved assessment of accuracy and performance using a rotational paper-based device for multiplexed detection of heavy metals. *Talanta* **2018**, *178*, 426–431. [CrossRef] [PubMed]
74. Liu, R.; Cao, H.; Nie, Z.; Si, S.; Zhao, X.; Zeng, X. A disposable expanded graphite paper electrode with self-doped sulfonated polyaniline/antimony for stripping voltammetric determination of trace Cd and Pb. *Anal. Methods* **2016**, *8*, 1618–1625. [CrossRef]
75. De Araujo, W.R.; Paixao, T.R.L.C. Fabrication of disposable electrochemical devices using silver ink and office paper. *Analyst* **2014**, *139*, 2742–2747. [CrossRef]

biosensors

MDPI

Review

Electrochemical (Bio)Sensors for Pesticides Detection Using Screen-Printed Electrodes

Beatriz Pérez-Fernández, Agustín Costa-García † and Alfredo de la Escosura- Muñiz *

NanoBioAnalysis Group-Department of Physical and Analytical Chemistry, University of Oviedo, Julián Clavería 8, 33006 Oviedo, Spain

* Correspondence: alfredo.escosura@uniovi.es; Tel.: +34-985-103-521

† In Memoriam to Prof. Agustín Costa-García.

Received: 12 March 2020; Accepted: 30 March 2020; Published: 2 April 2020

check for updates

Abstract: Pesticides are among the most important contaminants in food, leading to important global health problems. While conventional techniques such as high-performance liquid chromatography (HPLC) and mass spectrometry (MS) have traditionally been utilized for the detection of such food contaminants, they are relatively expensive, time-consuming and labor intensive, limiting their use for point-of-care (POC) applications. Electrochemical (bio)sensors are emerging devices meeting such expectations, since they represent reliable, simple, cheap, portable, selective and easy to use analytical tools that can be used outside the laboratories by non-specialized personnel. Screen-printed electrodes (SPEs) stand out from the variety of transducers used in electrochemical (bio)sensing because of their small size, high integration, low cost and ability to measure in few microliters of sample. In this context, in this review article, we summarize and discuss about the use of SPEs as analytical tools in the development of (bio)sensors for pesticides of interest for food control. Finally, aspects related to the analytical performance of the developed (bio)sensors together with prospects for future improvements are discussed.

Keywords: screen-printed electrodes; electrochemical (bio)sensors; pesticides; point-of-care; food control

1. Introduction

Pesticides are among the most used products in the agri-food industry for the control, prevention and elimination of pests. According to the target pest, they can be classified in insecticides, acaricides, fungicides, bactericides, herbicides, etc. The main pesticides are made of carbamates, dinitrocompounds, organochlorines, organophosphates, pyrethroids, neonicotinoids or triazines, among others [1]. However, such compounds have a high toxicity. In this line, according to the World Health Organization (WHO), they can be classified as carcinogenic, neurotoxic or teratogenic [2,3]. This makes necessary their strict control in wastewater, soil, food, animals and human beings. In the European Union, the Residual Maximum Limits (MRLs) allowed by the legislation are 0.1 μg/L for individual pesticides and 0.5 μg/L for total pesticides [4–6].

The United Nations (UN) estimates that 200,000 deaths from acute poisoning occur each year due to pesticides, 99% belonging to developing countries [7]. Continuous exposure to these compounds may cause cancer, Alzheimer's disease and Parkinson's disease, as well as neurological disorders, fertility issues, allergies and hypersensitivity.

The official methods for the determination of pesticides are based on chromatography, such as High Performance Liquid Chromatography (HPLC), HPLC-MS/MS and Gas Chromatography coupled to Mass Spectrometry (GC-MS/MS) [8–13]. Despite their high sensitivity, these techniques require very expensive equipment, long analysis times, high reagent sample volumes and qualified personnel. Due to these limitations, alternative methodologies for pesticides detection have been proposed in the last years, being probably the most relevant ones those based on electrochemical methods [14]. Electrochemical techniques have advantages over conventional ones related to their simplicity, the low sample volumes required (typically in the order of μL), the low cost of instruments and the short analysis time [15,16]. The main (bio)sensing routes for the electrochemical detection of pesticides are based on (Figure 1): (i) enzymatic sensors (inhibition and enzymatic catalysis); (ii) direct detection-based sensors (of electroactive pesticides); (iii) immunosensors (using specific antibodies as receptors); iv) aptasensors (using specific aptamers as receptors) and (v) biological sensors (using microorganisms as receptors).

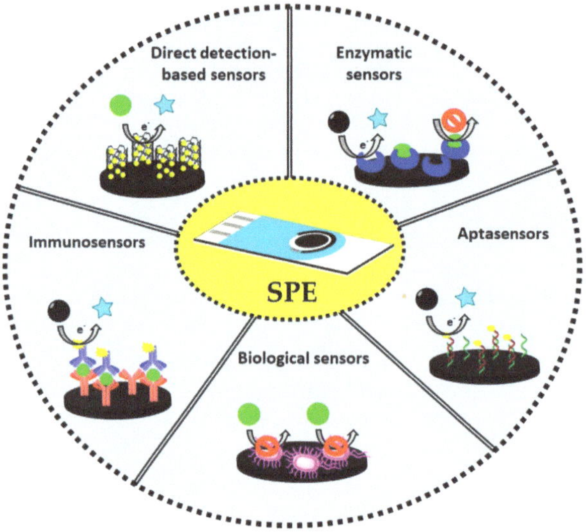

Figure 1. Main (bio)sensing routes followed for the electrochemical detection of pesticides.

However, the use of traditional electrodes requires relatively high sample volumes and quite complicated cell set-up, as they are not suitable for "in field" applications. In this sense, screen-printed electrodes (SPEs) have emerged as outstanding alternatives, overcoming the limitations of traditional electrodes. Screen printing is a well-developed technology widely used since the 1990s for the mass production of disposable and economical electrochemical sensors. This production process is carried out in several stages, as illustrated in Figure 2A [17].

Screen-printed electrodes (SPEs) are manufactured on ceramic or plastic substrates, in which different types of inks (typically carbon, graphite, silver and gold) are printed. In addition, these inks can be modified with nanomaterials or enzymes among other compounds, improving the analytical characteristics of the (bio)sensors developed from such electrodes.

SPEs satisfy the need for highly reproducible, sensitive and cost-effective detection methods, with additional advantages related to the low cost of production, flexibility in design, small size and ease of electrode surface modification [18–23]. The main methods used for the immobilization of (bio)receptors on the working electrode of SPEs for further pesticides detection are summarized in Figure 2B. The portability of the electrochemical instruments typically used also makes these systems ideal for point of care (POC) analysis [24–27]. The working electrode can be modified with various materials and recognition elements such as noble metal nanoparticles (i.e., Cu, Ni, Au, Pt, Ag) [28–37], nanotubes (CNT) [38–44], nanofibers (CNF) [43,45–49], graphene [50], graphene oxide (GO) [51–54], reduced graphene oxide (rGO) [55–58], quantum dots (QDs) [59–64], magnetic beads (MB) [65–68], enzymes

(AChE, ALP, GOD, HRP, FDH, OPH, Tyr) [69–80], antibodies [81–87], aptamers [88–93], DNA [94–98] and biological agents [99–101].

In this review, recent applications of SPEs for the electrochemical detection of pesticides are summarized, giving a critical vision on the advantages, drawbacks and perspectives.

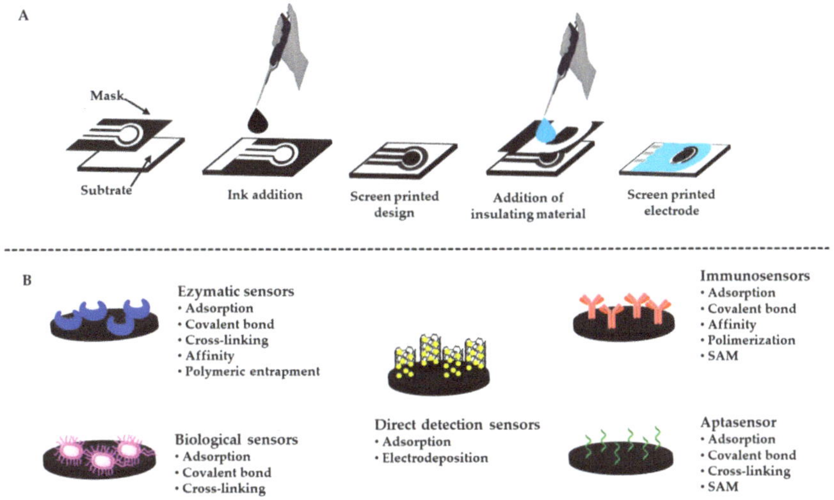

Figure 2. (**A**) Stages of the manufacturing process of screen-printed electrodes (SPEs). (**B**) Main methods used for the immobilization of (bio)receptors on the working electrode of SPEs for pesticides detection (SAM: Self Assembled Monolayers).

2. Electrochemical Techniques Used for Pesticides Detection

The typical electrochemical techniques used for pesticides detection, after following one of the (bio)sensing routes schematized in Figure 1, are: voltammetry (cyclic voltammetry, differential pulse voltammetry, square wave voltammetry), chronoamperometry and electrochemical impedance spectroscopy. The main characteristics of each technique are briefly described in this section.

2.1. *Cyclic Voltammetry*

Cyclic voltammetry (CV) is used to study the different electrochemical processes that take place when applying a potential scan. The measured peaks of current provide information on the oxidation and reduction processes of an electroactive specie. In addition, it provides information on the type of process object of study: (i) reversible; (ii) irreversible or (iii) quasi-reversible, depending on the separation between the anodic and cathodic peaks. Another characteristic that popularizes the CV is its ability to give information about the nature of a process in terms of adsorption and diffusion characteristics.

2.2. *Differential Pulse Voltammetry*

Differential pulse voltammetry (DPV) technique consists in applying a sequence of pulses of constant amplitude superimposed on a stepped potential increase. The current intensity is measured just before applying the pulse and at the end of it. The response obtained is the difference between the two current intensities, in relation to the potential at the start of the pulse, giving rise to a peak-like response.

This technique is used to determine oxidation or reduction processes depending on the analyte concentration. In general terms, DPV has better sensitivity than cyclic voltammetry.

2.3. *Square Wave Voltammetry*

In square wave voltammetry (SWV) technique, a large amplitude square wave potential sweep is applied with a stepped potential ramp. Current intensity is measured at the end of each applied pulse in the potential sweep cycle.

Generally, SWV is more sensitive, faster and more selective than DPV, since the background current is minimized.

2.4. *Chronoamperometry*

Chronoamperometry (AC) is a determination technique where a constant current intensity is applied for a certain time. During to the application of such potential, the electroactive analytes present in the solution are oxidized or reduced, generating and associated current, proportional to the amount of analyte. The main advantage of this technique compared with voltammetries is related to its simplicity.

2.5. *Electrochemical Impedance Spectroscopy*

Electrochemical impedance spectroscopy (EIS) is a technique commonly used to evaluate parameters of charge transfer, corrosion processes, double-layer formation, or modification processes of the electrode surface.

The EIS is given by the Nyquist diagram where values of the load transfer resistance and the resistance of the solution are obtained. Depending on the semicircle obtained in the Nyquist diagram, it can be determined if there is an impediment to the charge transfer or if it is favored by the modification of the working electrode. This technique is widely used in label-free biosensing formats, where the change in the impedance upon the analyte biorecognition allows its determination.

3. Enzymatic Sensors

Enzymatic sensors are the most widely used for the determination of pesticides on SPEs. Two are the main detection routes based on enzymes: (i) enzymatic inhibition route, where the pesticide inactivates the enzyme and (ii) catalytic route, where the pesticide is hydrolyzed by the enzyme generating an electroactive compound [102,103].

3.1. *Enzymatic Inhibition*

As stated above, inhibition reactions make enzymes inactive in the presence of pesticides. This is the case of acetylcholinesterase (AChE), butyrylcholinesterase (BChE), tyrosinase (Tyr) and alkaline phosphatase (ALP).

$$\text{Acetylcholine} + \text{H}_2\text{O} \overset{\text{AChE}}{\rightarrow} \text{Choline} + \text{Acetate} \quad (1)$$

AChE is one of the most commonly used enzymes for the determination of organophosphorus pesticides (OP). The reaction catalyzed by AChE is the following:

The presence of OP results in the inhibition of the enzymatic activity due to phosphorylation of the serine residue of the active center of the enzyme, which blocks the hydrolysis of acetylcholine (ACh). Therefore, the higher the concentration of OP, the more blocked the active center will be and the lower the signal of oxidation or reduction of the enzymatic products. Detection based on other enzymes relies in the same principles.

As shown in Table 1, many publications on enzymatic inhibition-based detection of pesticides on SPEs have been reported in the last years. For example, Solna et al. [74] developed a multi-analyte device for pesticides and phenols based on the immobilization of enzymes (AChE, BChE, Tyr, HRP) on different graphite and platinum working electrodes. AChE and BChe were immobilized on platinum working electrodes while Tyr and HRP did on graphite working electrodes. The limits of detection (LoDs) for the different pesticides studied depended on whether AChE or BChE were used, ranging from 0.8 to 130 nM for AChE and 2.8 to 2390 nM for BChE. Such difference in LoDs is related to the different affinity of each enzyme for the corresponding pesticide. Industrial wastewater was successfully analyzed with such system. Another biosensor based on the inhibition of AChE on screen-printed carbon electrodes (SPCEs) was the developed by Dou et al. [104], where after the manufacturing of the SPCEs, the enzyme was immobilized by polyacrylamide polymerization. Dichlorvos, Monocrotophs and Parathion pesticides were determined with LoD of 18.1, 26.4 and 14.4 nM respectively. In both works the enzymes were immobilized on unmodified working electrodes, giving rise to simple and fast sensors.

Table 1. Enzyme inhibition biosensors reported for pesticides detection on screen-printed electrodes.

Electrode	Enzyme	Analyte	Lineal Range	LoD	Ref
SPGE (array)	AChE	Carbaryl Heptenophos Fenitrothion Dichlorvos Phosphamide	0.8 nM–2.4 µM	0.80 nM 9.2 nM 85 nM 77 nM 130 nM	[74]
	BChE	Carbaryl Heptenophos Fenitrothion Dichlorvos Phosphamide		93 nM 2.8 nM 6.9 nM 14 nM 2390 nM	
Thick-film SPCE	Tyr	Diethyldithio carbamate	–	2 µM	[79]
SPCE	AChE	Dichlorvos Monocrotophs Parathion	16–28 nM	18.1 nM 26.4 nM 14.4 nM	[104]
PB/SPCE	AChE	Aldicarb Carbaryl	63–315 nM 124–497 nM	63 nM 124 nM	[105]
	BChE	Paraoxon Chlorpyrifos-methyl oxon	7–18 nM 1.6–6 nM	7 nM 1.6 nM	
PEDOT:PSS/SPGE	AChE	Chlorpyrifos-oxon	4–760 nM	4.4 nM	[106]
TCNQ/SPGE	AChE	Aldicarb Carbaryl Carbofuran Methomyl	10–500 nM 5–500 nM 1–750 nM 2.5–700 nM	8 nM 4 nM 1 nM 2 nM	[107]
CoPc/SWCNTs/ SPCE	AChE	Paraoxon Malaoxon	18–181 nM 6–159 nM	11 nM 6 nM	[108]
GA/ZnONPs/ SPCE	Tyr	Chlortoluron	1–100 nM	0.47 nM	[109]
Fe_3O_4/GR/SPCE	AChE	Chlorpyrifos	0.14–285 nM	0.06 nM	[110]
Al_2O_3/SPCE	AChE	Dichlorvos	1–60 µM	0.8 µM	[111]
CB/CoPc/SPCE	BChE	Paraoxon	Up to 100 nM	18 nM	[112]
CoPc/SPCE	AChE	Organophosphates	10^{-5}–10^{-9} M	–	[113]
Cyst/GA/AuSPE	AChE	Paraoxon	Up to 145 nM	7.3 nM	[114]
CoPc/CGCE	Tyr	Methyl parathion Diazinon Carbofuran Carbaryl	22.8–379.9 nM 62.4–164.3 nM 22.6–406.8 nM 49.7–248.5 nM	–	[115]
Nf/SPGE	BChE	Trichlorfon Coumaphos	4×10^{-7}–8×10^{-7} M 2×10^{-7}–5.5×10^{-6} M	3.5×10^{-7} M 1.5×10^{-7} M	[116]
PB/SPGE	ChO	Paraoxon	0.1–1 µM	0.1 µM	[117]
DEP-Au chips	AChE	Paraoxon Carbofuran	–	36.3 nM 36.1 nM	[118]
GA/IrOxNPs/ SPCE	Tyr	Chlorpyrifos	0.01–0.1 µM	3 nM	[119]
SPCE	AChE	Chlorpyrifos	1×10^{-6}–5×10^{-2} M	5 µM	[120]
DCHP/MWCNT/SPCE	AChE	Chlorpyrifos	0.14–2.85 nM	0.14 nM	[121]
Nf/PB/DSPCE	AChE	Isocarbophos Chlorpyrifos Trochlorfon	0.33–16.72 µM	0.33 µM	[122]

Table 1. *Cont.*

Electrode	Enzyme	Analyte	Lineal Range	LoD	Ref
SPCE	AChE	Permethrin	6.2–41 µM	8.1 µM	[123]
$Cu_3(PO_4)_2$/HNFs/ SPCE	AChE/ ChO	Paraoxon	2.18×10^{-5}–2.18 nM	21.8 fM	[124]
GA/Nf/BSA/ CBNPs/SPGE	BChE	Paraoxon	18.2–109 nM	18.2 nM	[125]
Nf/PB/ZrO_2/ CNT/SPCE	GMP-AChE	Dimethoate	0.004–43.6 nM	2 pM	[126]
TCNQ/SPCE	BChE	Chlorpyrifos-methyl Coumaphos Carbofuran	3×10^{-8}–3×10^{-7} M 1×10^{-7}–4×10^{-6} M 3×10^{-8}–1×10^{-7} M	20nM 50 nM 10 nM	[127]
PB/SPCE	AChE/ ChO	Chlorpyrifos-methyl Carbofuran	4×10^{-8}–5×10^{-7} M 1×10^{-8}–1×10^{-7} M	30 nM 8 nM	[127]
MWCNTs/SnO_2/ CHIT/SPCE	AChE	Chlorpyrifos	0.14–2.85×10^3 nM	< 0.14 nM	[128]
CS/PVA NFM/SPCE	AChE	Pirimiphos-methyl oxon	1×10^{-10}–8×10^{-9} M	0.2 nM	[129]
OMC-CHIT/ Fe_3O_4-CS/SPCE	AChE	Methamidophos Chlorpyrifos	–	7.09 nM 0.14 nM	[130]
SPSE	AChE	Chlorpyrifos	0–71.3 nM	7.13 nM	[131]
CS/CB/SPCE	AChE	Paraoxon	0.36–1.82 nM	0.18 nM	
MWCNT/SPCE	AChE	Paraoxon	Up to 6.9 nM	0.5 nM	[132]
ZnO/SPCE	AChE	Paraoxon	Up to 5 µM	0.13 µM	[133]
SPGE	AChE	Chlorpyrifos ethyl oxon	0–2×10^{-8} M 5×10^{-8}–2×10^{-7} M	3.6 pM	[134]
MWCNT/IL/ SPCE	AChE	Chlorpyrifos	0.14–2.85×10^5 nM	0.14 nM	[135]
PBNCs/rGO/ SPCE	AChE	Monocrotophos	4.5–2688 nM	0.45 nM	[136]
TCNQ/SPGE	AChE	Carbaryl Carbofuran Pirimicard	Up to 5×10^{-7} M Up to 1×10^{-7} M Up to 5×10^{-7} M	10 nM 0.8 nM 0.2 nM	[137]
CoPc/SPCE	AChE	Carbofuran	10^{-10}–10^{-8} M	0.5 nM	[138]
TCNQ/Nf/SPGE	AChE	Chlorpyrifos methyl	3–930 nM	68 nM	[139]
TCNQ/BSA/GA/ SPCE	AChE	Paraoxon	1.8×10^{-7}–5.4×10^{-5} M	0.18 µM	[140]
TCNQ/Nf/SPCE	AChE	Carbaryl Parathion methyl	9.9–447.3 nM 3.8–379.9 nM	9.9 nM 3.8 nM	[141]
CoPc/SPCE	AChE	Dichlorvos Parathion Azinphos	1×10^{-17}–1×10^{-4} M 1×10^{-16}–1×10^{-4} M 1×10^{-16}–1×10^{-4} M	fM 0.1 fM 0.1 fM	[142]

SPGE: Screen-printed graphite electrode; PB: Prussian Blue; PEDOT: Poly (3,4-ethylenedioxythiophene); PSS: Poly(styrene sulfonate); TCNQ:7,7′,8,8′-Tetracyanoquinodimethane; GR: Graphene; Cyst: Cysteamine; GA: Glutaraldehyde; AuSPE: Screen-printed gold electrode; CGCE: Acetylcelullose-graphite composite electrode; Nf: Nafion; ChO: Choline oxidase; DEP: Disposable electrochemical printed; DCHP: Dicyclohexyl phthalate; DSPCE: Dual-channel screen-printed carbon electrode; GMP-AChE: Gold magnetic particles-Acetylcholinesterase; HNF: Hybrid nanoflowers; CBNP: Carbon black nanoparticles; CNT: Carbon nanotube; CHIT: Chitosan; PVA: Poly (vinyl alcohol); NFM: nanofibrous membranes; OMC-CHIT: ordered mesoporous carbon–chitosan; SPSE: Screen-printed silver electrode; IL: Ionic liquid; PBNC: Prussian Blue Nanocubes; rGO: reduced Graphene Oxide; BSA: Bovine Serum Albumin.

However, in most cases working electrodes are modified with different materials so as to improve the efficiency of the enzyme immobilization and thus the pesticide analysis. An example of simple electrode modification is the reported by Arduini et al. [105] where Prussian Blue was immobilized before the enzyme (AChE and BChE) did. The concentration of several organophosphorus pesticides was determined, finding the highest sensitivity for Aldicarb and Carbaryl (LoDs 63 and 124 nM respectively) when AChE was used and for Paraoxon and Chlorpyrifos-methyl oxon (LoDs 7 and 1.6 nM, respectively) when using BChE. River water and wastewater samples were analyzed with such biosensor. Polymers like poly (3,4-ethylenedioxythiophene) polycation (PEDOT) and poly (styrenesulfonate) polyanion (PSS) were also used for the carbon electrode modification so as to increase its conductivity [106]. Organophosphorus pesticides such as Chlorpyrifos-oxon were determined (LoD of 4.4 nM) based on the thiocholine oxidation. In the case of the biosensor developed by Silva Nunes et al. [107], a screen-printed graphite electrode (SPGE) was modified with 7,7,8,8-tetracyanoquinodimethane (TCNQ) and photopolymerized with poly (vinyl alcohol) bearing styrylpyridinium groups (PVA-SbQ) to covalently immobilize AChE. This biosensor was applied for the determination of different carbamates such as Aldicarb (LoD 8 nM), Carbaryl (LoD 4 nM), Carbofuran (LoD 1 nM) and Methomyl (LoD 2 nM).

Different nanomaterials have also been proposed for the electrode's modification. For example, multi-walled carbon nanotubes (MWCNTs) were immobilized together with AChE and Co-phthalocyanine (Co-Pc; used as mediator) on the working electrode of a SPCE [108]. Such modification produces a decrease in the working potential, minimizing interferences and thus improving the selectivity of the biosensor. This device was developed for the determination of Paraoxon (LoD 11 nM) and Malaoxon (LoD 6 nM), even in tap and sparkling water. Metallic nanoparticles (NPs) have also been used as modifiers of the working electrode. This is the case of the Clortoluron sensing device developed by Haddaoui and Raouafi [109] using ZnO nNPs-modified electrodes for the immobilization of tyrosinase (Tyr) enzyme (Figure 3). Clortoluron is detected here with a LoD of 0.47 nM, having also a good performance in tap water, well water and river water. The stability of immobilized AChE as well as the electronic transference were also improved using magnetic nanoparticles (Fe_3O_4) coupled to a graphene (GR) film on a SPCE [110]. This biosensor was applied for the determination of Chlorpyrifos at levels of 0.06 nM even in vegetable samples (spinach and cabbage).

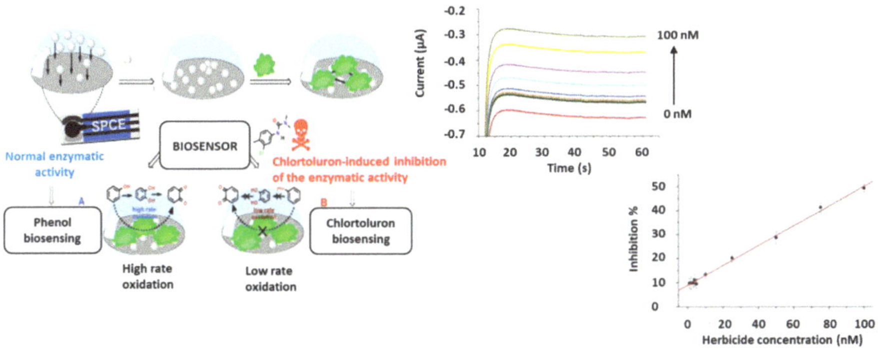

Figure 3. Enzyme inhibition biosensors for pesticides detection on screen-printed electrodes. (Left) scheme of a biosensor for Chlortoluron using a SPCE modified with ZnO NPs for Tyr enzyme immobilization; (Right) Chronoamperometric (CA) responses and calibration curve of inhibition % vs. concentration of herbicide. Reprinted from [109], Copyright 2015, with permission from Elsevier.

Alternatively, enzymes can also be trapped in gel matrices with which the electrode is subsequently modified. This is the case of the work developed by Shi et al. [111] where AChE is trapped in a sol-gel matrix of Al_2O_3. Such matrix not only increases the stability of AChE but also catalyzes the oxidation of thiocholine, decreasing the working potential and minimizing interferences. Dichlorvos pesticide was determined at levels of 0.8 μM in river water samples.

As summarized in Table 1, the modification of SPEs with different materials highly improves the analytical characteristics of the biosensors, lowering at levels as low as the femtomolar scale.

3.2. Catalytic Detection

As stated above, an alternative route for pesticides detection using enzyme receptors is based on the pesticide hydrolysis by the enzyme generating an easily detected electroactive compound. Organophosphorus anhydrolase acid (OPAA) and organophosphorus hydrolase (OPH) are the main enzymes used for such purpose [143–147]. The OPAA is a more restrictive enzyme with respect to organophosphorus compounds, being only able of catalyzing P-F bonds, while OPH also catalyzes organophosphorus containing P-O, P-S and P-CN bonds [148]. Therefore, OPH has a broader range of catalytic effect, being used to develop biosensors for total organophosphorus pesticides.

The catalytic reaction of OP through the use of OPH gives rise to the electroactive p-nitrophenol product. Consequently, the amount of OP is proportional to the production of p-nitrophenol. The generic reaction catalyzed by OPH is as follows:

$$\text{Aryldialkyl phosphate} + H_2O \overset{\text{OPH}}{\rightarrow} \text{Dialkyl phospahte} + \text{Aryl alcohol} \quad (2)$$

Table 2 summarizes the different biosensors reported for organophosphates detection based on enzymatic hydrolysis. As in the case of the enzymatic inhibition, different materials have been evaluated for the electrode modification so as to improve the performance of the biosensor.

Table 2. Enzymatic hydrolysis-based biosensors reported for pesticides detection on screen-printed electrodes.

Electrode	Enzyme	Analyte	Lineal Range	LoD	Ref
Nf/SPCE	OPH	Paraoxon Methyl parathion	4.6–46 µM Up to 5 µM	0.9 µM 0.4 µM	[148,149]
MWCNT/SPCE	OPH	Demeton-S	Up to 85 µM	1 µM	[150]
Fe_3O_4@Au-NC/ SPCE	MPH	Methyl parathion	1.9–3799 nM	0.38 nM	[151]
SPCE	PH	Parathion	34–343 nM	3.4 nM	[152]
BSA/GA/SPCE	OPH	Diazinon	–	0.59 µM	[153]

Nf: Nafion; OPH: Organophosphorus hydrolase; NC: Nanocomposite; MPH: Methyl parathion hydrolase; PH: Parathion hydrolase; GA: Glutaraldehyde.

As a representative example, Mulchandani et al. [149] modified the SPCE with Nafion (Nf) for the OPH immobilization and subsequent determination of Paraoxon and Methyl parathion in river water samples at 0.9 and 0.4 µM levels respectively. They and others [150], also evaluated the SPCE modification with MWCNT for the enzyme immobilization finding worse sensitivity than for the Nf-based modification.

Metallic NPs have also been proposed for the electrode modification. In particular, AuNPs surrounding the core of magnetic Fe_3O_4 NPs were used for the immobilization of methyl parathion hydrolase enzyme (MPH) which allowed the detection of Methyl parathion at levels as low as 0.38 nM (Figure 4) [151]. The use of the AuNP/ Fe_3O_4 NPs matrix also allowed to work at low potentials, minimizing interferences when analyzing river water samples.

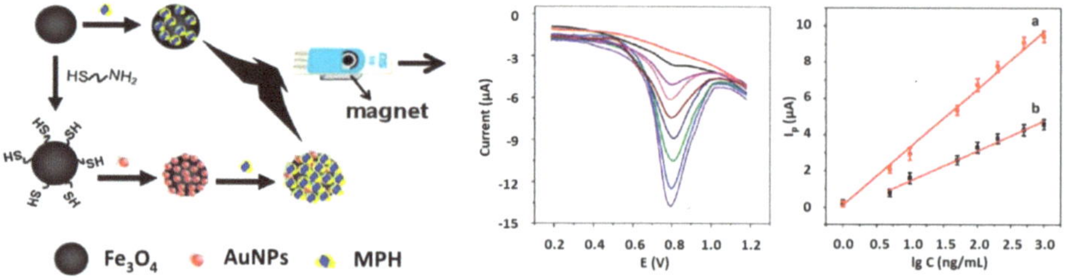

Figure 4. Enzymatic hydrolysis-based biosensors for pesticides detection on screen-printed electrodes. (Left) Scheme of an approach based on the SPCE modification with AuNPs/ Fe_3O_4 NPs for MPH immobilization and further determination of methyl parathion; (Right) SWV measurements of methyl parathion at different concentrations and calibration plots for the electrode with (**a**) and without (**b**) AuNPs. Reprinted from [151], Copyright 2013, with permission from Royal Society of Chemistry.

4. Direct Detection of Electroactive Pesticides

In spite of their advantages, the use of enzymes entails important limitations related to their low stability over time and their sensitivity to changes in temperature and pH [154]. Moreover, some pesticides are electroactive compounds able to be electrochemically detected without the need of enzymes [155,156]. For this reason, the growth of electrochemical sensors without enzymes is continuously increasing, benefiting of their lower cost, greater simplicity and faster analysis. However, in most cases the sensitivity of such direct detection on unmodified SPEs is not enough for detecting the pesticides at the maximum levels allowed by the legislation. Working electrode modification with different materials, mainly metallic nanoparticles (made of i.e., Ag, Zn, Cu, Ni) and carbon nanomaterials such as nanofibers (CNF), nanotubes (CNT) or graphene, etc. has been extensively studied so as to improve the electronic transference and thus the sensitivity of the pesticide detection.

Direct detection approaches for pesticides determination on SPEs are summarized in Table 3. As a representative example of direct detection without electrode modification, Geto et al. [157] self-prepared carbon made SPEs (SPCE) (Figure 5A) for the rapid determination of Bentazone (BTZN), an important herbicide used in agriculture. The voltammetric oxidation of the tertiary amine in the pesticide was selected as the analytical signal that allowed its quantification. A detection limit below the MRL (3.4 nM) was obtained, also demonstrating the good performance of the sensor in groundwater and lake water.

Table 3. Direct detection-based biosensors reported for pesticides determination on screen-printed electrodes.

Electrode	Analyte	Lineal range	LoD	Ref
SPCE	Bentazone	0.19–50 µM	34 mM	[157]
CB/SPCE	Carbofuran Isoprocarb Carbaryl Fenobucarb	0.1–100 µM 0.1–100 µM 0.1–100 µM 0.1–100 µM	49 nM 79 nM 48 nM 80 nM	[158]
Thick-film Bi/SPCE	Imidacloprid Thiamethoxam Dinotefuran Clothianidin Nitenpyram	0–110.26 µM	2.97 µM 2.68 µM 7.67 µM 4.12 µM 4.36 µM	[159]
NiO/SPCE	Parathion	0.1–5 µM and 5–30 µM	24 nM	[160]
AuNPs/SPCE	Thiram DEDMTDS Disulfiram	0.29–62.39 µM 0.15–26.62 µM 1.69–50.58 µM	90 nM 50 nM 550 nM	[161]
MIP/AuNPs/ERGO/ SPCE	Cyhexatin	2.60–1298.18 nM	0.52 nM	[162]
CoPc/SWCNT/SPGE	Thiocholine	0.07–0.45 mM	38 µM	[163]
ZnONPs/MWCNTs/ SPCE ZnONPs/Au-SPCE	Glyphosate AMPA	1–10 µM 10–100 µM	300 nM 3 µM	[164]
AG/AuNPs/SPCE	Hydrazine	0.002–936 µM	0.57 nM	[165]
CHIT/ZnO/SPCE	4-nitrophenol	0.5–400.6 µM	230 nM	[166]
Graphene/Nf/SPCE MWCNT/Nf/SPCE	4-nitrophenol	10–620 µM 25–620 µM	600 nM 1.3 µM	[167]
MWCNT-SPE	Sulfentrazone	1–30 µM	150 nM	[168]
AuNP/CeO_2/SPGE	Hydrazine	0.01–10 mM	–	[169]
CuONPs/SPCE	DCMU	0.5–2.5 µM	47 nM	[170]
NG-PVP/ AuNPs/SPCE	Hydrazine	2–300 µM	70 nM	[171]
Nafion/CNT/SPCE	Paraquat	0.54–4.30 µM	170 nM	[172]
AuNPs/GO/SPCE	Carbofuran	1–30 µM 30–250 µM	220 nM	[173]
Ag@GNRs/SPCE	Methyl parathion	0.005–2780 µM	0.5 nM	[174]
CoHCF/SPGE	Thiocholine	5×10^{-7}–1×10^{-5} M	500 nM	[175]
CB/CoPc/SPCE	Thiocholine	Up to 6 mM	4 µM	[112]

CB: Carbon-black; NP: Nanoparticles; DEDMTDS: N,N-diethyl-N′,N′-dimethylthiuram disulfide; CoPc: Cobalt phthalocyanine; SWCNT: Single-walled carbon nanotube; SPGE: Screen-printed graphite electrode; MWCNT: Multi-walled carbon nanotube; AMPA: Aminomethyl phosphoric acid; MIP: Molecularly imprinting polymer; ERGO: Electrochemical reduction graphene oxide; AG: Activ ated graphite; CHIT: Chitosan; Nf: Nafion; DCMU: 3-(3,4-dichlorophenyl)-1,1-dimethylurea; NG: Nitrogen-doped graphene; PVP: Polyvinylpyrrolidone; GNR: Graphene nanoribbons; CoHCF: Cobalt hexacyanoferrate.

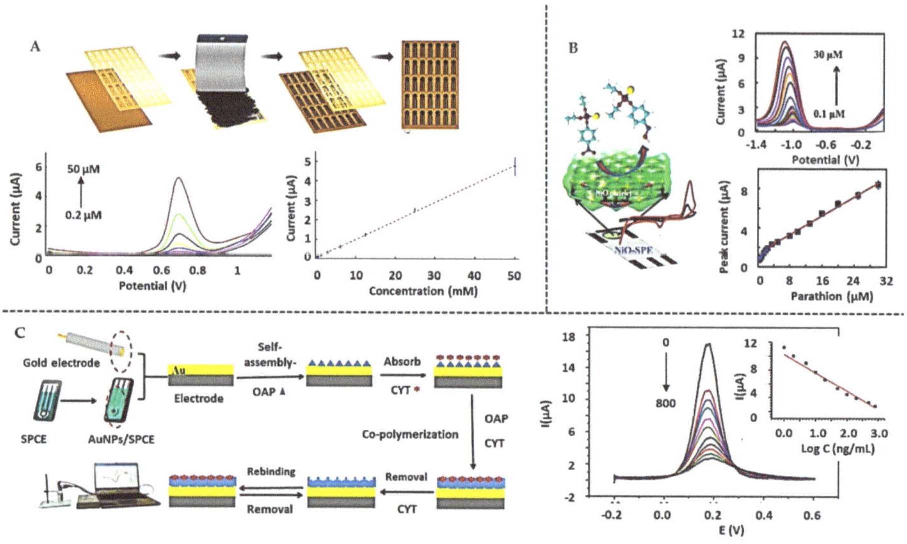

Figure 5. Direct detection-based biosensors for pesticides determination on screen-printed electrodes: (**A**) Scheme of the preparation of SPE for the determination of BTZN, SWV measurements and calibration curve. Reprinted from [157], Copyright 2019, with permission from Elsevier. (**B**) SPE modified with NiO NPs to determine Parathion, DPV signals and calibration curve. Reprinted from [160], Copyright 2018, with permission from Elsevier. (**C**) Use of MIP for the determination of CYT. Measurements by DPV and inset the calibration curve. Reprinted from [162], Copyright 2019, with permission from Elsevier.

However, as stated above, in most cases SPE modification is required for improving the sensor sensitivity. In this line, Della Pelle et al. [158] used black nanocarbon (CB) to develop a device for the determination of phenylcarbamates (i.e., Carbaryl, Carbofuran, Isoprocarb and Phenobucarb). In this case, the analytical signal corresponds to the voltammetric oxidation of the hydrolyzed forms of the pesticides, which allowed to reach LoD ranging from 48 to 80 nM. Wheat and corn samples were also analyzed after extraction and hydrolysis treatment.

Thick bismuth films have also been proposed as SPEs modifiers for the determination of neonicotinoid pesticides such as Clothianidin, Imidacloprid, Thiamethoxam and Nitenpyram [159]. The voltammetric reduction of the nitro group of the pesticides to hydroxylamine allowed the use of quantitative analysis, reaching LoDs at levels in the range 2.97–4.12 μM. Tap water, mineral water and samples from rivers and lakes were successfully analyzed.

NiO NPs have also been used to improve the sensitivity and stability for the determination of Parathion (Figure 5B) [160]. Such NPs catalyze the voltammetric reduction of the nitro group of the Parathion to hydroxylamine, allowing the detection of the pesticide at 24 nM levels. Tap water and human urine samples were analyzed without any pretreatment while a tomato juice sample only required a simple filtration.

AuNPs were also used to increase the electrode surface of the SPCE and catalyze the amperometric oxidation of Thiram, Disulfiram and N,N-diethyl-N′,N′-dimethylthiuram disulfide (DEDMTS) pesticides [161]. Interestingly, in this case, this sensor was coupled to an Ultra High-Performance Liquid Chromatography (UHPLC) system to perform a previous separation and improve the selectivity. LoDs ranging from 0.05 to 0.55 μM were obtained with such system, also analyzing samples of apple, grape and lettuce after extraction, filtration and centrifugation.

The high selectivity given by the use of molecularly imprinted polymeric membranes (MIP) has also been approached for the specific pesticide detection. As an example, a MIP combined with AuNPs and reduced graphene oxide (rGO), was used for the specific voltammetric determination of Cyhexatin (CYT) [162] (Figure 5C), reaching a LoD as low as 0.52 nM.

The SPEs electrodes used by Jubete et al. [163] were modified with single-wall carbon nanotubes (SWCNT) and cobalt phthalocyanine (CoPc) for the determination of Thiocholine (TCh), by monitoring its amperometric oxidation. This system gave a relatively low sensitivity with LoDs at 38 μM levels. In contrast, when zinc oxide nanoparticles (ZnONP) were used in combination with multi-walled carbon nanotubes (MWCNTs) [164] for the determination of Glyphosate and its hydrolysis product (aminomethylphosphonic acid, AMPA) the LoDs were significantly better (300 nM and 3 μM).

5. Immunosensors

Immunosensors for pesticides detection are based on the use of antibodies and antigens as recognition elements immobilized on the SPE. The main advantages of such biosensors over enzymatic ones rely on the higher stability of antibodies/antigens together with greater selectivity and specificity. However, the high cost and low availability of monoclonal antibodies is an important limitation that should be considered. The small size of the pesticides also usually avoids sandwich-based approaches, so competitive immunoassays are often required. Immunosensors for Imidacloprid, Parathion, Methyl Chlorpyrifos, Chlorsulfuron and Atrazine can be found in the literature, as shown in Table 4.

Table 4. Immunosensors reported for pesticides determination on screen-printed electrodes.

Electrode	Analyte	Lineal range	LoD	Ref
Ab/fG-SPCE	Parathion	0.3–3.43 × 10^3 pM	0.18 pM	[176]
Ab/NH_2-GQD/SPCE	Parathion	0.03–3.43 × 10^6 pM	0.16 pM	[177]
PO-SPCE	Chlorsulfuron	0.03–3.88 nM	30 pM	[178]
BSA-IMD/SPCE	Imidacloprid	50–10000 pM	24 pM	[179]
Ab/AuNP/SPCE	Imidacloprid	50–10000 pM	22 pM	[180]
BSA-Ag/Pt/SiO_2/SPCE	Chlorpyrifos methyl	1.24–62 nM	70 pM	[181]
Ab/PANI/PVS/SPCE	Atrazine	0.02–0.22 μM	4.6 nM	[182]
Ab/ATPh/GA/AuSPE	2,4-D	45 nM–0.45 mM	–	[183]

Ab: Antibody; fG: functionalized graphene; PO: Peroxidase; BSA: Bovine serum albumin; IMD: Imidacloprid; GQD: Graphene Quantum Dots; Ag: Antigen; ATPh: 4-aminothiophenol; GA: Glutaraldehyde; AuSPE: Screen-printed gold electrode; 2,4-D: 2,4-dichlorophenoxyacetic acid.

In most cases, SPEs are modified with different (nano)materials so as to improve the efficiency of the receptor immobilization. As example, graphene sheets [176] and graphene quantum dots (GQDs) [177] modified with amino groups have been proposed for the oriented immobilization of antibodies on SPCE. In both cases, anti-parathion antibodies were immobilized for the further Parathion recognition and final detection by means of electrochemical impedance spectroscopy (EIS), reaching LoDs as low as 0.16 pM, with high selectivity even in tomato and carrot samples after extraction (Figure 6A).

Representative examples of competitive immunosensors are those based on the use of enzymatic tags proposed for the determination of Chlorsulfuron [178] and Imidacloprid [179] at levels as low as 22 pM, also taking advantage of the use of AuNPs [180] for the immobilization of the antibody. Tap water, watermelon and tomato samples were analyzed here without the need of pretreatment (Figure 6B).

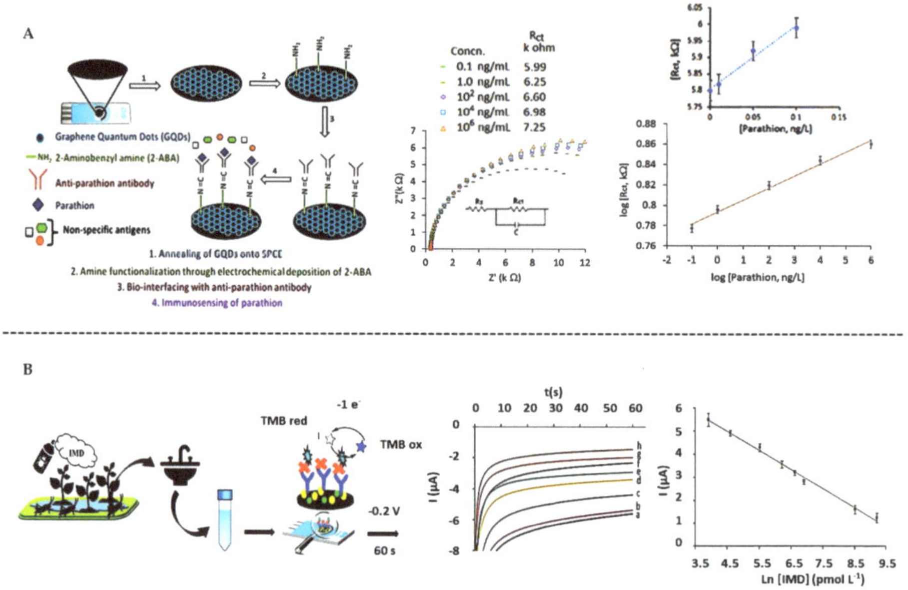

Figure 6. Immunosensors for the determination of pesticides on screen-printed electrodes. (**A**) Use of GQDs to determine Parathion, Niquist curves and calibration curve. Reprinted from [177], Copyright 2017, with permission from Elsevier. (**B**) Direct competitive immunosensor using AuNPs for the determination of Imidacoprid (IMD), chronoamperometroc measurements and calibration curve. Reprinted from [180], Copyright 2020, with permission from Elsevier.

6. Aptasensors

Aptamers are single-chain oligonucleotides that can be produced through the technique of "systematic evolution of ligands by exponential enrichment" (SELEX) [184]. These aptamers are able to fold into three-dimensional structures to bind small compounds such as pesticides and drugs or large organisms [185–190]. Aptamers and analytes are joined by Van der Waals forces, electrostatic interactions or hydrogen bonds [191], thus being able to reverse the aptamer/analyte bond. Some of the advantages of the use of aptamers are the lower cost with respect to enzymes and antibodies, higher stability, long service life, regeneration possibilities and simplicity and rapid response [185,192]. Aptamers can also be easily functionalized and immobilized on the SPE for the development of electrochemical aptasensors.

However, their use for the detection of pesticides has not been extensively reported, as shown in Table 5. Again, SPE modification with different (nano)materials seems to be crucial for improving the efficiency of both the aptamer immobilization and the electrochemical detection.

As example, polyaniline/AuNPs composites were proposed by Rapini et al. [193] for the immobilization of an acetamiprid-specific aptamer (Figure 7A). The pesticide detection was based on a competitive assay using an enzyme-tagged oligonucleotide complementary to the aptamer sequence, reaching a LOD of 86 nM with high selectivity. Good performance was also obtained when analyzing blackberry, apricot and peach juice samples.

Table 5. Aptasensors reported for pesticides determination on screen-printed electrodes.

Electrode	Analyte	Lineal range	LoD	Ref
Apt/PANI/AuNPs/SPGE	Acetamiprid	0.25–2 μM	86 nM	[193]
BSA/Apt/rGO-CuNPs/SPCE	Profenofos Phorate Isocarbophos Omethoate	0.01–100 nM 1–1000 nM 0.1–1000 nM 1–500 nM	3 pM 300 pM 30 pM 300 pM	[194]
Apt/MCH/AuNP/AuSPE	Diazinon	0.1–1000 nM	17 pM	[195]

Apt: Aptamer; PAMI: Polyaniline layer; SPGE: Screen-printed graphite electrode; MCH: 6-Mercapto-1-hexanol; AuSPE: Screen-printed gold electrode.

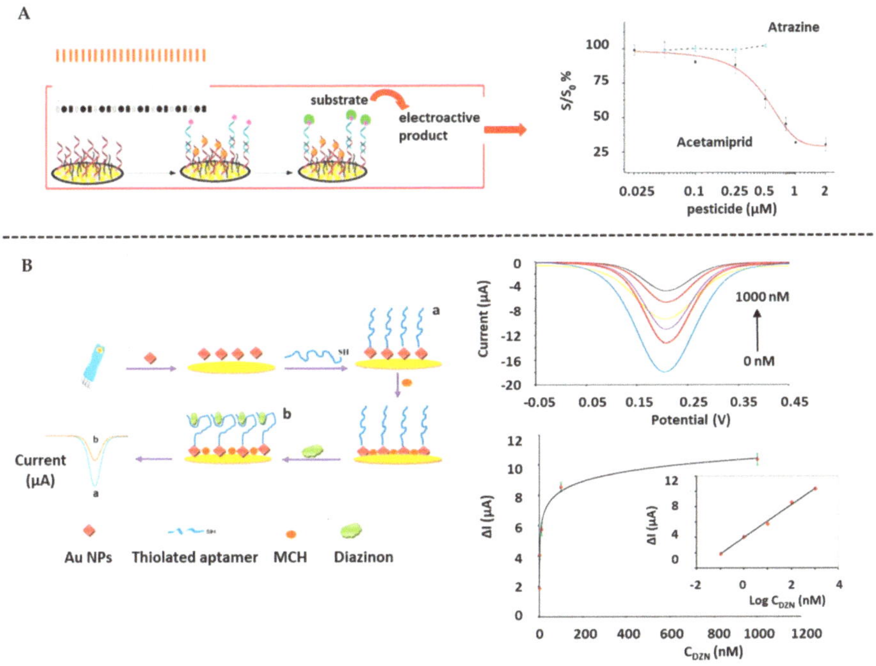

Figure 7. Aptasensors for pesticides determination on screen-printed electrodes: (**A**) Immobilization of aptamer on AuNPs-modified SPGE for the determination of Acetamiprid and the dose-response curve of Acetamiprid and Atrazine. Adapted from [193], Copyright 2016, with permission from Elsevier. (**B**) Immobilization of aptamer on AuNPs-modified gold SPE for Diazinon detection and DPV measurements at different concentrations together with the calibration curve. Reprinted from [195], Copyright 2018, with permission from Elsevier.

Aptamers have also been immobilized on SPE in composites with reduced graphene oxide (rGO) and Cu nanoparticles (CuNP). That is the case of the work reported by Fu et al. [194] for the determination of organophosphorus pesticides. The voltammetric signal decreases when increasing the pesticide concentration, since the complex aptamer-pesticide hindered the transfer electron of the $[Fe(CN)_6]^{3-/4-}$ ions. Under the optimal conditions LoDs ranging from of 0.003 to 0.3 nM were obtained for Profenofos, Phorate, Isocarbophos and Omethoate. The aptasensor was also successfully applied for rapeseed and spinach samples, after an extraction treatment.

AuNP-modified SPEs were also proposed for the immobilization of an aptamer specific for Diazinon (DZN) [195]. Increase in the impedance upon pesticide recognition was approached for its determination at levels as low as 17 fM, also taking advantage of the AuNPs as enhancers of the electronic transference (Figure 7B). Rat plasma samples were also analyzed with good sensor performance.

7. Biological Sensors

Cells and microorganisms can also be used as recognition elements in biosensors. In this case the analytical signal is commonly related to the activation or inactivation of cellular respiration upon analyte interaction, leading to the production of electroactive metabolites. A representative approach consists in the genetic modification of microorganisms with enzymes, such as OPH for the determination of organophosphorus pesticides by measuring the enzymatically produced p nitrophenol [102,196 198].

However, a key limitation of using microorganisms as a recognition element relies in their low sensitivity and long analysis time because of the slow transport of substrate and product through the cytoplasmic membranes of the cells. Due to this, few works on the use of microorganisms for the development of pesticide biosensors on SPE are found in the literature, as summarized in Table 6. From these works, the report by Touloupakis et al. deserves to be highlighted [199] which employed Photosystem II (PS II) that has an oxidoreductase-like behavior. In this work, they used the photosynthetic thylakoid of *Spinacia oleracea*, *Senecio vulgaris* and its atrazine resistant mutant immobilized with BSA-GA on the SPE for the detection of herbicides that selectively block the electronic transfer activity of PS II biomediators (Figure 8A). The developed multi-biosensor reached LoDs at levels ranging from 15 to 41 nM for Diuron, Atrazine, Simazine, Terbuthylazine and Deethylterbuthylazine, even in river water samples.

Table 6. Biological sensors reported for pesticides determination on screen-printed electrodes.

Bacteria	Analyte	Lineal range	LoD	Ref
Spinacia oleracea *Snecio vulgaris*	DIU	1×10^{-8}–1×10^{-6} M	15 nM	[199]
	ATR	1×10^{-8}–1×10^{-6} M	13 nM	
	SIM	1×10^{-8}–1×10^{-6} M	41 nM	
	TER	1×10^{-9}–1×10^{-6} M	25 nM	
	DET	1×10^{-8}–1×10^{-6} M	24 nM	
Rhodobacter sphaeroides	Terbutryn	0.001–10 µM	8 nM	[200]
Escherichia coli	Methyl parathion	2–80 µM	0.5 µM	[201]

DIU: Diuron; ATR: Atrazine; SIM: Simazine; TER: Terbuthylazine; DET: Deethylterbuthylazine.

Alternatively Chatzipetrou et al. [200] used as recognition elements the bacterial reaction centers (RC) of *Rhodobacter sphaeroides* immobilized on gold SPE for the determination of Terbutryn at 8 nM levels (Figure 8B).

Finally, the biosensor reported by Kumar and D'Souza [201] is based on the immobilization of whole cells of recombinant *Escherichia coli* on an SPCE for the detection of Methyl parathion. The organophosphorus hydrolase enzyme that catalyzes the hydrolysis of organosphorus pesticides such as Methyl parathion in p-nitrophenol is expressed in recombinant *Escherichia coli* cells, being an electroactive compound through which the concentration of Methyl parathion is directly determined. With this approach, Methyl parathion was detected at 0.5 µM levels.

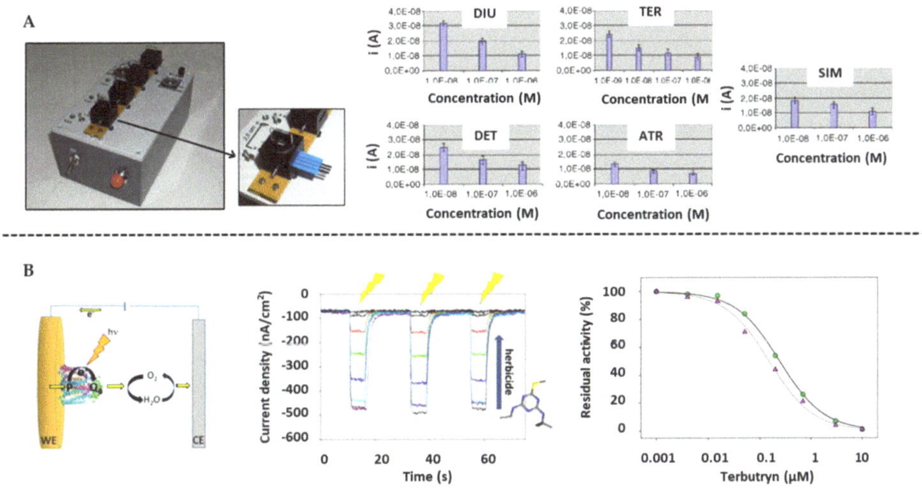

Figure 8. Biological sensors for pesticides determination on screen-printed electrodes: (**A**) Multi-flow detector device for the determination of herbicides using PS II as oxidoreductase. Representation of the current decrease with the herbicide's concentrations. Reprinted from [199], Copyright 2005, with permission from Elsevier. (**B**) Scheme of a biosensor using bacterial reaction centers on gold SPE for the determination of Terbutryn. Photocurrents detected by LIFT and Terbutryn curves in absence and presence of 80 µM of 2,3-dimethoxy-5-methyl-p-benzoquinone (UQ_0) (inset the Dixon plot). Reprinted from [200], Copyright 2016, with permission from Elsevier.

8. Conclusions

Screen-printed electrodes (SPEs) are emerging platforms with outstanding potential for their use as transducers in electrochemical (bio)sensing of pesticides. Their well-known advantages in terms of disposability, portability and low-volumes required, among others, make them ideal for the "in field" detection of pesticides at the point of need. Moreover, their versatility and easy modification with different materials is of key relevance for reaching ultralow detection limits that allow to detect the pesticides at the maximum levels allowed by the legislation. In this line, the use of nanomaterials, such as carbon-related ones (graphene, carbon nanotubes) and metallic nanoparticles is the object of an extensive research in recent years.

The (bio)sensing strategy to be followed for the pesticide detection must be carefully studied for each concrete case.

The simpler and faster strategy consists in the direct detection, taking advantage of the electroactivity of some pesticides, that is the presence of functional groups with red-ox properties. However, a limited group of pesticides can be sensitively detected through this route, typically at levels of µM–nM.

Enzymatic sensors are the most widely used, benefitting from the wide range of pesticides able to be detected and the high sensitivity reached, typically at levels of nM–pM. However, the use of enzymes entails important limitations related to their low stability over time and their sensitivity to changes in temperature and pH, among others, so alternative biosensing methods based on antibody receptors, are becoming popular in the last years.

The main advantages of such immunosensors over the enzyme-based ones rely on the higher stability of antibodies together with their superior selectivity and specificity. The detection limits reached are quite similar to the ones obtained through the enzymatic route, typically at pM levels. The high cost and low availability of monoclonal antibodies and the small size of the pesticides that usually avoids sandwich-based approaches are important limitations that should be considered before selecting this biosensing approach.

In conclusion, the selectivity and sensitivity/detection limit required, and the availability of specific enzymes and antibodies are the main parameters that should define the detection strategy more suitable for each particular application.

As far as we know, none of the reported electrochemical (bio)sensors for pesticides detection on SPEs are commercially available yet. In this line, some important issues should be solved for the implantation of such (bio)sensing systems for routine analysis, as alternative to centralized laboratory-based methods (HPLC-MS/MS; GC-MS/MS). Efficiency and long-term stability of the enzymes and the antibodies are crucial issues that are not addressed in most of the reviewed works. Moreover, multi-detection abilities should be strongly required for real applications in pesticides screening. Efforts in this sense should be the next at the current state of the art.

Overall, the higher potentiality of SPE-based pesticide (bio)sensors is in the decentralized "in field" analysis, in our opinion. The combination of such miniaturized electrochemical transducers, the cheap and portable electrochemical instruments and the stability of mainly the antibodies make altogether ideal for such applications. In the case of the immunosensors, the most challenging issue is related to the sampling, washing, etc. steps required, which limits their use by non-skilled people and consequently their commercial implantation. The combination with microfluidics seems to be of key relevance here, so high efforts in this sense are previewed for the coming years.

Author Contributions: B.P.-F.; writing—review and editing, A.C.-G.; funding acquisition, A.d.l.E.-M.; writing—review and editing and funding acquisition. All authors have read and agreed to the published version of the manuscript.

Funding: This research was funded by FC-GRUPIN-ID/2018/000166 project from the Asturias Regional Government and the CTQ2017-86994-R project from the Spanish Ministry Economy and Competitiveness (MINECO). A. de la Escosura-Muñiz acknowledges the Spanish Ministry of Science, Innovation and Universities (MICINN) for the "Ramón y Cajal" Research Fellow (RyC-2016-20299).

Conflicts of Interest: The authors declare no conflict of interest.

References

1. World Health Organization. *Human Biomonitoring: Facts and Figures*; WHO Regional Office for Europe: Copenhagen, Denmark, 2015.
2. WHO (World Health Organization). Available online: https://www.who.int/ (accessed on 30 December 2019).
3. United States Environmental Protection Agency (US EPA). Available online: https://www.epa.gov/ (accessed on 30 December 2019).
4. Consejo de la Unión Europea Directiva 98/83/CE del Consejo. *Off. J. Eur. Communities* **1998**, *L330*, 32–54.
5. Caso, E.L.; Plaguicidas, C.D.E.L.O.S. Análisis de la directiva europea 98/83/CE: Paradigma de la justificación y establecimiento de los valores paramétricos. *El caso concreto de los plaguicidas.* **2012**, *86*, 21–35.
6. Oficial, D.; Comisi, D.E.L.A.; Europeo, P.; Europeo, P.; Alimentarius, C. Scientific support for preparing an EU position for the 45th Session of the Codex Committee on Pesticide Residues (CCPR). *EFSA J.* **2013**, *11*, 1–5.
7. Human Resources Council. United Nations General Assembly. *Report of the Special Rapporteur on the Right to food*. Available online: https://www.ohchr.org/en/issues/food/pages/foodindex.aspx (accessed on 30 December 2019).
8. Anagnostopoulos, C.; Miliadis, G.E. Development and validation of an easy multiresidue method for the determination of multiclass pesticide residues using GC-MS/MS and LC-MS/MS in olive oil and olives. *Talanta* **2013**, *112*, 1–10. [CrossRef] [PubMed]
9. Xiao, Z.; Yang, Y.; Li, Y.; Fan, X.; Ding, S. Determination of neonicotinoid insecticides residues in eels using subcritical water extraction and ultra-performance liquid chromatography-tandem mass spectrometry. *Anal. Chim. Acta* **2013**, *777*, 32–40. [CrossRef] [PubMed]
10. Huang, Y.; Shi, T.; Luo, X.; Xiong, H.; Min, F.; Chen, Y. Determination of multi-pesticide residues in green tea with a modi fi ed QuEChERS protocol coupled to HPLC-MS/MS. *Food Chem.* **2019**, *275*, 255–264. [CrossRef] [PubMed]

11. Timofeeva, I.; Shishov, A.; Kanashina, D.; Dzema, D.; Bulatov, A. On-line in-syringe sugaring-out liquid-liquid extraction coupled with HPLC-MS/MS for the determination of pesticides in fruit and berry juices. *Talanta* **2017**, *167*, 761–767. [CrossRef] [PubMed]
12. Song, N.; Lee, J.Y.; Mansur, A.R.; Jang, H.W.; Lim, M.; Lee, Y. Determination of 60 pesticides in hen eggs using the QuEChERS procedure followed by LC-MS/MS and GC-MS/MS. *Food Chem.* **2019**, *298*, 125050. [CrossRef]
13. Zhu, B.; Xu, X.; Luo, J.; Jin, S.; Chen, W.; Liu, Z. Simultaneous determination of 131 pesticides in tea by on-line GPC-GC – MS/MS using graphitized multi-walled carbon nanotubes as dispersive solid phase extraction sorbent. *Food Chem.* **2019**, *276*, 202–208. [CrossRef]
14. Li, M.; Li, D.; Xiu, G.; Long, Y. Applications of screen-printed electrodes in current environmental analysis. *Curr. Opin. Electrochem.* **2012**, *3*, 137–143. [CrossRef]
15. Mullane, A.P.O. Electrochemistry. *Ref. Modul. Chem. Mol. Sci. Chem. Eng.* **2013**, *2*, 1–3.
16. Oja, S.M.; Wood, M.; Zhang, B. Nanoscale electrochemistry. *Anal. Chem.* **2013**, *85*, 473–486. [CrossRef] [PubMed]
17. Renedo, O.D.; Alonso-Lomillo, M.A.; Martínez, M.J.A. Recent developments in the field of screen-printed electrodes and their related applications. *Talanta* **2007**, *73*, 202–219. [CrossRef] [PubMed]
18. Fanjul-Bolado, P.; Hernández-Santos, D.; Lamas-Ardisana, P.J.; Martín-Pernía, A.; Costa-García, A. Electrochemical characterization of screen-printed and conventional carbon paste electrodes. *Electrochim. Acta* **2008**, *53*, 3635–3642. [CrossRef]
19. Lebik-elhadi, H.; Frontistis, Z.; Ait-amar, H.; Amrani, S. Electrochemical oxidation of pesticide thiamethoxam on boron doped diamond anode: Role of operating parameters and matrix effect. *Process Saf. Environ. Prot.* **2018**, *116*, 535–541. [CrossRef]
20. Chorti, P.; Fischer, J.; Vyskocil, V.; Economou, A.; Barek, J. Electrochimica Acta Voltammetric Determination of Insecticide Thiamethoxam on Silver Solid Amalgam Electrode. *Electrochim. Acta* **2014**, *140*, 5–10. [CrossRef]
21. Chu, Z.; Peng, J.; Jin, W. Advanced nanomaterial inks for screen-printed chemical sensors. *Sens. Actuators B Chem.* **2017**, *243*, 919–926. [CrossRef]
22. Martín-Yerga, D.; Costa Rama, E.; Costa García, A.A. Electrochemical study and determination of electroactive species with screen-printed electrodes. *J. Chem. Educ.* **2016**, *93*, 1270–1276. [CrossRef]
23. Arduini, F.; Micheli, L.; Moscone, D.; Palleschi, G.; Piermarini, S.; Ricci, F.; Volpe, G. Electrochemical biosensors based on nanomodified screen-printed electrodes: Recent applications in clinical analysis. *Trends Anal. Chem.* **2016**, *79*, 114–126. [CrossRef]
24. Fahem, D.K.; El, O.M.; El-rahman, M.K.A.; Zaazaa, H.E. A point of care screen printed potentiometric sensor for therapeutic monitoring of vecuronium. *Microchem. J.* **2019**, *147*, 532–537. [CrossRef]
25. Garcia, P.T.; Guimarães, L.N.; Dias, A.A.; Ulhoa, C.J.; Coltro, W.K.T. Amperometric detection of salivary α-amylase on screen-printed carbon electrodes as a simple and inexpensive alternative for point-of-care testing. *Sens. Actuators B Chem.* **2018**, *258*, 342–348. [CrossRef]
26. Moreira, F.T.C.; Dutra, R.A.F.; Noronha, J.P.C.; Fernandes, J.C.S.; Sales, M.G.F. Novel biosensing device for point-of-care applications with plastic antibodies grown on Au-screen printed electrodes. *Sens. Actuators B Chem.* **2013**, *182*, 733–740. [CrossRef]
27. Kampeera, J.; Pasakon, P.; Karuwan, C.; Arunrut, N. Point-of-care rapid detection of Vibrio parahaemolyticus in seafood using loop-mediated isothermal amplification and graphene-based screen-printed electrochemical sensor. *Biosens. Bioelectron.* **2019**, *132*, 271–278. [CrossRef] [PubMed]
28. Pérez-Fernández, B.; Martín-Yerga, D.; Costa-García, A. Electrodeposition of nickel nanoflowers on screen-printed electrodes and its application to non-enzymatic determination of sugars. *RSC Adv.* **2016**, *6*, 83748–83757. [CrossRef]
29. Pérez-Fernández, B.; Martín-Yerga, D.; Costa-García, A. Galvanostatic electrodeposition of copper nanoparticles on screen-printed carbon electrodes and their application for reducing sugars determination. *Talanta* **2017**, *175*, 108–113. [CrossRef] [PubMed]
30. Martínez-Paredes, G.; González-García, M.B.; Costa-García, A. In situ electrochemical generation of gold nanostructured screen-printed carbon electrodes. Application to the detection of lead underpotential deposition. *Electrochim. Acta* **2009**, *54*, 4801–4808. [CrossRef]
31. Kardaş, F.; Beytur, M.; Akyıldırım, O.; Yüksek, H.; Yola, M.L.; Atar, N. Electrochemical detection of atrazine in wastewater samples by copper oxide (CuO) nanoparticles ionic liquid modified electrode. *J. Mol. Liq.* **2017**, *248*, 360–363. [CrossRef]

32. Zhao, H.; Zhou, C.; Teng, Y.; Chen, C.; Lan, M. Applied Surface Science Novel Pt nanowires modified screen-printed gold electrode by electrodeposited method. *Appl. Surf. Sci.* **2011**, *257*, 3793–3797. [CrossRef]
33. Youse, A.; Babaei, A.; Delavar, M. Application of modified screen-printed carbon electrode with MWCNTs-Pt-doped CdS nanocomposite as a sensitive sensor for determination of natamycin in yoghurt drink and cheese. *J. Electroanal. Chem.* **2018**, *822*, 1–9. [CrossRef]
34. Song, Y.; Chen, J.; Sun, M.; Gong, C.; Shen, Y.; Song, Y.; Wang, L. A simple electrochemical biosensor based on AuNPs / MPS / Au electrode sensing layer for monitoring carbamate pesticides in real samples. *J. Hazard. Mater.* **2016**, *304*, 103–109. [CrossRef]
35. Tajik, S.; Safaei, M.; Beitollahi, H. A sensitive voltammetric sertraline nanosensor based on ZnFe2O4 nanoparticles modified screen printed electrode. *Measurement* **2019**, *143*, 51–57. [CrossRef]
36. Zia, S.; Beitollahi, H.; Allahabadi, H.; Rohani, T. Disposable electrochemical sensor based on modified screen printed electrode for sensitive cabergoline quantification. *J. Electroanal. Chem.* **2019**, *847*, 113223.
37. Chen, Y.; Kirankumar, R.; Kao, C.; Chen, P. Electrodeposited Ag, Au, and AuAg nanoparticles on graphene oxide-modified screen-printed carbon electrodes for the voltammetric determination of free sulfide in alkaline solutions. *Electrochim. Acta* **2016**, *205*, 124–131. [CrossRef]
38. Jeromiyas, N.; Elaiyappillai, E.; Senthil, A. Bismuth nanoparticles decorated graphenated carbon nanotubes modified screen-printed electrode for mercury detection. *J. Taiwan Inst. Chem. Eng.* **2019**, *95*, 466–474. [CrossRef]
39. Xuan, N.; Xuan, N.; Takamura, Y. Development of highly sensitive electrochemical immunosensor based on single-walled carbon nanotube modified screen-printed carbon electrode. *Mater. Chem. Phys.* **2019**, *227*, 123–129.
40. Tpra, R. Electrochemiluminescence study of AuNPs/CdTe-QDs/SWCNTs/chitosan nanocomposite modified carbon nanofiber screen-printed electrode with [Ru(bpy)3]2+/TPrA. *Inorg. Chem. Commun.* **2019**, *106*, 54–60.
41. Singh, M.; Tiwari, I.; Foster, C.W.; Banks, C.E. Highly sensitive and selective determination of dopamine using screen- printed electrodes modified with nanocomposite of N′-phenyl-p-phenylenediamine/multiwalled carbon nanotubes/nafion. *Mater. Res. Bull.* **2018**, *101*, 253–263. [CrossRef]
42. Lamas-Ardisana, P.A.; Queipo, P.; Fanjul-Bolado, P.; Costa-garcia, A. Multiwalled carbon nanotube modified screen-printed electrodes for the detection of p -aminophenol: Optimisation and application in alkaline phosphatase-based assays. *Anal. Chim. Acta* **2008**, *615*, 30–38. [CrossRef]
43. García-González, R.; Fernández-Abedul, M.T.; Costa-García, A. Nafion modified-screen printed gold electrodes and their carbon nanostructuration for electrochemical sensors applications. *Talanta* **2013**, *107*, 376–381. [CrossRef]
44. Calder, J.A.; Nagles, E.; García-beltr, O. Evaluation of the usefulness of a novel electrochemical sensor in detecting uric acid and dopamine in the presence of ascorbic acid using a screen-printed carbon electrode modi fi ed with single walled carbon nanotubes and ionic liquids. *Electrochim. Acta* **2017**, *258*, 512–523.
45. Gutiérrez, C.; Squella, J.A. Carbon nanofiber screen printed electrode joined to a flow injection system for nimodipine sensing. *Sens. Actuators B Chem.* **2015**, *220*, 456–462.
46. Nellaiappan, S.; Senthil, A. Electrocatalytic oxidation and flow injection analysis of isoniazid drug using a gold nanoparticles decorated carbon nanofibers-chitosan modified carbon screen printed electrode in neutral pH. *J. Electroanal. Chem.* **2017**, *801*, 171–178. [CrossRef]
47. Pérez-ràfols, C.; Serrano, N.; Díaz-cruz, J.M.; Ariño, C.; Esteban, M. Glutathione modified screen-printed carbon nanofiber electrode for the voltammetric determination of metal ions in natural samples. *Talanta* **2016**, *155*, 8–13. [CrossRef] [PubMed]
48. Lamas-ardisana, P.J.; Fanjul-bolado, P.; Costa-garcía, A. Manufacture and evaluation of cup-stacked carbon nanofiber-modified screen printed electrodes as electrochemical tools. *JEAC* **2016**, *775*, 129–134. [CrossRef]
49. Samie, H.A.; Arvand, M. RuO_2 nanowires on electrospun CeO_2 -Au nanofibers/functionalized carbon nanotubes / graphite oxide nanocomposite modified screen-printed carbon electrode for simultaneous determination of serotonin, dopamine and ascorbic acid. *J. Alloys Compd.* **2019**, *782*, 824–836. [CrossRef]
50. Chan, K.F.; Lim, H.N.; Shams, N.; Jayabal, S.; Pandikumar, A.; Huang, N.M. Fabrication of graphene / gold-modified screen-printed electrode for detection of carcinoembryonic antigen. *Mater. Sci. Eng. C* **2016**, *58*, 666–674. [CrossRef]

51. Sánchez Calvo, A.; Botas, C.; Martín-Yerga, D.; Álvarez, P.; Menéndez, R.; Costa-García, A. Comparative Study of Screen-Printed Electrodes Modified with Graphene Oxides Reduced by a Constant Current. *J. Electrochem. Soc.* **2015**, *162*, B282–B290. [CrossRef]
52. Jaiswal, N.; Tiwari, I.; Foster, C.W.; Banks, C.E. Highly sensitive amperometric sensing of nitrite utilizing bulk-modified MnO2 decorated Graphene oxide nanocomposite screen-printed electrodes. *Electrochim. Acta* **2017**, *227*, 255–266. [CrossRef]
53. Thunkhamrak, C.; Chuntib, P.; Ounnunkad, K.; Banet, P. Highly sensitive voltammetric immunosensor for the detection of prostate specific antigen based on silver nanoprobe assisted graphene oxide modified screen printed carbon electrode. *Talanta* **2020**, *208*, 120389. [CrossRef]
54. Li, S.; Zhang, Q.; Lu, Y.; Ji, D.; Zhang, D.; Wu, J.; Chen, X.; Liu, Q. One step electrochemical deposition and reduction of graphene oxide on screen printed electrodes for impedance detection of glucose. *Sens. Actuators B Chem.* **2017**, *244*, 290–298. [CrossRef]
55. Singh, M.; Jaiswal, N.; Tiwari, I.; Foster, C.W.; Banks, C.E. A reduced graphene oxide-cyclodextrin-platinum nanocomposite modified screen printed electrode for the detection of cysteine. *J. Electroanal. Chem.* **2018**, *829*, 230–240. [CrossRef]
56. Maity, D.; Minitha, C.R.; Rajendra Kumar, R.T. Glucose oxidase immobilized amine terminated multiwall carbon nanotubes/reduced graphene oxide/polyaniline/gold nanoparticles modified screen-printed carbon electrode for highly sensitive amperometric glucose detection. *Mater. Sci. Eng. C* **2019**, *105*, 110075. [CrossRef] [PubMed]
57. Ibáñez-Redín, G.; Wilson, D.; Gonçalves, D.; Oliveira, O.N., Jr. Low-cost screen-printed electrodes based on electrochemically reduced graphene oxide-carbon black nanocomposites for dopamine, epinephrine and paracetamol detection. *J. Colloid Interface Sci.* **2018**, *515*, 101–108. [CrossRef] [PubMed]
58. Wang, Y.; Huang, B.; Dai, W.; Ye, J.; Xu, B. Sensitive determination of capsaicin on Ag / Ag 2 O nanoparticles / reduced graphene oxide modified screen-printed electrode. *JEAC* **2016**, *776*, 93–100. [CrossRef]
59. Martín-Yerga, D.; Costa-García, A. Stabilization of electrogenerated copper species at quantum dots-modified electrodes. *Phys. Chem. Chem. Phys.* **2017**, *74*, 53–56.
60. Kokkinos, C.; Prodromidis, M.; Economou, A.; Petrou, P. Disposable integrated bismuth citrate-modified screen-printed immunosensor for ultrasensitive quantum dot-based electrochemical assay of C-reactive protein in human serum. *Anal. Chim. Acta* **2015**, *886*, 29–36. [CrossRef]
61. Martín-yerga, D.; González-garcía, M.B.; Costa-garcía, A. Electrochemical immunosensor for anti-tissue transglutaminase antibodies based on the in situ detection of quantum dots. *Talanta* **2014**, *130*, 598–602. [CrossRef]
62. García, C.; Navarro, F.; Celis, F.; Ruiz, D.; Toledo, S.; Sanhueza, L.; Qui, A. Electrochemical, spectroscopic and electrochemiluminescent characterization of self-assembled 3-aminopropyltriethoxysilane/CdTe quantum dots hybrids on screen-printed electrodes. *Electrochim. Acta* **2018**, *276*, 64–72. [CrossRef]
63. Roushani, M.; Jalilian, Z.; Nezhadali, A. Screen printed carbon electrode sensor with thiol graphene quantum dots and gold nanoparticles for voltammetric determination of solatol. *Heliyon* **2019**, *5*, e01984. [CrossRef]
64. Campuzano, S.; Paloma, Y.; Asadpour-zeynali, K. Ultrasensitive determination of receptor tyrosine kinase with a label- free electrochemical immunosensor using graphene quantum dots- modified screen-printed electrodes. *Anal. Chim. Acta* **2018**, *1011*, 28–34.
65. Centi, S.; Laschi, S.; Fr, M.; Mascini, M. A disposable immunomagnetic electrochemical sensor based on functionalised magnetic beads and carbon-based screen-printed electrodes (SPCEs) for the detection of polychlorinated biphenyls (PCBs). *Anal. Chim. Acta* **2005**, *538*, 205–212. [CrossRef]
66. Eguílaz, M.; Moreno-guzmán, M.; Campuzano, S.; González-cortés, A.; Pingarrón, J.M.; Yá, P. An electrochemical immunosensor for testosterone using functionalized magnetic beads and screen-printed carbon electrodes. *Biosens. Bioelectron.* **2010**, *26*, 517–522. [CrossRef] [PubMed]
67. Biscay, J.; Begoña, M.; García, G.; Costa, A. Electrochemical biotin detection based on magnetic beads and a new magnetic flow cell for screen printed electrode. *Talanta* **2015**, *131*, 706–711. [CrossRef] [PubMed]
68. Biscay, J.; García, M.B.G.; García, A.C. Electrochemical biotin determination based on a screen printed carbon electrode array and magnetic beads. *Sens. Actuators B Chem.* **2014**, *205*, 426–432. [CrossRef]
69. Panraksa, Y.; Siangproh, W.; Khampieng, T.; Chailapakul, O.; Apilux, A. Paper-based amperometric sensor for determination of acetylcholinesterase using screen-printed graphene electrode. *Talanta* **2018**, *178*, 1017–1023. [CrossRef]

70. Biscay, J.; Rama, E.C.; García, M.B.G.; Carrazón, J.M.P.; García, A.C. Enzymatic sensor using mediator-screen-printed carbon electrodes. *Electroanalysis* **2011**, *23*, 209–214. [CrossRef]
71. Wang, J.; Nascimento, V.B.; Kane, S.A.; Rogers, K.; Smyth, M.R.; Angnes, L. Screen-printed tyrosinase-containing electrodes for the biosensing of enzyme inhibitors. *Talanta* **1996**, *43*, 1903–1907. [CrossRef]
72. Alonso-lomillo, M.A.; Domínguez-renedo, O. Sensitive enzyme-biosensor based on screen-printed electrodes for Ochratoxin A. *Biosens. Bioelectron.* **2010**, *25*, 1333–1337. [CrossRef]
73. Alvarado-gámez, A.L.; Alonso-lomillo, M.A.; Domínguez-renedo, O.; Arcos-martínez, M.J. Vanadium determination in water using alkaline phosphatase based screen-printed carbon electrodes modified with gold nanoparticles. *J. Electroanal. Chem.* **2013**, *693*, 51–55. [CrossRef]
74. Biscay, J.; Costa Rama, E.; González García, M.B.; Julio Reviejo, A.; Pingarrón Carrazón, J.M.; García, A.C. Amperometric fructose sensor based on ferrocyanide modified screen-printed carbon electrode. *Talanta* **2012**, *88*, 432–438. [CrossRef]
75. Karim, N.; Jin, H. Amperometric phenol biosensor based on covalent immobilization of tyrosinase on Au nanoparticle modified screen printed carbon electrodes. *Talanta* **2013**, *116*, 991–996. [CrossRef] [PubMed]
76. Soln, R.; Ruzgas, T.; Skl, P. Amperometric screen-printed biosensor arrays with co-immobilised oxidoreductases and cholinesterases. *Anal. Chim. Acta* **2005**, *528*, 9–19. [CrossRef]
77. Sapelniko, S.; Dock, E.; Ruzgas, T.; Emne, J. Amperometric sensors based on tyrosinase-modified screen-printed arrays. *Talanta* **2003**, *61*, 473–483. [CrossRef]
78. Hatada, M.; Tsugawa, W.; Kamio, E.; Loew, N.; Klonoff, D.C. Development of a screen-printed carbon electrode based disposable enzyme sensor strip for the measurement of glycated albumin. *Biosens. Bioelectron.* **2017**, *88*, 167–173. [CrossRef]
79. Cerrato-alvarez, M.; Bernalte, E.; Bernalte-garcía, M.J.; Pinilla-gil, E. Fast and direct amperometric analysis of polyphenols in beers using tyrosinase-modified screen-printed gold nanoparticles biosensors. *Talanta* **2019**, *193*, 93–99. [CrossRef]
80. Dontsova, E.A.; Zeifman, Y.S.; Budashov, I.A.; Eremenko, A.V.; Kalnov, S.L.; Kurochkin, I.N. Screen-printed carbon electrode for choline based on MnO_2 nanoparticles and choline oxidase/polyelectrolyte layers. *Sens. Actuators B Chem.* **2011**, *159*, 261–270. [CrossRef]
81. Yan, M.; Zang, D.; Ge, S.; Ge, L.; Yu, J. A disposable electrochemical immunosensor based on carbon screen-printed electrodes for the detection of prostate specific antigen. *Biosens. Bioelectron.* **2012**, *38*, 355–361. [CrossRef]
82. Goud, K.Y.; Kumar, V.S.; Hayat, A.; Gobi, K.V.; Song, H.; Kim, K.; Louis, J. A highly sensitive electrochemical immunosensor for zearalenone using screen-printed disposable electrodes. *J. Electroanal. Chem.* **2019**, *832*, 336–342. [CrossRef]
83. Hern, D. Development of an immunosensor for the determination of rabbit IgG using streptavidin modified screen-printed carbon electrodes. *Talanta* **2005**, *65*, 565–573.
84. Deepthy, S.; Layek, K.; Mukherjee, R.; Kumar, K.; Ghosh, M. Development of screen-printed electrode based immunosensor for the detection of HER2 antigen in human serum samples. *Bioelectrochemistry* **2017**, *118*, 25–30.
85. Jampasa, S.; Siangproh, W.; Laocharoensuk, R.; Vilaivan, T. Electrochemical detection of c-reactive protein based on anthraquinone- labeled antibody using a screen-printed graphene electrode. *Talanta* **2018**, *183*, 311–319. [CrossRef] [PubMed]
86. Baradoke, A.; Jose, B.; Pauliukaite, R.; Forster, R.J. Properties of Anti-CA125 antibody layers on screen-printed carbon electrodes modified by gold and platinum nanostructures. *Electrochim. Acta* **2019**, *306*, 299–306. [CrossRef]
87. Ibáñez-redín, G.; Furuta, R.H.M.; Wilson, D.; Shimizu, F.M.; Materon, E.M.; Maria, L.; Batista, R.; Melendez, M.E.; Carvalho, A.L.; Manuel, R.; et al. Screen-printed interdigitated electrodes modified with nanostructured carbon nano-onion films for detecting the cancer biomarker CA19-9. *Mater. Sci. Eng. C* **2019**, *99*, 1502–1508. [CrossRef] [PubMed]
88. Eissa, S. A comparison of the performance of voltammetric aptasensors for glycated haemoglobin on different carbon nanomaterials-modified screen printed electrodes. *Mater. Sci. Eng. C* **2019**, *101*, 423–430. [CrossRef]
89. Erdem, A.; Congur, G.; Mayer, G. Aptasensor platform based on carbon nanofibers enriched screen printed electrodes for impedimetric detection of thrombin. *JEAC* **2015**, *758*, 12–19. [CrossRef]

90. Xie, D.; Li, C.; Shangguan, L.; Qi, H.; Xue, D.; Gao, Q. Click chemistry-assisted self-assembly of DNA aptamer on gold nanoparticles-modified screen-printed carbon electrodes for label-free electrochemical aptasensor. *Sens. Actuators B Chem.* **2014**, *192*, 558–564. [CrossRef]
91. Hashkavayi, A.B.; Raoof, J.B. Design an aptasensor based on structure-switching aptamer on dendritic gold nanostructures/Fe_3O_4@SiO_2/DABCO modified screen printed electrode for highly selective detection of epirubicin. *Biosens. Bioelectron.* **2017**, *91*, 650–657. [CrossRef]
92. Yeh, F.; Liu, T.; Tseng, I.; Yang, C.; Lu, L.; Lin, C. Gold nanoparticles conjugates-amplified aptamer immunosensing screen-printed carbon electrode strips for thrombin detection. *Biosens. Bioelectron.* **2014**, *61*, 336–343. [CrossRef]
93. Jo, H.; Her, J.; Lee, H.; Shim, Y.; Ban, C. Highly sensitive amperometric detection of cardiac troponin I using sandwich aptamers and screen-printed carbon electrodes. *Talanta* **2017**, *165*, 442–448. [CrossRef]
94. Ren, R.; Leng, C.; Zhang, S. A chronocoulometric DNA sensor based on screen-printed electrode doped with ionic liquid and polyaniline nanotubes. *Biosens. Bioelectron.* **2010**, *25*, 2089–2094. [CrossRef]
95. Evtugyn, G.; Mingaleva, A.; Budnikov, H.; Stoikova, E.; Vinter, V.; Eremin, S. Affinity biosensors based on disposable screen-printed electrodes modified with DNA. *Anal. Chim. Acta* **2003**, *479*, 125–134. [CrossRef]
96. Malecka, K.; Stachyra, A.; Góra-sochacka, A.; Sirko, A.; Zagórski-ostoja, W.; Radecka, H.; Radecki, J. Electrochemical genosensor based on disc and screen printed gold electrodes for detection of specific DNA and RNA sequences derived from Avian Influenza Virus H5N1. *Sens. Actuators B Chem.* **2016**, *224*, 290–297. [CrossRef]
97. Khairy, M.; Khorshed, A.A.; Rashwan, F.A.; Salah, G.A.; Abdel-wadood, H.M.; Banks, C.E. Sensitive determination of amlodipine besylate using bare/unmodified and DNA-modified screen-printed electrodes in tablets and biological fluids. *Sens. Actuators B Chem.* **2017**, *239*, 768–775. [CrossRef]
98. Zhang, Y.; Geng, X.; Ai, J.; Gao, Q.; Qi, H. Signal amplification detection of DNA using a sensor fabricated by one-step covalent immobilization of amino-terminated probe DNA onto the polydopamine-modified screen-printed carbon electrode. *Sens. Actuators B Chem.* **2015**, *221*, 1535–1541. [CrossRef]
99. Jiang, D.; Liu, Y.; Jiang, H.; Rao, S.; Fang, W.; Wu, M.; Yuan, L. A novel screen-printed mast cell-based electrochemical sensor for detecting spoilage bacterial quorum signaling molecules (N-acyl-homoserine-lactones) in freshwater fish. *Biosens. Bioelectron.* **2018**, *102*, 396–402. [CrossRef]
100. Chang, J.; Lien, C.; Vijayakumar, P.S.; Hsieh, P.; Zen, J. Electrochemical regulation of microbial growth on disposable screen printed carbon electrodes. *Electrochim. Acta* **2012**, *82*, 103–108. [CrossRef]
101. Wen, J.; He, D.; Yu, Z.; Zhou, S. In situ detection of microbial c-type cytochrome based on intrinsic peroxidase-like activity using screen-printed carbon electrode. *Biosens. Bioelectron.* **2018**, *113*, 52–57. [CrossRef]
102. Mulchandani, A.; Rajesh. Microbial biosensors for organophosphate pesticides. *Appl. Biochem. Biotechnol.* **2011**, *165*, 687–699. [CrossRef]
103. Kurbanoglu, S.; Ozkan, S.A.; Merkoçi, A. Nanomaterials-based enzyme electrochemical biosensors operating through inhibition for biosensing applications. *Biosens. Bioelectron.* **2017**, *89*, 886–898. [CrossRef]
104. Dou, J.; Fan, F.; Ding, A.; Cheng, L.; Sekar, R.; Wang, H.; Li, S. A screen-printed, amperometric biosensor for the determination of organophosphorus pesticides in water samples. *J. Environ. Sci.* **2012**, *24*, 956–962. [CrossRef]
105. Arduini, F.; Ricci, F.; Tuta, C.S.; Moscone, D.; Amine, A.; Palleschi, G. Detection of carbamic and organophosphorous pesticides in water samples using a cholinesterase biosensor based on Prussian Blue-modified screen-printed electrode. *Anal. Chim. Acta* **2006**, *580*, 155–162. [CrossRef] [PubMed]
106. Istamboulie, G.; Sikora, T.; Jubete, E.; Ochoteco, E.; Marty, J.; Noguer, T. Screen-printed poly (3, 4-ethylenedioxythiophene) (PEDOT): A new electrochemical mediator for acetylcholinesterase-based biosensors. *Talanta* **2010**, *82*, 957–961. [CrossRef] [PubMed]
107. Silva, G.; Jeanty, G.; Marty, J. Enzyme immobilization procedures on screen-printed electrodes used for the detection of anticholinesterase pesticides Comparative study. *Anal. Chim. Acta* **2004**, *523*, 107–115. [CrossRef]
108. Ivanov, A.N.; Younusov, R.R.; Evtugyn, G.A.; Arduini, F.; Moscone, D.; Palleschi, G. Acetylcholinesterase biosensor based on single-walled carbon nanotubes—Co phtalocyanine for organophosphorus pesticides detection. *Talanta* **2011**, *85*, 216–221. [CrossRef] [PubMed]
109. Haddaoui, M.; Raouafi, N. Chlortoluron-induced enzymatic activity inhibition in tyrosinase/ZnO NPs/SPCE biosensor for the detection of ppb levels of herbicide. *Sens. Actuators B Chem.* **2015**, *219*, 171–178. [CrossRef]

110. Wang, H.; Zhao, G.; Chen, D.; Wang, Z.; Liu, G. A sensitive acetylcholinesterase biosensor based on screen printed electrode modified with Fe3O4 nanoparticle and graphene for chlorpyrifos determination. *Int. J. Electrochem. Sci.* **2016**, *11*, 10906–10918. [CrossRef]
111. Shi, M.; Xu, J.; Zhang, S.; Liu, B.; Kong, J. A mediator-free screen-printed amperometric biosensor for screening of organophosphorus pesticides with flow-injection analysis (FIA) system. *Talanta* **2006**, *68*, 1089–1095. [CrossRef]
112. Cinti, S.; Neagu, D.; Carbone, M.; Cacciotti, I.; Moscone, D.; Arduini, F. Novel carbon black-cobalt phthalocyanine nanocomposite as sensing platform to detect organophosphorus pollutants at screen-printed electrode. *Electrochim. Acta* **2016**, *188*, 574–581. [CrossRef]
113. Crew, A.; Lonsdale, D.; Byrd, N.; Pittson, R.; Hart, J.P. A screen-printed, amperometric biosensor array incorporated into a novel automated system for the simultaneous determination of organophosphate pesticides. *Biosens. Bioelectron.* **2011**, *26*, 2847–2851. [CrossRef]
114. Arduini, F.; Guidone, S.; Amine, A.; Palleschi, G.; Moscone, D. Acetylcholinesterase biosensor based on self-assembled monolayer-modified gold-screen printed electrodes for organophosphorus insecticide detection. *Sens. Actuators B Chem.* **2013**, *179*, 201–208. [CrossRef]
115. De Albuquerque, Y.D.T.; Ferreira, L.F. Amperometric biosensing of carbamate and organophosphate pesticides utilizing screen-printed tyrosinase-modified electrodes. *Anal. Chim. Acta* **2007**, *596*, 210–221. [CrossRef]
116. Gogol, E.V.; Evtugyn, G.A.; Marty, J.; Budnikov, H.C.; Winter, V.G. Amperometric biosensors based on nafion coated screen-printed electrodes for the determination of cholinesterase inhibitors. *Talanta* **2000**, *53*, 379–389. [CrossRef]
117. Sajjadi, S.; Ghourchian, H.; Tavakoli, H. Choline oxidase as a selective recognition element for determination of paraoxon. *Biosens. Bioelectron.* **2009**, *24*, 2509–2514. [CrossRef]
118. Dounin, V.; Veloso, A.J.; Schulze, H.; Bachmann, T.T.; Kerman, K. Disposable electrochemical printed gold chips for the analysis of acetylcholinesterase inhibition. *Anal. Chim. Acta* **2010**, *669*, 63–67. [CrossRef]
119. Mayorga-Martinez, C.C.; Pino, F.; Kurbanoglu, S.; Rivas, L.; Ozkan, S.A.; Merkoçi, A. Iridium oxide nanoparticle induced dual catalytic/inhibition based detection of phenol and pesticide compounds. *J. Mater. Chem. B* **2014**, *2*, 2233–2239. [CrossRef]
120. Catalina, D.; Carvajal, S.; Peñuela, G. Effect of chlorpyrifos on the inhibition of the enzyme acetylcholinesterase by cross-linking in water-supply samples and milk from dairy cattle. *Talanta* **2013**, *111*, 1–7. [CrossRef]
121. Chen, D.; Liu, Z.; Fu, J.; Guo, Y.; Sun, X.; Yang, Q.; Wang, X. Electrochemical acetylcholinesterase biosensor based on multi-walled carbon nanotubes/dicyclohexyl phthalate modified screen-printed electrode for detection of chlorpyrifos. *J. Electroanal. Chem.* **2017**, *801*, 185–191. [CrossRef]
122. Shi, Q.; Teng, Y.; Zhang, Y.; Liu, W. Rapid detection of organophosphorus pesticide residue on Prussian blue modified dual-channel screen-printed electrodes combing with portable potentiostat. *Chin. Chem. Lett.* **2018**, *29*, 1379–1382. [CrossRef]
123. Domínguez-renedo, O.; Alonso-lomillo, M.A.; Recio-cebrián, P.; Arcos-martínez, M.J. Screen-printed acetylcholinesterase-based biosensors for inhibitive determination of permethrin. *Sci. Total Environ.* **2012**, *426*, 346–350. [CrossRef]
124. Jin, R.; Kong, D.; Zhao, X.; Li, H.; Yan, X.; Liu, F.; Sun, P.; Du, D.; Lin, Y.; Lu, G. Tandem catalysis driven by enzymes directed hybrid nanoflowers for on-site ultrasensitive detection of organophosphorus pesticide. *Biosens. Bioelectron.* **2019**, *141*, 111473. [CrossRef]
125. Arduini, F.; Forchielli, M.; Amine, A.; Neagu, D.; Cacciotti, I.; Nanni, F.; Moscone, D.; Palleschi, G. Screen-printed biosensor modified with carbon black nanoparticles for the determination of paraoxon based on the inhibition of butyrylcholinesterase. *Microchim. Acta* **2014**, *182*, 643–651. [CrossRef]
126. Gan, N.; Yang, X.; Xie, D.; Wu, Y.; Wen, W. A disposable organophosphorus pesticides enzyme biosensor based on magnetic composite nano-particles modified screen printed carbon electrode. *Sensors* **2010**, *10*, 625–638. [CrossRef]
127. Ivanov, A.; Evtugyn, G.; Budnikov, H.; Ricci, F.; Moscone, D.; Palleschi, G. Cholinesterase sensors based on screen-printed electrodes for detection of organophosphorus and carbamic pesticides. *Anal. Bioanal. Chem.* **2003**, *377*, 624–631. [CrossRef]
128. Chen, D.; Jiao, Y.; Jia, H.; Guo, Y.; Sun, X.; Wang, X.; Xu, J. Acetylcholinesterase biosensor for chlorpyrifos detection based on multi-walled carbon nanotubes-SnO_2-chitosan nanocomposite modified screen-printed electrode. *Int. J. Electrochem. Sci.* **2015**, *10*, 10491–10501.

129. El-Moghazy, A.Y.; Soliman, E.A.; Ibrahim, H.Z.; Marty, J.L.; Istamboulie, G.; Noguer, T. Biosensor based on electrospun blended chitosan-poly (vinyl alcohol) nanofibrous enzymatically sensitized membranes for pirimiphos-methyl detection in olive oil. *Talanta* **2016**, *155*, 258–264. [CrossRef]
130. Zhang, Q.; Xu, Q.; Guo, Y.; Sun, X.; Wang, X. Acetylcholinesterase biosensor based on the mesoporous carbon/ferroferric oxide modified electrode for detecting organophosphorus pesticides. *RSC Adv.* **2016**, *6*, 24698–24703. [CrossRef]
131. Tang, W.; Fan, K.; Zhou, J.; Wu, J.; Ji, F. Unmodified screen-printed silver electrode for facile detection of organophosphorus pesticide. *Ionics* **2015**, *21*, 587–592. [CrossRef]
132. Joshi, K.A.; Tang, J.; Haddon, R.; Wang, J.; Chen, W.; Mulchandani, A. A disposable biosensor for organophosphorus nerve agents based on carbon nanotubes modified thick film strip electrode. *Electroanalysis* **2005**, *17*, 54–58. [CrossRef]
133. Sinha, R.; Ganesana, M.; Andreescu, S.; Stanciu, L. AChE biosensor based on zinc oxide sol-gel for the detection of pesticides. *Anal. Chim. Acta* **2010**, *661*, 195–199. [CrossRef]
134. Bonnet, C.; Andreescu, S.; Marty, J. Adsorption: An easy and efficient immobilisation of acetylcholinesterase on screen-printed electrodes. *Anal. Chim. Acta* **2003**, *481*, 209–211. [CrossRef]
135. Chen, D.; Fu, J.; Liu, Z.; Guo, Y.; Sun, X.; Wang, X.; Wang, Z. A Simple acetylcholinesterase biosensor based on ionic liquid/multiwalled carbon nanotubes-modified screen-printed electrode for rapid detecting chlorpyrifos. *Int. J. Electrochem. Sci.* **2017**, *12*, 9465–9477. [CrossRef]
136. Zhang, L.; Zhang, A.; Du, D.; Lin, Y. Biosensor based on Prussian blue nanocubes/reduced graphene oxide nanocomposite for detection of organophosphorus pesticides. *Nanoscale* **2012**, *4*, 4674–4679. [CrossRef]
137. Bucur, B.; Fournier, D.; Danet, A.; Marty, J.L. Biosensors based on highly sensitive acetylcholinesterases for enhanced carbamate insecticides detection. *Anal. Chim. Acta* **2006**, *562*, 115–121. [CrossRef]
138. Laschi, S.; Ogończyk, D.; Palchetti, I.; Mascini, M. Evaluation of pesticide-induced acetylcholinesterase inhibition by means of disposable carbon-modified electrochemical biosensors. *Enzyme Microb. Technol.* **2007**, *40*, 485–489. [CrossRef]
139. Del Carlo, M.; Mascini, M.; Pepe, A.; Compagnone, D.; Mascini, M. Electrochemical bioassay for the investigation of chlorpyrifos-methyl in vine samples. *J. Agric. Food Chem.* **2002**, *50*, 7206–7210. [CrossRef]
140. Li, Y.G.; Zhou, Y.X.; Feng, J.L.; Jiang, Z.H.; Ma, L.R. Immobilization of enzyme on screen-printed electrode by exposure to glutaraldehyde vapour for the construction of amperometric acetylcholinesterase electrodes. *Anal. Chim. Acta* **1999**, *382*, 277–282. [CrossRef]
141. Del Carlo, M.; Mascini, M.; Pepe, A.; Diletti, G.; Compagnone, D. Screening of food samples for carbamate and organophosphate pesticides using an electrochemical bioassay. *Food Chem.* **2004**, *84*, 651–656. [CrossRef]
142. Law, K.A.; Higson, S.P.J. Sonochemically fabricated acetylcholinesterase micro-electrode arrays within a flow injection analyser for the determination of organophosphate pesticides. *Biosens. Bioelectron.* **2005**, *20*, 1914–1924. [CrossRef]
143. Di Sioudi, B.D.; Miller, C.E.; Lai, K.; Grimsley, J.K.; Wild, J.R. Rational design of organophosphorus hydrolase for altered substrate specificities. *Chem. Biol. Interact.* **1999**, *119–120*, 211–223. [CrossRef]
144. Efremenko, E.N.; Sergeeva, V.S. Organophosphate hydrolase—An enzyme catalyzing degradation of phosphorus-containing toxins and pesticides. *Russ. Chem. Bull.* **2001**, *50*, 1826–1832. [CrossRef]
145. Shimazu, M.; Mulchandani, A.; Chen, W. Simultaneous degradation of organophosphorus pesticides and p-nitrophenol by a genetically engineered Moraxella sp. with surface-expressed organophosphorus hydrolase. *Biotechnol. Bioeng.* **2001**, *76*, 318–324. [CrossRef]
146. Chough, S.H.; Mulchandani, A.; Mulchandani, P.; Chen, W.; Wang, J.; Rogers, K.R. Organophosphorus Hydrolase-Based Amperometric Sensor: Modulation of Sensitivity and Substrate Selectivity. *Electroanalysis* **2002**, *14*, 273–276. [CrossRef]
147. Lei, Y.; Mulchandani, P.; Wang, J.; Chen, W.; Mulchandani, A. Highly sensitive and selective amperometric microbial biosensor for direct determination of p-nitrophenyl-substituted organophosphate nerve agents. *Environ. Sci. Technol.* **2005**, *39*, 8853–8857. [CrossRef]
148. Mulchandani, A.; Chen, W.; Mulchandani, P.; Wang, J.; Rogers, K.R. Biosensors for direct determination of organophosphate pesticides. *Biosens. Bioelectron.* **2001**, *16*, 225–230. [CrossRef]
149. Mulchandani, A. Amperometric Thick-Film Strip Electrodes for Monitoring Organophosphate Nerve Agents Based on Immobilized Organophosphorus Hydrolase. *Anal. Chem.* **1999**, *71*, 2246–2249. [CrossRef]

150. Joshi, K.A.; Prouza, M.; Kum, M.; Wang, J.; Tang, J.; Haddon, R.; Chen, W.; Mulchandani, A. V-type nerve agent detection using a carbon nanotube-based amperometric enzyme electrode. *Anal. Chem.* **2006**, *78*, 331–336. [CrossRef]
151. Zhao, Y.; Zhang, W.; Lin, Y.; Du, D. The vital function of Fe_3O_4@Au nanocomposites for hydrolase biosensor design and its application in detection of methyl parathion. *Nanoscale* **2013**, *5*, 1121–1126. [CrossRef]
152. Sacks, V.; Eshkenazi, I.; Neufeld, T.; Dosoretz, C.; Rishpon, J. Immobilized Parathion Hydrolase: An Amperometric Sensor for Parathion. *Anal. Chem.* **2000**, *72*, 2055–2058. [CrossRef]
153. Mulyasuryani, A.; Dofir, M. Enzyme Biosensor for Detection of Organophosphate Pesticide Residues Base on Screen Printed Carbon Electrode (SPCE)-Bovine Serum Albumin (BSA). *Engineering* **2014**, *6*, 230–235. [CrossRef]
154. Primožič, M.; Čolnik, M.; Knez, Ž.; Leitgeb, M. Advantages and disadvantages of using SC CO_2 for enzyme release from halophilic fungi. *J. Supercrit. Fluids* **2019**, *143*, 286–293. [CrossRef]
155. Brahim, M.B.; Belhadj Ammar, H.; Abdelhédi, R.; Samet, Y. Electrochemical behavior and analytical detection of Imidacloprid insecticide on a BDD electrode using square-wave voltammetric method. *Chin. Chem. Lett.* **2016**, *27*, 666–672. [CrossRef]
156. Zhang, M.; Zhao, H.T.; Xie, T.J.; Yang, X.; Dong, A.J.; Zhang, H.; Wang, J.; Wang, Z.Y. Molecularly imprinted polymer on graphene surface for selective and sensitive electrochemical sensing imidacloprid. *Sens. Actuators B Chem.* **2017**, *252*, 991–1002. [CrossRef]
157. Geto, A.; Safaa, J.; Mortensen, J.; Svendsen, W.E.; Dimaki, M. Electrochemical determination of bentazone using simple screen-printed carbon electrodes. *Environ. Int.* **2019**, *129*, 400–407. [CrossRef] [PubMed]
158. Della Pelle, F.; Angelini, C.; Sergi, M.; Del Carlo, M.; Pepe, A.; Compagnone, D. Nano carbon black-based screen printed sensor for carbofuran, isoprocarb, carbaryl and fenobucarb detection: Application to grain samples. *Talanta* **2018**, *186*, 389–396. [CrossRef] [PubMed]
159. Lezi, N.; Economou, A. Voltammetric Determination of Neonicotinoid Pesticides at Disposable Screen-Printed Sensors Featuring a Sputtered Bismuth Electrode. *Electroanalysis* **2015**, *27*, 2313–2321. [CrossRef]
160. Khairy, M.; Ayoub, H.A.; Banks, C.E. Non-enzymatic electrochemical platform for parathion pesticide sensing based on nanometer-sized nickel oxide modified screen-printed electrodes. *Food Chem.* **2018**, *255*, 104–111. [CrossRef] [PubMed]
161. Charoenkitamorn, K.; Chailapakul, O.; Siangproh, W. Development of gold nanoparticles modified screen-printed carbon electrode for the analysis of thiram, disulfiram and their derivative in food using ultra-high performance liquid chromatography. *Talanta* **2015**, *132*, 416–423. [CrossRef]
162. Zhang, C.; Zhao, F.; She, Y.; Hong, S.; Cao, X.; Zheng, L.; Wang, S.; Li, T.; Wang, M.; Jin, M.; et al. A disposable molecularly imprinted sensor based on Graphe@AuNPs modified screen-printed electrode for highly selective and sensitive detection of cyhexatin in pear samples. *Sens. Actuators B Chem.* **2019**, *284*, 13–22. [CrossRef]
163. Jubete, E.; Zelechowska, K.; Loaiza, O.A.; Lamas, P.J.; Ochoteco, E.; Farmer, K.D.; Roberts, K.P.; Biernat, J.F. Derivatization of SWCNTs with cobalt phthalocyanine residues and applications in screen printed electrodes for electrochemical detection of thiocholine. *Electrochim. Acta* **2011**, *56*, 3988–3995. [CrossRef]
164. Habekost, A. Rapid and sensitive spectroelectrochemical and electrochemical detection of glyphosate and AMPA with screen-printed electrodes. *Talanta* **2017**, *162*, 583–588. [CrossRef]
165. Karuppiah, C.; Palanisamy, S.; Chen, S.; Kannan, S.; Periakaruppan, P. A novel and sensitive amperometric hydrazine sensor based on gold nanoparticles decorated graphite nanosheets modified screen printed carbon electrode. *Electrochim. Acta* **2014**, *139*, 157–164. [CrossRef]
166. Thirumalraj, B.; Rajkumar, C.; Chen, S.; Lin, K. Determination of 4-nitrophenol in water by use of a screen-printed carbon electrode modified with chitosan-crafted ZnO nanoneedles. *J. Colloid Interface Sci.* **2017**, *499*, 83–92. [CrossRef] [PubMed]
167. Arvinte, A.; Mahosenaho, M.; Pinteala, M.; Sesay, A.M.; Virtanen, V. Electrochemical oxidation of p-nitrophenol using graphene-modified electrodes, and a comparison to the performance of MWNT-based electrodes. *Microchim. Acta* **2011**, *174*, 337–343. [CrossRef]
168. De Oliveira, R.; Amorim, É.; Rogério, A. Electrochemically activated multi-walled carbon nanotubes modified screen-printed electrode for voltammetric determination of sulfentrazone. *J. Electroanal. Chem.* **2019**, *835*, 220–226.

169. Sun, H.; Zhao, S.; Qu, F. Gold nanoparticles modified ceria nanoparticles for the oxidation of hydrazine with disposable screen-printed electrode. *Measurement* **2012**, *45*, 1111–1113. [CrossRef]
170. Pino, F.; Mayorga-Martinez, C.C.; Merkoçi, A. High-performance sensor based on copper oxide nanoparticles for dual detection of phenolic compounds and a pesticide. *Electrochem. Commun.* **2016**, *71*, 33–37. [CrossRef]
171. Saengsookwaow, C.; Rangkupan, R.; Chailapakul, O. Nitrogen-doped graphene-polyvinylpyrrolidone/gold nanoparticles modified electrode as a novel hydrazine sensor. *Sens. Actuators B Chem.* **2016**, *227*, 524–532. [CrossRef]
172. Chuntib, P.; Themsirimongkon, S.; Saipanya, S.; Jakmunee, J. Sequential injection differential pulse voltammetric method based on screen printed carbon electrode modified with carbon nanotube/Nafion for sensitive determination of paraquat. *Talanta* **2017**, *170*, 1–8. [CrossRef]
173. Jirasirichote, A.; Punrat, E.; Suea-ngam, A.; Chailapakul, O. Voltammetric detection of carbofuran determination using screen-printed carbon electrodes modified with gold nanoparticles and graphene oxide. *Talanta* **2017**, *175*, 331–337. [CrossRef]
174. Govindasamy, M.; Mani, V.; Chen, S.M.; Chen, T.W.; Sundramoorthy, A.K. Methyl parathion detection in vegetables and fruits using silver@graphene nanoribbons nanocomposite modified screen printed electrode. *Sci. Rep.* **2017**, *7*, 46471. [CrossRef]
175. Arduini, F.; Cassisi, A.; Amine, A.; Ricci, F.; Moscone, D.; Palleschi, G. Electrocatalytic oxidation of thiocholine at chemically modified cobalt hexacyanoferrate screen-printed electrodes. *J. Electroanal. Chem.* **2009**, *626*, 66–74. [CrossRef]
176. Mehta, J.; Vinayak, P.; Tuteja, S.K.; Chhabra, V.A.; Bhardwaj, N.; Paul, A.K.; Kim, K.; Deep, A. Graphene modified screen printed immunosensor for highly sensitive detection of parathion. *Biosens. Bioelectron.* **2016**, *83*, 339–346. [CrossRef] [PubMed]
177. Mehta, J.; Bhardwaj, N.; Bhardwaj, S.K.; Tuteja, S.K.; Vinayak, P.; Paul, A.K.; Kim, K.; Deep, A. Graphene quantum dot modified screen printed immunosensor for the determination of parathion. *Anal. Biochem.* **2017**, *523*, 1–9. [CrossRef] [PubMed]
178. Dzantiev, B.B.; Yazynina, E.V.; Zherdev, A.V.; Plekhanova, Y.V.; Reshetilov, A.N.; Chang, S.; Mcneil, C.J. Determination of the herbicide chlorsulfuron by amperometric sensor based on separation-free bienzyme immunoassay. *Sens. Actuators B Chem.* **2004**, *98*, 254–261. [CrossRef]
179. Pérez-Fernández, B.; Mercader, J.V.; Checa-Orrego, B.I.; de la Escosura-Muñiz, A.; Costa-García, A. A monoclonal antibody-based immunosensor for the electrochemical detection of imidacloprid pesticide. *Analyst* **2019**, *144*, 2936–2941. [CrossRef]
180. Pérez-fernández, B.; Mercader, J.V.; Abad-fuentes, A.; Brenda, I.; De Escosura-muñiz, A.; Costa-garcía, A. Direct competitive immunosensor for Imidacloprid pesticide detection on gold nanoparticle-modified electrodes. *Talanta* **2019**, *209*, 120465. [CrossRef]
181. Wei, W.; Zong, X.; Wang, X.; Yin, L.; Pu, Y.; Liu, S. A disposable amperometric immunosensor for chlorpyrifos-methyl based on immunogen/platinum doped silica sol – gel film modified screen-printed carbon electrode. *Food Chem.* **2012**, *135*, 888–892. [CrossRef]
182. Grennan, K.; Strachan, G.; Porter, A.J.; Killard, A.J.; Smyth, M.R. Atrazine analysis using an amperometric immunosensor based on single-chain antibody fragments and regeneration-free multi-calibrant measurement. *Anal. Chim. Acta* **2003**, *500*, 287–298. [CrossRef]
183. Skla, P. The immunosensors for measurement of 2, 4-dichlorophenoxyacetic acid based on electrochemical impedance spectroscopy. *Bioelectrochemistry* **2004**, *62*, 11–18.
184. Stoltenburg, R.; Reinemann, C.; Strehlitz, B. SELEX—A (r) evolutionary method to generate high-affinity nucleic acid ligands. *Biomol. Eng.* **2007**, *24*, 381–403. [CrossRef]
185. Madianos, L.; Tsekenis, G.; Skotadis, E.; Patsiouras, L.; Tsoukalas, D. A highly sensitive impedimetric aptasensor for the selective detection of acetamiprid and atrazine based on microwires formed by platinum nanoparticles. *Biosens. Bioelectron.* **2018**, *101*, 268–274. [CrossRef]
186. Madianos, L.; Skotadis, E.; Tsekenis, G.; Patsiouras, L.; Tsigkourakos, M.; Tsoukalas, D. Impedimetric nanoparticle aptasensor for selective and label free pesticide detection. *Microelectron. Eng.* **2018**, *189*, 39–45. [CrossRef]
187. Jiao, Y.; Hou, W.; Fu, J.; Guo, Y.; Sun, X. A nanostructured electrochemical aptasensor for highly sensitive detection of chlorpyrifos. *Sens. Actuators B Chem.* **2017**, *243*, 1164–1170. [CrossRef]

188. Rapini, R.; Marrazza, G. Electrochemical aptasensors for contaminants detection in food and environment: Recent advances. *Bioelectrochemistry* **2017**, *118*, 47–61. [CrossRef] [PubMed]
189. Erdem, A.; Eksin, E.; Muti, M. Chitosan—Graphene oxide based aptasensor for the impedimetric detection of lysozyme. *Colloids Surfaces B Biointerfaces* **2014**, *115*, 205–211. [CrossRef]
190. Wang, C.; Chen, B.; Zou, M.; Cheng, G. Cyclic RGD-modified chitosan / graphene oxide polymers for drug delivery and cellular imaging. *Colloids Surfaces B Biointerfaces* **2014**, *122*, 332–340. [CrossRef]
191. Liu, M.; Khan, A.; Wang, Z.; Liu, Y.; Yang, G.; Deng, Y.; He, N. Aptasensors for pesticide detection. *Biosens. Bioelectron.* **2019**, *130*, 174–184. [CrossRef]
192. Fei, A.; Liu, Q.; Huan, J.; Qian, J.; Dong, X.; Qiu, B.; Mao, H.; Wang, K. Label-free impedimetric aptasensor for detection of femtomole level acetamiprid using gold nanoparticles decorated multiwalled carbon nanotube-reduced graphene oxide nanoribbon composites. *Biosens. Bioelectron.* **2015**, *70*, 122–129. [CrossRef]
193. Rapini, R.; Cincinelli, A.; Marrazza, G. Acetamiprid multidetection by disposable electrochemical DNA aptasensor. *Talanta* **2016**, *161*, 15–21. [CrossRef]
194. Fu, J.; An, X.; Yao, Y.; Guo, Y.; Sun, X. Electrochemical aptasensor based on one step co-electrodeposition of aptamer and GO-CuNPs nanocomposite for organophosphorus pesticide detection. *Sens. Actuators B Chem.* **2019**, *287*, 503–509. [CrossRef]
195. Hassani, S.; Akmal, M.R.; Salek-Maghsoudi, A.; Rahmani, S.; Ganjali, M.R.; Norouzi, P.; Abdollahi, M. Novel label-free electrochemical aptasensor for determination of Diazinon using gold nanoparticles-modified screen-printed gold electrode. *Biosens. Bioelectron.* **2018**, *120*, 122–128. [CrossRef] [PubMed]
196. Mulchandani, A.; Mulchandani, P.; Chauhan, S.; Kaneva, I.; Chen, W. A Potentiometric Microbial Biosensor for Direct Determination of Organophosphate Nerve Agents. *Electroanalysis* **1998**, *10*, 733–737. [CrossRef]
197. Chen, W.; Mulchandani, A. Flow Injection Amperometric Enzyme Biosensor for Direct Determination of Organophosphate Nerve Agents. *Environ. Sci. Technol.* **2001**, *35*, 2562–2565.
198. Mulchandani, P.; Mulchandani, A.; Kaneva, I.; Chen, W. Biosensor for direct determination of organophosphate nerve agents. 1. Potentiometric enzyme electrode. *Biosens. Bioelectron.* **1999**, *14*, 77–85. [CrossRef]
199. Touloupakis, E.; Giannoudi, L.; Piletsky, S.A.; Guzzella, L.; Pozzoni, F.; Giardi, M.T. A multi-biosensor based on immobilized Photosystem II on screen-printed electrodes for the detection of herbicides in river water. *Biosens. Bioelectron.* **2005**, *20*, 1984–1992. [CrossRef] [PubMed]
200. Chatzipetrou, M.; Milano, F.; Giotta, L.; Chirizzi, D.; Trotta, M.; Massaouti, M.; Guascito, M.R.; Zergioti, I. Functionalization of gold screen printed electrodes with bacterial photosynthetic reaction centers by laser printing technology for mediatorless herbicide biosensing. *Electrochem. Commun.* **2016**, *64*, 46–50. [CrossRef]
201. Kumar, J.; Souza, S.F.D. Microbial biosensor for detection of methyl parathion using screen printed carbon electrode and cyclic voltammetry. *Biosens. Bioelectron.* **2011**, *26*, 4289–4293. [CrossRef]

biosensors

MDPI

Review

Screen-Printed Electrodes Modified with Metal Nanoparticles for Small Molecule Sensing

Daniel Antuña-Jiménez, María Begoña González-García, David Hernández-Santos and Pablo Fanjul-Bolado *

Metrohm DropSens S.L., Edificio CEEI-Parque Tecnológico de Asturias, 33428 Llanera, Spain; daniel.antuna@metrohm.com (D.A.-J.); begona.gonzalez@metrohm.com (M.B.G.-G.); david.hernandez@metrohm.com (D.H.-S.)

* Correspondence: pablo.fanjul@metrohm.com; Tel.: +34-985-277-685

Received: 17 January 2020; Accepted: 29 January 2020; Published: 1 February 2020

check for updates

Abstract: Recent progress in the field of electroanalysis with metal nanoparticle (NP)-based screen-printed electrodes (SPEs) is discussed, focusing on the methods employed to perform the electrode surface functionalization, and the final application achieved with different types of metallic NPs. The ink mixing approach, electrochemical deposition, and drop casting are the usual methodologies used for SPEs' modification purposes to obtain nanoparticulated sensing phases with suitable tailor-made functionalities. Among these, applications on inorganic and organic molecule sensing with several NPs of transition metals, bimetallic alloys, and metal oxides should be highlighted.

Keywords: metal nanoparticles; bimetallic alloys; metal oxides; screen printed electrode; ink-mixing; drop casting; electrodeposition; electrocatalysis; enzymatic sensor; enzyme-free sensor

1. Introduction

Screen-printed electrodes are well-known suitable platforms for sensing devices' development. Technology involving its preparation implies the use of many different substrates, such as ceramic, plastic, paper, or glass. In addition, the possibility of producing patterns of cells or electrodes with different architectures, such as single electrodes or conventional three-or-more electrode cells, interdigitated electrodes, flow cells, etc., affords a wide range of different applications. Its easy preparation and cost-effective production allows mass production of customized electrode configurations with devices made of different substrates, geometries, shapes, and sizes.

In addition to its principal advantage that is customable manufacturing, screen-printed electrodes (SPEs) also offer the possibility of producing tailor-made surfaces to achieve desirable applications for the detection of specific analytes in several fields, such as industry, clinical, or academic research.

Modification of the electrode surface in these devices is usually achieved by three well-known methods represented in Figure 1: Ink mixing with the modifying agent, electrochemical deposition of a metallic precursor, or drop casting of a preformed nanoparticulated material. The first method is carried out before ink curing and has more critical parameters, such as the curing temperature and mixing recipes, that need to be very well supervised in order to achieve batch reproducibility [1]. The other two methods are performed, after electrode preparation, on their surface, so they are more suitable for working with commercially available SPEs.

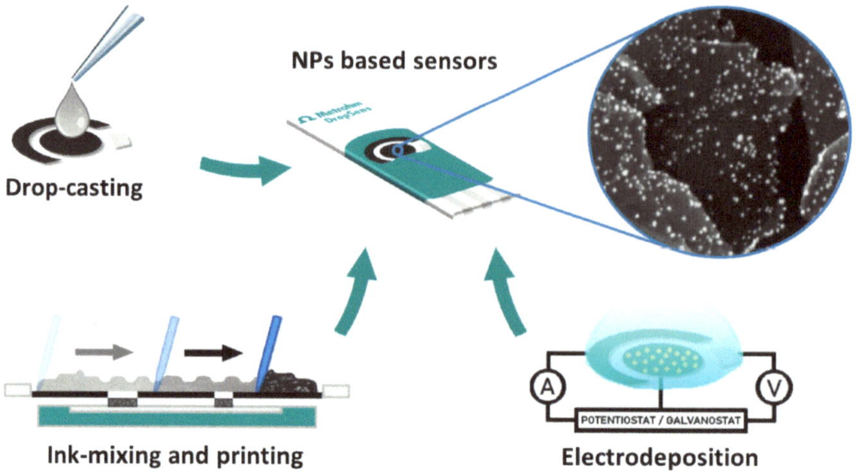

Figure 1. Schematic representation of the three main methodologies usually employed to modify SPEs with metal NPs.

There is a vast literature on the use of metal NPs for the detection of heavy metal ions, and several reviews have been published [2–6]. So, the scope of this review was only centered on applications with the aim of detecting inorganic and organic molecules and the strategies employed to improve the analytical signal when using NPs as the sensing phase.

2. Modification Methodologies

2.1. Ink Mixing Method

The first approach used for screen-printed electrode modification with nanoparticles copies the usual methodology previously explored with modifiers onto carbon paste electrodes. Functionalization in these devices was achieved by mixing modifying material with the carbon-based paste that was subsequently pressed and polished.

In this way, ink mixing SPE modification consists in the preparation of an ink where three main components are always presented: Conductive particles usually made of carbonous material, a solvent/binder mixture that allows to transfer particulated matter onto the substrate, and the modifying agent, metal NPs in this case. Depending on the final application, the main parameters that should be optimized are the recipe of the ink, its rheology, substrate selection, and thermal curing.

Screen printing starts with the positioning of the printing media upon the mesh screen followed by the application of pressure with a squeegee that forces printing medium through the previously designed pattern. Finally, a curing temperature is applied to dry the ink. This procedure can be applied several times in order to obtain more layers of the material onto the substrate, either conductive ink or dielectric material. Further details of the procedure were reviewed elsewhere [7].

Being the first approach explored, original papers appear in the early 1990s. These publications employed conductive material attached directly to metal particles, so the modification consists of metallized carbon with platinum [8], palladium [9], or iridium [10,11] mixed with conventional inks used for screen printing to obtain second generation enzymatic sensors, where metallic particles are used as a catalyst. The main advantage of this methodology is in the fact that commercially available materials were simply mixed with printing binders and solvents, so only the mixing recipe was optimized.

Later, pristine metallic particles previously separated from carbon were mixed with conductive material and binder [12,13]. Although the percentage of metal dispersion can modulate the reactivity of the supported metal substrates, when the ink mixing approach is used, the analytical response is not equally affected. This behavior can be explained by the nature of the mixing process. As nanoparticulated material is mixed together, agglomeration occurs, diminishing the degree of dispersion, so, finally, bigger particles in the micro range are obtained with the same metal loading. This is the reason why only a few publications [14,15] are based on ink mixing with metal NPs, since the main advantage of NPs, a high surface area due to interparticle spacing and their nanometer size, is lost.

Recently, an interesting paper deals with suitable ink mixing modification procedures to achieve better results regarding the catalytic performance [14]. Due to the higher surface having more active sites, an inkjet-printing approach was used with capped silver NPs. The inkjet approach is based in the same principle, like screen printing, as an ink that is previously mixed is required. In this case, the prepared ink passes through a nozzle so th printing material is applied directly onto the surface and mild thermal treatment compared to heat curing is required. Moreover, environmentally friendly organic solvents like glycerol are employed, so the printing conditions are less aggressive.

Although this technique is more suitable for printing detailed patterns onto flexible substrates, inkjet printing affords thinner conductive substrates so more metallic particles are exposed to the analyte solution, diminishing the aggregation effect when compared to the ink mixing method. Moreover, inhibition is avoided due to the lack of paste additives used for ink mixing that can decrease the catalytic activity. The results obtained with this methodology with silver NPs show a greater catalytic response against hydrogen peroxide. One more improvement can be made by using a decapping method, where hydrochloric acid is used to dissolve the capped agent of silver NPs. Following this protocol, agglomerated particles can be interconnected, decreasing the resistance and affording better electron transfer and, consequently, improving the analytical performance [14]. This is the only approach, until now, that overcomes ink mixing drawback with metal NPs in a practical manner, but no further metals were tested.

2.2. *Drop Casting Method*

This is the easiest method employed to modify screen-printed electrodes. Only one parameter should be optimized: The final amount drop casted onto the electrode. This can be modulated by changing the size of the drop and the concentration of the metal NPs dispersed in the solution.

Sensors developed with this methodology are based in two main strategies: Direct modification of selected NPs onto the working electrode or ex-situ fabrication of composites made of NPs attached with carbonous nanomaterials.

The first strategy is easier to carry out, as no functionalization step is necessary. So, many metal NP solutions were used, such as bismuth [16], platinum [17–19], rhodium [20], gold [21], silver [22], copper [23], and nickel [24]. Sensors, thus prepared, are stable, even in flow injection systems, making easier and affordable sensing devices possible when compared to those obtained by the ink mixing method. On the other hand, agglomeration is the main drawback observed in these devices. As NPs were simply casted and dried onto the working electrode, NPs tends to aggregate while drying, so the final material attached to the surface has a microparticulated appearance, even when a nanosized material was casted initially.

In order to overcome this issue, the second strategy is more commonly used since more reproducible nano-sized metallic centers are obtained. NPs are synthesized in this way onto a conductive carbonous substrate, in a similar manner to the particulated metallized carbonous materials firstly employed with the ink mixing method. When compared to the successive casting in the two steps of carbon nanomaterial followed by NPs [21], these novel composites afford not only a "real" nanoparticulated substrate but also an increased electroactive area due to the nature of the carbon nanomaterials employed. Taking this advantage, platinum [25], silver [26], nickel [27], and gold [28] were used in combination with carbon nanotubes [25], nanoporous carbon [27], and reduced graphene oxide [26] or graphene oxide [28] to obtain novel composite sensing devices with improved catalytic performance.

Recently, bimetallic clusters made of Cu-Ti, where porous titanium phosphate NPs were used as thee carrier for copper NPs, were also tested in a similar manner to carbonous materials, offering novel sensing platforms based only on metallic substrates [29]. All these published works are mainly focused on the previously synthesized composite than on the methodology itself, since the main novelty of these sensors is due to the use of a casted composite responsible for the improved catalysis.

2.3. Methods based on Electrochemical Deposition

This is the most common method used to modify SPEs with metal NPs because it is the best method to control the morphology of NPs in an accurate way. This methodology is based on the reduction of oxidized species, typically metallic water-soluble salts, at a fixed potential or current to obtain tailor-made metal particles grown on conductive substrates.

The parameters usually optimized are separated in two main groups: The ones related to the precursor solution where the salt type and concentration are involved, and the conditions of electrochemical deposition. A list of usual salts employed are $AgNO_3$ [30–34] for silver NPs; $HAuCl_4$ [31,35,36] and $AuCl_3$ [37] for gold NPs; $Bi(NO_3)_3$ [16,38,39] for bismuth NPs; $CoCl_2$ [40] for cobalt NPs; $CuCl_2$ [41], $CuSO_4$ [42], and $CuNO_3$ [43,44] for copper NPs; $NiCl_2$ [40,45] and $NiSO_4$ [46,47] for nickel NPs; $PdCl_2$ [48–50] for palladium NPs; H_2PtCl_4 [51], H_2PtCl_6 [30,52–56], and $PtCl_2$ [37] for platinum NPs; $RhCl_3$ [57] for rhodium NPs, etc. Although higher concentrations of precursor allow bigger particles to be obtained, the size and shape are usually controlled electrochemically; so, the precursor concentration usually tends to be high enough to have sufficient material susceptible to be deposited and it is not often optimized [33]. With this in mind, two parameters are crucial to control the size and shape of growth NPs: The potential or current applied and the time of deposition. The last one modulates the amount and size of metal onto the electrode in such a way that a longer time gives rise to higher amounts of NPs with a bigger particle size [37]. The first parameter will be discussed separately in subsequent sections since potential changes are the basis of potentiostatic techniques while current changes are the basis of galvanostatic techniques.

2.3.1. Electrochemical Methods based on Potentiostatic Techniques

These methods are based on the application of a fixed potential. When a specific potential to deposit NPs is applied, their size and shape can be modulated. As the potential becomes more negative, the nucleation rate increases, so more NPs are obtained with a smaller particle size, increasing the electroactive surface area. Focusing on the shape, negative potentials around −0.1 V (vs. a silver pseudo-reference electrode) give rise to more spherical NPs since growth is favored more than nucleation while a more negative potential affords an heterogeneous morphology [53]. Generally, the sensitivity of sensors made by potentiostatic deposition increases as the deposition potential is more negative in combination with a higher deposition time. In this way, when applying a deposition potential equal or more negative than −0.5 V (vs. an Ag/AgCl pseudo-reference electrode), platinum NPs' fast growth creates a heterogeneous concentration of platinum ions around NPs, making the formation of polyhedron shapes that grow up faster than the rest of the facets possible, forming flower-like nanostructures [51]. Similar structures are obtained with bismuth [38] at −1 V and gold [58] at −0.2 V vs. Ag/AgCl pseudo-reference electrodes.

Although morphology is important, "best photo" is not often the goal but improved analytical performance, so the optimization has to be carried out to achieve a better response depending on the application goal. For example, more negative potentials (−0.6 V vs. Ag/AgCl) and shorter deposition times (190 s) than the ones employed to achieve nanoflowers are employed for the deposition of platinum onto ultramicroelectrodes [52]. Mild reduction conditions are employed to obtain a wider linearity range and higher sensitivity for bromide detection with rhodium NPs [57] and for ascorbic acid detection with gold NPs [35].

Deposition is usually performed in one potentiostatic reduction step by applying a potential negative enough to afford a reduction of the metal species to the zero-valence state, but sometimes, two steps [32] or more [54] are employed. In these cases, the objective is to more precisely control both the nucleation and growth rates. NPs, thus prepared, show improved homogeneity and the possibility of having more accurate results.

Examples of a well-controlled electrodeposition process are bimetallic alloys. These systems are based on particles that combine two different metals in one single particle as a core shell [54] approach or simply by co-deposition [59]. The main goal of nanoparticulated alloys is the combination of the advantages from the different metals involved and the minimization of the disadvantages of each one [60]. A good example of the accuracy level that is needed when these NPs are fabricated is the nanoflower-like morphology obtained with the Pt-Pd system by using a multi-step potentiostatic protocol [54]. A mixture of platinum and palladium salts are reduced under an initial constant potential, provoking the formation of numerous crystal seeds. A subsequent multi-potential step electrodeposition repeated for 50 cycles promotes precise growth of the colloidal nanostructures onto preformed seeds. This protocol allows porous nanostructures with a large surface area and abundant catalytic sites against peroxide reduction to be obtained. The application of only one of the steps from the whole protocol affords particles with serious agglomeration that reduces the surface active sites and electrocatalytic properties dramatically, so a very well controlled growth has to be achived with these bialloy systems.

Electrodeposition protocol is not the only parameter that should be taken into account when bimetallic alloys are developed. The selection of metals deposited, and the order of deposition are also crucial to achieve the best results. For example, a hydrazine sensor prepared with core-shell Cu-Pd NPs shows better results when copper is deposited first [61].

Cyclic voltammetry is also employed for electrodeposition, but control is more difficult because more parameters need to be optimized, such as the scan rate, scan cycles, and potential window. Published works that use this technique with noble metals are mainly focused on the novelty of the substrates employed like graphite nanosheets [49], carbon nanotubes [55], fullerenes [48], graphene [46], and graphene oxide [59] than electrodeposition itself.

Bimetallic alloys were also obtained when using cyclic voltammetry as the electrochemical technique. Gold and silver have been deposited together onto the carbonous working electrode, improving the sensitivity by slightly giving up the dynamic range [59]. Silver as a substrate was also employed in another bimetallic system, where bismuth was electrodeposited onto silver SPE, creating an alloy capable of catalyzing hydrogen peroxide oxidation, but the mechanism involved remain unknown [39].

Cyclic voltammetry is a useful technique when copper as a metal is employed, since passivated copper with an oxide/hydroxide layer is the species responsible for the catalysis of sugars and amino acids. Since the catalytic properties of CuNPs towards these analytes depend on the size, shape, and nature of CuNPs, several studies have been carried out to clarify the complex mechanism of electrocatalysis [44,62]. The first approach consists of cyclic scans successively applied in several cycles to achieve the deposition of copper onto the SPE and also passivation of the reduced metal to obtain catalytic centers in a two-step protocol. This methodology affords well-controlled cubic copper NPs without the aid of protective agents [43] with good sensing properties. Copper nanobelt can be synthesized via the potentiostatic method by applying a high potential and longer time of deposition. When nanobelts are compared with NPs, the response against sugar increases dramatically due to a higher oxidized surface being obtained as a greater area is exposed to ambient self-oxidation [41].

Amino acids can also be detected with copper NPs because their carboxylic and amine terminals act like a chelating agent in a bidentate ligand. Their complexation with oxidized species of copper is capable of decreasing the detection potential at 0 V, while electrocatalysis increases this potential from the +0.4 to +0.8 V range. Interestingly, this phenomenon only occurs when 100-nm-sized copper NPs are electrodeposited onto SPEs, so accurate control of the particle size needs to be controlled. For this reason, a photo-irradiated electrodeposition method was developed based on potentiostatic electrodeposition applied under a xenon light source, because the modulation intensity of light is capable of controlling the size growth of copper NPs [44].

2.3.2. Methods based on the Galvanostatic Technique

These methods apply a constant negative current capable of reducing precursor metallic salt. The more negative the current applied, the higher the nucleation rate achieved, similar to potentiostatic methods. The deposition time is also a crucial parameter.

Although this technique is not extensively used, the employment of current instead of potentials is more convenient when using SPEs [63]. Due to the pseudo-reference electrode, potentials can change when oxidizing media are used in the deposition step. As the morphology of the nanoparticle surface is responsible for the final analytical response of the sensor, small variations in the reference potential give rise to less accurate results. The application of a constant current minimizes the effects of pseudo-referenced systems [42] but good control of the current applied is still necessary. In this way, negative current densities (around -0.2 mA cm^{-2}) afford homogeneous nanoflower-like structures when nickel is reduced [47] while a more negative current (around -1.8 mA cm^{-2}) affords a more heterogeneous particle size distribution when copper is deposited [42]. This phenomenon observed with copper can be explained because nucleation and growth seem to take place at different times so two different particle size are obtained. Taking into account the deposition time, a suitable time window has to be applied when using sputtered paper as the substrate, as a larger time can produce gold detachment [64].

The galvanostatic technique also offers the possibility of synthesizing porous metallic substrates with a high surface area via hydrogen evolution-assisted electrodeposition. This methodology consists in the application of a very large current density (around 1.4 A cm^{-2}) to the electrode system, provoking a quick reduction of metal ions at the working electrode of the SPE combined with the arising of numerous bubbles of hydrogen, hindering normal diffusion of the remaining ions of the precursor salt. Since no ion can occupy the space of hydrogen bubbles, the deposition is only achieved in the inner space among these bubbles so the final result is the formation of highly porous architectures where pores of hundreds of nanometers are easily obtained. This method has been employed to achieve 3D porous nickel structures with an extremely large electroactive area [45]. A greater surface area to increase the electroactivity can be obtained with subsequent electrodeposition. A combination of hydrogen evolution-assisted galvanostatic reduction followed by potentiostatic electrodeposition was assayed in a two-step protocol [36]. Firstly, a large current density affords a highly porous gold substrate, and secondly, several cyclic voltammetry cycles are applied to obtain conventional gold NPs deposit. The final electrode is made of microporous and nanoparticulated gold, offering a dramatic increase in the analytical signal.

Bimetallic alloys can also be obtained using galvanostatic methods to combine the advantages of different metals. Nickel and cobalt can be co-deposited together, providing similar current values for glucose detection than those obtained with cobalt NPs but at a lower applied potential [40].

Galvanic displacement was also explored for bimetallic alloys' electrosynthesis [30]. It is based on the phenomenon observed when a moderately active metal is partially replaced (e.g., oxidized) by a less active or more noble metal. Based on this fact, silver NPs were deposited first to serve as the active metal. Secondly, SPE modified with silver NPs is immersed into platinum salt solution and left without the application of any potential or current. Platinum displaces silver on the surface so the final particle is made of a core of silver and a shell of platinum, with no electrochemical step applied. Platinum shell extension can be modulated by changing the time of contact. More time provides more shell growth and a better response against peroxide catalysis.

2.4. Other Methods for NPs Modification

Although the majority of SPEs modified with metallic NPs are obtained with the above-mentioned methods, there are others that have been used to fabricate these devices.

Chemical deposition has been assayed for nanostructurated sensing phase synthesis. Reducing precursor salt onto the SPE allows hydroxilated copper nanowires that can be oxidized via thermal treatment to be obtained, affording a high catalytic surface area of copper oxide suitable for glucose detection [65]. The main drawback is that the extension of the reduced material cannot be controlled with accuracy with this approach. Chemical deposition of iridium onto graphene was also tested as a pH sensor in a cheap and portable pHmeter [66].

In a more aggressive way, spark discharge was also applied onto SPEs. It is based on the application of an electric field high enough to create an ionized electrically conductive channel through air between two electrodes. Bismuth NPs were synthesized in this way by using a +1.2 kV DC power supply, where bismuth wire is connected to the positive pole and the working electrode is connected to the negative pole. Applying several electric discharge (sparking) cycles under atmospheric conditions, homogeneously distributed bismuth NPs of 2–5 nm separated and covered with a carbonous layer are produced in the surface of the working electrode. With this approach, it is possible to detect nanomolar concentrations of riboflavin without deaeration of the sample due to the presence of the carbon shell layer formed around the bismuth oxide NPs provoked by the own nature of the sparking process. The presence of this layer restricts oxygen interaction with bismuth NPs during voltammetric detection, affording lower blanks that allow measurements at lower concentrations [67].

3. Roles and Applications of Metal Nanoparticles

In this section, the main roles of the SPEs modified with NPs are revised. As mentioned above, heavy metal detection is the main application of metallic particles, but the extensive bibliography exceeds the scope of this review. Despite this, a brief mention is made of bismuth NPs that are the most commonly sensing phase employed in the detection of heavy metals since they are considered the most popular "green" substitutes for the classical mercury drop or film electrodes. Due to its wide cathodic range and its low toxicity, this allows the detection of heavy metal ions in a similar manner as mercury, but without the environmental risk of mercury waste disposal, such as bioaccumulation and acute toxicity by vapor inhalation [68,69]. Electrodes modified with bismuth can be achieved by the use of a precursor or simply by nanoparticle deposition using different modification methods [3].

In this section, the applications achieved with other metal NPS will be discussed, focusing on the specific role developed in the sensing device. In this way, three main different roles are considered. Metal NPs can act as a catalyst in enzymatic and non-enzymatic devices, as sensing phases for direct detection of several analytes, or as anchorage substrates for sensing platforms. In addition, two tables summarizing the real applications of SPEs based on metallic NPs (see Table 1) and bimetallic alloys (see Table 2) are included for clarification.

3.1. *As catalyst in Enzymatic and Non-Enzymatic Devices*

Metallic NPs are capable of catalyzin relevant processes, such as peroxide reduction or carbohydrates oxidation, making the detection of enzymatic products, sugars, and amino acids in real samples possible. The following discussion is focused on the catalytical effect of several NPs against these and other analytes.

3.1.1. Hydrogen Peroxide Monitoring

Hydrogen peroxide monitoring is essential because it participates in classical enzymatic reactions extensively used in the biosensing field and it is an additive in many food, pharmaceutical, and environmental goods. Among enzymatic sensing devices, the most commonly used ones are based on peroxidases and oxidases due to their high selectivity and sensitivity. The reaction pathways of these enzymes involve hydrogen peroxide as the reagent or product, respectively, so it is the target analyte when detecting substrates with this two kind of enzymes. The monitoring is achieved directly in first generation biosensors, by the use of mediators in second generation biosensors or by the catalytic center of the enzyme in third generation biosensors. Although these successive improvements increase the use of enzymatic devices, they usually have poor chemical and long-term stability that limits their fabrication and increases the cost of the final device.

Platinum group metallic NPs are a well-known electrocatalyst towards the oxidation or reduction of hydrogen peroxide [70]. The mechanism involved [71] is shown in Figure 2a. Taking into account its catalytical properties, it was possible to develop disposable glucose biosensor made by platinized carbon particles [8]. Further studies based on platinum nanoflowers electrodeposited potentiostatically affords a sensor with a wide linear range capable of monitoring glucose in serum samples [51].

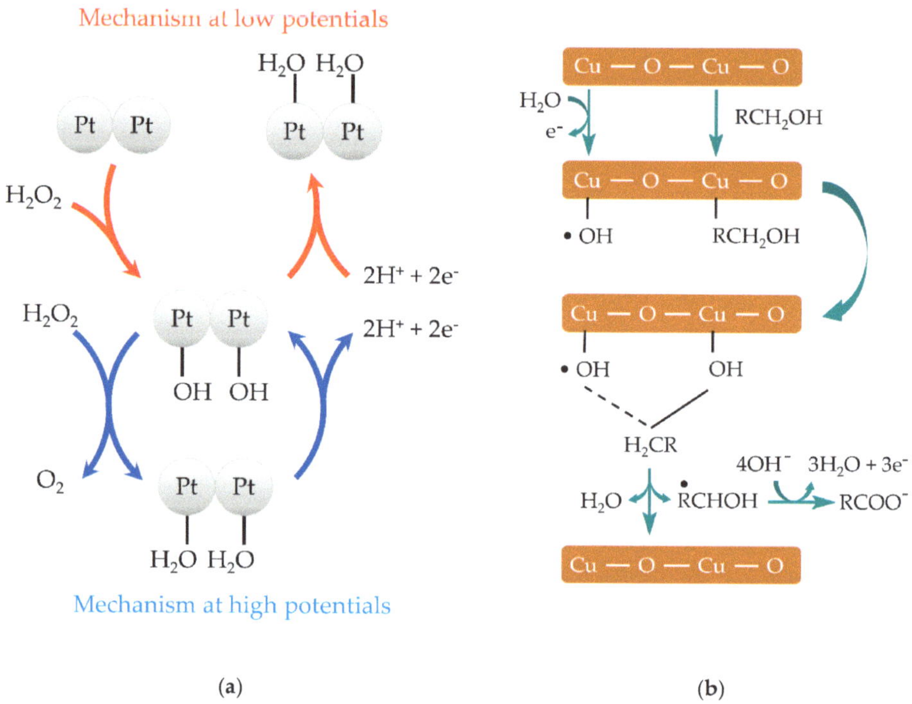

Figure 2. Schemes showing the proposed mechanism of (**a**) hydrogen peroxide oxidation onto platinum NPs [71] and (**b**) sugar oxidation onto copper NPs [62].

Similar to platinum, palladium shows high catalytic activity towards several electrochemical processes, making it possible to perform sensors based on screen-printing technologies [12]. Although being more expensive than platinum, oxygen reduction reaction onto palladium is only a slightly lower potential than that onto platinum, so it is possible to monitor dissolved oxygen with good reproducibility in ground and tap water [50]. As a peroxide catalyst, palladium NPs were employed for monitoring glucose with an SPE strip based on palladium-dispersed carbon ink [9].

Iridium NP-modified SPEs by the ink mixing approach were also used as transducers for enzymatic reactions. By means of peroxide detection, uric acid was detected via uricase reaction [10].

As noble metal NPs are capable of detecting peroxide at low overpotential when compared to bare classical electrodes, they were also employed in enzyme-free devices. There are many advantages in the electrochemical detection of peroxide without the presence of an enzyme, such as the improvement in stability and reproducibility and the possibility of obtaining simple and inexpensive devices [72]. Moreover, it is now possible to prepare noble metal NPs with a highly controllable size, shape, surface charge, and physicochemical characteristics for specific electrocatalytic applications [73–75]. Combining all these facts with other capabilities, such as low toxicity, high surface area, wide surface functionalization chemistry, and colloidal stability, these nanomaterials were extensively used in recent years to perform new biosensing platforms and devices by themselves.

Platinum is the most common noble metal employed for non-enzymatic sensors. Due to its good electrocatalytic activity towards peroxide, platinum can be used to detect this molecule in several real samples, such as cosmetics [53], household goods [30], or food and beverages [25], with good recoveries. A simple electrodeposition step is needed [37] to perform sensors with this method, affording a facile and robust methodology to monitor peroxide "in-situ" and making it even susceptible to academic demonstrations with hair lightener [56] and whitening strips [18] as real samples.

Like platinum, rhodium NPs were also used for peroxide detection. By a simple drop casting method, SPEs modified with rhodium NPs were used to detect hydrogen peroxide produced by autoxidation of polyphenols in tea extracts [20].

In a similar manner to platinum-like metals, other noble metals, such as gold, silver, and copper, are employed in sensing devices. Sensors developed with these NPs show less catalytic activity towards hydrogen peroxide and they are very dependent on the particle size and preparation steps but offer cheaper devices compared to platinum-modified SPEs.

Silver NP-based SPEs offer better catalytic properties towards peroxide detection than bulk silver electrodes. Unfortunately, sensors are only stable for a week because the main drawback of silver-based nanomaterials relies on its inherent and fast oxidation, so the stability is compromised for mass production purposes [14]. Drop casting methodologies were also applied in combination with carbon nanomaterials attached directly to silver NPs [26] or by the ink mixing method [22] to enhance the sensitivity.

Focusing on the field of biosensing and organic molecules, only a few works employing bismuth NPs were published [76] because the detection capabilities with enzymes or organic compounds were not comparable to those obtained with heavy metal stripping assays. Moreover, due to high background limitations, these electrodes need to be deaerated before using in many practical applications when very low detection limits are necessary.

Bismuth can act as a catalyst for hydrogen peroxide detection but only when it is combined with silver. Electrodeposition of bismuth NPs onto silver electrodes forms an alloy capable of catalyzing the reduction of hydrogen peroxide, including in real cosmetic samples. The existence of these alloys can open the door to a new metallic catalyst for sensing devices [39].

Metal oxide nanoparticles were extensively used for sensing gases [77], but the applications related to that field exceed the scope of this review. When focusing on electrochemical sensing and biosensing devices, not many applications were developed when employing screen printing technology, and much less in real samples [78]. Although transition metal oxides are sensitive to the same analytes as their reduced counterparts, and similar modification methods were employed, the preparation procedures do not allow nanoparticulated electrode surfaces to be obtained.

Acting as a catalyst for peroxide oxidation, metallic oxides can be classified depending on their nature. The first group corresponds to common oxides, such as CuO [65,79,80] and NiO [81], with MnO_2 [82] as the most representative one, and the second group correspond to oxides of platinum group metals, such as RuO_2 [83], RhO_2 [84] and PtO_2, PdO, or IrO_2 [85]. The latter afford more expensive sensors but are more chemically stable. Electrodes modified with these compounds are prepared by ink mixing [86], the electrodeposition method [80,87], or ex-situ growth of copper with graphene and subsequent drop casting methodology [79]. Assays with real samples were only achieved in the past years in combination with enzymes for glucose monitoring purposes in fruits [83] and food [84].

Metal alloys prepared with several elements, such as Pt-Ag [30] and Pt-Pd [54], afford a better response towards peroxide [30] and glucose [54] than each metal separately. Good detection is achieved in real samples, such as antiseptic and laundry boosters [30].

3.1.2. Carbohydrate Monitoring

Several metallic elements are capable of catalyzing the hydrolysis of carbohydrates, making the detection of these analytes in real samples possible.

Copper-modified SPEs were evaluated for sugar detection. The proposed mechanism for sugars' oxidation [62] onto copper NPs is schematized in Figure 2b. For this purpose, electrodeposition of copper nanospheres [42] or nanobelts [41] and chemically synthesized copper nanowires [65] have been prepared. Nickel NPs show good electrocatalytical properties towards sugar oxidation in a similar way. They have been applied in non-enzymatic devices against glucose in food [24,42,47], blood [45], or urine [46] as real samples. It is worth mentioning that a voltammetric or amperometric pretreatment must be previously done with these NPs to obtain metal oxide species, such as oxyhydroxides, which are responsible for the catalytic oxidation of carbohydrates.

Gold surfaces are also capable of catalyzing sugar oxidation due to the presence of chemisorpted hydroxyl anions forming hydrous gold oxides, which are believed to be the catalytic component of gold electrodes. The main applications of this behavior relies on the development of novel non-enzymatic sensing in the health science field. In this way, glycated hemoglobin can be monitored in serum with electrodeposited gold nano-flowers [58]. Glucose levels in serum and blood can also be analyzed with electrodeposited porous gold nanostructures [36] or drop-casted gold NPs onto graphene nanocomposites [28]. Sugars were also quantified in beverages with gold NPs electrodeposited onto a gold-sputtered paper, obtaining similar results to those obtained using a commercial enzymatic kit [64]. In addition, advanced devices were also recently tested, like an enzymatic fuel cell to monitor glucose and oxygen in human saliva [88].

Although metal oxides are capable of detecting hydroxide [81] directly and can be used in non-enzymatic sensing of carbohydrates via commercial oxide powder modification [62] or electrodeposition of nanoparticulated material [80], no real samples were assayed.

Bimetallic alloys based on NPs capable of catalyzing carbohydrate oxidation were prepared. Cu-Ti [29] and Ni-Co [40] offer an improved response towards glucose, combining the capabilities of both metals involved.

Table 1. Application of NPs with real samples.

NPs	Modification		Analyte	Detection		Performance		Sample	Year	Ref.
	Tech.	Parameters		Tech.	Parameters	Linear range	LOD			
Ag	DC	12 µL AgNP-rGO composite, RT	H_2O_2	AD	−0.3 V	0.5 µM to 12 mM	0.21 µM	Contact lens care solution	2016	[26]
	PE	CA, −1.2 V, 10 s	Sulfite	AD	+0.4 V	1.96 to 16.66 mM	1.99 mM	Beverages	2013	[31]
	PE	*Step 1*: CA, 0.13 V, 5 ms *Step 2*: CA, 0.24 V, 25 s	Metronidazole	DPV	E_{amp}: −0.1 V 0.075 Vs^{-1}	3.1 to 310 µM	0.4 µM	Serum, Urine, and Tablets	2012	[32]
	PE	CA, −1.2 V, 120 s	Lamotrigine	DPCSV (CA+DPV)	A: −0.90, 147 s	0.33 to 1.50 µM	0.372 µM	Pharmaceuticals	2007	[33]
	PE	CA, −1.2 V, 20 s	Chloride Bromide Iodide	LSV	−0.2 to 0.6 V, 0.01 V s^{-1}	3 µM to 100 µM 5 µM to 90 µM 5 µM to 80 µM	3 µM 5 µM 5 µM	Synthetic sweat	2018	[34]
Au	IM	Ionophore based ink	Trazodone	OCP	-	10 µM to 10 mM	6.8 µM	Pharmaceuticals	2018	[15]
	DC	-	Carbofuran	DPCSV (CA+DPV)	A: 0 V, 60 s DPV: −0.2 to 0.35 V, E_p: 0.15 V, t_p: 0.3 s, E_{step}: 0.01 V	1–250 µM	0.22 µM	Food	2017	[21]
	DC	1.8 µL AuNPs Graphene composite, RT, 12 h	H_2O_2 Glucose	AD	−0.2V	0.2 to 4.2 mM 2 to 10mM	– 180 µM	Blood	2010	[28]
	PE	CA, +0.18 V, 10 s	Sulfite	AD	+0.3 V	9.8 to 83.33 µM	9.79 µM	Beverages	2013	[31]
	PE	CA, +0.18 V, 50 s	Ascorbic acid	DPV	−0.2 to 0.8 V, 0.1 V s^{-1} E_p: 0.012 V, t_p: 0.07 s, E_{step}: 0.025 V	1.9 to 16.6 µM	0.99 µM	Serum	2017	[35]
	HEA-GE	*Step 1*: CP, 3 Acm^{-2}, 100 s, RT *Step 2*: CV, 10 cycles, −0.7 to 0.4 V, 0.05 Vs^{-1}	Glucose	AD	−0.2 V	1.5 and 16 mM	25 µM	Serum	2018	[36]
	PE	CA, −0.2 V, 150 s	Glycated hemoglobin	CV	0 to −0.6 V, 0.1 Vs^{-1} Calibrated at −0.45 V	2 to 20%	0.65%	Serum	2019	[58]

Table 1. *Cont.*

NPs	Modification		Analyte	Detection		Performance		Sample	Year	Ref.
	Tech.	Parameters		Tech.	Parameters	Linear range	LOD			
Bi	PE	CV, 5 cycles +0.4 to −0.6 V, 0.05 Vs^{-1}	Sulfide	DPCSV (CA+DPV)	A, +0.4 V, 60 s DPV: +0.4 to −0.9 V, E_p: 0.008 V, t_p: 0.05 s, E_{step}: 0.1 V	0.05 to 1.5 μM	0,2 uM	Tap water	2016	[59]
	GE	CP, −100 μA, 6000 s	Glucose	CV	−0.3 to +0.5 V, 0.1 Vs^{-1}	0.01 to 5 mM	6 μM	Beverages	2017	[64]
	PE	CA, −1 V, 4 min	Phenol	CA	+0.8 V, 150 s	5 to 100 μM	480 nM	Wastewater	2010	[38]
	PE	CV, 20 cycles −0.6 to 0.3 V	H_2O_2	CV	−0.3 to −1.3 V	100 μM to 5 mM	57 μM	Cosmetic	2011	[39]
	SD	1.2 kV, 20 cycles	Riboflavin	SWV	0 to −0.8 V, Freq: 50 Hz, E_{amp}: 0.05 V, E_{step}: 0.0015 V	1 to 100 nM	0.7 nM	Multivitamin	2015	[67]
Cu	GE	CP, −225 μA, 60 s	Glucose Fructose Arabinose Galactose Mannose Xylose	CA	+0.65 V, 100 s	1 μM to 10 mM	0.57 μM 0.61 μM 1.0 μM 0.89 μM 1.3 μM 1.04 μM	Honey and beverages	2017	[42]
	IM	Ink with 50% of $Cu(OH)_2$ nanorods	Ascorbic acid	CA	0 V, 25 s	0.0125 to 10 mM	6 mM	Tablets Urine	2017	[89]
Ir	IM	Ink with 0,9:5 of Ir-C powder (5 % Ir)	Triglyceride	CA	+0.15 V, 30 s	Up to 10 mM	-	Serum	2008	[11]
Ni	DC	15 μL (10 g L^{-1}) *Activation:* A, −1.5 V, 600 s in NaOH 0.1 M	Glucose Fructose Mix 1:1	AD, FIA	+0.7 V, 2 mL min^{-1}	0.05 to 1 mM	0.06 mM 0.04 mM 0.04 mM	Honey	2012	[24]
	HEA-GE	CP, 0.1 A, 30 s	Glucose	CA	+0.5 V, 100 s	0.5 μM to 4 mM	0.07 μM	Blood	2013	[45]
	PE	CV, 40 cycles, 0.05 Vs^{-1}, 0 to −1.5 V *Activation:* CV, 40 cycles, 0.1 Vs^{-1}, 0 to +0.8 V in 0.1 M NaOH	Glucose	AD	+0.6 V	0.2 to 9 mM	4.1 μM	Urine	2013	[46]

Table 1. *Cont.*

NPs	Modification		Analyte	Detection		Performance		Sample	Year	Ref.
	Tech.	Parameters		Tech.	Parameters	Linear range	LOD			
	GE	CP, −25 µA, 60 s *Activation:* CV, 50 cycles, 0.1 Vs^{-1}, +0.2 to +0.7 V in 0.1 M NaOH	Glucose Fructose	CA	+0.6 V, 120 s	25 to 1000 µM	Between 8 µM and 20 µM	Food	2016	[47]
Pd	PE	CV, 10 cycles, 0.05 Vs^{-1} −0.25 to +1.2 V	Dopamine	DPV	−0.1 to +0.6 V	0.35 to 135.35 µM	0.056 µM	Injection	2015	[48]
Pd	PE	CV, 20 cycles, 0.02 Vs^{-1} +1.2 to −0.25 V	Hydrazine	AD	−0.05 V	0.05 to 1415 µM	4 nM	Drainage water	2016	[49]
Pd	PE	CA, −0.6 V, 180 s	Dissolved O_2	CV	0.5 to −0.3 V, 0.02 V s^{-1}	Up to 250 µM	-	Ground and tap water	2006	[50]
Pt	DC	12 µL, RT, 24 h	H_2O_2	AD	−0.3 V	1 µM to 10 mM	0.43 µM	Contact lens care solution	2016	[17]
Pt	DC	20 µL, dried at 80 °C, 10 min	H_2O_2	AD	0.345 V	Up to 0.1 mM	6.6 µM	Whitening Strips	2015	[18]
Pt	DC	10 µL (2 g L^{-1}), dried at 40 °C, 180 min	Ethanol	LSV	−1 to 1 V, 0.05 V s^{-1}	15 to 102 mM	15 mM	Beverages	2017	[19]
Pt	DC	0.5 µL PtNP-MWCNT composite, RT	H_2O_2	CA	+0.3 V, 60 s	10 to 100 µM	10 µM	Green tea	2018	[25]
Pt	PE	CA, −0.5 V, 300 s	H_2O_2	CA	−0.7 V, 30 s	500 µM to 20 mM	32.8 µM	Serum	2017	[51]
Pt	PE	CA, −0.4 V, 900 s	H_2O_2	AD	+0.7 V	6 to 215 µM	7.6 µM	Hair lightener Antiseptic Plant extract	2017	[53]
Pt	PE	CA, 12.4 V, 12 min	H_2O_2	AD	+0.7 V	Up to 6.5 mM	80 µM	Hair lightener	2018	[56]
Rh	DC	15 µL, RT	H_2O_2	AD	0 V	5 to 600 µM	2 µM	Tea extracts	2015	[20]
Rh	PE	CA, −0.25 V, 480 s	Bromide	CSV (CA+LSV)	A: +1.25 V, 20 s LSV: +1 to −0.25	Up to 40 mM	39 µM	Seawater Pharmaceuticals	2019	[57]

AD: Amperometric detection; **CA**: Chronoamperometry; **CP**: Chronopotentiommetry; **CV**: Cyclic Voltammetry; **CSV**: Cathodic Stripping Voltammetry; **DC**: Drop-Casting; **DPCSV**: Differential-Pulse Cathodic Stripping Voltammetry; **DPV**: Differential Pulse Voltammetry; **FIA**: Flow Injection Analysis; **GE**: Galvanostatic electrodeposition; **HEA-GE**: Hydrogen-Evolution-Assisted Galvanostatic Electrodeposition; **IM**: Ink-Mixing; **LSV**: Linear Sweep Voltammetry; **OCP**: Open Circuit Potential; **PE**: Potentiostatic electrodeposition; **RT**: Room Temperature; **SD**: Spark Discharge; **SWV**: Square Wave Voltammetry; **Tech.**: electrochemical technique.

Table 2. Application of bimetallic NPs with real samples.

NPs	Modification		Analyte	Detection		Performance		Sample	Year	Ref.
	Tech.	Parameters		Tech.	Parameters	Linear range	LOD			
Cu-Ti	DC	4 µL dried at RT	Glucose	CA	+0.5 V, 60 s	25 µM to 2 mM	7 µM	Honey Plasma	2017	[29]
Pt-Ag	PE-GD	*Step 1:* CA, −0.3 V, 900 s *Step 2:* 0.2 mM H_2PtCl_6 at pH 3.4 with 0.2 mM AA for 2.5 hours	H_2O_2	AD	+0.7 V	2.2 to 67 µM	0.34 µM	Antiseptic and Laundry boosters	2019	[30]
Pt-Pd	PE	*Step 1:* CP, 0.4 V, 20 s *Step 2:* 50 cycles of: CP, 0.5 V, 0.2 s CP, 0.4 V, 10 s	H_2O_2 Glucose	AD	−0.4 V	0.005 to 6 mM Up to 16 mM	0.87 µM 10 µM	Simulative blood	2012	[54]
Au-Ag	PE	CV, 5 cycles, +0.4 to −0.6 V 0.05 Vs^{-1}	Sulfide	LSCSV (CA+LSV)	A, +0.2 V, 30 s LSV, +0.2 to −0.9 V 0.05 Vs^{-1}	0.5 to 12.5 µM	0,2 µM	Water	2016	[59]
Cu-Pd	PE	*Step 1, Cu:* CA, −0.7 V, 300 s *Step 2, Pd:* CA, −0.6 V, 180 s	Hydrazine	AD, FIA	+0.2 V 0.5 mL min^{-1}	2 to 100 µM	270 nM	Cigarette tobacco	2005	[61]

AD: Amperometric detection; **CA**: Chronoamperometry; **CP**: Chronopotentiommetry; **CV**: Cyclic Voltammetry; **DC**: Drop-Casting; **FIA**: Flow Injection Analysis; **GD**: Galvanic Displacemenet; **GE**: Galvanostatic electrodeposition; **LCSV**: Linear-Scan Cathodic Stripping Voltammetry; **LSV**: Linear Sweep Voltammetry; **PE**: Potentiostatic electrodeposition; **RT**: Room Temperature; **Tech.**: electrochemical technique.

3.2. As sensing Phase for Other Analytes

Metal NPs are also capable of monitoring other analytes, opening the path for new applications. Platinum NPs are capable of catalyzing the oxidation of small organic molecules, such as ethanol and formaldehyde, making the development of an alcohol sensor for wine and beer [19] and a gas sensor against formaldehyde possible [52].

Palladium nanoparticles are also capable of detecting formaldehyde and other analytes, such as hydrazine and sulphuric acid, with the classical approach of ink mixing. In addition, hydrogen was also tested in a proof-of-concept assay, making possible the application of palladium-based SPE sensors as gas detectors [12]. Although no real samples were assayed in this work, subsequent publications based on carbon nanomaterial substrates were successfully applied to monitor dopamine in pharmaceuticals [48] and hydrazine in wastewater samples [49]. Electrodeposition of rhodium NPs also gives rise to a good platform to detect bromide in seawater and pharmaceuticals [57].

Iridium NPs are also capable of monitoring NADH produced in the dehydrogenase oxidation of glycerol. With this approach, a triglyceride biosensor was developed with a good correlation in bovine and human serum [11]. In a similar way to other proton-responsive metals, iridium oxide NPs in combination with graphene were tested as pH sensor [66].

The affinity of sulphur against gold offers the possibility of performing direct detection of sulphur species, obtaining devices capable of detecting sulfite in beverages [31] and free sulfide in tap water [59]. Other molecules were also detected directly onto gold NPs or by mediation of the ion association complex [15], making the monitorization of ascorbic acid in serum [35], carbofuran in food [21], or trazodone in a potentiometric sensor stable up to 7 months [15].

Silver affinity towards sulfur and halide species was also explored for sensor development purposes. Sulfite was detected amperometrically with silver NPs in drinking water, pickle juice, and vinegar [31], and bromide, chloride, and iodide were detected voltammetricallly in synthetic sweat [34]. Focusing on organic molecules, direct detection with electrodeposited silver NPs is possible against metronidazole [32] and lamotrigine [33] in pharmaceutical tablets.

Direct detection with bismuth NPs was only achieved when using deposition onto SPEs by sparking discharge. With this approach, it is possible to detect riboflavin in a nanomolar concentration in multivitamin real samples [67].

Copper NPs are capable of a complex amino acid as previously mentioned, so α-,β- [44] and γ-amino acids [43] can be monitored in flow systems with good correlation and detectability. Although the potential employed for the detection of amino acid can be decreased dramatically when using copper NPs via complexation, no real samples were assayed in these reports. Other analyte, such as ascorbic acid, was tested with copper NPs via drop-casting methodology [23], but again, no real sample was assayed. Only by using hydroxilated copper nanorods with an ink-mixing approach can ascorbic acid be detected in urine with good recoveries [89].

Alloys of Cu-Pd [61] and Au-Ag [59] were developed to increase the sensibility for hydrazine [61] and free sulfide [59] detection in real samples of tap water [59] and cigarettes [61].

Metal oxides were successfully tested for direct sensing of pharmaceuticals in real samples by using other metal-transition oxides not previously mentioned like ZrO alone [90] or doped with rare earth metals [91].

3.3. As platforms for Sensing Phases

Metallic NPs can also serve as carriers for sensing phases in other applications not mentioned before.

Gold NPs were extensively used as an anchoring platform. Easy functionalization and tailor-made size, shape, and nature are advantages that offer a wide range of applications in the electrochemical sensors field [92]. A recent review deals with different approaches and novelties published until now with gold-based sensors for the detection of small molecules, DNA, and aptamers [93]. Another advantage in the use of gold NPs consists in the strong binding of thiol ligands onto their surface due to the soft character of both gold and sulfur. Monolayers onto gold NPs were extensively used to stabilize them and also for functionalization purposes in many different applications [94].

Bismuth NPs were also explored as a biosensing platform in mediator-free enzymatic SPE devices. These NPs are capable of complex oxidase activity tissues, affording a substrate capable of detecting phenolic compounds, thus direct monitoring the product of the enzymatic reaction instead of hydrogen peroxide was achieved. Deposition of cationic bismuth and negatively charged mushroom tissue together offers immobilization of polyphenol oxidase enzyme onto carbon SPEs, making the determination of phenol possible [38]. With a similar approach, the same group developed a tyrosinase-based biosensor against phenol and cathecol [16].

Only a few more metallic NPs were studied as platforms for sensing devices with a rich related bibliography. Iron and cobalt NPs are one of the most assayed ones due to its low-cost production, tailor-made magnetic properties, easy functionalization processes, and biocompatibility. Iron NPs are often used as a magnetic carrier more than a catalyst [95] while cobalt NPs are mostly employed to achieve inks for electronic 3D printing [96], so the applications and SPE-based devices achieved exceed the scope of this review.

Bimetallic clusters can also serve for biotin labelling detection in multivitamin tablets with Cd-Ti functionalized with neutravidin [97] and SERS enhancement effect to study 4-mercaptopyridine adsorption-desorption onto electrodeposited dendritic Au-Ag NPs [98].

4. Conclusions

After practically 25 years from the first publication, metal NPs are still nowadays employed as modifying agents in sensing devices based on screen printing technology. Ink mixing was the first approach explored capable of offering metallized SPE for sensing development. The main drawbacks of this method is the agglomeration of NPs, complex ink recipes, and bad reproducibility among batches. Drop casting methodologies offer a suitable approach to modify SPEs since the modification is carried out after ink preparation. NPs casted onto working electrodes offer a high active surface against analyte but agglomeration is also achieved. Nanoparticulated material with accurate tailor-made sizes and shapes is only obtained when using electrodeposition. Potentiostactic techniques are employed more due to their use for years in the classical three-electrode cell with a well-controlled reference electrode. However, galvanostatic techniques are more appropriate when working with SPEs. Deposition under a controlled current is not affected by potential variations from the pseudo-reference electrode, offering similar capabilities to control the nucleation and growth of metallic NPs as potentiostatic techniques. Large-scale synthesis is the main drawback of this method due to electrodeposition having to be done with each sensor separately, and the deposition step can be a large time-consuming process when thinking in large batch preparation.

Recent scientific publications dealing with these sensing phases are focused on the development of novel strategies capable of affording a better analytical performance based on an increasing surface area or novel modifying agent synthesis. Newly developed electrochemical techniques that can increase the electroactive area in a great amount are based on previous knowledge not applied before with SPEs like hydrogen-evolution-assisted galvanostatic electrodeposition and spark discharge, or the use of carbonous nanomaterials. On the other hand, bimetallic clusters as modifying agents can be electrodeposited with an accurate growth control, affording novel surfaces with advantages from several metals in one particle, offering unknown metal synergies suitable for future screen printing devices based on metal NPs.

Author Contributions: The authors contributed equally. All authors have read and agreed to the published version of the manuscript.

Funding: This research received no external funding.

Conflicts of Interest: The authors declare no conflict of interest.

References

1. Chu, Z.; Peng, J.; Jin, W. Advanced nanomaterial inks for screen-printed chemical sensors. *Sens. Actuatorsb Chem.* **2017**, *243*, 919–926. [CrossRef]
2. Hernández-Santos, D.; González-García, M.B.; García, A.C. Metal-nanoparticles based electroanalysis. *Electroanalysis* **2002**, *14*, 1225–1235. [CrossRef]
3. Serrano, N.; Alberich, A.; Díaz-Cruz, J.M.; Ariño, C.; Esteban, M. Coating methods, modifiers and applications of bismuth screen-printed electrodes. *Trac. Trends Anal. Chem.* **2013**, *46*, 15–29. [CrossRef]
4. Serrano, N.; Díaz-Cruz, J.M.; Ariño, C.; Esteban, M. Antimony- based electrodes for analytical determinations. *Trac. Trends Anal. Chem.* **2016**, *77*, 203–213. [CrossRef]
5. Jovanovski, V.; Hočevar, S.B.; Ogorevc, B. Bismuth electrodes in contemporary electroanalysis. *Curr. Opin. Electrochem.* **2017**, *3*, 114–122. [CrossRef]
6. Economou, A. Screen-printed electrodes modified with "green" metals for electrochemical stripping analysis of toxic elements. *Sensors* **2018**, *18*, 1032. [CrossRef]
7. Foster, C.W.; Kadara, R.O.; Banks, C.E. *Fundamentals of Screen-Printing Electrochemical Architectures*; Springer: Cham, Switzerland, 2016; ISBN 9783319251912.
8. Cardosi, M.F.; Birch, S.W. Screen printed glucose electrodes based on platinised carbon particles and glucose oxidase. *Anal. Chim. Acta* **1993**, *276*, 69–74. [CrossRef]
9. Wang, J.; Chen, Q. Screen-printed glucose strip based on palladium-dispersed carbon ink. *Analyst* **1994**, *119*, 1849–1851. [CrossRef]
10. Luo, Y.C.; Do, J.S.; Liu, C.C. An amperometric uric acid biosensor based on modified Ir-C electrode. *Biosens. Bioelectron.* **2006**, *22*, 482–488. [CrossRef]
11. Liao, W.Y.; Liu, C.C.; Chou, T.C. Detection of triglyceride using an iridium nano-particle catalyst based amperometric biosensor. *Analyst* **2008**, *133*, 1757–1763. [CrossRef]
12. Metters, J.P.; Tan, F.; Banks, C.E. Screen-printed palladium electroanalytical sensors. *J. Solid State Electrochem.* **2013**, *17*, 1553–1562. [CrossRef]
13. Jeon, W.Y.; Choi, Y.B.; Kim, H.H. Disposable non-enzymatic glucose sensors using screen-printed nickel/carbon composites on indium tin oxide electrodes. *Sensors* **2015**, *15*, 31083–31091. [CrossRef] [PubMed]
14. Shi, L.; Layani, M.; Cai, X.; Zhao, H.; Magdassi, S.; Lan, M. An inkjet printed Ag electrode fabricated on plastic substrate with a chemical sintering approach for the electrochemical sensing of hydrogen peroxide. *Sens. Actuatorsb Chem.* **2018**, *256*, 938–945. [CrossRef]
15. Salama, F.M.; Attia, K.A.; Said, R.A.; El-Olemy, A.; Abdel-Raoof, A.M. Disposable gold nanoparticle functionalized and bare screen-printed electrodes for potentiometric determination of trazodone hydrochloride in pure form and pharmaceutical preparations. *RSC Adv.* **2018**, *8*, 11517–11527. [CrossRef]
16. Mayorga-Martinez, C.C.; Cadevall, M.; Guix, M.; Ros, J.; Merkoçi, A. Bismuth nanoparticles for phenolic compounds biosensing application. *Biosens. Bioelectron.* **2013**, *40*, 57–62. [CrossRef]
17. Yang, X.; Ouyang, Y.; Wu, F.; Hu, Y.; Zhang, H.; Wu, Z. In situ & controlled preparation of platinum nanoparticles dopping into graphene sheets@cerium oxide nanocomposites sensitized screen printed electrode for nonenzymatic electrochemical sensing of hydrogen peroxide. *J. Electroanal. Chem.* **2016**, *777*, 85–91. [CrossRef]
18. Popa, A.; Abenojar, E.C.; Vianna, A.; Buenviaje, C.Y.A.; Yang, J.; Pascual, C.B.; Samia, A.C.S. Fabrication of Metal Nanoparticle-Modified Screen Printed Carbon Electrodes for the Evaluation of Hydrogen Peroxide Content in Teeth Whitening Strips. *J. Chem. Educ.* **2015**, *92*, 1913–1917. [CrossRef]
19. Pereira Silva Neves, M.M.; González-García, M.B.; Bobes-Limenes, P.; Pérez-Junquera, A.; Hernández-Santos, D.; Vidal-Iglesias, F.J.; Solla-Gullón, J.; Fanjul-Bolado, P. A non-enzymatic ethanol sensor based on a nanostructured catalytic disposable electrode. *Anal. Methods* **2017**, *9*, 5108–5114. [CrossRef]
20. Gatselou, V.A.; Giokas, D.L.; Vlessidis, A.G.; Prodromidis, M.I. Rhodium nanoparticle-modified screen-printed graphite electrodes for the determination of hydrogen peroxide in tea extracts in the presence of oxygen. *Talanta* **2015**, *134*, 482–487. [CrossRef]
21. Jirasirichote, A.; Punrat, E.; Suea-Ngam, A.; Chailapakul, O.; Chuanuwatanakul, S. Voltammetric detection of carbofuran determination using screen-printed carbon electrodes modified with gold nanoparticles and graphene oxide. *Talanta* **2017**, *175*, 331–337. [CrossRef]

22. Shamkhalichenar, H.; Choi, J.W. An inkjet-printed non-enzymatic hydrogen peroxide sensor on paper. *J. Electrochem. Soc.* **2017**, *164*, B3101–B3106. [CrossRef]
23. Shabalina, A.V.; Svetlichnyi, V.A.; Ryzhinskaya, K.A.; Lapin, I.N. Copper nanoparticles for ascorbic acid sensing in water on carbon screen-printed electrodes. *Anal. Sci.* **2017**, *33*, 1415–1419. [CrossRef] [PubMed]
24. García, M.; Escarpa, A. A class-selective and reliable electrochemical monosaccharide index in honeys, as determined using nickel and nickel-copper nanowires. *Anal. Bioanal. Chem.* **2012**, *402*, 945–953. [CrossRef] [PubMed]
25. Chou, T.; Wu, K.; Hsu, F.; Lee, C. ScienceDirect Pt-MWCNT modified carbon electrode strip for rapid and quantitative detection of H_2O_2 in food. *J. Food Drug Anal.* **2017**, *26*, 662–669. [CrossRef] [PubMed]
26. Yao, Z.; Yang, X.; Wu, F.; Wu, W.; Wu, F. Synthesis of differently sized silver nanoparticles on a screen-printed electrode sensitized with a nanocomposites consisting of reduced graphene oxide and cerium(IV) oxide for nonenzymatic sensing of hydrogen peroxide. *Microchim. Acta* **2016**, *183*, 2799–2806. [CrossRef]
27. Hjiri, M.; Dhahri, R.; Ben Mansour, N.; El Mir, L.; Bonyani, M.; Mirzaei, A.; Leonardi, S.G.; Neri, G. Electrochemical properties of a novel Ni-doped nanoporous carbon. *Mater. Lett.* **2015**, *160*, 452–455. [CrossRef]
28. Shan, C.; Yang, H.; Han, D.; Zhang, Q.; Ivaska, A.; Niu, L. Graphene/AuNPs/chitosan nanocomposites film for glucose biosensing. *Biosens. Bioelectron.* **2010**, *25*, 1070–1074. [CrossRef]
29. Martín-Yerga, D.; Carrasco-Rodríguez, J.; Fierro, J.L.G.; García Alonso, F.J.; Costa-García, A. Copper-modified titanium phosphate nanoparticles as electrocatalyst for glucose detection. *Electrochim. Acta* **2017**, *229*, 102–111. [CrossRef]
30. Gómez-Monedero, B.; González-Sánchez, M.I.; Iniesta, J.; Agrisuelas, J.; Valero, E. Design and characterization of effective Ag, Pt and agpt nanoparticles to H_2O_2 electrosensing from scrapped printed electrodes. *Sensors* **2019**, *19*, 1685. [CrossRef]
31. Molinero-Abad, B.; Alonso-Lomillo, M.A.; Domínguez-Renedo, O.; Arcos-Martínez, M.J. Amperometric determination of sulfite using screen-printed electrodes modified with metallic nanoparticles. *Microchim. Acta* **2013**, *180*, 1351–1355. [CrossRef]
32. Sadeghi, S.; Hemmati, M.; Garmroodi, A. Preparation of Ag-Nanoparticles/Ionic-Liquid Modified Screen-Printed Electrode and Its Application in the Determination of Metronidazole. *Electroanalysis* **2013**, *25*, 316–322. [CrossRef]
33. Burgoa Calvo, M.E.; Domínguez Renedo, O.; Arcos Martínez, M.J. Determination of lamotrigine by adsorptive stripping voltammetry using silver nanoparticle-modified carbon screen-printed electrodes. *Talanta* **2007**, *74*, 59–64. [CrossRef] [PubMed]
34. Bujes-Garrido, J.; Izquierdo-Bote, D.; Heras, A.; Colina, A.; Arcos-Martínez, M.J. Determination of halides using Ag nanoparticles-modified disposable electrodes. A first approach to a wearable sensor for quantification of chloride ions. *Anal. Chim. Acta* **2018**, *1012*, 42–48. [CrossRef] [PubMed]
35. Alonso-Lomillo, M.A.; Domínguez-Renedo, O.; Saldaña-Botín, A.; Arcos-Martínez, M.J. Determination of ascorbic acid in serum samples by screen-printed carbon electrodes modified with gold nanoparticles. *Talanta* **2017**, *174*, 733–737. [CrossRef] [PubMed]
36. Viet, N.X.; Chikae, M.; Ukita, Y.; Takamura, Y. Enzyme-free glucose sensor based on micro-nano Dualporous gold-modified screen-printed carbon electrode. *Int. J. Electrochem. Sci.* **2018**, *13*, 8633–8644. [CrossRef]
37. Chikae, M.; Idegami, K.; Kerman, K.; Nagatani, N.; Ishikawa, M.; Takamura, Y.; Tamiya, E. Direct fabrication of catalytic metal nanoparticles onto the surface of a screen-printed carbon electrode. *Electrochem. Commun.* **2006**, *8*, 1375–1380. [CrossRef]
38. Merkoçi, A.; Anik, U.; Çevik, S.; Çubukçu, M.; Guixa, M. Bismuth film combined with screen-printed electrode as biosensing platform for phenol detection. *Electroanalysis* **2010**, *22*, 1429–1436. [CrossRef]
39. Chiu, M.H.; Kumar, A.S.; Sornambikai, S.; Chen, P.Y.; Shih, Y.; Zen, J.M. Cosmetic hydrogen peroxide detection using nano bismuth species deposited built-in three-in-one screen-printed silver electrode. *Int. J. Electrochem. Sci.* **2011**, *6*, 2352–2365.
40. Lien, C.H.; Chen, J.C.; Hu, C.C.; Wong, D.S.H. Cathodic deposition of binary nickel-cobalt hydroxide for non-enzymatic glucose sensing. *J. Taiwan Inst. Chem. Eng.* **2014**, *45*, 846–851. [CrossRef]
41. Huang, T.K.; Lin, K.W.; Tung, S.P.; Cheng, T.M.; Chang, I.C.; Hsieh, Y.Z.; Lee, C.Y.; Chiu, H.T. Glucose sensing by electrochemically grown copper nanobelt electrode. *J. Electroanal. Chem.* **2009**, *636*, 123–127. [CrossRef]

42. Pérez-Fernández, B.; Martín-Yerga, D.; Costa-García, A. Galvanostatic electrodeposition of copper nanoparticles on screen-printed carbon electrodes and their application for reducing sugars determination. *Talanta* **2017**, *175*, 108–113. [CrossRef] [PubMed]
43. Lee, M.Y.; Peng, J.; Wu, C.C. Geometric effect of copper nanoparticles electrodeposited on screen-printed carbon electrodes on the detection of α-, β- And γ-amino acids. *Sens. Actuatorsb Chem.* **2013**, *186*, 270–277. [CrossRef]
44. Zen, J.M.; Hsu, C.T.; Kumar, A.S.; Lyuu, H.J.; Lin, K.Y. Amino acid analysis using disposable copper nanoparticle plated electrodes. *Analyst* **2004**, *129*, 841–845. [CrossRef] [PubMed]
45. Niu, X.; Lan, M.; Zhao, H.; Chen, C. Highly sensitive and selective nonenzymatic detection of glucose using three-dimensional porous nickel nanostructures. *Anal. Chem.* **2013**, *85*, 3561–3569. [CrossRef] [PubMed]
46. Yang, J.; Yu, J.H.; Rudi Strickler, J.; Chang, W.J.; Gunasekaran, S. Nickel nanoparticle-chitosan-reduced graphene oxide-modified screen-printed electrodes for enzyme-free glucose sensing in portable microfluidic devices. *Biosens. Bioelectron.* **2013**, *47*, 530–538. [CrossRef]
47. Pérez-Fernández, B.; Martín-Yerga, D.; Costa-García, A. Electrodeposition of nickel nanoflowers on screen-printed electrodes and their application to non-enzymatic determination of sugars. *RSC Adv.* **2016**, *6*, 83748–83757. [CrossRef]
48. Palanisamy, S.; Thirumalraj, B.; Chen, S.M.; Ali, M.A.; Al-Hemaid, F.M.A. Palladium nanoparticles decorated on activated fullerene modified screen printed carbon electrode for enhanced electrochemical sensing of dopamine. *J. Colloid Interface Sci.* **2015**, *448*, 251–256. [CrossRef]
49. Karuppiah, C.; Velmurugan, M.; Chen, S.M.; Devasenathipathy, R.; Karthik, R.; Wang, S.F. Electrochemical Activation of Graphite Nanosheets Decorated with Palladium Nanoparticles for High Performance Amperometric Hydrazine Sensor. *Electroanalysis* **2016**, *28*, 808–816. [CrossRef]
50. Yang, C.C.; Kumar, A.S.; Zen, J.M. Electrocatalytic reduction and determination of dissolved oxygen at a preanodized screen-printed carbon electrode modified with palladium nanoparticles. *Electroanalysis* **2006**, *18*, 64–69. [CrossRef]
51. Fu, L.; Wu, K.; Ji, J.; Zhang, J.; Guo, X. A highly sensitive disposable glucose biosensor based on platinum nanoflowers decorated screen printed carbon electrode. In Proceedings of the IEEE Sensors 2017, Glasgow, Scotland, UK, 29 October–1 November 2017; pp. 1–3. [CrossRef]
52. Chou, C.; Chang, J.; Zen, J. Sensors and Actuators B: Chemical Effective analysis of gaseous formaldehyde based on a platinum-deposited screen-printed edge band ultramicroelectrode coated with Nafion as solid polymer electrolyte. *Sens. Actuators B. Chem.* **2010**, *147*, 669–675. [CrossRef]
53. Agrisuelas, J.; González-Sánchez, M.I.; Valero, E. Hydrogen peroxide sensor based on in situ grown Pt nanoparticles from waste screen-printed electrodes. *Sens. Actuatorsb Chem.* **2017**, *249*, 499–505. [CrossRef]
54. Niu, X.; Chen, C.; Zhao, H.; Chai, Y.; Lan, M. Biosensors and Bioelectronics Novel snowflake-like Pt–Pd bimetallic clusters on screen-printed gold nanofilm electrode for H_2O_2 and glucose sensing. *Biosens. Bioelectron.* **2012**, *36*, 262–266. [CrossRef] [PubMed]
55. Niu, X.; Zhao, H.; Chen, C.; Lan, M. Electrochimica Acta Platinum nanoparticle-decorated carbon nanotube clusters on screen-printed gold nanofilm electrode for enhanced electrocatalytic reduction of hydrogen peroxide. *Electrochim. Acta* **2012**, *65*, 97–103. [CrossRef]
56. González-Sánchez, M.I.; Gómez-Monedero, B.; Agrisuelas, J.; Valero, E. Recycling Metals from Spent Screen-Printed Electrodes while Learning the Fundamentals of Electrochemical Sensing. *J. Chem. Educ.* **2018**, *95*, 847–851. [CrossRef]
57. Cunha-Silva, H.; Arcos-Martinez, M.J. A disposable rhodium nanoparticle-modified screen-printed sensor for direct determination of bromide anions. *Sens. Actuatorsb Chem.* **2019**, *282*, 603–608. [CrossRef]
58. Wang, X.; Su, J.; Zeng, D.; Liu, G.; Liu, L.; Xu, Y.; Wang, C.; Liu, X.; Wang, L.; Mi, X. Gold nano-flowers (Au NFs) modified screen-printed carbon electrode electrochemical biosensor for label-free and quantitative detection of glycated hemoglobin. *Talanta* **2019**, *201*, 119–125. [CrossRef]
59. Chen, Y.H.; Kirankumar, R.; Kao, C.L.; Chen, P.Y. Electrodeposited Ag, Au, and AuAg nanoparticles on graphene oxide-modified screen-printed carbon electrodes for the voltammetric determination of free sulfide in alkaline solutions. *Electrochim. Acta.* **2016**, *205*, 124–131. [CrossRef]

60. Rick, J.; Tsai, M.C.; Hwang, B.J. Biosensors incorporating bimetallic nanoparticles. *Nanomaterials* **2015**, *6*, 5. [CrossRef]
61. Yang, C.C.; Kumar, A.S.; Kuo, M.C.; Chien, S.H.; Zen, J.M. Copper-palladium alloy nanoparticle plated electrodes for the electrocatalytic determination of hydrazine. *Anal. Chim. Acta* **2005**, *554*, 66–73. [CrossRef]
62. Choudhry, N.A.; Kampouris, D.K.; Kadara, R.O.; Jenkinson, N.; Banks, C.E. Next generation screen printed electrochemical platforms: Non-enzymatic sensing of carbohydrates using copper(ii) oxide screen printed electrodes. *Anal. Methods* **2009**, *1*, 183–187. [CrossRef]
63. Martínez-Paredes, G.; González-García, M.B.; Costa-García, A. In situ electrochemical generation of gold nanostructured screen-printed carbon electrodes. Application to the detection of lead underpotential deposition. *Electrochim. Acta* **2009**, *54*, 4801–4808. [CrossRef]
64. Núnez-Bajo, E.; Blanco-López, M.C.; Costa-García, A.; Fernández-Abedul, M.T. In situ gold-nanoparticle electrogeneration on gold films deposited on paper for non-enzymatic electrochemical determination of glucose. *Talanta* **2018**, *178*, 160–165. [CrossRef] [PubMed]
65. Espro, C.; Donato, N.; Galvagno, S.; Aloisio, D.; Leonardi, S.G.; Neri, G. CuO nanowires-based electrodes for glucose sensors. *Chem. Eng. Trans.* **2014**, *41*, 415–420. [CrossRef]
66. Yang, J.; Kwak, T.-J.; Zhang, X.; McClain, R.; Chang, W.J.; Gunasekaran, S. Iridium Oxide-reduced Graphene Oxide Nanohybrid Thin Film Modified Screen-printed Electrodes as Disposable Electrochemical Paper Microfluidic pH Sensors. *J. Vis. Exp.* **2016**, e53339. [CrossRef]
67. Riman, D.; Avgeropoulos, A.; Hrbac, J.; Prodromidis, M.I. Electrochimica Acta Sparked-bismuth oxide screen-printed electrodes for the determination of ribo fl avin in the sub-nanomolar range in non-deoxygenated solutions. *Electrochim. Acta* **2015**, *165*, 410–415. [CrossRef]
68. Niu, P.; Fernández-Sánchez, C.; Gich, M.; Navarro-Hernández, C.; Fanjul-Bolado, P.; Roig, A. Screen-printed electrodes made of a bismuth nanoparticle porous carbon nanocomposite applied to the determination of heavy metal ions. *Microchim. Acta* **2016**, *183*, 617–623. [CrossRef]
69. Barton, J.; García, M.B.G.; Santos, D.H.; Fanjul-Bolado, P.; Ribotti, A.; McCaul, M.; Diamond, D.; Magni, P. Screen-printed electrodes for environmental monitoring of heavy metal ions: a review. *Microchim. Acta* **2016**, *183*, 503–517. [CrossRef]
70. Waryo, T.; Kotzian, P.; Begić, S.; Bradizlova, P.; Beyene, N.; Baker, P.; Kgarebe, B.; Turkušić, E.; Iwuoha, E.; Vytřas, K.; et al. Amperometric Hydrogen Peroxide Sensors with Multivalent Metal Oxide-Modified Electrodes for Biomedical Analysis. In Proceedings of the 13th International Conference on Biomedical Engineering, Singapore, 3–6 December 2008; pp. 829–833.
71. Katsounaros, I.; Schneider, W.B.; Meier, J.C.; Benedikt, U.; Biedermann, P.U.; Auer, A.A.; Mayrhofer, K.J.J. Hydrogen peroxide electrochemistry on platinum: Towards understanding the oxygen reduction reaction mechanism. *Phys. Chem. Chem. Phys.* **2012**, *14*, 7384–7391. [CrossRef]
72. Dhara, K. Recent advances in electrochemical nonenzymatic hydrogen peroxide sensors based on nanomaterials: A review. *J. Mater. Sci.* **2019**, *54*, 12319–12357. [CrossRef]
73. García-Cruz, L.; Montiel, V.; Solla-Gullón, J. Shape-controlled metal nanoparticles for electrocatalytic applications. *Phys. Sci. Rev.* **2018**, *4*, 1–34. [CrossRef]
74. Cao, S.; Tao, F.F.; Tang, Y.; Li, Y.; Yu, J. Size- and shape-dependent catalytic performances of oxidation and reduction reactions on nanocatalysts. *Chem. Soc. Rev.* **2016**, *45*, 4747–4765. [CrossRef] [PubMed]
75. Leong, G.J.; Schulze, M.C.; Strand, M.B.; Maloney, D.; Frisco, S.L.; Dinh, H.N.; Pivovar, B.; Richards, R.M. Shape-directed platinum nanoparticle synthesis: Nanoscale design of novel catalysts. *Appl. Organomet. Chem.* **2014**, *28*, 1–17. [CrossRef]
76. Lezi, N.; Vysko, V.; Economou, A. Electroanalysis of Organic Compounds at Bismuth Electrodes: A Short Review. *Sens. Electroanal.* **2012**, *7*, 71–78.
77. Dey, A. Semiconductor metal oxide gas sensors: A review. *Mater. Sci. Eng. B Solid-State Mater. Adv. Technol.* **2018**, *229*, 206–217. [CrossRef]
78. George, J.M.; Antony, A.; Mathew, B. Metal oxide nanoparticles in electrochemical sensing and biosensing: A review. *Microchim. Acta* **2018**, 185. [CrossRef]
79. Sun, C.L.; Cheng, W.L.; Hsu, T.K.; Chang, C.W.; Chang, J.L.; Zen, J.M. Ultrasensitive and highly stable nonenzymatic glucose sensor by a CuO/graphene-modified screen-printed carbon electrode integrated with flow-injection analysis. *Electrochem. Commun.* **2013**, *30*, 91–94. [CrossRef]

80. Leonardi, S.G.; Marini, S.; Espro, C.; Bonavita, A.; Galvagno, S.; Neri, G. In-situ grown flower-like nanostructured CuO on screen printed carbon electrodes for non-enzymatic amperometric sensing of glucose. *Microchim. Acta* **2017**, *184*, 2375–2385. [CrossRef]
81. Hallam, P.M.; Kampouris, D.K.; Kadara, R.O.; Jenkinson, N.; Banks, C.E. Nickel oxide screen printed electrodes for the sensing of hydroxide ions in aqueous solutions. *Anal. Methods* **2010**, *2*, 1152–1155. [CrossRef]
82. Beyene, N.W.; Kotzian, P.; Schachl, K.; Alemu, H.; Turkušić, E.; Čopra, A.; Moderegger, H.; Švancara, I.; Vytřas, K.; Kalcher, K. (Bio)sensors based on manganese dioxide-modified carbon substrates: Retrospections, further improvements and applications. *Talanta* **2004**, *64*, 1151–1159. [CrossRef]
83. Kotzian, P.; Brázdilová, P.; Kalcher, K.; Vytřas, K. Determination of hydrogen peroxide, glucose and hypoxanthine using (bio)sensors based on ruthenium dioxide-modified screen-printed electrodes. *Anal. Lett.* **2005**, *38*, 1099–1113. [CrossRef]
84. Kotzian, P.; Brázdilová, P.; Řezková, S.; Kalcher, K.; Vytřas, K. Amperometric glucose biosensor based on rhodium dioxide-modified carbon ink. *Electroanalysis* **2006**, *18*, 1499–1504. [CrossRef]
85. Kotzian, P.; Brázdilová, P.; Kalcher, K.; Handlíř, K.; Vytřas, K. Oxides of platinum metal group as potential catalysts in carbonaceous amperometric biosensors based on oxidases. *Sens. Actuatorsb. Chem.* **2007**, *124*, 297–302. [CrossRef]
86. Ahmad, R.; Vaseem, M.; Tripathy, N.; Hahn, Y.-B. Wide Linear-Range Detecting Nonenzymatic Glucose Biosensor Based on CuO Nanoparticles Inkjet-Printed on Electrodes. *Anal. Chem.* **2013**, *85*, 10448–10454. [CrossRef]
87. Kovalyk, A.; Tananaiko, O.; Borets, A.; Etienne, M.; Walcarius, A. Voltammetric and microscopic characteristics of MnO2 and silica-MnO2hybrid films electrodeposited on the surface of planar electrodes. *Electrochim. Acta.* **2019**, *306*, 680–687. [CrossRef]
88. Bollella, P.; Fusco, G.; Stevar, D.; Gorton, L.; Ludwig, R.; Ma, S.; Boer, H.; Koivula, A.; Tortolini, C.; Favero, G.; et al. A Glucose/Oxygen Enzymatic Fuel Cell based on Gold Nanoparticles modified Graphene Screen-Printed Electrode. Proof-of-Concept in Human Saliva. *Sens. Actuatorsb Chem.* **2018**, *256*, 921–930. [CrossRef]
89. Raveendran, J.; Krishnan, R.G.; Nair, B.G.; Satheesh Babu, T.G. Voltammetric determination of ascorbic acid by using a disposable screen printed electrode modified with Cu(OH)2nanorods. *Microchim. Acta.* **2017**, *184*, 3573–3579. [CrossRef]
90. Hendawy, H.A.M.; Eldin, G.M.G.; Fekry, A.M. A Zirconium Oxide Nanoparticle Modified Screen-printed Electrode for Anodic Stripping Determination of Daclatasvir Dihydrochloride. *Electroanalysis* **2019**, *31*, 858–866. [CrossRef]
91. Rezaei, R.; Foroughi, M.M.; Beitollahi, H.; Tajik, S.; Jahani, S. Synthesis of Lanthanium-doped ZnO Nanoflowers: Supported on Graphite Screen Printed Electrode for Selective and Sensitive Detection of Hydrochlorothiazide. *Int. J. Electrochem. Sci.* **2019**, *14*, 2038–2048. [CrossRef]
92. Neves, M.M.P.S.; González-García, M.B.; Hernández-Santos, D.; Fanjul-Bolado, P. Screen-Printed Electrochemical 96-Well Plate: A High-Throughput Platform for Multiple Analytical Applications. *Electroanalysis* **2014**, *26*, 2764–2772. [CrossRef]
93. Xiao, T.; Huang, J.; Wang, D.; Meng, T.; Yang, X. Au and Au-Based nanomaterials: Synthesis and recent progress in electrochemical sensor applications. *Talanta* **2020**, *206*, 120210. [CrossRef]
94. Pezzato, C.; Maiti, S.; Chen, J.L.Y.; Cazzolaro, A.; Gobbo, C.; Prins, L.J. Monolayer protected gold nanoparticles with metal-ion binding sites: Functional systems for chemosensing applications. *Chem. Commun.* **2015**, *51*, 9922–9931. [CrossRef] [PubMed]
95. Hasanzadeh, M.; Shadjou, N.; de la Guardia, M. Iron and iron-oxide magnetic nanoparticles as signal-amplification elements in electrochemical biosensing. *Trac Trends Anal. Chem.* **2015**, *72*, 1–9. [CrossRef]
96. Tan, H.W.; An, J.; Chua, C.K.; Tran, T. Metallic Nanoparticle Inks for 3D Printing of Electronics. *Adv. Electron. Mater.* **2019**, *5*, 1800831. [CrossRef]

97. Martín-Yerga, D.; Carrasco-Rodríguez, J.; García Alonso, F.J.; Costa-García, A. Competitive electrochemical biosensing of biotin using cadmium-modified titanium phosphate nanoparticles and 8-channel screen-printed disposable electrodes. *Anal. Methods* **2017**, *9*, 3983–3991. [CrossRef]
98. Teng, Y.; Ding, G.; Liu, W.; Liu, J.; Nie, Y.; Li, P. Electrodeposition of dendritic gold/silver nanaoparticles on disposable screen-printed carbon electrode and its application of 4-mercaptopyridine in in situ electrochemical surface-enhanced Raman scattering. *Surf. Interface Anal.* **2016**, *48*, 990–994. [CrossRef]

biosensors

MDPI

Review

Screen-Printed Electrode-Based Sensors for Food Spoilage Control: Bacteria and Biogenic Amines Detection †

Ricarda Torre [1], Estefanía Costa-Rama [1,2,*], Henri P. A. Nouws [1] and Cristina Delerue-Matos [1,*]

1 REQUIMTE/LAQV, Instituto Superior de Engenharia do Porto, Instituto Politécnico do Porto, Dr. António Bernardino de Almeida 431, 4200-072 Porto, Portugal; rdvdt@isep.ipp.pt (R.T.); han@isep.ipp.pt (H.P.A.N.)
2 Departamento de Química Física y Analítica, Universidad de Oviedo, Av. Julián Clavería 8, 33006 Oviedo, Spain
* Correspondence: costaestefania@uniovi.es (E.C.-R.); cmm@isep.ipp.pt (C.D.-M.)
† This article is dedicated to the memory of Professor Agustín Costa-García.

Received: 31 August 2020; Accepted: 28 September 2020; Published: 30 September 2020

check for updates

Abstract: Food spoilage is caused by the development of microorganisms, biogenic amines, and other harmful substances, which, when consumed, can lead to different health problems. Foodborne diseases can be avoided by assessing the safety and freshness of food along the production and supply chains. The routine methods for food analysis usually involve long analysis times and complex instrumentation and are performed in centralized laboratories. In this context, sensors based on screen-printed electrodes (SPEs) have gained increasing importance because of their advantageous characteristics, such as ease of use and portability, which allow fast analysis in point-of-need scenarios. This review provides a comprehensive overview of SPE-based sensors for the evaluation of food safety and freshness, focusing on the determination of bacteria and biogenic amines. After discussing the characteristics of SPEs as transducers, the main bacteria, and biogenic amines responsible for important and common foodborne diseases are described. Then, SPE-based sensors for the analysis of these bacteria and biogenic amines in food samples are discussed, comparing several parameters, such as limit of detection, analysis time, and sample type.

Keywords: screen-printed electrode; electroanalysis; electrochemical sensor; biosensor; immunosensor; food analysis; bacteria; biogenic amines; histamine

1. Introduction

The impact of food contamination by microorganisms and other poisonous substances is considered a major public health and safety concern. According to the World Health Organization (WHO), each year 600 million people (almost 1 in 10) fall ill because of contaminated food [1]. Pathogens have the ability to adapt to various environments, causing contaminations in different stages of the food production and supply chains. Thus, they can appear in raw food but also at any point of the food production process and even after the consumer acquires the food if the necessary precaution to transport and store is not taken. Many microorganisms are affected by heat and can be destroyed or inactivated after cooking [2,3]. However, some of them, and substances such as histamine (the main biogenic amine), are not affected by cooking, freezing, or canning processes [3,4]. Taking this into account, the importance of the control of contamination along the whole food chain is clear. Analytical methods and devices for real-time control of food safety and quality provide immediate information that allows corrective actions to be taken before the food products are made available for consumption.

Among the microorganisms that cause foodborne illnesses, bacteria are the most important because of their high occurrence [1,5]. These bacteria can be detected by appropriate techniques and methods such as cell culture and colony counting, polymerase chain reaction (PCR) and immunological assays [3,6]. Biogenic amines are nitrogenous species usually present in different foods that, at normal levels, do not entail health risks. However, their levels increase when food, especially fish, is stored for a long time and/or at an inadequate temperature (>4 °C) [7–9]. Therefore, the quantification of biogenic amines, especially histamine, is included in the routine analysis of many food industries. The analysis of biogenic amines is often performed through chromatographic methods (mainly liquid) coupled to different detectors [10]. Enzymatic kits are also employed since they are simpler and require cheaper instrumentation [11].

Although the above-mentioned methods (i.e., cell culture and colony counting, PCR, chromatography) are very useful, robust and provide accurate results, they are time consuming, involve complex processing steps and require highly trained analysts and expensive/complex instrumentation (Figure 1A). Therefore, the analyses have to be performed in centralized laboratories and the results are not available in real-time. Taking into account the time the different steps of the analytical process take (sampling, sample preparation, analysis, results interpretation, and communication) and the short shelf life of food products, the development of analytical methods that allow rapid screening of pathogens and spoilage indicators is critical to ensure food safety.

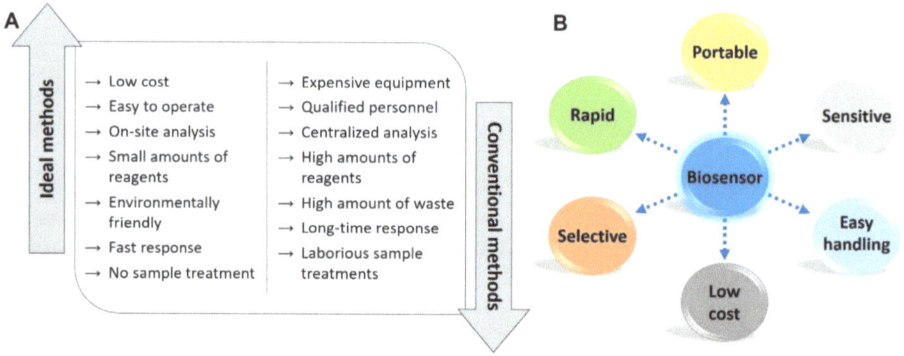

Figure 1. Schematic representation of (**A**) the advantageous features of ideal analytical methods vs. disadvantageous features of conventional ones and (**B**) the advantages of screen-printed based biosensors. The last one adapted from [12] with permission from Elsevier.

In this context, electrochemical (bio)sensors based on screen-printed electrodes (SPEs) have gained increasing interest as analytical tools for food analysis since SPEs provide great advantages that make these kind of sensors have the important characteristics of ideal biosensors (Figure 1B) [12]: ease of use, low-cost, and portability [13,14]. So, the screen-printed technology has significantly contributed to the transition from the traditional unwieldy electrochemical cells to miniaturized and portable electrodes that meet the needs for on-site analysis [12,15]. Although a screen-printed electrode (SPE) is not as robust as a conventional electrode, such as glassy carbon or gold disk, and the surface of its working electrode is not as perfect as the one of a mirror-like polished solid electrode, the advantages of SPEs regarding cost and size led to their increasing use in the last years as transducers in (bio)sensing. The use of SPE-based sensors in the control of food spoilage as complementary analytical tools to the conventional methods allows a rapid screening at any point of the food production chain, preventing the occurrence of foodborne illnesses and the reduction of food waste.

The purpose of this article is to review SPE-based biosensors for the analysis of bacteria and biogenic amines related with food spoilage, focusing on the analyte, and discussing the different approaches and trends in the development of these sensors. The main characteristics of SPEs as transducers and the main challenges on improved SPE-based biosensors are also highlighted.

2. Screen-Printed Electrodes as Transducers

2.1. Production and Design of Screen-Printed Electrodes

The screen-printing technology was adapted from the microelectronics industry and is used, among others, to produce screen-printed electrodes (SPEs) (Figure 2A,B). These electrodes offer the main characteristics required to obtain electrochemical sensing platforms for on-site analysis. Although this technology exists in its present form since the 20th century [16], it began to be used for the fabrication of electrochemical cells in the 1990's. Since then, the use of SPEs as transducers for many different electrochemical sensors has steadily increased (Figure 2C). Nowadays, the screen-printing technology is a common and well-established technique for the conception of electroanalytical devices with assorted applications: from point-of-care (POC) devices for biomedical applications [17–19] to portable sensors for food analysis [13,14] and detection of environmental contaminants [12,20,21]. SPEs usually contain an electrochemical cell composed of three electrodes (working-WE-, reference-RE-, and counter-CE-electrodes) printed on a solid substrate (Figure 2A). Different inks (the most common are carbon and metallic inks) to print the electrodes [22] and different substrates (often ceramic or plastic) can be used. The SPE's fabrication process is fast and allows large scale and highly reproducible production of small-sized, cheap, and disposable electrodes. Therefore, it is not necessary to clean and/or polish them, avoiding tedious pretreatment steps, saving a lot of time. In contrast, the robustness of the printed electrodes and their electrochemical features are not as good as those of conventional electrodes (e.g., glassy carbon, gold disk, etc.). However, SPEs show adequate electroanalytical features for sensing applications and this, together with their low-cost and ease of use (which avoids the need of highly skilled analysts) make SPEs clearly advantageous as transducers for applications in which on-site one-point measurements are required. Moreover, the miniaturized design of SPEs not only allows to transport them to perform on-site measurements for real-time analysis, but also avoids the use of high amounts of reagents and samples. All these characteristics are in accordance with the principles of Green Analytical Chemistry [23,24].

Briefly, the fabrication of SPEs consists of the following steps: (i) design of the screen or mesh that will define the geometry and size of the SPE; (ii) selection and preparation of the conductive inks and selection of the substrate material; (iii) layer-by-layer deposition of the chosen inks on the solid substrate and (iv) drying and curing [17,20]. By covering the electrical circuits with an insulating material it is possible to perform the analytical measurement by depositing a single drop of the reagent/sample solution onto the SPE, by immersing it into a solution or by including it in a flow system. Regarding the inks for the WE, as mentioned before, the most popular ones are based on carbon (graphite, graphene, fullerene, carbon nanomaterials, etc.) because of their suitable features for electroanalysis (i.e., good conductivity, chemical inertness, ease of modification, low background currents, and a wide potential range) and their low costs [17,25]. Besides carbon inks, conductive metallic inks have increasingly been used; among them gold ink is the most common due to its high affinity with thiol moieties that allows easy surface modification with proteins by the formation of self-assembled monolayers (SAMs). SPEs with a WE made of other metallic inks such as silver, platinum or palladium are also available on the market [26] but their use is scarce and limited to specific applications. The use of SPEs with an optically transparent WE, made of indium tin oxide (ITO), PEDOT or even gold (obtained by sputtering process) or carbon (made of carbon nanotubes), is increasing because of the growing interest in spectroelectrochemistry [26–31]. The RE is often made of silver or silver/silver chloride ink. This is considered a pseudo-reference or quasi-reference electrode since its potential is not as stable as that of an ideal reference electrode (e.g., conventional silver/silver chloride RE, which is the most common). Therefore, the applied potential is not as exact and reproducible as when an Ag/AgCl electrode is used. This can be problematic for electrochemical studies in which the control of the potential is essential;

however, for sensing applications, this is rarely a problem. The CE is usually made of the same ink as the WE. Because the composition of the inks defines the electrochemical characteristics of the electrode, SPEs are highly versatile. However, the versatility of SPEs is not only due to the use of different inks, but also because of the ease of modification of the WE. The purpose of these modifications is to enhance the electroanalytical characteristics of the SPEs (such as sensitivity, precision, operational stability) and to improve the immobilization of the recognition element (which are often biological elements (e.g., proteins, DNA, etc.), but can also be synthetic (e.g., molecularly imprinted polymers (MIPs), see Section below)) [17]. For example, great enhancements of the analytical features have been achieved by using carbon nanomaterials (nanotubes, nanofibers, graphene, among others) and metallic nanoparticles (primarily gold nanoparticles, since they are cheaper than a WE made of gold ink) [18,32]. Besides these nanomaterials many other materials can be used: redox mediators, polymers, complexing agents, metallic oxides, etc. The simplest procedure to modify SPEs is by deposition of the modifying agent onto the WE; this procedure is facilitated because of the planar nature of the SPE, so it can be performed through an automatic dispenser in a mass-producible way. However, the WE of an SPE can also be modified by adding the modifier to the ink before printing, by chemical adsorption or by electrochemical deposition (a good example is the in-situ generation of metallic nanoparticles) [32–34].

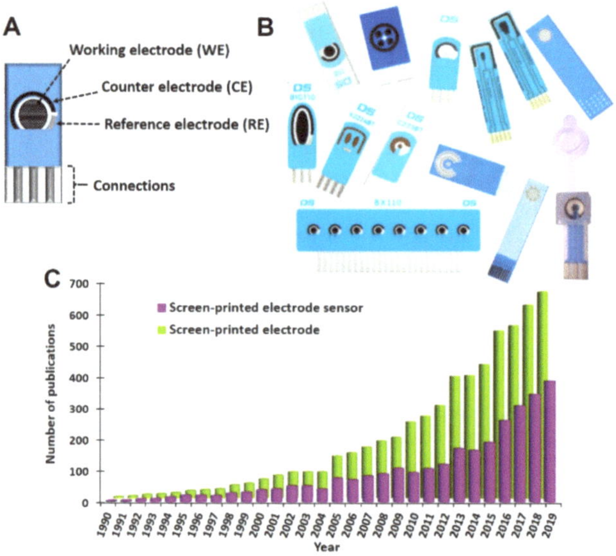

Figure 2. (**A**) Scheme of the most common configuration of a screen-printed electrode. (**B**) Examples of commercial screen-printed electrodes with different configurations and designs. Reproduced from [19] with permission from Wiley. (**C**) Number of publications per year when searching "screen-printed electrode" and "screen-printed electrode sensor" in Scopus database for the last 30 years (1990–2019).

Another source of the SPE's versatility is the possibility of printing the electrochemical cell on a wide variety of substrates. The choice of the substrate will determine the stability, robustness, disposability, and applicability of the SPE. As mentioned before, the most common are rigid substrates such as ceramics. However, although printing the electrodes on non-planar and non-rigid surfaces is not so easy as on rigid ones, there are several works describing SPEs that were fabricated using paper sheets, cloths, stretch and foldable films, and even epidermis [18,20,25,35–38]. To choose the correct substrate, it is important to keep the final application in mind: for example, ceramics are easy to print on and are highly robust but are more expensive than paper. Paper is light and easy to transport but its flexibility and moisture tolerance is limited. Polymeric substrates, especially flexible ones, are interesting for wearable sensors; in these cases, the limitation is related to the bending endurance of the printed electrodes.

So, SPEs offer numerous advantages, but the most important one is their high adaptability. This adaptability is not confined to the materials to fabricate them (inks and substrates); it also covers their design. As said before, the most common option is printing one electrochemical cell (with three electrodes) on the substrate, but many others configurations are possible: SPEs with more than one WE sharing the same RE and CE, platforms with several complete electrochemical cells, 96-well SPE plate or even SPEs with an integrated micro-well/reactor [20,26,39,40] (Figure 2B). Thus, their high versatility together with their ease of use and portability make SPEs one of the main transducers for the development of electroanalytical devices.

2.2. *(Bio)Sensors Based on SPEs*

As mentioned before, there are a great amount and variety of (bio)sensors based on SPEs with applications in very different fields. A biosensor is a type of chemical sensor; it can be defined as an analytical device able to provide (bio)chemical information, usually the concentration of a substance in a complex matrix, which consists of two main parts: a biological recognition element that selectively identifies the analyte of interest, and a transducer that transforms that recognition event into an measurable signal (Figure 3A) [41,42]. A biosensor should therefore contain biological elements that can be, mainly, (i) enzymes (catalytic biosensors) and (ii) proteins (antibody or antigen), or DNA or RNA strands (affinity biosensors) (Figure 3C). However, because of the advantages of artificial biomimetic receptors, such as MIPs and aptamers, regarding physical and chemical stability, it is increasingly accepted to include them in the "biosensor" category [25].

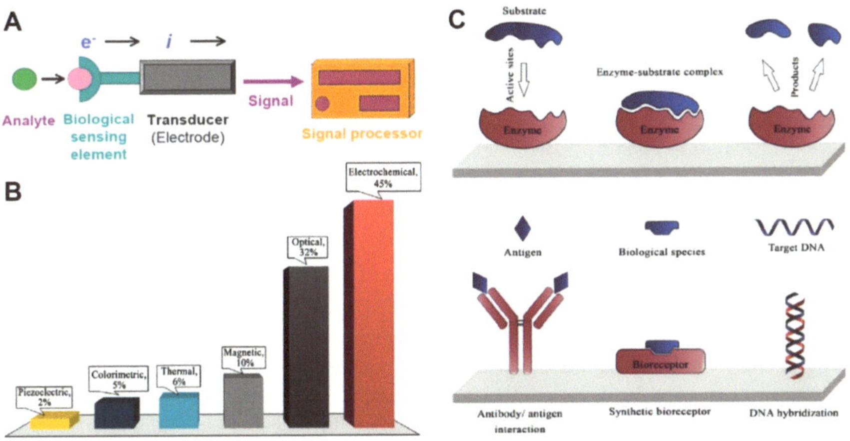

Figure 3. (**A**) A schematic representation of a biosensor with electrochemical transduction. Reproduced from [43] with permission from The Royal Society of Chemistry. (**B**) Distribution of different types of techniques for signal transduction using biosensors (data retrieved from the Scopus database from 2017 to August 2019). Reproduced from [25] with permission from Springer 2019. (**C**) Schematic illustration of enzymatic reaction on catalytic-based biosensors (top) and three different types of affinity-based biosensors (bottom). Reproduced from [25] with permission from Springer 2019.

The most common SPE-based biosensors for food analysis are enzymatic- and immunosensors. Enzymatic biosensors are based on the highly selective interaction of the target analyte with an enzyme through its active sites, forming a complex that transforms the analyte into a (or more than one) product(s) [25,44]. The determination of the analyte is usually carried out by measuring the amount of generated product. Nevertheless, since co-factors or other co-reagents are sometimes needed, their consumption can be also used to monitor the analyte–enzyme interaction. Immunosensors are based on antibody–antigen interactions and take advantage of the high specificity of an antibody towards the corresponding antigen. In these sensors, the target analyte can either be the antibody or the antigen. Briefly, there are two main ways of following the immunoreaction: (i) using a label attached to a one of the immunoreagents, e.g., an enzyme or a nanoparticle that produces a detectable signal, and (ii) through label-free detection; in this case, the formation of the immunocomplex (antibody–antigen) produces a detectable physical/chemical change [14,19]. Immunosensors are highly specific and can be applied to a wide variety of analytes provided that an antibody that interacts with the analyte is available. Moreover, different strategies (e.g., the use of different labels or nanomaterials) can be used to improve their sensitivity. However, compared to enzymatic sensors, immunosensors are usually more labour intensive and less robust since several steps with long incubation times are required.

Independent of the type of recognition element, they have to be immobilized on the surface of the WE. The versatility of SPEs allows to choose between many different immobilization procedures: from the simplest one, the direct adsorption of the receptor by incubating it on the WE surface, to others that require more steps such as crosslinking, SAM formation, covalent binding, entrapment, or affinity binding (e.g., using the avidin-biotin system). By taking advantage of the transducer, immobilization of the recognition element through electrodeposition is also possible (a good example is the case of electrogenerated MIPs [45]). These immobilization methods are extensively described in several previous reviews [32,44,46–48].

When using biosensors, mainly electrochemical techniques are used for signal transduction, but colorimetric (without instrumentation), optical, magnetic, piezoelectric, and thermal techniques can also be employed (Figure 3B) [25,49]. In electrochemical biosensors, the analytical signal can be provided by different techniques: amperometry and voltammetry (based on current measurement), electrochemical impedance spectroscopy (EIS), potentiometry or conductometry [43,50,51]. The amperometric and voltammetric sensors are the most widely used because of their simplicity and applicability. Nevertheless, EIS sensors are gaining interest since there is no need for labels (especially used in immunosensing), but their sensitivities are often lower than the amperometric and voltammetric sensors [43].

3. Parameters Related to Food Spoilage

The production of safe and high-quality foodstuffs requires the control of several parameters at different points of the food production and supply chain: the quality of raw ingredients, the hygienic conditions of food production, the suitability of storage conditions, and the nutritional properties of the finished products [2]. Inadequate conditions at any stage of this chain often lead to food spoilage, involving chemical and physical changes (e.g., oxidation, colour changes, nasty smells, physical damages, etc.). Although food spoilage can be originated by various causes, the growth of microorganisms is the most common [1,5]. Many factors can contribute or accelerate food deterioration, such as exposure to inadequate levels of oxygen, moisture, light, or temperature. The microorganisms responsible for food spoilage include several bacteria, viruses, moulds, and yeasts. Among these, the bacteria *Salmonella*, *Escherichia coli*, *Campylobacter*, and *Listeria* are the most common foodborne pathogens [1,34]. Besides microorganisms, mycotoxins, which are toxic metabolites produced by fungi, are also important causes of foodborne illnesses [1]. Another common parameter to evaluate food safety and freshness is the level of biogenic amines, which are produced by the microbial decarboxylation of amino acids present in some foodstuffs such as fish, meat, and fermented foods [10,52,53].

3.1. Bacteria

Bacteria are the most common cause of foodborne illnesses. It is often difficult to notice their presence at low but harmful levels since visual or olfactory changes are not always easy to observe. Consequently, the consumer can ingest food contaminated with bacteria without realising it, causing illnesses with important implications. Although there are a great variety of bacteria responsible for foodborne illnesses, *Campylobacter*, *Escherichia coli*, *Salmonella*, and *Listeria* are the most common causes [1,5,42,54,55]. Therefore, these bacteria will be focussed on in this review.

Campylobacter are Gram-negative bacteria that live as commensals organisms in the gastrointestinal tract of humans and many domestic animals [56,57]. Campylobacteriosis is the most commonly reported gastrointestinal infection in humans in the EU [54,55]. It normally produces symptoms such as diarrhoea and vomiting that can last from 2 to 10 days. Its main food sources are undercooked meat (specially poultry), unpasteurized milk and vegetables [3,58]. Within the genus *Campylobacter*, the species *C. jejuni* is responsible for more than 80% of *Campylobacter* infections [3].

Escherichia coli (*E. coli*) are Gram-negative bacteria belonging to the *Enterobacteriaceae* family that inhabit the gastrointestinal tract of humans and warm-blooded animals. As commensal microorganism, *E. coli* live in mutually beneficial association with its host without causing diseases [59,60]. However, there are several *E. coli* strains with virulent attributes associated mainly with three clinical syndromes: diarrhoea, urinary tract infections or meningitis. [3,60]. Its ease of handling and the availability of its complete genome sequence makes *E. coli* an important microorganism in biotechnological, medical and industrial applications [59]. Among the intestinal pathogenic *E. coli* there are six well-describes variants known as pathovars or pathotypes: Enteropathogenic *E. coli* (EPEC), Enterotoxigenic *E. coli* (ETEC), Enteroaggregative *E. coli* (EAEC), Enteroinvasive *E. coli* (EIEC) and Diffusely adherent *E. coli* (DAEC) and Shiga toxin-producing *E. coli* (STEC, which includes the Enterohemorrhagic *E. coli*-EHEC-) [3,60]. Foodborne illness outbreaks related with *E. coli* can be associated with many types of food: from meats and unpasteurized milk or fruit juice to vegetables such as lettuce and spinach [3]. Symptoms of *E. coli* infections can be minor for some people; however, sometimes the infection may become a life-threatening illness causing serious problems such as kidney failure. STEC is the third most common cause of foodborne zoonotic illness [54]. *E. coli* O157:H7 currently accounts for most of the EHEC infections worldwide [3].

Salmonella are Gram-negative bacteria that belong to the *Enterobacteriaceae* family and is classified into two species: *S. bongori* and *S. enterica*. The latter are associated with the main public health concern [3,61]. Based on the Kaufmann-White scheme, *Salmonella* spp. are subdivided into serotypes and consequently, they are usually referred to by their serotype names [61]. Within *S. enterica*, which includes more than 2500 serotypes, *S.* Typhi and *S.* Paratyphi are responsible for typhoid illness (typhoid and paratyphoid fever, respectively) characterized by fever, headache, abdominal pain, and diarrhoea which can be fatal if suitable treatment is not provided [3,42,61,62]. Besides typhoidal illness, the other *Salmonella* serotypes can cause gastrointestinal illness (salmonellosis) that is less serious and its symptoms normally last for a few days [3]. In 2018, nearly 30% of the total foodborne illness outbreaks reported in the EU (5146 outbreaks affecting 48,365 people) were caused by *Salmonella* [54]. These outbreaks were mainly linked to eggs [54], however salmonellosis can also occur by the ingestion of other animal-derived contaminated foods such as milk, meat, or poultry, or even of contaminated fruits or raw vegetables [3,63].

Listeria are Gram-positive bacteria that comprise seventeen species, including *Listeria monocytogenes*, which is responsible for Listeriosis that, although presenting a low incidence, leads to high hospitalizations and mortality rates [50,55,64]. *L. monocytogenes* is highly persistent: it is salt-tolerant and can survive, and even grow, at temperatures below 1°C unlike many other pathogens [3]. *Listeria* can grow in several kinds of foods: raw milk, smoked fish, meats, and raw vegetables [3,50].

The main methods for the detection of these pathogens in foods are based on culturing and colony counting, which are characterized by laborious and time-consuming procedures, consumption of high amounts of reagents and the need for highly-trained personnel [3,5]. Alternative methods are those based on polymerase chain reaction (PCR) or real-time (quantitative) PCR that considerably reduce the analysis time (24 h or 3–6 h, respectively) but also involve laborious procedures [3,5,6,42]. Other detection methods are those based on immunoassays. Among them, enzyme-linked immunosorbent assays (ELISA) are the most common since their commercialization as kits facilitate their use and large-scale application [5,50,51].

3.2. Biogenic Amines

Biogenic amines (BAs) are nitrogenous low-molecular weight compounds that are mainly produced by the microbial decarboxylation of amino acids. There are eight BAs commonly present in animals, plants, and foods and can be classified in three groups based on their structure: (i) aliphatic (putrescine, cadaverine, spermine, and spermidine); (ii) aromatic (tyramine and phenylethylamine); and (iii) heterocyclic (histamine and tryptamine) (Table 1) [7–9,52]. BAs are important in several physiological processes, such as neuromodulating functions, and each one of them has key roles in organisms [7]. For example, histamine acts as a neurotransmitter, is related with intestinal physiological functions, and is involved in allergic reactions; tyramine has antioxidant effects, and putrescine is an important constituent of all mammalian cells [52,65].

Table 1. Classification and basic information of the eight most common biogenic amines. Adapted from [52] with permission from Elsevier.

Classification	Name	Molecular Formula	Structure	Molecular Weight (g/mol)
Heterocyclic	Histamine (HIS)	$C_5H_9N_3$	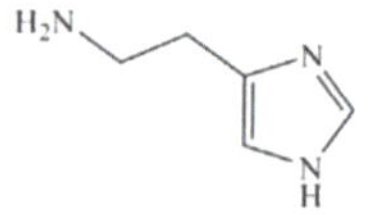	111.15
	Tryptamine (TRYP)	$C_{10}H_{12}N_2$	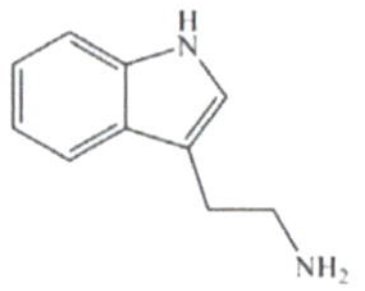	160.21
Aromatic	Phenylethylamine (PHEN)	$C_8H_{11}N$	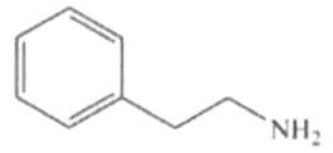	121.18
	Tyramine (TYR)	$C_8H_{11}NO$	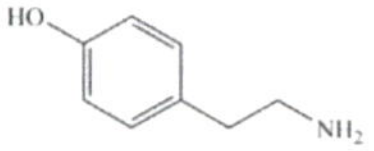	137.18
Aliphatic	Spermidine (SPD)	$C_{10}H_{26}N_4$	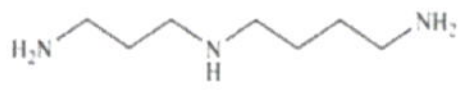	145.25
	Spermine (SPM)	$C_7H_{19}N_3$		202.34
	Cadaverine (CAD)	$C_5H_{14}N_2$		102.18
	Putrescine (PUT)	$C_4H_{12}N_2$	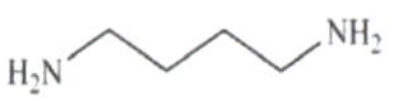	88.15

In suitable levels BAs have beneficial effects, but the consumption of an excessive amount of BAs can be toxic to humans. The most common example is histamine fish poisoning (also known as scombroid poisoning) which is generally caused by the consumption of fish with high levels of histamine. The symptoms are headache, gastrointestinal, and skin problems, and their severity depends on the dosage [4,7,9]. The rapid increase of the concentration of histamine in fish is induced by unsuitable storage conditions (mainly temperatures >4 °C and long storage times) [4,7]. Fish with high levels of histidine (such as sardine or tuna) are more prone to develop histamine than histidine-poor fish. Moreover, histamine shows a high temperature stability, so it is not affected by cooking or freezing nor by sterilization or canning processes [4,7]. Thus, the concentration of histamine is a common parameter that is used in the fish industry as a quality and freshness indicator.

Although histamine is the main BA of concern due to its toxicity, the other BAs can also induce harmful effects on human health; for example, tyramine, phenylethylamine, and tryptamine cause hypertension, and putrescine and cadaverine can cause hypotension and bradycardia, and potentiate the toxicity of other amines, especially of histamine [7,8,66,67].

Besides fish and sea-food, BAs are found in several daily-life foodstuffs (wine, beer, cheese, other fermented foods and meat) [9,10,66]. Because of their microbiological origin, the concentration of BAs has been used for the assessment of the freshness of certain foods. With this aim, the biogenic amine index (BAI) has been proposed; this parameter can include different BAs depending on the type of food to be evaluated [10]. The most widely used BAI includes histamine, cadaverine, putrescine, and tyramine. A BAI lower than 5–10 mg/kg indicates a good quality and fresh food [8,52,66].

Hence, the concentration of BAs in food is an important parameter to control. The main methods for their quantification are based on chromatographic techniques combined with different extraction techniques such as solid phase extraction, ultrasound-assisted extraction or dispersive liquid–liquid microextraction [10,52,66,67]. Regarding chromatographic techniques, the most common is liquid chromatography (LC) combined with ultraviolet or fluorescence detectors (in which the BAs need to be derivatized since they exhibit neither UV absorption nor fluorescence emission), or tandem mass spectrometry [8,10,52,65–67].

4. SPE-Based (Bio)Sensors for the Determination of Food Spoilage Parameters

When a bibliographic search for articles about SPE-based sensors for determining parameters related to food spoilage is performed, a considerable number of works is found. However, this number is lower than for clinical or biomedical applications. Therefore, the development of these kind of sensors in this area will surely continue to be explored in the next years. Among the SPE-based sensors, there are several enzymatic- and immunosensors, but aptasensors were also described. The electroactivity of some of the analytes is also explored in some studies, avoiding the use of a recognition element. In the following section, different SPE-based sensors for the determination of important food spoilage parameters are discussed. The classification of these sensors is based on the analytes and mainly focuses on those with applications in food analysis.

4.1. SPE-Based (Bio)Sensors for Bacteria Detection

As mentioned before, among the bacteria responsible for foodborne diseases the main contributors, because of their incidence and the illnesses they cause, are *Campylobacter*, *Salmonella*, *E. coli*, and *Listeria*. Because of this, numerous biosensors have been developed for the determination of these microorganisms in food, as a whole or through target indicators of their presence (for example, specific cell membrane proteins or toxins). The wide incidence of salmonellosis has led to the development of many biosensors for the determination of *Salmonella* in foods such as milk or chicken meat (Table 2): most of them are immunosensors, for the serotype *Salmonella* Typhimurium, and based on SPEs with a carbon WE, both unmodified [68,69] or modified with nanomaterials [70–72], polymers [70], or an ionic liquid [71]. Although immunosensors are the main type of sensors, an aptasensor for *Salmonella* detection in apple juice is also reported [73] (Figure 4B). This is a label-free impedimetric sensor that used an SPCE modified with diazonium salt through chemical grafting on which the aminated-aptamer is immobilized. With this approach a concentration range between 10^1 and 10^8 CFU/mL and a limit of detection (LOD) of 6 CFU/mL was achieved. Several sensors use magnetic beads on which either the capture or the detection antibody is immobilized. A good example of this is the one developed for Ngoensawat et al. [74] in which the monoclonal capture antibody is immobilized on carboxylic acid-modified Fe_2O_3 magnetic particles on which multiwalled carbon nanotubes (MWCNT) modified with Methylene blue (the detection label) are immobilized. Once the immunomagnetic separation of *Salmonella* from the sample is performed, the detection is carried out through a sandwich type assay on an avidin-modified SPCE on which a biotin-labelled polyclonal antibody is immobilized. Using DPV as detection technique, a good LOD in milk samples is obtained: 17.3 CFU/mL. As in this case, using labels for monitoring the immunoaffinity event is the most common: the enzyme horseradish peroxidase is widely used. However, the use of nanomaterials such as gold nanoparticles (AuNP) [69,75,76] and CdS nanocrystals [68] is also frequent. A remarkable label is the one based on a polymeric dendron modified with CdTe Quantum Dots (QD) which was recently developed by Murasova et al. [77]. Using a specific anti-*Salmonella* antibody modified with this label, a sandwich immunoassay is performed using an antibody attached to magnetic beads. The detection is carried out through square-wave anodic stripping voltammetry (SWASV) on an SPCE modified with an on-site generated bismuth film obtaining a LOD of 4 CFU/mL. Viswanathan et al. explored the fact that metals show different redox potentials by using different metallic nanoparticles (CuS, CdS and PbS) to construct a multiplexed immunosensor for the simultaneous determination of *Salmonella*, *E. coli*, and *Campylobacter* [78]. The sensor consisted of a MWCNT-polyallylamine-modified SPCE on which specific antibodies for each one of the bacteria are immobilized. The sandwich is formed with detection antibodies specifically labelled with each one of the three different nanocrystals (Figure 4A). Using SWASV as technique detection, calibration curves in the range 10^3–5×10^5 cell/mL, and LODs of 400 cell/mL for *Salmonella* and *Campylobacter* and 800 cell/mL for *E. coli* are obtained. Another approach to develop sensors able to perform simultaneous measurements is the use of SPEs with more than one WE or more than one electrochemical cell. Examples of these sensors for *Salmonella* detection are also reported: from SPEs with two WEs [79], to a microfluidic system with eight WEs [75] (Figure 4C) or a 96-well-SPE plate [80].

Table 2. SPE-based sensors for *Salmonella* and *Listeria* detection in foods.

Serotype	Sensor Construction	Detect. Tech.	Conc. Range	LOD	Analysis Time	Sample	Ref.
Salmonella Pullorum and *Salmonella* Gallinarum	Immunoassay; HRP as indirect label; ERGO/PVA-PDMS/SPCE	CV	10–10^9 CFU/mL	1.61 CFU/mL	≈31 min	Chicken, eggs	[70]
Salmonella Pullorum and *Salmonella* Gallinarum	Immunoassay (sandwich); HRP as label; IL/Ab/AuNP/SPCE	CV	10^4–10^9 CFU/mL	3×10^3 CFU/mL	≈81 min	Chicken, eggs	[71]
Salmonella Typhimurium	Immunoassay (sandwich); Ab-coated MB; Ag measurement; Avidin-SPCE	DPASV	10–10^6 CFU/mL	12.6 CFU/mL	≈105 min	Milk, green bean sprouts, eggs	[81]
Salmonella Typhimurium	Immunoassay; Au-coated MB/SAM/Ab; CdSNP as label; SPCE	SWASV	10–10^6 cell/mL	13 cells/mL	≈40 min	Milk	[68]
Salmonella Typhimurium	Immunoassay (sandwich); Ab on MB-MWCNT-Methylene Blue (which is the label); Avidin-SPCE	DPV	10–10^6 CFU/mL in buffer and milk	7.9 CFU/mL in buffer; 17.3 CFU/mL in milk	≈55 min	Milk	[74]
Salmonella Typhimurium	Immunoassay (sandwich); Capture Ab on MB; AuNP as label; SPCE	DPV	10^3–10^6 cell/mL	143 cells/mL	≈95 min	Milk	[69]
Salmonella Typhimurium	Immunoassay; Label free; Ferrocyanide measurement; rG-GO/SPCE	EIS	–	10^1 CFU/mL in samples	≈15 min	Water, orange juice	[72]
Salmonella Typhimurium	Aptasensor; Label free; diazonium salt- modified SPCE	EIS	10–10^8 CFU/mL	6 CFU/mL	≈45 min	Apple juice	[73]
Salmonella Typhimurium	Paper-based immunoassay (sandwich); AuNP as label; SPCE	C	10–10^8 CFU/mL	10 CFU/mL	≈35 min	Water	[76]

Table 2. *Cont.*

Serotype	Sensor Construction	Detect. Tech.	Conc. Range	LOD	Analysis Time	Sample	Ref.
Salmonella Typhimurium	Immunoassay (sandwich); HRP as label; SPAuE	CA	10–10^7 CFU/mL	≈20 CFU/mL	≈150 min	Chicken	[82]
Salmonella Typhimurium	Immunoassay (sandwich); HRP as label; SAM/Protein A/SPAuE	CA	–	10 CFU/mL	≈125 min	Milk	[62]
Salmonella (no serotype)	Immunoassay (sandwich); Capture Ab on MB; QD (CdTe) dendron as label; BiSPCE	SWASV	–	4 CFU/mL	≈80 min	Milk	[77]
Salmonella Typhimurium	Immunoassay; Label free; SAM/GA/Ab/2-SPAuE	EIS	10^3–10^8 CFU/mL	10^3 CFU/mL	≈20 min	Milk	[79]
Salmonella. Typhimurium (and *E. coli* O157:H7)	Immunoassay (sandwich); Capture biotinylated Ab on stretavidin-MB; GOX-as label; SP-IDME (gold)	EIS	10^2–10^6 CFU/mL for both	1.66 × 10^3 CFU/mL (3.90 × 10^2 CFU/mL for *E. coli*)	≈180 min	Chicken carcass (ground beef for *E. coli*)	[83]
Salmonella Typhimurium and *Salmonella aureus* (and *E. coli*)	Antimicrobial petide melittin on MB; SP-IDME (silver)	EIS	10–10^4 CFU/mL; 10–10^6 CFU/mL (1–10^6 CFU/mL for *E. coli*)	10 CFU/mL for both (1 CFU/mL for *E. coli*)	≈30 min	Water, apple juice	[84]
Salmonella Pullorum and *Salmonella* Gallinarum	Immunoassay (sandwich); Capture Ab on AuNP-modified MB (SiO_2/Fe_3O_4); HRP as label; 4-SPCE	CV	10^2–10^6 CFU/mL	32 CFU/mL	≈70 min	Chicken	[85]
Salmonella Typhimurium	Immunoassay (sandwich); Capture Ab on MB; AuNP as label; μFD-8-SPCE	DPV	10.0–100.0 cell/mL in milk	7.7 cells/mL	≈75 min	Milk	[75]
Salmonella (no serotype)	Immunoassay (sandwich); HRP as label; 96-well SPCE plate	IPA	5 × 10^6–5 × 10^8 CFU/mL	2 × 10^6 CFU/mL	≈100 min	Pork, chicken, beef	[80]
Salmonella (no serotype) (Multiplexed: *E. coli*, *Campylobacter*)	Immunoassay (sandwich); specific nanolabel for each specie (CuS, CdS, PbS); MWCNT-PAH/SPCE	SWASV	10^3–5 × 10^5 cell/mL	400 cells/mL for *Salmonella* and *Campylobacter*; 800 cells/mL for *E. coli*	≈70 min	Milk	[78]
Listeria monocytogenes	Immunoassay (sandwich); Ab capture on MB; Ab detection/urease (as label) modified AuNP; SP-IDE (gold)	EIS	1.9 × 10^3–1.9 × 10^6 CFU/mL	1.6 × 10^3 CFU/mL	≈115 min	Lettuce	[86]
Listeria innocua Serovar 6b	Label-free; Bacteriophage endolysin CBD500 covalent immobilized on SPAuE	EIS	10^4–10^9 CFU/mL	1.1 × 10^4 CFU/mL	≈25 min	Milk	[87]

4-SPCE: screen-printed carbon electrode with 4 working electrodes; 2-SPE: screen-printed electrode with 2 working electrodes; 8-SPCE: screen-printed carbon electrode with 8 working electrodes; μFD: microfluidic device; Ab: antibody; AP: alkaline phosphatase; AuNP: gold nanoparticles; BiSPCE: Bi film-modified screen-printed carbon electrode; C: conductometry; CA: chronoamperometry; CdSNP: CdS nanoparticles; CV: cyclic voltammetry; DPASV: differential pulse anodic stripping voltammetry; DPV: differential pulse voltammetry; EIS: electrochemical impedance spectroscopy; ERGO: electrochemically reduced graphene oxide; GA: Glutaraldehyde; HRP: horseradish peroxidase; GOX: glucose oxidase; IL: ionic liquid; IPA: intermittent pulse amperometry; MB: magnetic beads; MWCNT: multiwalled carbon nanotube; n.r: not reported; LSV: linear sweep voltammetry; PAH: polyallylamine; PDMS: polydimethylsiloxane; PVA: polyvinyl alcohol; QD: Quantum Dot; rG-GO: reduced graphene-graphene oxide; SPAuE: screen-printed gold electrode; SPCE: screen-printed carbon electrode; SP-IDME: screen-printed interdigitated electrode; SWASV: square wave anodic stripping voltammetry.

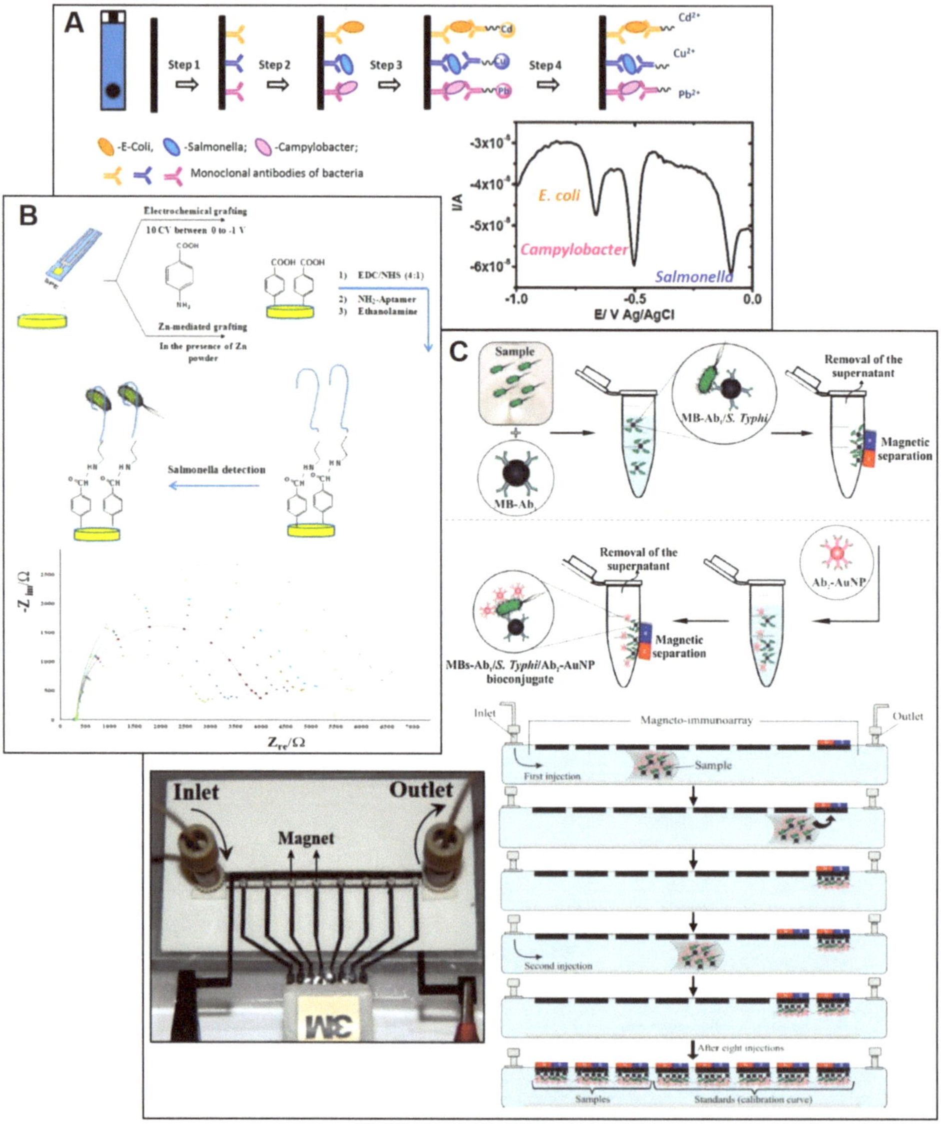

Figure 4. (**A**) Schematic representation of the multiplexed immunosensor developed by Viswanathan et al. for *E. coli*, *Salmonella*, and *Campylobacter* detection and the analytical signals obtained for the three bacteria by square-wave anodic stripping voltammetry (SWASV). Reproduced from [78] with permission from Elsevier. (**B**) Schematic representation of the aptasensor developed by Bagheryan et al. for *Salmonella* detection and the electrochemical impedance spectroscopy (EIS) signals obtained for different *Salmonella* concentrations included in the calibration curve. Reproduced from [73] with permission from Elsevier. (**C**) Schematic representation of the immunoassay based on magnetic beads developed by T.R. de Oliveira et al. using the microfluidic multiplex system shown in the picture. Reproduced from [75] with permission from Elsevier.

For *Listeria* detection in food samples scarce works were found. A noteworthy example is the one developed by Tolba et al. [87] that used the cell wall binding domain (CBD) of bacteriophage-encoded peptidoglycan hydrolases (endolysin) as biorecognition element. CBD500 was immobilized by covalent binding on an SPE with a gold WE. After the reaction with the *Listeria* present in the sample, the analytical signal was obtained by EIS using $[Fe(CN)_6]^{3-/4-}$ as redox probe.

As in the case of *Salmonella*, the SPE-based sensors for *E. coli* detection in foods are mostly based on immunoassays (Table 3). A simple biosensor was developed by Yueh-Hui Lui et al. that consisted of a sandwich immunoassay. The capture antibody was immobilized on an SPCE modified with AuNP and ferrocene dicarboxylic acid (FeDC) [88]. The detection antibody was labelled with HRP and H_2O_2 was used as substrate. The combination of AuNP and FeDC resulted in a significant improvement of the current intensity (studied by CV) when compared with an SPCE that was modified only with AuNP or FeDC. The SPCE contained two electrodes, both made of carbon ink: one acting as working and the other acting as both reference and counter electrodes. Using chronoamperometry (at 300 mV vs. carbon counter/reference electrode) as detection technique, the obtained immunosensor showed a quite good LOD of 600 CFU/mL. A lower LOD (309 CFU/mL in tap water and 457 CFU/mL in minced beef) was obtained by Hassan et al. [89] who immobilized the capture antibody on magnetic beads and carried out a sandwich immunoassay using a AuNP-modified detection antibody (Figure 5A). The quantification of the bacteria was performed through the Hydrogen Evolution Reaction (HER) catalysed by the AuNP using chronoamperometry (applying +1.35 V for 60 s and then, −1.00 V for 100 s) and an SPCE as transducer. Wenchao Dou et al. used a nanocomposite consisting of gold-platinum core/shell nanoparticles, neutral red, and reduced graphene oxide (rGO-NR-Au@Pt) to develop different sensors for *E. coli* determination [90–92]. Using this nanocomposite to label the detection antibody, a sandwich-type immunosensor was developed by immobilizing the capture antibody on an AuNP/polyaniline-SPCE [90]. *E. coli* was quantified by taking advantage of the catalytic effect of the Au@Pt particles on the reduction of H_2O_2, achieving a high LOD of 2840 CFU/mL. Using a similar immunoassay and detection system (measurement of the reduction of H_2O_2 by CV on SPCE), Wenchao Dou et al. achieved a much better LOD (450 CFU/mL) by immobilizing the capture antibody on magnetic beads and using thionine as electron mediator [91]. They achieved an even lower LOD (91 CFU/mL), introducing HRP as a label (to catalyse the H_2O_2 reaction) in the rGO-NR-Au@Pt-detection antibody composite [92] (Figure 5B).

Table 3. SPE-based sensors for *E. coli* O157:H7 detection in foods.

Sensor Construction	Detect. Tech.	Conc. Range	LOD	Analysis Time	Sample	Ref.
Immunoassay (sandwich); HRP as label; AuNP/FeDC-SPCE	CA	10^2 to 10^7 CFU/mL	600 CFU/mL	≈35 min	Milk	[88]
Immunoassay (sandwich); Ab capture on MB; AuNP as label (catalysing HER); SPCE	CA	10^2–10^5 CFU/mL in samples	309 CFU/mL in tap water, 457 CFU/mL in minced beef	≈70 min	Water, minced beef	[89]
Immunoassay (sandwich); rGO-NR-Au@Pt nanocomposite-detection Ab (measurement of H_2O_2 reduction); AuNP/PANI-SPCE	CV	8.9×10^3–8.9×10^9 CFU/mL	2840 CFU/mL	≈110 min	Milk, pork	[90]
Immunoassay (sandwich); Capture Ab on MB; rGO-NR-Au@Pt nanocomposite-detection Ab (measurement of H_2O_2 reduction); Thionine as mediator; SPCE	CV	4×10^3–4×10^8 CFU/mL	450 CFU/mL	≈115 min	Milk, pork	[91]

Table 3. *Cont.*

Sensor Construction	Detect. Tech.	Conc. Range	LOD	Analysis Time	Sample	Ref.
Immunoassay (sandwich); Capture Ab on MB; rGO-NR-Au@Pt nanocomposite HRP-modified detection-Ab; HRP as label; Thionine as mediator; 4-SPCE	CV	4×10^2–4×10^8 CFU/mL	91 CFU/mL	≈135 min	Milk, pork	[92]
Immunoassay; Label-free (measurement of $Fe(CN)_6^{3-/4-}$); AuNP-SPCE	CV	1.19×10^3–1.19×10^9 CFU/mL	594 CFU/mL	≈55 min	Milk powder	[93]
Immunoassay; Label-free (measurement of $Fe(CN)_6^{3-/4-}$); AuNP/PANI-SPCE	DPV	4×10^4–4×10^9 CFU/mL	7980 CFU/mL	≈45 min	Milk	[94]
Immunoassay (sandwich); Ab photochemical immobilization; Label free; SPAuE	EIS	10^2–10^3 CFU/mL in drinking water	30 CFU/mL	≈70 min	Drinking water	[95]
Immunoassay; Capture Ab on MB; Label free; SP-IDME of gold	EIS	10^4–10^7 CFU/mL	$10^{4.45}$ CFU/mL	≈60 min	Ground beef	[96]
Immunoassay; Ab on AuNP/MB-GOX@PDA; Filtration step; GOX as label; Prussian Blue-modified SP-IDME of gold	A	10^3–10^6 CFU/g in ground beef	190 CFU/g	≈75 min	Ground beef	[97]

4-SPCE: screen-printed carbon electrode with 4 working electrodes; A: amperometry; Ab: antibody; AuNP: gold nanoparticles; CA: chronoamperometry; CV: cyclic voltammetry; DPV: differential pulse voltammetry; EIS: electrochemical impedance spectroscopy; FeDC: ferrocene dicarboxylic acid; HER: hydrogen evolution reaction; HRP: horseradish peroxidase; ITO: indium tin oxide; MB: magnetic beads; NP: nanoparticles; NR: neutral red; PANI: polyaniline; PDA: polydopamine; rGO: reduced graphene oxide; SPAuE: screen-printed gold electrode; SPE: screen-printed electrode; SP-IDME: screen-printed interdigitated microelectrode.

Among the label-free sensors for *E. coli* detection, it is worthy to note the one recently developed by Cimafonte et al. for drinking water [95]. It consisted of a sandwich-type immunosensor in which the capture antibody was immobilized with a suitable orientation on a SPAuE by a photochemical technique. The determination was performed by EIS using the $[Fe(CN)_6]^{3-/4-}$ redox probe, achieving a very low LOD of 30 CFU/mL.

Studies using screen-printed interdigitated electrodes were also found. An interesting one is the sensor developed by Xu et al. [97] that used a Prussian blue (PB)-modified screen-printed interdigitated gold microelectrode achieving a LOD of 190 CFU/g. An anti-E.coli antibody was immobilized on magnetic beads that were coated with polydopamine and modified with glucose oxidase (GOX) and AuNP. After the immunoreaction with the bacteria, a filtration was performed (through a paper with 0.8 μm pores) to separate the immunocomplex formed with the bacteria from the free nanocomposite-Ab. The analytical signal was recorded by amperometry dropping the filtered solution onto the PB-modified electrode together with a glucose solution (enzymatic substrate for GOX).

4.2. SPE-Based (Bio)Sensors for Biogenic Amines Detection

As mentioned before, histamine is the main BA. This explains the large number of sensors developed for its determination when compared to those developed for the other BAs. The most frequently reported SPE-based sensors for the determination of BAs in food samples (mainly fish) are enzymatic, although some immunosensors [98,99] or sensors based on the electroactivity of the amines [100,101] can also be found (Table 4). An interesting example among the immunosensors for histamine determination is the one recently developed by Shkodra et al. [98] using a flexible SPE with a WE made of a silver polymeric paste. The three-electrode cell is screen-printed on a polyethylene terephthalate (PET) flexible substrate to obtain a sensor that withstands frequent bending without signal loss (Figure 6A). To perform the immunoassay, an anti-histamine antibody is immobilized on the WE, previously modified with oxygen plasma-treated carbon nanotubes. Then, the competitive immunoassay is carried out using HRP-labelled histamine to compete with the histamine of the sample. Using 3,3′,5,5′-tetramethylbenzidine (TMB) as enzymatic substrate and chronoamperometry as detection technique, a sensor with a very low LOD (0.022 nM) and a high selectivity (tested using other BAs (cadaverine, putrescine, and tyramine) was obtained.

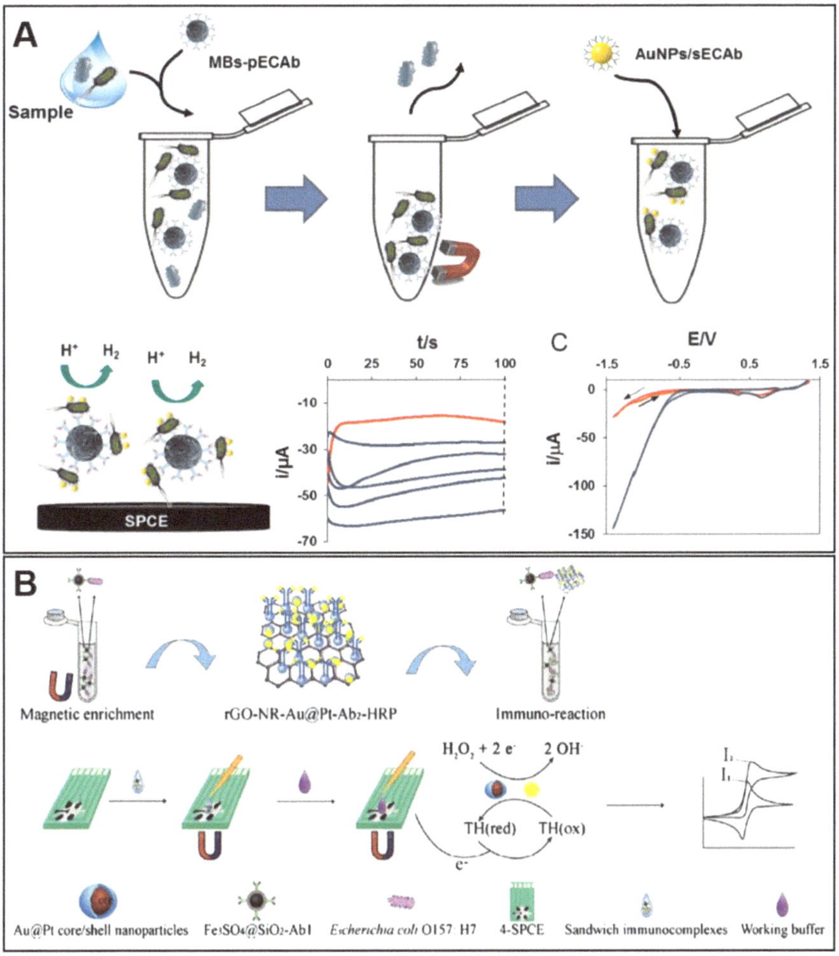

Figure 5. (**A**) Schematic representation of the magneto immunoassay developed by Hassan et al. for *E.coli* O157:H7 detection based on the Hydrogen Evolution Reaction electrocatalyzed by AuNP; chronoamperograms for different bacteria concentration; cyclic voltammograms in absence (red line) and presence of bacteria (blue line). Reproduced from [89] with permission from Elsevier. (**B**) Schematic representation of the magneto immunoassay, using rGO-NR-Au@Pt nanocomposite and HRP as label, developed by Wenchao Dou et al. *E. coli* O157:H7 detection. Reproduced from [92] with permission from Springer 2018.

Table 4. SPE-based sensors for biogenic amines detection in food samples.

Biogenic Amines	Sensor Construction	Detect. Tech.	Conc. Range	LOD	Analysis Time	Sample	Ref.
HIS	Rhenium (IV) oxide-SPCE	A	4.5–90 μM	1.8 μM	≈3 min	Fish sauce	[100]
HIS	Nafion/$Cu_3(PO_4)_2$NP/SPCE	A	0.045–4.5 mM	0.027 mM	≈3 min	Fish	[101]
HIS	Immunoassay (competitive); Histamine labelled with HRP; Capture Ab on SWCNT/SPE (flexible with a silver WE)	CA	0.045–450 nM	0.022 nM	≈140 min	Fish	[98]
HIS	Immunoassay (competitive); HRP-labelled detection Ab; Histamine-ovalbumin conjugate on PB/chitosan/AuNP/SPCE	CV	0.09–900 μM	0.01 nM	≈130 min	Fish	[99]
HIS	DAO on SPCE	CA	9–675 μM	4.5 μM	≈1 min	Fish (hake, mackerel)	[102]
HIS	DAO on SPCE; $[Fe(CN)_6]^{3-}$ in solution as mediator	CA	45–675 μM	8.7 μM	≈7 min	Fish (tuna, mackerel)	[103]
HIS	DAO and HRP on polysulfone/ MWCNT/ferrocene membrane/SPCE; SPCE with two WE, ferrocene as mediator	A	0.3–20 μM	0.17 μM	≈2 min	Fish (anchovy, tuna, sardine, mackerel, shrimp, grater weever)	[104]
HIS	DAO on PtNP/rGO/chitosan/SPCE	A	0.1–300 μM	25.4 nM	≈2 min	Fish (carp, tench, catfish, perch)	[105]
PUT	MAO on TTF-SPCE; TTF as mediator	A	16–101 μM	17.2 μM	≈2 min	Anchovy, Courgette	[106]
PUT	PUO on TTF-SPCE; TTF as mediator	A	10–74 μM	10.1 μM	≈2 min	Octopus, courgette	[107]
TYR	DAO on GO/PVF-modified SPCE MAO on GO/PVF-modified SPCE	A	0.99–120 μM 0.9–110 μM	0.41 μM 0.61 μM	≈2 min	Cheese	[108]
TYR	Ty on SWCNT/SPCE	A	5–180 μM	0.62 μM	≈2 min	Fish	[109]
TYR	1-methyl-4-mercaptopyridine/AuNP/PEDOT:PSS/SPCE	DPV	5–100 nM	2.31 nM	≈6 min	Milk	[110]
TYR	Nafion/Ty/Fe_3O_4-chitosan/poly-L-lysine/SPCE	A	0.49–63 μM	0.075 μM	≈2 min	Cheese	[111]
TYR	PAO on SPCE (hydroxymethylferrocene in cell solution as mediator)	A	2–164 μM	2.0 μM	≈2 min	Cheese	[112]
TYR	HRP on SPCE	A	2–456 μM	2.1 μM	≈2 min	Cheese	[113]

Table 4. *Cont.*

Biogenic Amines	Sensor Construction	Detect. Tech.	Conc. Range	LOD	Analysis Time	Sample	Ref.
HIS CAD	DAO on PB/ITO nanoparticles/SPCE MAO on PB/ITO nanoparticles/SPCE	A	6.0–690 µM 3–1000 µM	1.9 µM 0.9 µM	≈2 min	Cheese	[114]
HIS PUT	HMD and PUO respectively on TTF-SPCE (with 4 WE); TTF as mediator	A	– –	8.1 µM 10 µM	≈2 min	Octopus	[115]
PUT CAD	MAO (for PUT) or MAO/AuNPs (for PUT and CAD) on TTF-SPCE (with two WE); TTF as mediator	A	9.9–74.1 µM 19.6–107.1 µM	9.9 µM 19.9 µM	≈2 min	Octopus	[116]
Total biogenic amines (calibration with HIS, PUT, CAD)	DAO on MB; PB-SPCE	CA	0.01–1 mM for HIS, PUT, CAD	4.8 µM for HIS; 0.9 µM for PUT; 0.67 µM for CAD	≈15 min	Fish (sea bass)	[117]
Total biogenic amines (calibration with HIS)	DAO and HRP on aryl diazonium salt/SPCE	A	0.2–1.6 µM	0.18 µM	≈2 min	Fish (anchovy)	[118]
Total biogenic amines (calibration with PUT)	DAO on polyazetidine prepolimer/SPE (with two WE of gold)	A	8–227 µM	2.3 µM	≈2 min	Wine, beer	[119]
Total biogenic amines (calibration with CAD, PUT, TYR, HIS)	Nafion/DAO/MnO_2-SPCE (MnO_2 as mediator)	A	1–50 µM for CAD and PUT; 10–300 µM for TYR and HIS	0.3 µM for CAD and PUT; 3.0 µM for TYR and HIS	≈5 min	Chicken meat	[120]

A: amperometry; AuNP: gold nanoparticles; BSA: bovine serum albumin; CA: chronoamperometry; CAD: cadaverine; CV: cyclic voltammetry; DAO: diamine oxidase; DPV: differential pulse voltammetry; GO: graphene oxide; HIS: histamine; HMD: histamine dehydrogenase; HRP: horseradish peroxidase; ITO: indium tin oxide; MAO: monoamine oxidase; MB: magnetic beads; MWCNT: multi-walled carbon nanotubes; NP: nanoparticles; PAO: plasma amine oxidase; PB: Prussian blue; PEDOT:PSS: poly(3,4-ethylenedioxythiophene):poly-styrene sulfonate: PtNP: platinum nanoparticles; PUO: putrescine oxidase; PUT: putrescine; PVF: polyvinylferrocene; rGO: reduced graphene oxide; SPCE: screen-printed carbon electrode; SPE: screen-printed electrode; SWCNT: single-walled carbon nanotubes; TTF: tetrathiafulvalene; Ty: tyrosinase; TYR: tyramine; WE: working electrode.

Among the large number of enzymatic sensors reported for the determination of BAs in food, most are based on the use of the enzymes monoamine oxidase (MAO) or diamine oxidase (DAO). These enzymes catalyse the oxidation of BAs, producing hydrogen peroxide [102,105,106,116–118]. The detection in these sensors is usually carried out by amperometric techniques and the use of redox mediators, such as $[Fe(CN)_6]^{3-/4-}$ [103], ferrocene [104,112], or tetrathiafulvalene (TTF) [106,107,115,116] is very common. The use of these mediators decreases the detection potential, improving the selectivity of the sensor. An interesting work was reported by S. Leonardo et al. in which different mono- (DAO) and bienzymatic (DAO and HRP) sensors using magnetic beads and different mediators (Co(II)-phthalocyanine (CoPh), Prussian Blue (PB), and Os-polyvinylpyridine (Os-PVP)) were developed and compared [117] (Figure 6C). Although calibration curves for histamine, putrescine and cadaverine were obtained for each one of the sensors (DAO-MB/CoPh-SPCE, DAO-MB/PB-SPCE, and DAO-MB/Os-PVP-HRP/SPCE), obtaining LODs from 0.47 μM to 5.13 μM, the one that included Prussian Blue as mediator was chosen for the determination of BAs in sea bass.

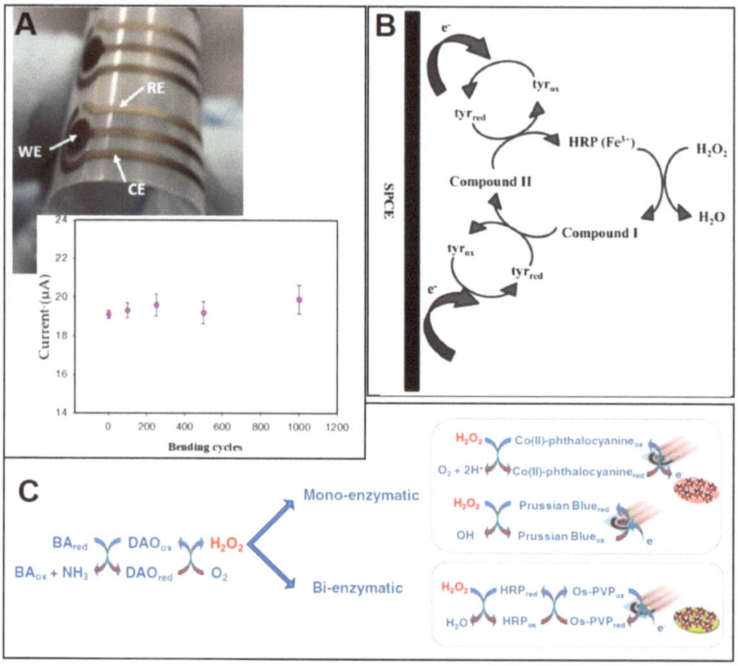

Figure 6. (**A**) Photograph of a flexible three-electrode SPE with silver working electrode used as transducer of an immunosensor for histamine, and flexibility test of that sensor (current intensity obtained after bending it). Reproduced from [98]). (**B**) Schematic representation of the enzymatic reaction occurring at the surface of the HRP/SPCE sensor for tyramine detection; Compound I and compound II are reaction intermediates (compound I (oxidation state +5) comprising a ferryl species (Fe^{4+}=O) and a porphyrin radical cation; compound II (oxidation state +4) is formed by the first reduction of the porphyrin radical cation). Reproduced from [113] with permission from Wiley. (**C**) Scheme of the enzymatic and electrochemical reaction occurring on DAO-MB mono- and bi-enzymatic sensors for biogenic amines (BAs) detection. Reproduced from [117] from Springer 2016.

The use of nanomaterials is also frequent: single or multi-walled carbon nanotubes [104,109], graphene [108], nanoparticles [110,111,114,116], or the combination of different kind of nanomaterials [105]. An example is the sensor developed by Pérez et al. [104] that combines the use of two enzymes, DAO and horseradish peroxidase (HRP), with MWCNT and ferrocene as mediator. In this case, an SPCE with two

WEs was used: one contained the enzymes immobilized on a polysulfone/MWCNT/ferrocene membrane and the other only contained the membrane. With this strategy, the response towards any electroactive species present in the samples that could interfere in the determination is eliminated, improving the selectivity of the sensor.

Sensors based on DAO and MAO or oxidase enzymes such as putrescine oxidase or plasma amine oxidase sometimes show problems regarding selectivity when just one biogenic amine is the target analyte [102,106,107,112,121]. Since for many food applications the objective is obtain the BAI (biogenic amine index), this fact could be not a problem and indeed, several sensors based on DAO and MAO are focused on the quantification of the total amount of BAs [117–120]. With the aim of obtaining more selective enzymatic sensors, the use of other enzymes as recognition element has been reported: e.g., tyrosinase [109] and histamine dehydrogenase [115]. A noteworthy example of a selective SPE-based enzymatic sensor for tyramine is the one developed by Calvo-Pérez et al. [113]. In this work HRP was used as recognition element for tyramine, which is not among the common substrates for this enzyme. The recognition of tyramine through HRP is based on the oxidation of the –OH group present in the molecular structure of tyramine (Figure 6B). Two immobilization procedures were assessed: (i) cross-linking with glutaraldehyde and bovine serum albumin and (ii) mixing the carbon ink used for screen-printing the WE with HRP. Since the second procedure was easier and provided better reproducibility, the sensor obtained in that way was the chosen for its application in real samples. A high selectivity of this HRP-sensor was demonstrated; no response was observed when calibration curves for other BAs (putrescine, cadaverine, histamine, tryptamine, spermine and spermidine were evaluated) in the same concentration range than for tyramine were performed. Another approach to greatly improve the selectivity is to add a separation step before the measurement with the sensor as reported by Li et al. [110]. In this work, a sensor based on an SPCE modified with a conductive polymer (PEDOT.PSS), AuNP and 1-methyl-4-mercaptopyridine (1m-4-MP) was developed to detect tyramine using DPV as technique detection. Before the electrochemical sensing, a sample was treated through a solid-phase extraction based on MIP technology (using a MIP synthetized with methacrylic acid as monomer). The combination of the MIP-based solid phase extraction with the 1-m-4-MP/AuNP/PEDOT:PSS/SPCE provided a LOD of 2.31 nM.

5. Conclusions

Nowadays, food safety is a key concern because it is directly related to public health. Therefore, the development of methods that allow rapid and on-site analysis has gained special relevance in food safety and quality assurance. Disposable electrodes, such as screen-printed electrodes (SPEs), have attracted attention worldwide since they allow the development of easy-to-handle and cost-efficient biosensors. The easy mass-production of reproducible SPEs allows the use of SPE-based sensors as one-shot devices. Besides the concern from the food industry and public-health-related administration about food safety and quality, the growing consumer concern about the security and healthiness of the food they eat enormously increases the interest in point-of-need sensors that can be used by untrained people.

Although there are many published papers on biosensors for food applications, the number of those commercially available is scare since the knowledge transfer from research laboratories to the market is hard. The main challenge for the commercialization of biosensors (for any kind of application) is often the low stability of their recognition element since they are biological compounds that requires special storage conditions. In the case of biosensors for food applications, another important difficulty is related to the sample since it is usually solid, and the measurements normally have to be performed in aqueous medium. This is an important limitation when compared with biosensors for clinical application that are typically applied to bodily fluids. Although the development of multiplex biosensors is increasing, multi-analyte detection is still a big challenge. In the case of food sensors, this is a key issue since, for example, a biosensor can be able to selectively determine a single bacteria serotype (i.e., *S. typhimurium*) but does not provide any information about the presence of other serotypes that can also be harmful. In those cases, the selection of a biological recognition able to detect several serotypes (or kinds of analytes) or the design of a multiplex devices is of paramount importance.

Therefore, it is obvious that biosensors cannot replace the conventional methods (e.g., PCR or HPLC-MS), since these show better features in terms of accuracy, selectively, sensitivity, or multi-analyte detection ability. However, the advantages of SPE-based biosensors mentioned in this review make them exceptional devices for on-site screening. The hard work of electroanalytical researchers to make portable sensors a suitable alternative to the centralized analysis, together the great advances in digital communication networks is leading to promising tools for food control and analysis. Nowadays, in a growing number of situations, it is much more advantageous to have simple tools for fast and on-site screening than sophisticated instrumentation in centralized laboratories.

Funding: This work was supported by UIDB/50006/2020 and UIDP/50006/2020 with funding from the Fundação para a Ciência e a Tecnologia (FCT)/the Ministério da Ciência, Tecnologia e Ensino Superior (MCTES) through national funds. The authors also thank FCT and the EU for funding through the projects: FishBioSensing—Portable electrochemical (bio)sensing devices for safety and quality assessment of fishery products (02/SAICT/2016, POCI-01-0145-FEDER-023817), PTDC/QUI-QAN/30735/2017—TracAllerSens—Electrochemical sensors for the detection and quantification of trace amounts of allergens in food products (POCI-01-0145-FEDER-030735), and PTDC/ASP-PES/29547/2017—CECs(Bio)Sensing—(Bio)sensors for assessment of contaminants of emerging concern in fishery commodities (POCI-01-0145-FEDER-029547), supported by national funds by FCT/MCTES and co-supported by Fundo Europeu de Desenvolvimento Regional (FEDER) through COMPETE 2020—Programa Operacional Competitividade e Internacionalização. E. Costa-Rama thanks the Government of Principado de Asturias and Marie Curie-Cofund Actions for the post-doctoral grant "Clarín-Cofund" ACA17-20. R. Torre is grateful to FCT for her PhD grant (SFRH/BD/143753/2019), financed by POPH–QREN–Tipologia 4.1–Formação Avançada, subsidized by FSE and MCTES.

Conflicts of Interest: The authors declare no conflict of interest. The funders had no role in the design of the study; in the collection, analyses, or interpretation of data; in the writing of the manuscript, or in the decision to publish the results.

References

1. World Health Organization. Available online: https://www.who.int/news-room/fact-sheets/detail/food-safety (accessed on 29 September 2020).
2. Den Besten, H.M.W.; Wells-Bennik, M.H.J.; Zwietering, M.H. Natural Diversity in Heat Resistance of Bacteria and Bacterial Spores: Impact on Food Safety and Quality. *Annu. Rev. Food Sci. Technol.* **2018**, *9*, 383–410. [CrossRef] [PubMed]
3. Abraham, A.; Al-Khaldi, S.; Assimon, S.A.; Beuadry, C.; Benner, R.A.; Bennett, R.; Binet, R.; Cahill, S.M.; Burkhardt, W., III. Bad Bud Book. In *Handbook of Foodborne Pathogenic Microorganisms and Natural Toxins Introduction*, 2nd ed.; Food and Drug Administration: Silver Spring, MD, USA, 2012; ISBN 9780323401814.
4. Naila, A.; Flint, S.; Fletcher, G.; Bremer, P.; Meerdink, G. Control of biogenic amines in food—Existing and emerging approaches. *J. Food Sci.* **2010**, *75*, R139–R150. [CrossRef] [PubMed]
5. Velusamy, V.; Arshak, K.; Korostynska, O.; Oliwa, K.; Adley, C. An overview of foodborne pathogen detection: In the perspective of biosensors. *Biotechnol. Adv.* **2010**, *28*, 232–254. [CrossRef] [PubMed]
6. Zhao, X.; Lin, C.W.; Wang, J.; Oh, D.H. Advances in rapid detection methods for foodborne pathogens. *J. Microbiol. Biotechnol.* **2014**, *24*, 297–312. [CrossRef] [PubMed]
7. European Food Safety Authority (EFSA). BIOHAZ Scientific Opinion on risk based control of biogenic amine formation in fermented foods. *EFSA J.* **2011**, *9*, 2393. [CrossRef]
8. Biji, K.B.; Ravishankar, C.N.; Venkateswarlu, R.; Mohan, C.O.; Gopal, T.K.S. Biogenic amines in seafood: A review. *J. Food Sci. Technol.* **2016**, *53*, 2210–2218. [CrossRef] [PubMed]
9. Jairath, G.; Singh, P.K.; Dabur, R.S.; Rani, M.; Chaudhari, M. Biogenic amines in meat and meat products and its public health significance: A review. *J. Food Sci. Technol.* **2015**, *52*, 6835–6846. [CrossRef]
10. Papageorgiou, M.; Lambropoulou, D.; Morrison, C.; Kłodzińska, E.; Namieśnik, J.; Płotka-Wasylka, J. Literature update of analytical methods for biogenic amines determination in food and beverages. *TrAC Trends Anal. Chem.* **2018**, *98*, 128–142. [CrossRef]
11. Köse, S.; Kaklikkaya, N.; Koral, S.; Tufan, B.; Buruk, K.C.; Aydin, F. Commercial test kits and the determination of histamine in traditional (ethnic) fish products-evaluation against an EU accepted HPLC method. *Food Chem.* **2011**, *125*, 1490–1497. [CrossRef]

12. Mishra, R.K.; Nunes, G.S.; Souto, L.; Marty, J.L. Screen printed technology—An application towards biosensor development. In *Encyclopedia of Interfacial Chemistry*; Elsevier: Amsterdam, The Netherlands, 2018; pp. 487–498.
13. Smart, A.; Crew, A.; Pemberton, R.; Hughes, G.; Doran, O.; Hart, J.P. Screen-printed carbon based biosensors and their applications in agri-food safety. *TrAC Trends Anal. Chem.* **2020**, *127*, 115898. [CrossRef]
14. Vasilescu, A.; Nunes, G.; Hayat, A.; Latif, U.; Marty, J.L. Electrochemical affinity biosensors based on disposable screen-printed electrodes for detection of food allergens. *Sensors* **2016**, *16*, 1863. [CrossRef] [PubMed]
15. Díaz-Cruz, J.M.; Serrano, N.; Pérez-Ràfols, C.; Ariño, C.; Esteban, M. Electroanalysis from the past to the twenty-first century: Challenges and perspectives. *J. Solid State Electrochem.* **2020**. [CrossRef] [PubMed]
16. Roberts, G.; Age, S.; Simon, S. History's Influence on Screen Printing's Future Explore How Screenprinting's Past Will Shape Its Future. *Screen Print.* February 2006. Available online: https://www.screenweb.com/content/historys-influence-screen-printings-future (accessed on 30 September 2020).
17. Couto, R.A.S.; Lima, J.L.F.C.; Quinaz, M.B. Recent developments, characteristics and potential applications of screen-printed electrodes in pharmaceutical and biological analysis. *Talanta* **2016**, *146*, 801–814. [CrossRef] [PubMed]
18. Arduini, F.; Micheli, L.; Moscone, D.; Palleschi, G.; Piermarini, S.; Ricci, F.; Volpe, G. Electrochemical biosensors based on nanomodified screen-printed electrodes: Recent applications in clinical analysis. *Trends Anal. Chem.* **2016**, *79*, 114–126. [CrossRef]
19. Rama, E.C.; Costa-García, A. Screen-printed Electrochemical Immunosensors for the Detection of Cancer and Cardiovascular Biomarkers. *Electroanalysis* **2016**, *28*, 1700–1715. [CrossRef]
20. Li, M.; Li, Y.T.; Li, D.W.; Long, Y.T. Recent developments and applications of screen-printed electrodes in environmental assays-A review. *Anal. Chim. Acta* **2012**, *734*, 31–44. [CrossRef]
21. Hayat, A.; Marty, J.L. Disposable screen printed electrochemical sensors: Tools for environmental monitoring. *Sensors* **2014**, *14*, 10432–10453. [CrossRef]
22. Cano-Raya, C.; Denchev, Z.Z.; Cruz, S.F.; Viana, J.C. Chemistry of solid metal-based inks and pastes for printed electronics–A review. *Appl. Mater. Today* **2019**, *15*, 416–430. [CrossRef]
23. Yáñez-Sedeño, P.; Campuzano, S.; Pingarrón, J.M. Electrochemical (bio)sensors: Promising tools for green analytical chemistry. *Curr. Opin. Green Sustain. Chem.* **2019**, *19*, 1–7. [CrossRef]
24. Gałuszka, A.; Migaszewski, Z.; Namieśnik, J. The 12 principles of green analytical chemistry and the SIGNIFICANCE mnemonic of green analytical practices. *TrAC Trends Anal. Chem.* **2013**, *50*, 78–84. [CrossRef]
25. Sanati, A.; Jalali, M.; Raeissi, K.; Karimzadeh, F.; Kharaziha, M.; Mahshid, S.S.; Mahshid, S. A review on recent advancements in electrochemical biosensing using carbonaceous nanomaterials. *Microchim. Acta* **2019**, *186*, 773. [CrossRef] [PubMed]
26. Metrohm DropSens. Available online: http://www.dropsens.com/ (accessed on 29 September 2020).
27. Micrux Technologies. Available online: http://www.micruxfluidic.com/ (accessed on 29 September 2020).
28. Pine Research. Available online: https://pineresearch.com/ (accessed on 29 September 2020).
29. Gwent Group. Available online: http://www.gwent.org/ (accessed on 29 September 2020).
30. PalmSens. Available online: https://www.palmsens.com/ (accessed on 29 September 2020).
31. Rusens. Available online: http://www.rusens.com/indexeng.html (accessed on 29 September 2020).
32. Putzbach, W.; Ronkainen, N.J. Immobilization techniques in the fabrication of nanomaterial-based electrochemical biosensors: A review. *Sensors (Basel)* **2013**, *13*, 4811–4840. [CrossRef] [PubMed]
33. Antuña-Jiménez, D.; González-García, M.B.; Hernández-Santos, D.; Fanjul-Bolado, P. Screen-printed electrodes modified with metal nanoparticles for small molecule sensing. *Biosensors* **2020**, *10*, 9. [CrossRef]
34. Duffy, G.F.; Moore, E.J. Electrochemical Immunosensors for Food Analysis: A Review of Recent Developments. *Anal. Lett.* **2017**, *50*, 1–32. [CrossRef]
35. Windmiller, J.R.; Bandodkar, A.J.; Parkhomovsky, S.; Wang, J. Stamp transfer electrodes for electrochemical sensing on non-planar and oversized surfaces. *Analyst* **2012**, *137*, 1570–1575. [CrossRef]
36. Mishra, R.K.; Hubble, L.J.; Martín, A.; Kumar, R.; Barfidokht, A.; Kim, J.; Musameh, M.M.; Kyratzis, I.L.; Wang, J. Wearable flexible and stretchable glove biosensor for on-site detection of organophosphorus chemical threats. *ACS Sens.* **2017**, *2*, 553–561. [CrossRef]
37. Desmet, C.; Marquette, C.A.; Blum, L.J.; Doumèche, B. Paper electrodes for bioelectrochemistry: Biosensors and biofuel cells. *Biosens. Bioelectron.* **2016**, *76*, 145–163. [CrossRef]

38. Moro, G.; Bottari, F.; Van Loon, J.; Du Bois, E.; De Wael, K.; Moretto, L.M. Disposable electrodes from waste materials and renewable sources for (bio)electroanalytical applications. *Biosens. Bioelectron.* **2019**, *146*. [CrossRef]
39. Neves, M.M.P.S.; González-García, M.B.; Hernández-Santos, D.; Fanjul-Bolado, P. Screen-Printed Electrochemical 96-Well Plate: A High-Throughput Platform for Multiple Analytical Applications. *Electroanalysis* **2014**, *26*, 2764–2772. [CrossRef]
40. Piermarini, S.; Micheli, L.; Ammida, N.H.S.; Palleschi, G.; Moscone, D. Electrochemical immunosensor array using a 96-well screen-printed microplate for aflatoxin B1 detection. *Biosens. Bioelectron.* **2007**, *22*, 1434–1440. [CrossRef]
41. Thévenot, D.R.; Toth, K.; Durst, R.A.; Wilson, G.S. Electrochemical biosensors: Recommended definitions and classification1International Union of Pure and Applied Chemistry: Physical Chemistry Division, Commission I.7 (Biophysical Chemistry); Analytical Chemistry Division, Commission V.5 (Electroanalytical). *Biosens. Bioelectron.* **2001**, *16*, 121–131. [CrossRef]
42. Sharma, H.; Mutharasan, R. Review of biosensors for foodborne pathogens and toxins. *Sens. Actuators B Chem.* **2013**, *183*, 535–549. [CrossRef]
43. Ronkainen, N.J.; Halsall, H.B.; Heineman, W.R. Electrochemical biosensors. *Chem. Soc. Rev.* **2010**, *39*, 1747–1763. [CrossRef] [PubMed]
44. Sassolas, A.; Blum, L.J.; Leca-Bouvier, B.D. Immobilization strategies to develop enzymatic biosensors. *Biotechnol. Adv.* **2012**, *30*, 489–511. [CrossRef] [PubMed]
45. Crapnell, R.D.; Hudson, A.; Foster, C.W.; Eersels, K.; van Grinsven, B.; Cleij, T.J.; Banks, C.E.; Peeters, M. Recent advances in electrosynthesized molecularly imprinted polymer sensing platforms for bioanalyte detection. *Sensors (Switzerland)* **2019**, *19*, 1204. [CrossRef]
46. Tudorache, M.; Bala, C. Biosensors based on screen-printing technology, and their applications in environmental and food analysis. *Anal. Bioanal. Chem.* **2007**, *388*, 565–578. [CrossRef]
47. Ricci, F.; Adornetto, G.; Palleschi, G. A review of experimental aspects of electrochemical immunosensors. *Electrochim. Acta* **2012**, *84*, 74–83. [CrossRef]
48. Cesewski, E.; Johnson, B.N. Electrochemical biosensors for pathogen detection. *Biosens. Bioelectron.* **2020**, *159*, 112214. [CrossRef]
49. Wang, Y.; Duncan, T.V. Nanoscale sensors for assuring the safety of food products. *Curr. Opin. Biotechnol.* **2017**, *44*, 74–86. [CrossRef]
50. Silva, N.F.D.; Neves, M.M.P.S.; Magalhães, J.M.C.S.; Freire, C.; Delerue-Matos, C. Emerging electrochemical biosensing approaches for detection of *Listeria monocytogenes* in food samples: An overview. *Trends Food Sci. Technol.* **2020**, *99*, 621–633. [CrossRef]
51. Silva, N.F.D.; Magalhães, J.M.C.S.; Freire, C.; Delerue-Matos, C. Electrochemical biosensors for *Salmonella*: State of the art and challenges in food safety assessment. *Biosens. Bioelectron.* **2018**, *99*, 667–682. [CrossRef] [PubMed]
52. Zhang, Y.-J.; Zhang, Y.; Zhou, Y.; Li, G.-H.; Yang, W.-Z.; Feng, X.-S. A review of pretreatment and analytical methods of biogenic amines in food and biological samples since 2010. *J. Chromatogr. A* **2019**, *1605*, 360361. [CrossRef] [PubMed]
53. Prabhakar, P.K.; Vatsa, S.; Srivastav, P.P.; Pathak, S.S. A comprehensive review on freshness of fish and assessment: Analytical methods and recent innovations. *Food Res. Int.* **2020**, *133*, 109157. [CrossRef] [PubMed]
54. European Food Safety Authority (EFSA). Available online: http://www.efsa.europa.eu/en/news/salmonella-most-common-cause-foodborne-outbreaks-european-union (accessed on 29 September 2020).
55. European Food Safety Authority (EFSA). ECDC The European Union summary report on trends and sources of zoonoses, zoonotic agents and food-borne outbreaks in 2015. *EFSA J.* **2016**, *14*, 4364. [CrossRef]
56. Bolton, D.J. *Campylobacter* virulence and survival factors. *Food Microbiol.* **2015**, *48*, 99–108. [CrossRef]
57. Silva, J.; Leite, D.; Fernandes, M.; Mena, C.; Gibbs, P.A.; Teixeira, P. *Campylobacter* spp. As a foodborne pathogen: A review. *Front. Microbiol.* **2011**, *2*, 1–12. [CrossRef]
58. Fabiani, L.; Delibato, E.; Volpe, G.; Piermarini, S.; De Medici, D.; Palleschi, G. Development of a sandwich ELIME assay exploiting different antibody combinations as sensing strategy for an early detection of *Campylobacter*. *Sens. Actuators B Chem.* **2019**, *290*, 318–325. [CrossRef]

59. Allocati, N.; Masulli, M.; Alexeyev, M.F.; Di Ilio, C. *Escherichia coli* in Europe: An overview. *Int. J. Environ. Res. Public Health* **2013**, *10*, 6235–6254. [CrossRef]
60. Kaper, J.B.; Nataro, J.P.; Mobley, H.L.T. Pathogenic *Escherichia coli*. *Nat. Rev. Microbiol.* **2004**, *2*, 123–140. [CrossRef]
61. Eng, S.K.; Pusparajah, P.; Ab Mutalib, N.S.; Ser, H.L.; Chan, K.G.; Lee, L.H. *Salmonella*: A review on pathogenesis, epidemiology and antibiotic resistance. *Front. Life Sci.* **2015**, *8*, 284–293. [CrossRef]
62. Alexandre, D.L.; Melo, A.M.A.; Furtado, R.F.; Borges, M.F.; Figueiredo, E.A.T.; Biswas, A.; Cheng, H.N.; Alves, C.R. A Rapid and Specific Biosensor for *Salmonella* Typhimurium Detection in Milk. *Food Bioprocess Technol.* **2018**, *11*, 748–756. [CrossRef]
63. Cinti, S.; Volpe, G.; Piermarini, S.; Delibato, E.; Palleschi, G. Electrochemical biosensors for rapid detection of foodborne *Salmonella*: A critical overview. *Sensors (Switzerland)* **2017**, *17*, 1910. [CrossRef] [PubMed]
64. Buchanan, R.L.; Gorris, L.G.M.; Hayman, M.M.; Jackson, T.C.; Whiting, R.C. A review of *Listeria monocytogenes*: An update on outbreaks, virulence, dose-response, ecology, and risk assessments. *Food Control* **2017**, *75*, 1–13. [CrossRef]
65. De Jong, W.H.A.; De Vries, E.G.E.; Kema, I.P. Current status and future developments of LC-MS/MS in clinical chemistry for quantification of biogenic amines. *Clin. Biochem.* **2011**, *44*, 95–103. [CrossRef]
66. Ahmad, W.; Mohammed, G.I.; Al-Eryani, D.A.; Saigl, Z.M.; Alyoubi, A.O.; Alwael, H.; Bashammakh, A.S.; O'Sullivan, C.K.; El-Shahawi, M.S. Biogenic Amines Formation Mechanism and Determination Strategies: Future Challenges and Limitations. *Crit. Rev. Anal. Chem.* **2019**, *0*, 1–16. [CrossRef]
67. Ordóñez, J.L.; Troncoso, A.M.; García-Parrilla, M.D.C.; Callejón, R.M. Recent trends in the determination of biogenic amines in fermented beverages—A review. *Anal. Chim. Acta* **2016**, *939*, 10–25. [CrossRef]
68. Freitas, M.; Viswanathan, S.; Nouws, H.P.A.; Oliveira, M.B.P.P.; Delerue-Matos, C. Iron oxide/gold core/shell nanomagnetic probes and CdS biolabels for amplified electrochemical immunosensing of *Salmonella* typhimurium. *Biosens. Bioelectron.* **2014**, *51*, 195–200. [CrossRef]
69. Afonso, A.S.; Pérez-López, B.; Faria, R.C.; Mattoso, L.H.C.; Hernández-Herrero, M.; Roig-Sagués, A.X.; Maltez-da Costa, M.; Merkoçi, A. Electrochemical detection of *Salmonella* using gold nanoparticles. *Biosens. Bioelectron.* **2013**, *40*, 121–126. [CrossRef]
70. Wang, D.; Dou, W.; Chen, Y.; Zhao, G. Enzyme-functionalized electrochemical immunosensor based on electrochemically reduced graphene oxide and polyvinyl alcohol-polydimethylsiloxane for the detection of *Salmonella* pullorum & *Salmonella* gallinarum. *RSC Adv.* **2014**, *4*, 57733–57742. [CrossRef]
71. Fei, J.; Dou, W.; Zhao, G. A sandwich electrochemical immunosensor for *Salmonella* pullorum and *Salmonella* gallinarum based on a screen-printed carbon electrode modified with an ionic liquid and electrodeposited gold nanoparticles. *Microchim. Acta* **2015**, *182*, 2267–2275. [CrossRef]
72. Mutreja, R.; Jariyal, M.; Pathania, P.; Sharma, A.; Sahoo, D.K.; Suri, C.R. Novel surface antigen based impedimetric immunosensor for detection of *Salmonella* typhimurium in water and juice samples. *Biosens. Bioelectron.* **2016**, *85*, 707–713. [CrossRef] [PubMed]
73. Bagheryan, Z.; Raoof, J.B.; Golabi, M.; Turner, A.P.F.; Beni, V. Diazonium-based impedimetric aptasensor for the rapid label-free detection of *Salmonella* typhimurium in food sample. *Biosens. Bioelectron.* **2016**, *80*, 566–573. [CrossRef] [PubMed]
74. Ngoensawat, U.; Rijiravanich, P.; Surareungchai, W.; Somasundrum, M. Electrochemical Immunoassay for *Salmonella* Typhimurium Based on an Immuno-magnetic Redox Label. *Electroanalysis* **2018**, *30*, 146–153. [CrossRef]
75. De Oliveira, T.R.; Martucci, D.H.; Faria, R.C. Simple disposable microfluidic device for *Salmonella* typhimurium detection by magneto-immunoassay. *Sens. Actuators B Chem.* **2018**, *255*, 684–691. [CrossRef]
76. Wonsawat, W.; Limvongjaroen, S.; Supromma, S.; Panphut, W.; Ruecha, N.; Ratnarathorn, N.; Dungchai, W. A paper-based conductive immunosensor for the determination of *Salmonella* Typhimurium. *Analyst* **2020**, *145*, 4637–4645. [CrossRef] [PubMed]
77. Murasova, P.; Kovarova, A.; Kasparova, J.; Brozkova, I.; Hamiot, A.; Pekarkova, J.; Dupuy, B.; Drbohlavova, J.; Bilkova, Z.; Korecka, L. Direct culture-free electrochemical detection of *Salmonella* cells in milk based on quantum dots-modified nanostructured dendrons. *J. Electroanal. Chem.* **2020**, *863*, 114051. [CrossRef]
78. Viswanathan, S.; Rani, C.; Ho, J.A.H.A. Electrochemical immunosensor for multiplexed detection of food-borne pathogens using nanocrystal bioconjugates and MWCNT screen-printed electrode. *Talanta* **2012**, *94*, 315–319. [CrossRef]

79. Farka, Z.; Juřík, T.; Pastucha, M.; Kovář, D.; Lacina, K.; Skládal, P. Rapid Immunosensing of *Salmonella* Typhimurium Using Electrochemical Impedance Spectroscopy: The Effect of Sample Treatment. *Electroanalysis* **2016**, *28*, 1803–1809. [CrossRef]
80. Delibato, E.; Volpe, G.; Stangalini, D.; De Medici, D.; Moscone, D.; Palleschi, G. Development of SYBR-green real-time PCR and a multichannel electrochemical immunosensor for specific detection of *Salmonella* enterica. *Anal. Lett.* **2006**, *39*, 1611–1625. [CrossRef]
81. Pratiwi, F.W.; Rijiravanich, P.; Somasundrum, M.; Surareungchai, W. Electrochemical immunoassay for *Salmonella* Typhimurium based on magnetically collected Ag-enhanced DNA biobarcode labels. *Analyst* **2013**, *138*, 5011–5018. [CrossRef]
82. Salam, F.; Tothill, I.E. Detection of *Salmonella* typhimurium using an electrochemical immunosensor. *Biosens. Bioelectron.* **2009**, *24*, 2630–2636. [CrossRef] [PubMed]
83. Xu, M.; Wang, R.; Li, Y. Rapid detection of *Escherichia coli* O157:H7 and *Salmonella* Typhimurium in foods using an electrochemical immunosensor based on screen-printed interdigitated microelectrode and immunomagnetic separation. *Talanta* **2016**, *148*, 200–208. [CrossRef]
84. Wilson, D.; Materón, E.M.; Ibáñez-Redín, G.; Faria, R.C.; Correa, D.S.; Oliveira, O.N. Electrical detection of pathogenic bacteria in food samples using information visualization methods with a sensor based on magnetic nanoparticles functionalized with antimicrobial peptides. *Talanta* **2019**, *194*, 611–618. [CrossRef] [PubMed]
85. Fei, J.; Dou, W.; Zhao, G. A sandwich electrochemical immunoassay for Salmonella pullorum and *Salmonella* gallinarum based on a AuNPs/SiO_2/Fe3O_4 adsorbing antibody and 4 channel screen printed carbon electrode electrodeposited gold nanoparticles. *RSC Adv.* **2015**, *5*, 74548–74556. [CrossRef]
86. Wang, D.; Chen, Q.; Huo, H.; Bai, S.; Cai, G.; Lai, W.; Lin, J. Efficient separation and quantitative detection of *Listeria monocytogenes* based on screen-printed interdigitated electrode, urease and magnetic nanoparticles. *Food Control* **2017**, *73*, 555–561. [CrossRef]
87. Tolba, M.; Ahmed, M.U.; Tlili, C.; Eichenseher, F.; Loessner, M.J.; Zourob, M. A bacteriophage endolysin-based electrochemical impedance biosensor for the rapid detection of *Listeria* cells. *Analyst* **2012**, *137*, 5749–5756. [CrossRef]
88. Lin, Y.H.; Chen, S.H.; Chuang, Y.C.; Lu, Y.C.; Shen, T.Y.; Chang, C.A.; Lin, C.S. Disposable amperometric immunosensing strips fabricated by Au nanoparticles-modified screen-printed carbon electrodes for the detection of foodborne pathogen *Escherichia coli* O157:H7. *Biosens. Bioelectron.* **2008**, *23*, 1832–1837. [CrossRef]
89. Hassan, A.R.H.A.A.; de la Escosura-Muñiz, A.; Merkoçi, A. Highly sensitive and rapid determination of *Escherichia coli* O157:H7 in minced beef and water using electrocatalytic gold nanoparticle tags. *Biosens. Bioelectron.* **2015**, *67*, 511–515. [CrossRef]
90. Mo, X.; Wu, Z.; Huang, J.; Zhao, G.; Dou, W. A sensitive and regenerative electrochemical immunosensor for quantitative detection of: *Escherichia coli* O157:H7 based on stable polyaniline coated screen-printed carbon electrode and rGO-NR-Au@Pt. *Anal. Methods* **2019**, *11*, 1475–1482. [CrossRef]
91. Zhu, F.; Zhao, G.; Dou, W. A non-enzymatic electrochemical immunoassay for quantitative detection of *Escherichia coli* O157:H7 using Au@Pt and graphene. *Anal. Biochem.* **2018**, *559*, 34–43. [CrossRef]
92. Zhu, F.; Zhao, G.; Dou, W. Electrochemical sandwich immunoassay for *Escherichia coli* O157:H7 based on the use of magnetic nanoparticles and graphene functionalized with electrocatalytically active Au@Pt core/shell nanoparticles. *Microchim. Acta* **2018**, *185*. [CrossRef] [PubMed]
93. Huang, Y.; Wu, Z.; Zhao, G.; Dou, W. A Label-Free Electrochemical Immunosensor Modified with AuNPs for Quantitative Detection of *Escherichia coli* O157:H7. *J. Electron. Mater.* **2019**, *48*, 7960–7969. [CrossRef]
94. Mo, X.; Zhao, G.; Dou, W. Electropolymerization of Stable Leucoemeraldine Base Polyaniline Film and Application for Quantitative Detection of *Escherichia coli* O157:H7. *J. Electron. Mater.* **2018**, *47*, 6507–6517. [CrossRef]
95. Cimafonte, M.; Fulgione, A.; Gaglione, R.; Papaianni, M.; Capparelli, R.; Arciello, A.; Censi, S.B.; Borriello, G.; Velotta, R.; Ventura, B. Della Screen printed based impedimetric immunosensor for rapid detection of *Escherichia coli* in drinking water. *Sensors* **2020**, *20*, 274. [CrossRef] [PubMed]
96. Wang, R.; Lum, J.; Callaway, Z.; Lin, J.; Bottje, W.; Li, Y. A label-free impedance immunosensor using screen-printed interdigitated electrodes and magnetic nanobeads for the detection of *E. coli* O157:H7. *Biosensors* **2015**, *5*, 791–803. [CrossRef] [PubMed]

97. Xu, M.; Wang, R.; Li, Y. An electrochemical biosensor for rapid detection of: *E. coli* O157:H7 with highly efficient bi-functional glucose oxidase-polydopamine nanocomposites and Prussian blue modified screen-printed interdigitated electrodes. *Analyst* **2016**, *141*, 5441–5449. [CrossRef]
98. Shkodra, B.; Abera, B.D.; Cantarella, G.; Douaki, A.; Avancini, E.; Petti, L.; Lugli, P. Flexible and printed electrochemical immunosensor coated with oxygen plasma treated SWCNTs for histamine detection. *Biosensors* **2020**, *10*, 35. [CrossRef]
99. Dong, X.X.; Yang, J.Y.; Luo, L.; Zhang, Y.F.; Mao, C.; Sun, Y.M.; Lei, H.T.; Shen, Y.D.; Beier, R.C.; Xu, Z.L. Portable amperometric immunosensor for histamine detection using Prussian blue-chitosan-gold nanoparticle nanocomposite films. *Biosens. Bioelectron.* **2017**, *98*, 305–309. [CrossRef]
100. Veseli, A.; Vasjari, M.; Arbneshi, T.; Hajrizi, A.; Švorc, L.; Samphao, A.; Kalcher, K. Electrochemical determination of histamine in fish sauce using heterogeneous carbon electrodes modified with rhenium(IV) oxide. *Sens. Actuators B Chem.* **2016**, *228*, 774–781. [CrossRef]
101. Lee, M.-Y.; Wu, C.-C.; Sari, M.I.; Hsieh, Y. A disposable non-enzymatic histamine sensor based on the nafion-coated copper phosphate electrodes for estimation of fish freshness. *Electrochim. Acta* **2018**, *283*, 772–779. [CrossRef]
102. Torre, R.; Costa-Rama, E.; Lopes, P.; Nouws, H.P.A.; Delerue-Matos, C. Amperometric enzyme sensor for the rapid determination of histamine. *Anal. Methods* **2019**, *11*, 1264–1269. [CrossRef]
103. Torre, R.; Costa-rama, E.; Nouws, H.P.A.; Delerue-Matos, C. Diamine oxidase-modified screen-printed electrode for the redox-mediated determination of histamine. *J. Anal. Sci. Technol.* **2020**, *3*, 4–11. [CrossRef]
104. Pérez, S.; Bartrolí, J.; Fàbregas, E. Amperometric biosensor for the determination of histamine in fish samples. *Food Chem.* **2013**, *141*, 4066–4072. [CrossRef] [PubMed]
105. Apetrei, I.M.; Apetrei, C. Amperometric biosensor based on diamine oxidase/platinum nanoparticles/graphene/chitosan modified screen-printed carbon electrode for histamine detection. *Sensors* **2016**, *16*, 422. [CrossRef]
106. Henao-Escobar, W.; Domínguez-Renedo, O.; Alonso-Lomillo, M.A.; Arcos-Martínez, M.J. A screen-printed disposable biosensor for selective determination of putrescine. *Microchim. Acta* **2013**, *180*, 687–693. [CrossRef]
107. Henao-Escobar, W.; Domínguez-Renedo, O.; Alonso-Lomillo, M.A.; Cascalheira, J.F.; Dias-Cabral, A.C.; Arcos-Martínez, M.J. Characterization of a Disposable Electrochemical Biosensor Based on Putrescine Oxidase from Micrococcus rubens for the Determination of Putrescine. *Electroanalysis* **2015**, *27*, 368–377. [CrossRef]
108. Erden, P.E.; Erdoğan, Z.Ö.; Öztürk, F.; Koçoğlu, İ.O.; Kılıç, E. Amperometric Biosensors for Tyramine Determination Based on Graphene Oxide and Polyvinylferrocene Modified Screen-printed Electrodes. *Electroanalysis* **2019**, *31*, 2368–2378. [CrossRef]
109. Apetrei, I.M.; Apetrei, C. The biocomposite screen-printed biosensor based on immobilization of tyrosinase onto the carboxyl functionalised carbon nanotube for assaying tyramine in fish products. *J. Food Eng.* **2015**, *149*, 1–8. [CrossRef]
110. Li, Y.; Hsieh, C.H.; Lai, C.-W.; Chang, Y.-F.; Chan, H.-Y.; Tsai, C.-F.; Ho, J.A.; Wu, L. Tyramine detection using PEDOT:PSS/AuNPs/1-methyl-4-mercaptopyridine modified screen-printed carbon electrode with molecularly imprinted polymer solid phase extraction. *Biosens. Bioelectron.* **2017**, *87*, 142–149. [CrossRef] [PubMed]
111. Dalkıran, B.; Erden, P.E.; Kaçar, C.; Kılıç, E. Disposable Amperometric Biosensor Based on Poly-L-lysine and Fe_3O_4 NPs-chitosan Composite for the Detection of Tyramine in Cheese. *Electroanalysis* **2019**, *31*, 1324–1333. [CrossRef]
112. Calvo-Pérez, A.; Domínguez-Renedo, O.; Alonso-Lomillo, M.A.; Arcos-Martínez, M.J. Disposable amperometric biosensor for the determination of tyramine using plasma amino oxidase. *Microchim. Acta* **2013**, *180*, 253–259. [CrossRef]
113. Calvo-Pérez, A.; Domínguez-Renedo, O.; Alonso-Lomillo, M.A.; Arcos-Martínez, M.J. Disposable Horseradish Peroxidase Biosensors for the Selective Determination of Tyramine. *Electroanalysis* **2013**, *25*, 1316–1322. [CrossRef]
114. Kaçar, C.; Erden, P.E.; Dalkiran, B.; İnal, E.K.; Kiliç, E. Amperometric biogenic amine biosensors based on Prussian blue, indium tin oxide nanoparticles and diamine oxidase—Or monoamine oxidase–modified electrodes. *Anal. Bioanal. Chem.* **2020**, *412*, 1933–1946. [CrossRef] [PubMed]

115. Henao-Escobar, W.; Román, L.D.T.-D.; Domínguez-Renedo, O.; Alonso-Lomillo, M.A.; Arcos-Martínez, M.J. Dual enzymatic biosensor for simultaneous amperometric determination of histamine and putrescine. *Food Chem.* **2016**, *190*, 818–823. [CrossRef] [PubMed]
116. Henao-Escobar, W.; Domínguez-Renedo, O.; Asunción Alonso-Lomillo, M.; Julia Arcos-Martínez, M. Simultaneous determination of cadaverine and putrescine using a disposable monoamine oxidase based biosensor. *Talanta* **2013**, *117*, 405–411. [CrossRef]
117. Leonardo, S.; Campàs, M. Electrochemical enzyme sensor arrays for the detection of the biogenic amines histamine, putrescine and cadaverine using magnetic beads as immobilisation supports. *Microchim. Acta* **2016**, *183*, 1881–1890. [CrossRef]
118. Alonso-Lomillo, M.A.; Domínguez-Renedo, O.; Matos, P.; Arcos-Martínez, M.J. Disposable biosensors for determination of biogenic amines. *Anal. Chim. Acta* **2010**, *665*, 26–31. [CrossRef]
119. Di Fusco, M.; Federico, R.; Boffi, A.; MacOne, A.; Favero, G.; Mazzei, F. Characterization and application of a diamine oxidase from Lathyrus sativus as component of an electrochemical biosensor for the determination of biogenic amines in wine and beer. *Anal. Bioanal. Chem.* **2011**, *401*, 707–716. [CrossRef]
120. Telsnig, D.; Kalcher, K.; Leitner, A.; Ortner, A. Design of an Amperometric Biosensor for the Determination of Biogenic Amines Using Screen Printed Carbon Working Electrodes. *Electroanalysis* **2013**, *25*, 47–50. [CrossRef]
121. Lange, J.; Wittmann, C. Enzyme sensor array for the determination of biogenic amines in food samples. *Anal. Bioanal. Chem.* **2002**, *372*, 276–283. [CrossRef]

biosensors

MDPI

Review

Screen-Printed Electrodes: Promising Paper and Wearable Transducers for (Bio)Sensing

Paloma Yáñez-Sedeño *, Susana Campuzano and José Manuel Pingarrón

Departamento de Química Analítica, Facultad de CC. Químicas, Universidad Complutense de Madrid, E-28040 Madrid, Spain; susanacr@quim.ucm.es (S.C.); pingarro@quim.ucm.es (J.M.P.)

* Correspondence: yseo@quim.ucm.es

Received: 19 June 2020; Accepted: 7 July 2020; Published: 9 July 2020

check for updates

Abstract: Screen-printing technology has revolutionized many fields, including that of electrochemical biosensing. Due to their current relevance, this review, unlike other papers, discusses the relevant aspects of electrochemical biosensors manufactured using this technology in connection to both paper substrates and wearable formats. The main trends, advances, and opportunities provided by these types of devices, with particular attention to the environmental and biomedical fields, are addressed along with illustrative fundamentals and applications of selected representative approaches from the recent literature. The main challenges and future directions to tackle in this research area are also pointed out.

Keywords: screen-printed; electrochemical (bio)sensing; paper; wearable; environmental monitoring; clinical analysis

1. Screen-Printed Paper Electrodes for (Bio)Sensing

Electrochemical paper-based analytical devices (e-PADs) combine the inherent advantages of electrochemical detection—such as high sensitivity and low detection limits (LODs), the possibility of enhancing selectivity by applying different potential values or using modified electrodes, and low cos—with those of paper—such as porosity, allowing liquid transport by capillarity, high surface area/volume ratio, and easy waste disposal by incineration [1]. Paper-based platforms are interesting alternatives to develop disposable, eco-friendly, and inexpensive electrochemical sensors. Lightness and flexibility are additional characteristics of these sensors, which confer unique exploitable properties for application in electroanalysis. Since 2009, when Dungchai et al. introduced e-PADs [2], research in this field has been intensive. Different fabrication procedures [3,4], materials [5], and various practical aspects [6–8] have been reviewed. Akyazi et al. [9] reported a critical overview on the fabrication techniques, production limitations, and the commercialization of paper devices.

In addition to general reports, a variety of methods using different configurations of (bio)sensors and paper-based microfluidic designs as detection platforms have been proposed. With the aim of providing comprehensive information, Tables 1 and 2 summarize the fundamentals and main characteristics of relevant electroanalytical methods involving screen-printed paper-based devices applied to environmental and clinical monitoring, respectively.

2. Screen-Printed Paper Electrochemical (Bio)Sensors

2.1. Environmental Applications

An area in which screen-printed paper electrodes have shown particular relevance is environmental monitoring (Table 1). Different types of paper impregnated with suitable reagents [10], modified with metal nanoparticles or carbon nanostructures both in the absence or in the presence [11] of specific enzymes, have been used to determine contaminants such as heavy metals [12], anions [13], and gases [14,15]. For instance, the electrocatalytic activity of polyoxymetalates (POMs) toward the electrochemical reduction of chlorate was employed to prepare a vanadium-containing POM ($[PMO_{11}VO_{40}]^{5-}$) for the determination of ClO_3^- in soils. Just like in other paper-based designs, the SPCE brought the electrodes and the paper together to create the appropriate cell volume for the electrolyte solution, resulting in a thin layer cell for electrochemical detection [10]. Moreover, in this particular application, paper also acted as a filter for soil analysis. Using chronocoulometry, a LOD of 0.31 mg mL^{-1} ClO_3^- was achieved. Using a similar configuration, a disposable gas-sensing paper-based device (gPAD) was fabricated in origami design, integrating in a single device activated carbon as the gas adsorbent and the electrochemical detection consisting of a screen-printed graphene electrode modified with copper nanoparticles. Both NO and NO_2 (as NOx) were detected with the same current responses measured by differential pulse voltammetry (DPV) achieving LODs of 0.23 vppm and 0.03 vppm with exposure times of 25 min and 1 h, respectively. Relative standard deviation (RSD) values less than 5.1% (n = 7 devices) for 25, 75, and 125 vppm NO_2 were reported, and the gPAD was applied to detect NOx in air and exhaust gases from cars [14].

The nerve agent VX gas is prohibited as a chemical warfare agent. Since it cannot be used in research experiments, dimethyl methylphosphonate (DMMP) is utilized as a model. A conductive paper prepared with poly(aniline) (PANI) nanofiber and graphene sheet was used to detect DMMP at parts per billion within few seconds. The intermolecular forces between the cellulosic paper and the conductive additives were improved by using the copolymer poly(vinylbutyral-co-vinyl-alcohol-co-vinyl acetate) (Figure 1). The resulting sensor exhibited a detectable level of 3 ppb and a response time of 2 s [15]. In this field, strategies consisting of electrodes screen-printed onto a filter paper support allows enzymes and other reagents to be pre-loaded into the cellulose network. An illustrative example is the paper-based wearable electrode constructed for the detection of mustard agent, one of the most dangerous chemical warfare agents (CWAs). It is an origami-like device where the detection is based on the inhibitory effects of the analyte toward the enzyme choline oxidase. The amperometric responses were measured at a carbon black/Prussian blue nanocomposite distributed on the electrode surface and profiting its electrocatalytic activity for H_2O_2 reduction. A Keithley 2400 current source meter (Keithley Co., Cleveland, OH, USA) was used to measure the electrical properties and sensing performance of the conductive papers. The LOD attained in the aerosol phase was 0.019 g min m^{-3} [11].

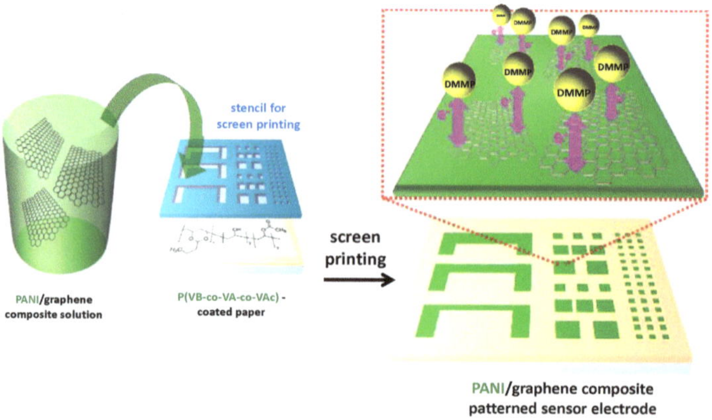

Figure 1. Fabrication process of conductive paper containing sensor patterns for dimethyl methylphosphonate (DMMP) based on poly(aniline) (PANI)/graphene composite. Reproduced and adapted with permission of American Chemical Society [15].

A carbon-nanotubes-based ink prepared with sodium dodecyl sulfate (SDS) and chitosan (CS) absorbed onto cellulose fibers was used to prepare a paper electrode for the determination of Pb^{2+} trace levels in water samples. Square-wave anodic stripping voltammetry (SWASV) using a bismuth film prepared by in situ plating of Bi into the CNT-CS-SDS paper electrodes made the determination of Pb^{2+} in the presence of Bi (10–200 ppb) with a LOD of 6.74 ppb possible [12]. Furthermore, a paper-based, disposable electrochemical platform was developed for the determination of nitrite. Graphene nanosheets and gold nanoparticles were assembled to form a three-dimensional structure onto mixed cellulose ester (MCE) filter papers, leading to thin layer rather than planar diffusion behavior of nitrite at the paper-based electrode. The resulting platform provided larger currents compared to conventional gold or glassy carbon electrodes and, consequently, allowed an improved sensitivity. Importantly, this design effectively avoided the fouling arising from the adsorption of oxidation products thus allowing the determination of nitrite in environmental samples such as waters and industrial sewage. The calibration curve at +0.74 V covered a wide concentration range of 0.3–720 μM, and the LOD was 0.1 μM (S/N = 3) [13].

Contamination of water by microbial pathogens leading to water-borne diseases requires strict controls in drinking water resources, particularly in poor regions, to reduce mortality incidence. Although not included in Table 1, various methods related to the detection of microorganisms should be highlighted. For instance, a simple low-cost paper-based impedimetric sensor for the detection of bacteria in water was prepared using carbon electrodes screen-printed with a conductive ink onto a commercial hydrophobic paper. Concanavalin A covalently immobilized onto the carboxylated electrode surface was used as the biorecognition element due to its ability to selectively interact with mono- and oligo-saccharides on bacteria. In this method, the hydrophobicity of the cellulose paper used as a substrate prevented any unspecific adsorption. The biosensor was applied to bacterial cultures from sewage sludge that were grown in synthetic water, then filtered and enumerated for defining the stock solution. The calibration plot showed an increase in the charge transfer resistance (R_{CT}) over the 10^3 to 10^6 colony forming unit (CFU) mL^{-1} range, with an estimated LOD of $1.9{\times}10^3$ CFU mL^{-1} [16]. A fast-flow paper-based electrochemical sensor was developed by Channon et al. for the label-free detection of virus particles [17]. West Nile viruses were detected by electrochemical impedance spectroscopy using antibody functionalized Au microwires, achieving a LOD of 10.2 particles in 50 μL of cell culture media. The sensing approach is easily controllable by means of a smartphone and may presumably be applied to a range of biological targets. Toxins secreted from pathogens can also be detected in bacterial cultures using paper-based electrochemical sensors. This is the case of pyocyanin, a toxin solely produced by *Pseudomonas aeruginosa*, whose detection was performed using an inexpensive approach involving electrode printing of carbon ink on photo paper and square wave voltammetry [18]. It is worth noting that some authors have also exploited the bacterial enzyme activity (expression of β-glucosidase by *Enterococcus* spp. and the production of β-galactosidase and β-glucuronidase by *E. coli*) for their determination at paper-based electrochemical sensors [19].

The use of biological indicators to determine the toxicological effect of environment pollutants is an interesting research area where microorganisms combined with paper-based electrochemical sensors provide important advantages. In this context, a μPAD for highly integrated biotoxicity measurements was prepared involving screen-printing with conductive carbon ink and chromatographic paper. The μPAD contained three functional units for injection, separation, and detection zones with hydrophobic barriers. The *E. coli* cell incubation and the fluid-cell separation were integrated as special microfluidic units, and an interesting scheme for determination making use of inhibition of the microorganism respiratory chain was utilized. The procedure involved the addition of benzoquinone (BQ) to react with the electron or hydrogen carriers including enzymes, co-enzymes, prosthetic groups, or co-factors, which compose the respiratory chain, to form hydroquinone (HQ). When toxic environment inhibits the cellular respiratory chain, the quantity of HQ decreases leading to lower electrochemical current. The relationship between toxicant and HQ production provided the fundamentals for the biotoxicity assay. As proofs of concept; two heavy metals, Cu^{2+} and Pb^{2+} in water and soil; the antibiotic penicillin in soil; and the pesticides acetamiprid, triazolone, and acephate in vegetable juices were detected [20].

Table 1. Screen-printed paper electrochemical (bio)sensors for environmental applications.

Configuration	Technique and Method	Detection	Analyte/Sample	Analytical Characteristics	Ref.
Origami gas-sensing paper-based with CuNPs/SPGE	Gas absorption and electrocatalytic oxidation of NO_2 reduced form in the presence of CuNPs	DPV (nitrite)	NOx/air, exhaust gases from cars	0.03 vppm	[14]
Origami paper based multiple biosensor with BChE, AP or Tyr	Detection of TCh, 1-naphthol or 1,2-benzoquinone	Differential amperometry	paraoxon, 2,4-DCPA, atrazine	L.R: 2–100 ppb LOD: 2 ppb	[21]
PANI/G/PEO/p(VB-co-VA-co-VAc) on cellulosic paper	Conductive paper with printed sensor patterns	Resistance changes	nerve gas (DMMP)	L.R.: 3–30,000 ppb LOD: 3 ppb	[15]
ChOx/PB/CBNPs/office paper SPE	Inhibition of ChOx activity	Amperometry (H_2O_2)	Sulphur mustard (Yprite)	L.R: 1–4 mM LOD: 0.9 mM	[11]
Microfluidic device with chromatographic paper/CE	BQ mediated *E. coli* respiration	Amperometry (HQ)	pesticides/soils, vegetables	LOD: 37.5 μg g^{-1} (triazolone)	[20]
G/AuNPs/mixed cellulose ester filter paper	Direct electrochemical oxidation	DPV (NO_2^-)	nitrite/waters	L.R: 0.3–720 μM LOD: 0.1 μM	[13]
CNTs/Chit/SDS/cellulosic paper with electrodeposited Bi	Anodic stripping previous accumulation at −1.2 V for 240 s	SWASV	Pb^{2+}/waters	L.R: 10–200 ppb LOD: 6.74 ppb	[12]
G/CNTs/ionic liquid/cellulosic paper with electroplated Bi	Anodic stripping previous accumulation at −1.3 V for 300 s	SWASV	Cd^{2+}, Pb^{2+}/wood	L.R: 1–50 μg L^{-1} LOD: 0.2 μg L^{-1}	[22]
$[PMo_{11}VO_{40}]^{5-}$/Whatman #4 filter paper/SPE	Direct electrochemical reduction	CV	ClO_3^-/soil	L.R: 0.312–2.5 mg mL^{-1} LOD: 0.15 mg mL^{-1}	[10]
CB/Prussian Blue paper-based SPE	Reagent-free nitrocellulose membrane with enzyme substrate BTCh	Differential amperometry	nerve agents (paraoxon)	L.R: up to 25 μg L^{-1} LOD: 3 μg L^{-1}	[23]
CNFs or rGO/AuNPs	Whatman Grade 1 cellulose paper modified by ink (bottom side) and nanomaterials (upper side)	LSV after preconcentration at +0.2 V vs Ag for 600 s	Hg(II)/river waters	L.R: up to 1.2 μM LOD: 30 nM	[24]
SiNs/paper/rGO/SPCE	Paper-based immunocapture assay with anti-EE2	SWV	EE2/ waters	L.R: 0.5–120 ng L^{-1} LOD: 0.1 ng L^{-1}	[25]
carbon black ink/filter paper SPE	Direct electrochemical oxidation	SWV	BPA/waters	L.R: 0.1–0.9; 1–50 μM LOD: 0.03 μM	[26]

AP: alkaline phosphatase; BChE: butyrylcholinesterase, CB: carbon black; Chit: chitosan; CFU: colony forming unit; ChOx: choline oxidase; CNF: carbon nanofibers; DCPA: 2,4-dichloro-phenoxyacetic acid; DMMP: dimethyl methylphosphomate; *E. coli*: *Escherichia coli*; EE2: ethinyl estradiol; EIS: electrochemical impedance spectroscopy; G: graphene; HQ: hydroquinone; LOD: limit of detection; L.R: linear range; CNTs: carbon nanotubes; PANI, polyaniline; PB, Prussian Blue; PEO, polyethylene oxide; p(VB-co-VA-co-VAc): poly(vinylbutyral-co-vinyl alcohol-co-vinyl acetate); rGO: reduced graphene oxide; SPE: screen-printed electrode; SPGE: screen-printed gold electrode; SDS: sodium docecylsulfate; SiNs: silica nanoparticles; SWASV: square-wave anodic stripping voltammetry.

2.2. Clinical Applications

Sensitive and selective sensors constructed with screen-printed electrode (SPEs) have been developed for different analytes of clinical relevance. These sensors exhibit great advantages allowing fabrication of attractive designs for single and multiple determination even in the absence of biological elements. In this context, the special features of paper as support material for the preparation of diagnostic devices, together with those of screen-printed platforms, represents an important advance for easy self-testing and point-of-care (POC) assessment. Table 2 summarizes the analytical characteristics and the main properties of some recent and representative methods applied to analytes of clinical interest in biological samples [27–60]. Some selected examples are discussed below.

An illustrative example is a wax-printed paper-based device reported by Martins et al. [27] for the electrochemical detection of 3-nitrotyrosine (3-NT), a biomarker of oxidative stress. The paper was modified to become a hydrophobic support, and then carbon and silver conductive inks were applied to generate a three electrode-system on a small spot. Square wave voltammetry (SWV) was employed to determine 3-NT in a range from 500 nM to 1 mM with a low LOD of 49.2 nM. More recently, a disposable paper-based printed electroanalytical strip has been reported for the rapid and high-throughput detection of glutathione in blood [28]. The detection involved a thiol-disulfide exchange reaction giving an electroactive product easily oxidizable at a Prussian Blue/carbon black nanocomposite screen-printed onto a wax-patterned filter paper. The resulting configuration, where the paper provides a reagents-free device, allowed the detection of glutathione up to a concentration 10 mM, with a LOD value of 60 µM, and was employed to quantify blood glutathione at physiological levels.

Hydrogen peroxide is an important biomarker associated with respiratory and pulmonary diseases such as asthma and lung cancer. A disposable cellulose paper-based electrochemical sensor integrated into a commercial respiratory mask was reported for on-site testing of H_2O_2 in exhaled breath (Figure 2) [29]. The device involved a Prussian-Blue-mediated carbon electrode for H_2O_2 detection and a carbon blank electrode for subtracting the background currents. In the presence of the analyte, the oxidation product formed from Prussian Blue was electrochemically reduced providing amperometric responses related to the H_2O_2 concentration. This configuration did not exhibit influence from environmental conditions or interferents due to differential measurements. In addition, the use of paper as flexible substrate and hygroscopic porous support eliminated the need for additional membranes.

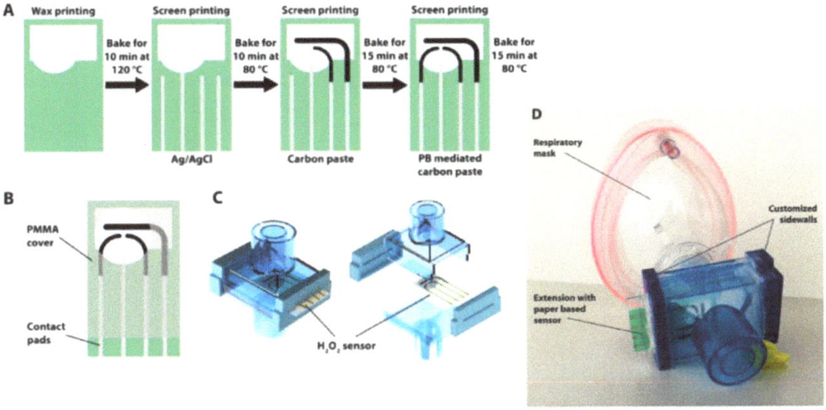

Figure 2. (**A**) Schematics of fabrication steps and (**B**) computer-aided design (CAD) drawing of the disposable cellulose paper-based electrochemical sensor for on-site testing of H_2O_2 in exhaled breath with poly-methylmethacrylate (PMMA) carrier. (**C**) Model of a filter extension for respiratory mask. (**D**) Image of respiratory mask with the commercial filter extension with customized sidewalls, containing the sensor chip. Reproduced and adapted with permission of American Chemical Society [29].

Table 2. Screen-printed paper electrochemical (bio)sensors for clinical applications.

Configuration	Technique and Method	Detection	Analyte/Sample	Analytical Characteristics	Ref.
$Fe(CN)_6^{3-}$/banana peel tissue/SN-MPTS/paper	L-Tyr oxidation catalyzed by tyrosinase and mediated by $Fe(CN)_6^{3-}$	DPV	L-Tyr/plasma	L.R: 0.05–600 µM LOD: 0.02 µM	[39]
MWCNTs/THI/AuNPs/SPE	Label-free microfluidic paper based immunosensor with immobilized anti-E2	DPV (THI)	17β-estradiol (E2)/serum	L.R: 0.01–100 ng mL^{-1} LOD: 10 pg mL^{-1}	[40]
$(NH_2$-G)/THI/AuNPs/SPE	Label-free microfluidic paper based immunosensor with immobilized anti-BNP	Amperometry	BNP/serum	L.R: 0.05–30 ng mL^{-1} LOD: 12 pg mL^{-1}	[41]
rGO/THI/AuNPs/SPE	Label-free microfluidic paper based immunosensor with immobilized anti-FSH	DPV (THI)	FSH/serum	L.R: 1–100 mIU mL^{-1} LOD: 1 mIU mL^{-1}	[42]
rGO-TEPA/AuNPs/SPE	Microfluidic paper-based immunosensor with immobilized anti-AFP; HRP-GNRs-dAb as signal probe	SWV (H_2O_2/OPD)	AFP/serum	L.R: 0.01–100 ng mL^{-1} LOD: 0.005 ng mL^{-1}	[35]
L-Cys-AuNPs/G/SPE	Label-free origami paper based immunosensor with immobilized anti-CRP	EIS ($Fe(CN)_6^{3-/4-}$)	CRP/serum	L.R: 50–10^5 ng mL^{-1} LOD: 15 ng mL^{-1}	[43]
Q-MA/SPGE	Label-free microfluidic paper based immunosensor with immobilized anti-CEA	DPV	CEA/serum	L.R: 1–100 ng mL^{-1} LOD: 0.33 ng mL^{-1}	[44]
DPA/Ag/G/SPCE	Microfluidic plastic-paper based immunosensor with immobilized anti-AFP	EIS	AFP	L.R: 1–10^4 ng mL^{-1} LOD: 1 ng mL^{-1}	[37]
OPANI/G/SPE	Label-free microfluidic paper based immunosensor with immobilized anti-IFN-γ	EIS ($Fe(CN)_6^{3-/4-}$)	IFN-γ/serum	L.R: 5–10^3 pg mL^{-1} LOD: 3.4 pg mL^{-1}	[45]
rGO/THI/AuNPs	Label-free microfluidic paper based immunosensor with immobilized anti-CA125	DPV (THI)	CA125/serum	L.R: 0.1–200 U mL^{-1} LOD: 0.01 U mL^{-1}	[46]
cMWCNTs/cellulose paper/SPE	Label-free paper based immunosensor with immobilized anti-cTnI	EIS ($Fe(CN)_6^{3-/4-}$)	cTnI/serum	L.R: 0.05–50 ng mL^{-1} LOD: 0.05 ng mL^{-1}	[47]
Fe_3O_4@AuNPs@SiO_2 MIP /Whatman paper/CPE	3D-ePAD Direct oxidation	LSV	serotonin/capsules, urine	L.R: 0.01–1,000 mM LOD: 0.002 mM	[30]
C/Ag/paper/SPE	Label-free detection, current decrease	SWV	3-nitrotyrosine	L.R: 500 nM–1 mM LOD: 49.2 nM	[27]

Table 2. *Cont.*

Configuration	Technique and Method	Detection	Analyte/Sample	AnalyticalCharacteristics	Ref.
Patterned waxed paper screen-printed with silver ink	Electrochemical oxidation in the presence of silver	CV	chloride/serum, sweat	L.R: up to 200 mM LOD: 1 mM	[49]
Prussian Blue/paper/SPEs	Differential current measurements	amperometry	H_2O_2/simulated exhaled breath	L.R: 5–320 µM LOD: —	[29]
Prussian Blue/C black/wax patterned wax filter paper	Thiol-disulfide exchange reaction	Amperometry +0.3 V	glutathione	L.R: up to 10 mM LOD: 60 µM	[28]
CoPc/G/IL/paper/SPCE	Non-enzymatic detection	Amperometry +0.7 V	glucose/serum, honey, wine	L.R: 0.01–1.3–5.0 mM LOD: 0.67 µM	[50]
AuNPs/porous paper/SPE	Non-enzymatic detection	CV	glucose	L.R: 0.01–5 mM LOD: 6 µM	[51]
ATCh/G/Ag/AgCl ink/wax printing paper	ATCh hydrolysis by AChE giving TCh directly oxidized	Amperometry/TCh	AChE	L.R: 0.1–15 U mL^{-1} LOD: 0.1 U mL^{-1}	[34]
PheDH/paper/ERGO/SPCE	Phe hydrolysis by PheDH in the presence of NAD^+	Amperometry/NADH	Phe/neonatal blood	L.R: 1–600 µM LOD: 0.2 µM	[33]
paper-based wax printing/CB/SPCE	BTCh as substrate of BChE	Amperometry/TCh/PB	BChE activity/serum	L.R: up to 12 IU /mL LOD: 0.5 IU/mL	[52]
MBs/paper microfluidic/SPCE	On-chip single-step magneto-immunoassay with cAb-MBs and poly-HRP-biotin-dAb	Amperometry/H_2O_2/TMB	MMP-9/plasma	L.R: 0.03–2 ng mL^{-1} LOD: 0.01 ng mL^{-1}	[53]
AQ-PNA/G-PANI/paper/SPCE	PNA-DNA duplexes obstruct electron transfer from AQ label	SWV/AQ	HPV/DNA from SiHa cell line	L.R: 10–200 nM LOD: 2.3 nM	[54]
MB-tagged TFO/AuNPs/paper/SPCE	filter and copy papers compared for detection of ssDNA or dsDNA	SWV/MB	HIV/serum	L.R: 3–3,000 nM LOD: 3 nM ssDNA;	[55]
CuO/IL/ERGO/SPCE/PAD	CuO/IL delivered from a HP D300 digital dispenser	Amperometry	Creatinine/human serum	L.R: 0.01–2.0 µM LOD: 0.22 µM	[56]
GOx-rGO-TEPA/PB-paper/SPE	3D paper-based microfluidic SPE	Amperometry H_2O_2/PB	Glucose/human sweat, blood	L.R: 0.1–25 mM LOD: 25 µM	[57]
rGO/AuNPs-paper-SPE	Wax-patterning on filter paper Whatman No1; rGO prepared from GO and dopamine	SWV	uric acid/urine	L.R: 2.5–1,000 µM LOD: 0.74 µM	[58]

Table 2. *Cont.*

Configuration	Technique and Method	Detection	Analyte/Sample	AnalyticalCharacteristics	Ref.
Wax printed amino-functional graphene (NG)/THI/AuNPs and PB/PEDOT/AuNPs/SPE PADs	Label-free aptasensors	DPV	CEA, NSE/serum	L.R: 0.01–500 ng mL^{-1} (CEA); 0.05–500 ng mL^{-1} (NSE); LOD: 2 pg mL^{-1} (CEA); 10 pg mL^{-1} (NSE)	[59]
Wax screen printing patterns on cellulose paper/Nafion/Chit/GOx/PB/SPE	3D paper-based microfluidic SPE	Amperometry H_2O_2/PB	glucose/sweat	L.R.: up to 1.9 mM LOD: 5 μM	[60]

ATCh: acetylthiocholine chloride; AF: alpha-fetoprotein; AP: alkaline phosphatase; AQ-PNA:anthraquinone-labeled pyrrolidinyl peptide nucleic acid; AuNPs: gold nanoparticles; BChE: butyrylcholinesterase, BNP: B-type natriuretic peptide; BTCh: butyrylthiocholine; cAb: capture antibody; CB: carbon black; CEA: carcinoembryonic antigen; Chit: chitosan; CNFs: carbon nanofibers; CoPc: cobalt phthalocyanine; CRP: C-reactive protein; dAb: detector antibody; DCPA: 2,4-dichlorophenoxyacetic acid; DPA: diphenylalanine; EIS: electrochemical impedance spectroscopy; ERGO: elecrochemically reduced graphene oxide; FSH: follicle stimulating hormone; G: graphene; GNR: gold nanorods; HIV: human immunodeficiency virus HPV: human papillomavirus; HRP: horseradish peroxidase; IL: ionic liquid; MAQ: mercapto-amine quinone-functionalized receptor; MB: methylene blue; MWCNT: multi-walled carbon nanotubes; NSE: neuronspecific enolase; OPD: o-phenylenediamine; PANI: polyaniline; PB: Prussian Blue; Phe: phenylalanine; Q: quinone; Q-MA: quinone-based mercapto amine; rGO: reduced graphene oxide; SN-MPTS: 3-mercaptopropyl trimethoxysilane functionalized silica nanoparticles; SPE: screen-printed electrode; SPGE: screen-printed gold electrode; TCh: thiocholine; TEPA: tetraethylene pentamine; TFO: triple forming oligonucleotides; THI: thionine; TMB: tetramethylbenzidine.

Molecularly imprinted polymers (MIPs) used as electrode surface modifiers allow high selective recognition, although they sometimes lack the required sensitivity due to the poor conductivity of building materials. To improve the analytical performance, nanostructured configurations yielding larger currents and fast responses have been proposed. The resulting nano-sized MIPs have been combined with paper-based analytical devices to obtain three-dimensional electrochemical PADs (3D-ePADs), which provide additional advantages such as lower cost and smaller sample and reagents volumes. An interesting example is the method involving filter papers prepared by alkyl ketene dimer (AKD)-inkjet printing of a circular hydrophobic detection zone coupled with screen-printed graphite electrodes drop coated with Fe_3O_4@Au@SiO_2-MIP nanocomposites (Figure 3), for the voltammetric determination of serotonin. Linear sweep voltammetry at +0.39 V provided a linear range from 0.01 to 1,000 mM with a LOD of 0.002 mM. The resulting MIPs exhibited strong affinity for the analyte, and the electrochemical sensor showed electrocatalytic activity toward the oxidation of serotonin. The sensor was successfully applied to the analysis of pharmaceutical capsules and urine samples [30].

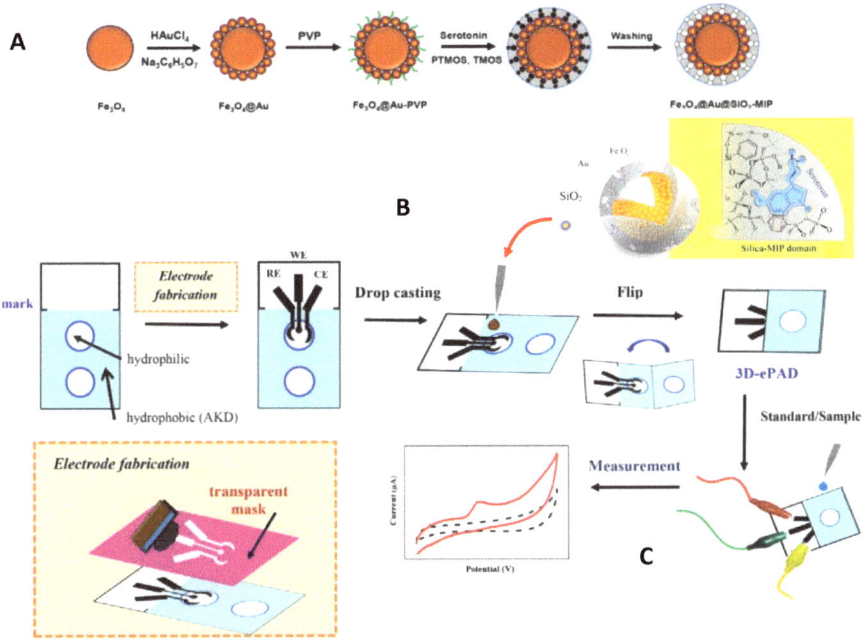

Figure 3. (**A**) Synthesis of Fe_3O_4@Au@SiO_2-MIP, (**B**) preparation of the sensor for serotonin, and (**C**) electrochemical detection using the 3D-ePAD. Reproduced and adapted with permission of Elsevier [30].

Human C-reactive protein (CRP) is a nonspecific pentameric protein produced by hepatocytes in the liver upon stimulation by endogenous proinflammatory cytokines. CRP is an important biomarker for various cardiovascular diseases and its determination requires sensitive and accurate methods with high selectivity for application in complex clinical samples. Pinyorospathum et al. [31] developed a single step method for the determination of CRP in human serum involving a AuNP-modified SPCE self- assembled with PADs tethered with a biomimetic polymer consisting of thiol-terminated poly(2-methacryloyloxyethyl phosphorylcholine) (PMPC-SH). The approach took advantage of the specific binding of protomers subunits forming the CRP structure to the phosphorylcholine group in the presence of calcium ion [32]. Figure 4A shows that PMPC-SH copolymer reacts with AuNPs; then, $[Fe(CN)_6]^{3-/4-}$ current at the resulting PMPC-SH/AuNPs-SPCE is measured by DPV to further subtracting it from the response in the presence of CRP. Figure 4B–F shows the preparation of the PAD in three parts—the middle where SPCE is placed, the green flap for the storage of calcium ions and dropping the sample, and the purple flap used for the detection with the $[Fe(CN)_6]^{3-/4-}$ redox probe. The current decreased in the presence of CRP and Ca^{2+} over the 5 to 5,000 ng mL^{-1} CRP concentration range with a LOD value of 1.6 ng mL^{-1}. The use of a PMPC-modified surface reduced the nonspecific adsorption of proteins, and the sensor response was not interfered by bilirubin, myoglobin, or albumin. The sensor was successfully applied to the determination of CRP in certified human serum.

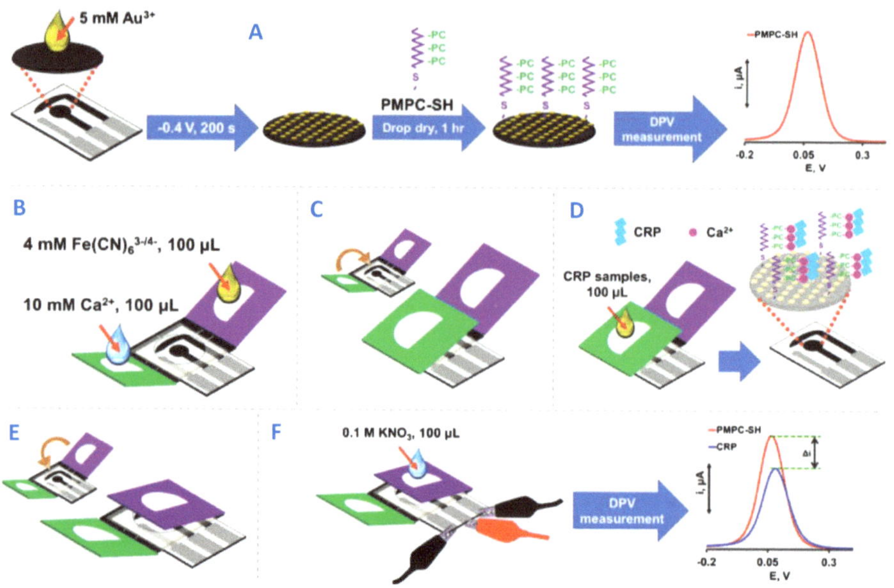

Figure 4. (**A**) Preparation of thiol-terminated poly(2-methacryloyloxyethyl phosphorylcholine) (PMPC-SH)-AuNPs/SPCE. (**B–F**) Steps for preparation of the PMPC-SH-AuNPs/SPCE/PAD sensor for the differential pulse voltammetry (DPV) determination of C-reactive protein (CRP). Reproduced and adapted with permission of Springer [31].

Electrochemical enzyme paper biosensors (EPADs) have been shown to provide an adequate microenvironment for direct measurements, while physical adsorption of the enzyme did not affect its native structure, function, and electrocatalytic activity [32]. An illustrative example is a biosensor for phenylketonuria (PKU) screening based on the determination of phenylalanine (Phe). The biosensor was implemented by immobilization of phenylalanine dehydrogenase (PheDH) over paper microzones placed onto an electrochemically reduced graphene oxide (ERGO)-modified SPCE. The detection of the NADH formed in the presence of NAD^+ provided a sensitive, low-cost, and fast method for PKU monitoring in neonatal blood samples [33]. An original paper-based biosensor was developed for the detection of acetylcholinesterase (AChE) [34]. This enzyme catalyzes the hydrolysis of acetylcholine neurotransmitter and its abnormal function can promote and accelerate the aggregation of amyloid-betapeptides closely related to Alzheimer's disease. The bioelectrode was fabricated by immobilization of acetylthiocholine on a sheet prepared with double adhesive tape. The enzyme samples were dropped on the backside of the electrode where, after hydrolysis, amperometric detection was performed and provided a LOD value of 0.1 U mL^{-1} AChE.

The combination of SPEs with simple paper-based microfluidics (µ-PEI) exhibits several advantages for the preparation of electrochemical biosensors compared with conventional analytical devices fabricated with other substrates (glass, silicon, or polymers). The resulting devices are inexpensive, easy to fabricate, and compatible with a variety of chemical or biochemical applications [35]. Cellulose papers with high surface area have become useful substrates in combination with SPEs for prototyping new point of care testing devices (POCTs) involving microfluidic systems in clinical diagnostics. Among the recent designs, those involving immunoassays constitute a challenge where the stability of the immunoreagents and the preparation of a surface capable of promoting electronic transfer to effectively enhance the assay sensitivity and selectivity, are critical factors. An illustrative method is that reported for the determination of alpha-fetoprotein (AFP). The method used paper-based microfluidic channels to integrate sampling, detection, and adsorption zones, as well as an rGO-tetraethylene pentamine (TEPA)/AuNPs nanocomposite for immobilization of specific AFP antibodies and sensitive detection (Figure 5) [35]. AFP is one of the most important biomarkers in diagnosing hepatocellular carcinoma, and in the case of pregnant women, it is the first serologic biomarker to detect birth defects in a developing baby [36,37]. The immunoreaction was performed by applying the tested solution to the sample zone and letting it elute slowly to the detection zone where AFP was captured. Then, gold nanorods decorated with horseradish peroxidase and detector antibodies (HRP-GNRs-Ab_2) were dropped onto the sample zone to form a sandwich-type configuration on the working electrode and SWV detection was carried out in the presence of H_2O_2. The calibration plot showed a wide linear range (0.01–100.0 ng mL^{-1}) with a low LOD value of 0.005 ng mL^{-1}.

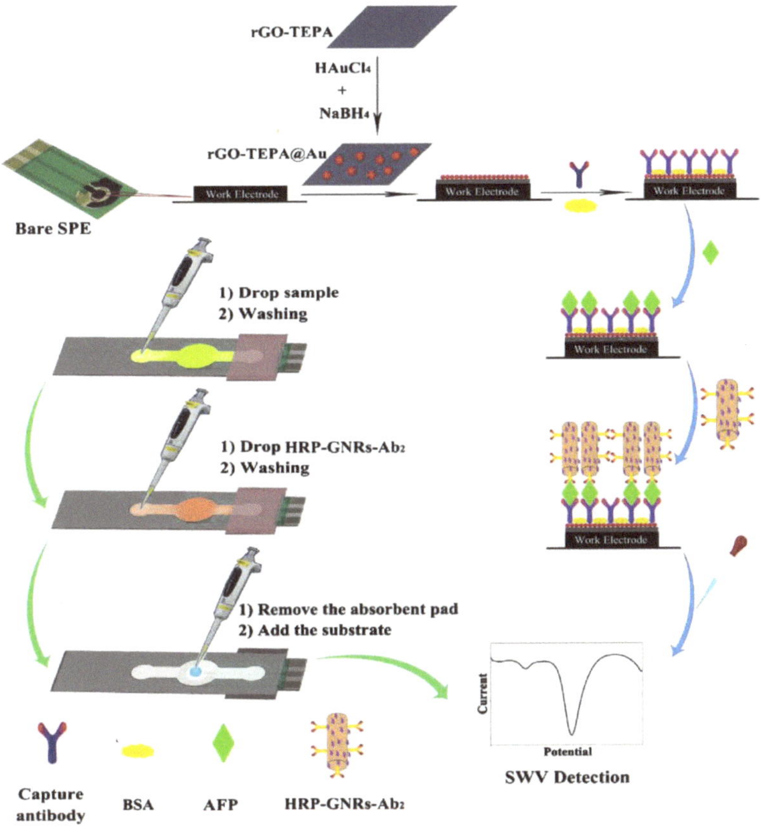

Figure 5. Schematic displays of the modification and assay procedure for the implementation of an immunosensor for the determination of alpha-fetoprotein (AFP) using paper-based microfluidic channels to integrate sampling, detection and adsorption zones, and a reduced graphene oxide (rGO)-tetraethylene pentamine (TEPA)/AuNPs nanocomposite for immobilization of specific AFP antibodies. Reproduced with permission of Elsevier [35].

Oher configurations involving label-free paper-based immunosensors with immobilized capture antibodies have been reported for the detection of hormones [40,42], CRP [43], interferon (IFN-γ) [45], cancer biomarkers [44,46], and cardiac biomarkers [41,47]. Among them, it is worth mentioning the platform described by Ruecha et al. [45] involving a paper-based microfluidic device coupled with a label-free electrochemical impedimetric immunosensor for the detection of IFN-γ in serum. This multifunctional cytokine, originally characterized by its viral activities, is primarily secreted by natural killer T cells as a part of the innate immune response to intracellular pathogens [48] and plays a crucial role related to inflammatory and antoimmune diseases. In Ruecha's method, a wax-printing strategy was implemented to fabricate the paper electrode, which was screened with graphene modified ink to deposit polyaniline (PANI) and further covalent immobilization of the specific anti- IFN-γ antibodies. The increase of the charge transfer resistance with the cytokine loading provided a linear relationship with logarithmic concentrations of IFN-γ in the 5–1000 pg mL^{-1} range with a LOD of 3.4 pg mL^{-1}.

Although the number of paper-based electrochemical biosensing platforms has increased in the last few years, the vast majority of the reported methods involve the use of enzymes and antibodies as recognition elements, and so far, they have not been expanded to nucleic acid-based assays. In this context, Liu et al. [38] prepared a paper modified with signal molecule-labelled DNA and a screen-printed electrode along with target recognition solutions to achieve the detection of multiple biomarkers. The method is based on the target-induced synthesis of Mg^{2+}- dependent DNAzyme for catalyzing the cleavage of substrate DNA from paper, taking advantage of the high specific target-triggered polymerization/nicking and DNAzyme-catalyzed signal amplification. The performance of this method was evaluated using a microRNA recognition probe for lung cancer-specific miR-21, a phosphorylated hairpin probe for targeting alkaline phosphatase (ALP), and a DNA aptamer for carcinoembryonic antigen (CEA). Ferrocene-labeled DNA (Fc-DNA) was immobilized on paper by functionalizing it with aldehyde groups and further Schiff-based reaction (Figure 6A). Then, the paper-electrochemical biosensor was prepared by sticking the Fc-DNA modified paper onto a plastic slide and carbon nanotubes modified SPE. As an example, the fundamentals of miR-21 detection are illustrated in Figure 6B. After incubation with the recognition solution that contains the ssDNA probe (P1), KF polymerase and nicking endonuclease Nt.BbvCI, the polymerization via KF activity is initiated to extend the 3′-end, providing dsDNA with recognition site for endonuclease whose activity to cut one strand of dsDNA generates new replication sites. Then, the Mg^{2+}-dependent DNAzyme strand is displaced and released. This cycle produces a large amount of DNAzyme strands that fold into the catalytically active loop structure and bind to immobilized Fc-DNA resulting in the release of DNAzyme strands and cleaved Fc-shorter ssDNA from the paper, which diffuses to the surface of CNTs-SPE giving a DPV response.

Among paper-based electrochemical DNA sensors, configurations developed for the detection of human papillomavirus (HPV) [54] and human immunodeficiency virus (HIV) [55] are particularly relevant. Teengam et al. [54] reported a graphene-PANI modified electrode with immobilized anthraquinone-labeled pyrrolidinyl peptide nucleic acid probe (AQ-PNA) for the detection of a synthetic 14-base oligonucleotide target with the sequence of HPV type 16 DNA by electrochemical measurement of the AQ response by SWV. A linear range of 10–200 nM and a LOD value of 2.3 nM were obtained. The performance of this biosensor was tested with the detection of PCR-amplified DNA. On the other hand, Cinti et al. [55] developed a series of paper-based strips for the electrochemical detection of single and double stranded DNA, which were successfully applied to a synthetic PCR amplified dsDNA sequence related to HIV in serum samples. Paper-based AuNPs-SPE platforms and triplex forming oligonucleotides (TFO) including Methylene Blue (MB) were used as the recognizing probes.

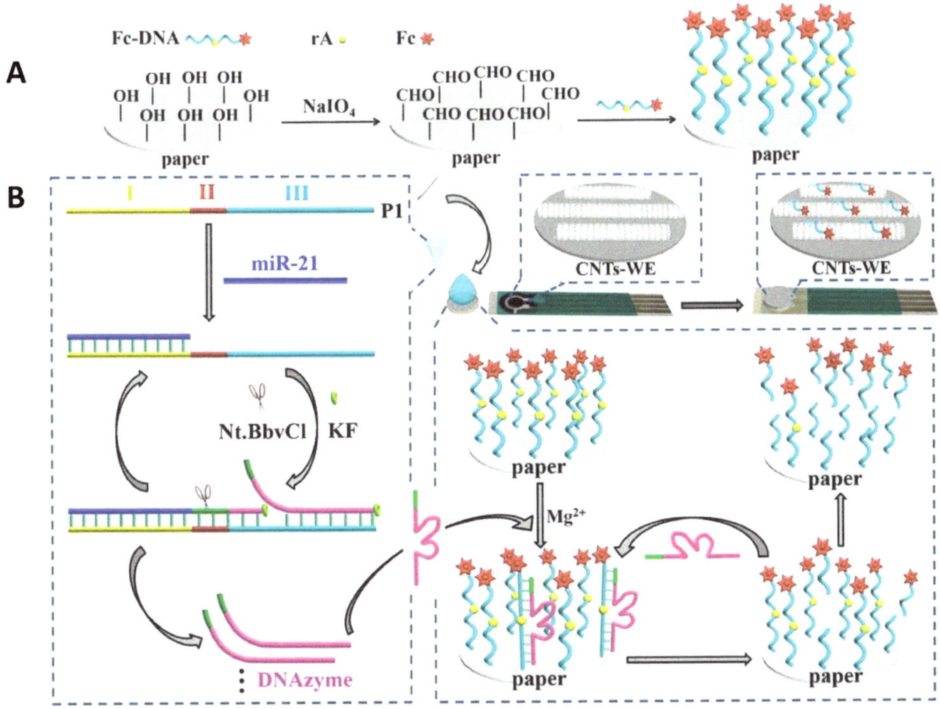

Figure 6. Schemes of (**A**) Ferrocene-labeled DNA (Fc-DNA) immobilization on paper and CNTs-SPEs, and (**B**) miRNA assay for the recognition of miR-21 (left), and the electrochemical response to the released DNAzymes (right). Reproduced and adapted with permission of American Chemical Society [38].

3. Wearable Printed Electrodes for Biosensing Applications

Due to the booming research activity in the field of wearable and/or flexible printed electrodes, excellent reviews have been recently reported highlighting the versatility and tremendous potential of these devices [61–69]. Therefore, this section is just limited to give a rough overview of late advances and prospects to draw the current landscape of wearable and flexible printed electrodes (not implantable) for biosensing. Accordingly, only a few of the most representative methods reported during 2018 and 2019, applied mainly to clinical diagnosis and environment monitoring, are critically discussed. Table 3 summarizes the main features of these methods.

SPEs can be easily printed in a variety of shapes (flower, skull, marijuana, panda bear, etc., Figure 7a–d) and sizes and can be modified with different biological elements and nanomaterials. Leveraging on these advantages, screen printing has been employed to construct affordable wearable printed electrochemical sensors to provide real-time information on both the wearer's health and performance (opening the door to individualized medicine) and the surrounding environment. In the physiological monitoring field, the active sensor surface is in close contact with the epidermis (oral mucosa in the mouth, stratum corneum or skin) to detect relevant biomarkers such as glucose [70] and ethanol [71] in different informative biofluids (saliva, sweat, tears). However, for monitoring the wearer's environment, the sensor faces away from the epidermis in the direction of the surrounding to detect risk of exposure to chemicals [63]. Moreover, the recently explored robotic assisted strategy implies that the robot fingertips are kept in close contact with the target sample [72].

The spectacular growth witnessed in wearable printed sensors is largely due to the development of novel materials that imparted the resulting sensors the capabilities to fold, bend, stretch, and repair, ensuring their performance during on-body applications under extreme tensile stress [73]. Wearable electrochemical sensors have been implemented on head-to-toe wearable platforms and in connection to different biofluids, environments and analytes [64]. For the realization of wearable applications matching the non-planarity and mechanical properties of the human body, electrochemical sensors have been printed on temporary tattoo, bendable bandage, gloves, contact lens, water-soluble silk thin-film substrates (transferred to tooth enamel) or textile substrates (GORE-TEX and Neoprene) (Figure 7e–j) [63] or incorporated in mouthguards, eyeglasses, or rings (Figure 7k–m). The great progress experienced by electronics in terms of flexibility and miniaturization [65,74–76], in the development of effective methods for stimulating/controlling of non-invasive bio-fluids collection and the proliferation of smart-phones and connected devices, together with a growing consumer demand for health awareness, and the imperative need for doctors to obtain as much objective and quality data from their patients as possible, have been crucial aspects in the development of fully implementable wearable electrochemical devices and in opening up new avenues for body-integrated electronics previously unattainable [74]. However, powering is still the main Achilles' heel of these devices and the size and weight of the power source may limit the wearability of the biodevices and hinder the wearer's activity. Therefore, the rational integration of power sources with biosensors is a desperate requirement and additional efforts are required to develop anatomically compliant, miniaturized, stretchable and flexible power sources [64,77].

In general, there are three different modes of integrating wearable biosensors and power supplies: (1) an external circuit connection which is bulky and cumbrous; (2) a flexible substrate-based integration; and (3) all-in-one integration [78]. The last two strategies, made possible by advances in device designs and micro/nanofabrication technologies, are more widely used. The second strategy implies each component is relatively independent and can be considered a general integration strategy applicable to diverse sensing systems without having to worry about structural compatibility between components but difficult to allow the level of miniaturization required. The third strategy is effective for miniaturized designs in which all the components suffer from similar deformations simultaneously. However, the endurability difference among them should be minimized to guarantee the normal function of the respective component.

Currently, wearable biodevices are powered mainly by (i) safe high energy wearable batteries; (ii) energy conversion devices (piezoelectric and triboelectric nanogenerators, which harvest the mechanical energy in human motions, such as walking, breathing, and waving arms; solar cells, which harness light energy; thermoelectric supercapacitors; biofuel; and water-voltage cells); (iii) energy storage devices (mechanically flexible energy storage elements, mainly supercapacitor, and lithium-ion battery); (iv) hybrid power supplies combining energy conversion with energy storage; and (v) wireless energy transfer (wireless coils, like RF antennas). Significant progress have also been made in self-power and energy-efficient or even energy-free systems devices, fueled by the development of high-efficiency energy acquisition approach and ultra-low power consumption technique [68,69,78–80].

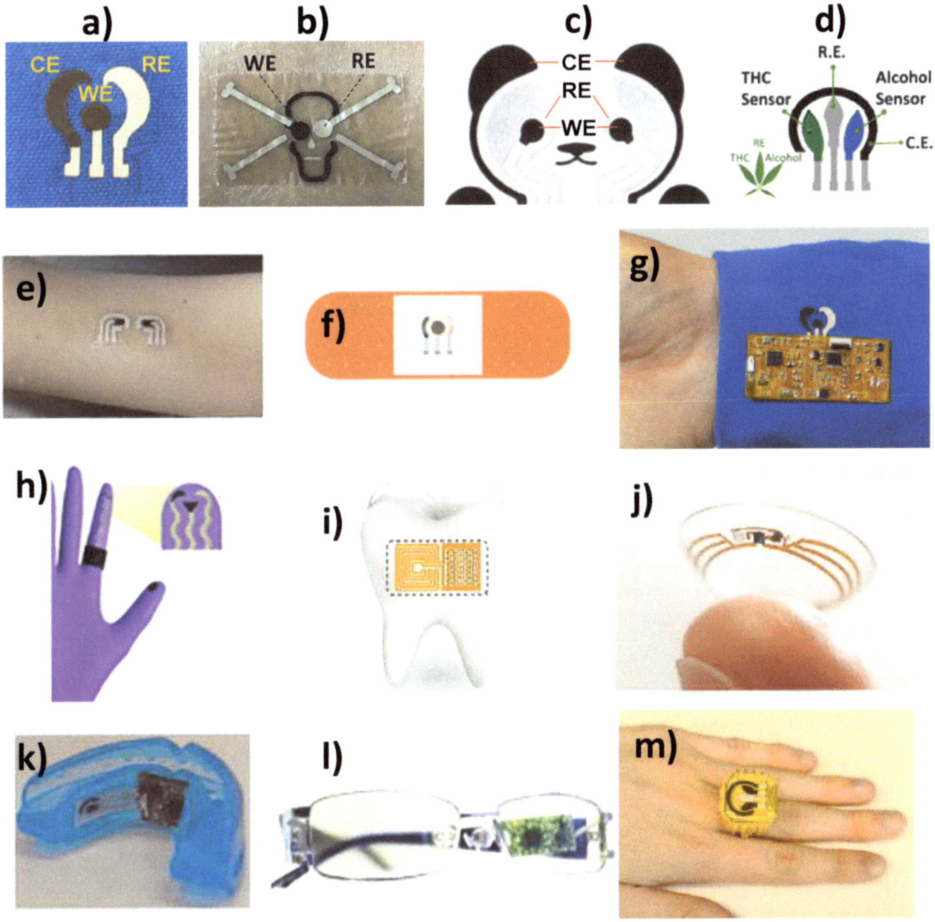

Figure 7. SPEs printed on flower (**a**), skull (**b**), panda bear (**c**), and marijuana (**d**) shapes. SPEs fabricated on a temporary tattoo (**e**), bendable bandage (**f**), textile substrate (**g**), glove (**h**), water-soluble silk thin-film substrates (transferred to tooth enamel) (**i**), contact lens (**j**), or incorporated in a mouthguard (**k**), eyeglasess (**l**) or ring (**m**). Reprinted and adapted with permission of Springer [68] (**e**,**j**,**k**) Wiley [81] (**f**), Elsevier [82] (**a**,**g**), Elsevier [82](**b**), Wiley [83] (**c**), Wiley [65] (**h**), Nature Research [84] (**i**), Elsevier [85] (**l**), and Elsevier [86] (**c**,**m**).

As a previous step to on-body measurements, Payne et al. made an exhaustive study to characterize the effects of five different salts in physiologically relevant concentration ranges on the performance of a printed, flexible, wearable biosensor involving lactate oxidase and tetrathiafulvalene for the amperometric detection of lactate in sweat [87].

The extensive and pioneering work performed by Wang's group in the development of wearable and flexible printed electrodes for biosensing in healthcare, food, and security fields should be noted. Wang's team proposed the use of fully integrated wearable bendable bandage and minimally invasive microneedle-based sensors modified with catechol (CAT) for rapid and decentralized screening of skin melanoma through the amperometric detection of the benzoquinone (BQ) generated in the presence of the tyrosinase (TYR) biomarker (Figure 8). The bandage sensor exhibited high resiliency against mechanical strains due to the use of stress-enduring inks for its printing. These skin-worn sensors were interfaced to flexible electronic board that controlled the electrochemical operation and transmitted data wirelessly to a mobile device, and were used to screen biomarkers both on the skin surface (bandage sensor) or under the skin (microneedle device). They were applied to analyze TYR-containing agarose phantom gel and porcine skin [81]. These epidermal sensors allow skin cancer screening in less than 4 min obviating the need of painful solid biopsies and the associated delays and anxiety.

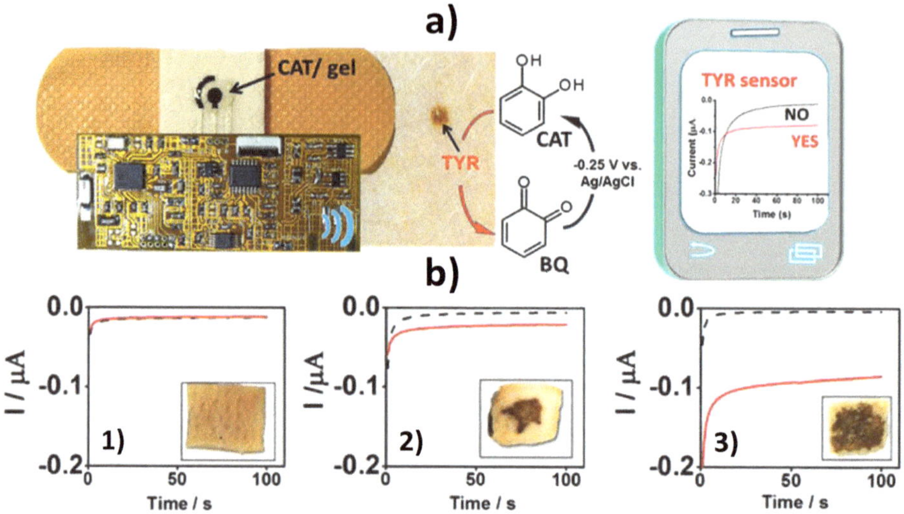

Figure 8. TYR biosensing using a bandage electrochemical sensor modified with a CAT-containing agarose gel and involving wireless chronoamperometric data transmission to a smart device (**a**). Chronoamperometric responses provided by the bandage sensors before (black line) and after (red line) 2 min interaction with skin pork samples untreated (1) and treated with 0.5 (2), and 2.5 mg mL^{-1} TYR (3) (**b**). Reprinted and adapted with permission of Wiley [81].

The same group reported a strategy using a single wearable and flexible epidermal platform for the simultaneous yet independent noninvasive sampling and analysis of two different epidermal biofluids (sweat and skin interstitial fluid (ISF) with a blood-like composition) at two physically separate locations. This approach involves parallel operation of iontophoretic delivery of the sweat-inducing pilocarpine into the skin and reverse iontophoretic ISF extraction across the skin at anode and cathode, respectively (Figure 9a,b) [83]. The developed wearable device was implemented using a cost-effective screen-printing technique with body-compliant temporary tattoo materials for disposable single use and conformal wireless readout circuits, and integrated amperometric GOx and AOx biosensors (Figure 9c). It was used for real-time monitoring of alcohol and glucose levels in sweat and ISF, respectively, from individuals consuming food and alcoholic drinks.

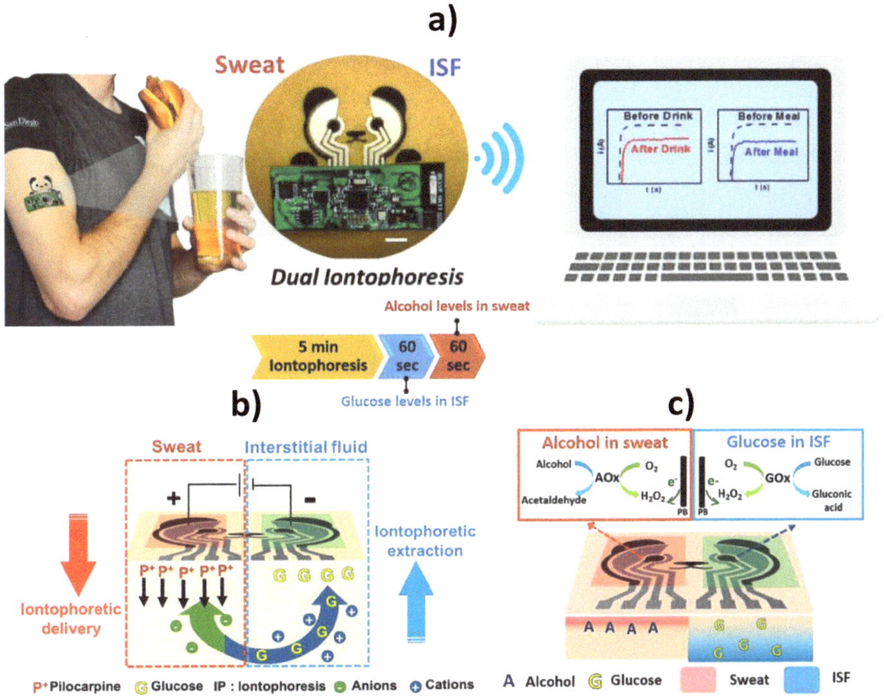

Figure 9. Wearable iontophoretic biosensing device developed on a printed tattoo platform for simultaneous glucose and alcohol monitoring in interstitial fluid (ISF) and sweat, respectively, and wireless real-time transmission of the recorded response (**a**). Schematic display of the iontophoretic operation to simultaneously induce generation of alcohol-containing sweat by iontophoretic delivery of pilocarpine at the anode and sampling of ISF glucose at the cathode by reverse iontophoretic (**b**); biosensing operations to detect amperometrically alcohol in the stimulated sweat and of glucose in the extracted ISF by measuring the hydrogen peroxide generated in the AOx and GOx enzymatic reactions (**c**). Reprinted and adapted with permission of Wiley [83].

Flexible epidermal tattoo and textile-based electrochemical biosensors using stretchable organophosphorus hydrolase (OPH) enzyme electrodes have been developed for continuous vapor-phase detection of organophosphorus (OP) threats [82] (Figure 10a). These wearable sensors were fabricated with elastomeric inks and displayed resiliency toward mechanical stress expected from the wearer's activity without compromising the biosensing performance. They were coupled with a fully integrated conformal flexible electronic interface providing square-wave voltammetry (SWV) detection of the enzymatically-generated nitrophenol product (Figure 10b) and wireless data transmission. The sensor achieved a LOD of 12 mg L^{-1} in terms of OP air density. The same group proposed also a wearable tattoo OPH–pH biosensor for real-time on-body potentiometric monitoring of G-type nerve agent simulants (using fluorine-containing OP nerve agent simulant diisopropyl fluorophosphate, DFP, as model) in both liquid and vapor phases. The OPH biocatalytic recognition phase was coupled on a flexible printed transducer with a pH-responsive poly(aniline) PANI layer for monitoring the proton release during the enzymatic hydrolysis of DFP by OPH. This skin-worn OP potentiometric sensor withstands severe mechanical strains without compromising the analytical performance and displays a wide dynamic range, fast response, high selectivity towards DFP and good reproducibility. These wearable OP biosensing devices hold considerable promise for real-time on-body detection and warning exposure to chemical threats such CWAs and pesticides in a variety of scenarios for triggering timely countermeasure actions, changing dramatically the protection of civilians, farmers, and military personnel.

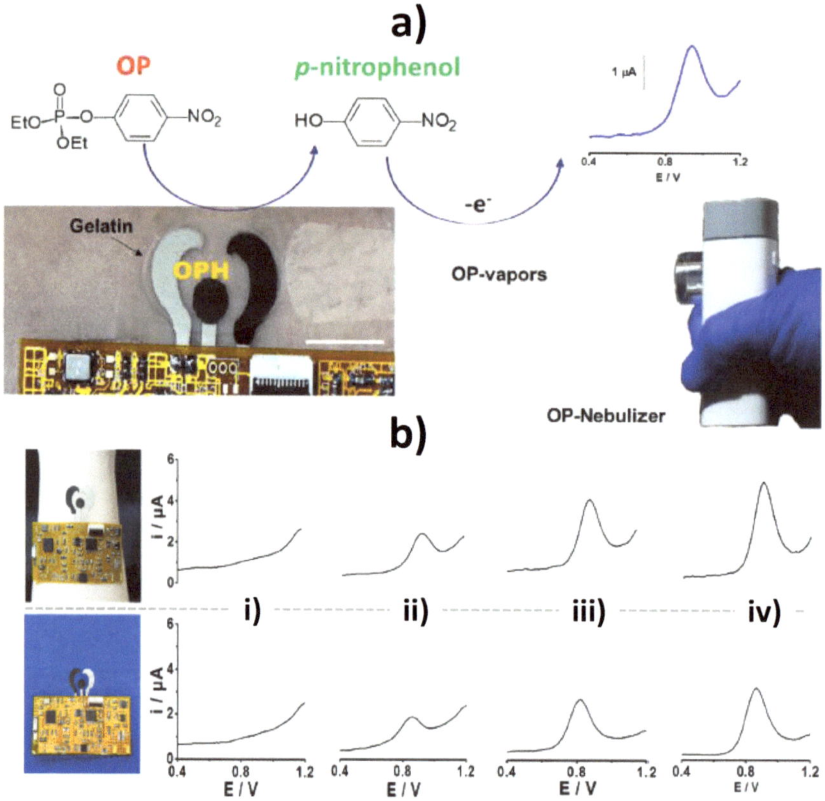

Figure 10. Epidermal tattoo organophosphorus hydrolase (OPH)-based biosensor for vapor-phase detection of OP through SWV measurements of the p-nitrophenol generated after interaction of the MPOx micro-droplets released from the nebulizer with the OPH layer (**a**). Pictures of the OPH based epidermal tattoo (up) and textile (down) biosensors integrated with the flexible electronic interface and SWV responses they provide upon spraying 0 (i), 5 (ii), 10 (iii), and 15 (iv) mM MPOx in the vapor phase (**b**). Reprinted and adapted with permission of Elsevier [82].

Wang's team reported recently a wearable eyeglasses-based tear biosensing system for non-invasive monitoring of key biomarkers [85]. The wearable tear bioelectronic platform integrates a microfluidic electrochemical detector into an eyeglasses nose-bridge pad and the wireless electronic circuitry into the eyeglasses frame and yielded a fully portable, convenient-to-use fashionable sensing device placed outside the eye region addressing drawbacks of sensor systems involving direct contact with the eye as the contact lenses platform (Figure 11a). The concept was used for real-time non-invasive detection of alcohol, glucose and multiple vitamins in tears in connection with enzymatic (AOx and GOx) biosensing fluidic system (alcohol and glucose) and rapid voltammetric scanning (vitamins). This tear alcohol sensing strategy exhibited good correlation to concurrent blood alcohol concentration (BAC) in the monitoring of alcohol intake in individuals over multiple drinking courses.

A novel sensor ring concept, comprising a powerful wireless electronic board embedded into a ring platform and a readily replaceable printed dual-sensor electrode cap, was developed for the simultaneous, direct and rapid field detection of D9-tetrahydrocannabinol (THC) and alcohol in diluted saliva [86]. The ring sensing platform contained a voltammetric THC sensor based on a multi-walled carbon nanotubes MWCNTs/carbon electrode and an amperometric alcohol biosensor involving Prussian-blue (PB) mediator, coated with AOx/chitosan reagent layer on the ring cap (Figure 11b). The dual-analyte THC/alcohol ring sensor system showed no cross talk and high sensitivity (0.5 μM THC and 0.2 mM alcohol). THC and alcohol were determined simultaneously in the same diluted saliva sample within 3 min without any interference from the matrix. This new wearable THC/alcohol ring sensor, readily expanded to detecting other drugs of abuse, is very promising for rapid testing of suspected drivers and for alerting users to their own levels before driving.

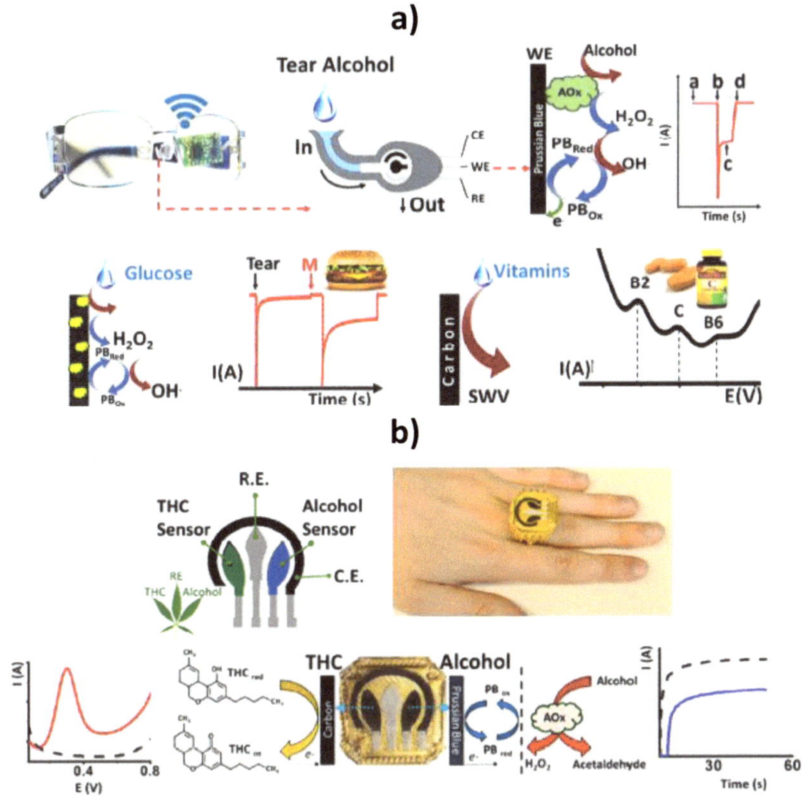

Figure 11. Schematic cartoon of the fluidic device, wireless electronics integrated into the eyeglasses platform, enzymatic detections of alcohol and glucose by chronoamperometry and non-enzymatic determination of vitamins by SWV in collected tears (**a**). Ring-based sensor platform embedded with marijuana designed sensor for detecting THC and alcohol in undiluted saliva samples using SWV and chronoamperometry (**b**). Reprinted and adapted with permission of Elsevier [85] (**a**) and [86] (**b**).

The same group have also reported in a pioneering way advances in wearable chemical sensor technology and flexible electronics to develop chemical sensing robotic fingers (printed on the robotic glove) for rapid discrimination between sweetness, sourness, and spiciness, via electrochemical monitoring of glucose, ascorbic acid, and capsaicin in different drinks (juices, sport and soft drinks and coffee) and extracts (green chili, red paprika and red pepper) (Figure 12) [72]. It is worth remarking that although it was out of the period to which this section has been restricted this group proposed also glove type wearable devices for use in forensic analysis (gunshot residues and nitroaromatic explosive compounds) [88]. This chemical sensing robotic skin is a key demonstration to spur future development of wearable printed electrodes, which offers great opportunity for automated chemical sensing machinery, facilitating robotic decision in a wide range of applications and even in potentially hazardous environments for human counterparts.

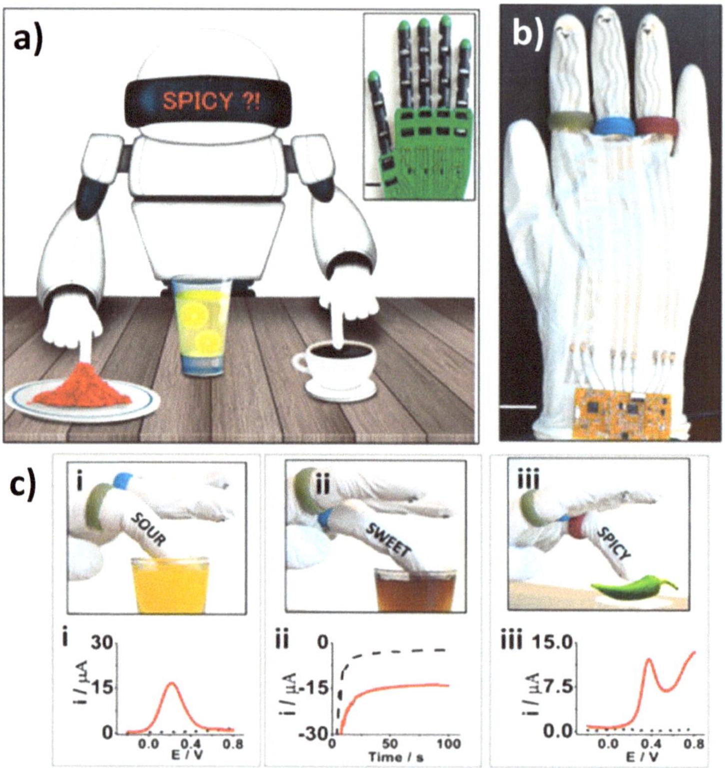

Figure 12. Automated taste discrimination in food samples through chemical sensing at the robot fingertips (**a**) Prototype of the screen-printed robotic sense fingers (carbon-printed sour-finger in green, GOx PB-printed sweet-finger in blue and carbon-printed spicy-finger in red) with long connections to the electronic interface (**b**). Images and corresponding electrochemical responses (in red) of: robotic sour-finger dipped in orange juice and the square wave voltammetry (SWV) signature of ascorbic acid (i), robotic sweet-finger in cherry juice and amperometry data of glucose (ii), spicy-finger on green-pepper, and SWV feedback response to the presence of capsaicin (iii). For comparison purposes the response obtained in phosphate buffer saline (PBS) response are displayed in black dotted lines (**c**). Reprinted and adapted with permission of American Chemical Society [72].

Wang's group has reported very recently an epidermal AAOx biosensor able to monitor the dynamics of vitamin C in sweat after the intake of vitamin C pills and fruit juices [89]. This method combines the use of a flexible vitamin C tattoo patch fabricated on a polyurethane substrate with a localized iontophoretic sweat stimulation system. Chrono-amperometric cathodic detection of the oxygen cosubstrate consumption during the enzymatic reaction, demonstrates very interesting potential for personalized nutrition solutions.

Table 3. Wearable and flexible printed electrodes reported during 2018–2020 for biosensing applications.

Type of Wearable Sensor	Methodology	Analyte	Detection Technique	LOD	Application, Samples and Assay Time	Ref.
Flexible, wearable lactate sweat sensor	Biosensors using LOx and TTF	Lactate	Chrono-amperometry	—	Detection in artificial sweat	[87]
Bendable bandage and microneedle based sensors	In the presence of the surface TYR biomarker, its catechol substrate, immobilized on the transducer surface is rapidly converted to benzoquinone	TYR (Melanome biomarker)	Chono-amperometry	—	Melanoma screening in skin and tissues/Tyr-containing agarose phantom gel and porcine skin in less than 4 min (2 min of incubation and 100 s for the measurement)	[81]
Tattoo-like flexible iontophoretic platform integrated with electrochemical biosensors	Parallel operation of reverse iontophoretic ISF extraction across the skin and iontophoretic delivery of the sweat-inducing pilocarpine into the skin at separate locations and GOx and AOx-based biosensors	Glucose and alcohol	Chrono-amperometry	—	Simultaneous and real-time determination of alcohol and glucose levels on demand localized sampled sweat and ISF biofluids	[83]
Flexible epidermal tattoo and textile-based electrochemical biosensors	OPH-based skin- and textile-worn biosensors for continuous vapor-phase detection of OP threats integrated with a soft, flexible, skin-conforming electronic interface	Vapor-phase detection of OP nerve agents.	SWV	12 mg L^{-1} in terms of OP air density	Continuous and real-time vapor-phase detection of MPOx	[82]
Tattoo paper biosensor	Epidermal OPH–pH biosensor printed onto a temporary tattoo paper coated with PANi (for monitoring the proton release during the enzymatic hydrolysis of DFP by OPH) and with a PVA-acrylamide hydrogel which ensures surface distribution of the target DFP vapors	DFP in both liquid and vapor phases	Potentiometry	—	Real-time detection of DFP in both liquid and vapor phases	[90]
Eyeglasses platform for biosensing in tears	Enclosing the electrochemical biosensor within a microfluidic chamber, with the supporting electronics embedded onto the eyeglasses' inner frame	Ethanol, glucose and multiple vitamins (B_2, C and B_6)	Chronoamperometry (ethanol and glucose) SWV (vitamins)	—	Real-time detection of alcohol intake and glucose and vitamins in human subjects	[85]

Table 3. *Cont.*

Type of Wearable Sensor	Methodology	Analyte	Detection Technique	LOD	Application, Samples and Assay Time	Ref.
Ring-based dual sensing platform	Wireless electronic board embedded into a ring platform, along with a printed dual-sensor electrode cap comprising a voltammetric THC sensor based on a MWCNTs/carbon electrode and an amperometric alcohol biosensor based on a Prussian-blue transducer, coated with *AOx*/chitosan reagent layer	THC and ethanol	SWV (THC) and chrono-amperometry (ethanol)	THC: 0.5 μM; alcohol: 0.2 mM	Simultaneous detection of THC and ethanol in undiluted saliva sample within 3 min	[86]
Finger devices printed on the robotic glove	Robotic assisted automated taste sweetness, sourness, and spiciness discrimination in food samples	Glucose, ascorbic acid, and capsaicin.	Chrono-amperometry (ethanol and glucose) SWV (vitamins)	—	Ascorbic acid in orange juice, cola, lemon juice, sports drink, and pineapple juice; Glucose in: apple cider, sugar-free sports drink, cola, sugar-free energy drink, and apple juice; Capsaicin in: green chili extract, coffee, red paprika extract, watermelon juice and red pepper extract	[72]
Flexible printable tattoo electrodes	Flexible AAOx enzymatic biosensing tattoo patch fabricated on a polyurethane substrate and combined with a localized iontophoretic sweat stimulation system	Ascorbic acid	Chronoamperometry (Oxygen cosustrate depletion)	—	Sweat from subjects taking varying amounts of commercial vitamin C pills or vitamin C-rich beverages	[89]

AAOx: ascorbate oxidase; AOx: alcohol oxidase; GOx: glucose oxidase; DFP: diisopropyl fluorophosphate; ISF: skin interstitial fluid; LOx: lactate oxidase; MPOx: methyl paraoxon; MWCNTs: multi-walled carbon nanotubes; OP: organophosphorus; OPH: organophosphorus hydrolase; PANi: polyaniline; PVA: polyvinyl alcohol; SWV: square wave voltammetry; TTF: tetrathiafulvalene; THC: D9-tetrahydrocannabinol; Tyr: tyrosinase.

Wang's group has reported very recently an epidermal AAOx biosensor able to monitor the dynamics of vitamin C in sweat after the intake of vitamin C pills and fruit juices [89]. This method combines the use of a flexible vitamin C tattoo patch fabricated on a polyurethane substrate with a localized iontophoretic sweat stimulation system. Chrono-amperometric cathodic detection of the oxygen cosubstrate consumption during the enzymatic reaction, demonstrates very interesting potential for personalized nutrition solutions.

4. General Considerations, Challenges to Face, and Future Prospects

The advances that have occurred in recent years in screen-printing technology provide unimaginable possibilities in electrochemical sensing and biosensing. Paper-based and wearable (bio)sensors are two of the areas that benefit from such advances.

e-PADs show excellent opportunities for sensing and biosensing mainly in the environmental and clinical fields. For these purposes, different types of paper impregnated with the suitable reagents, modified with different nanomaterials (metal nanoparticles, carbon nanostructures, and nano-sized MIPs) either in the absence or in the presence of specific bioreceptors (enzymes, lectins, antibodies, and, to a much lesser extent, nucleic acids) have been proposed. The resulting devices exploit the interesting features offered by nanomaterials in terms of electrocatalytic properties, biocompatibility and high surface area, and the strong selectivity of biological molecules or MIPs, with the advantages of electrochemical detection. In the environmental field, e-PADs have been applied to the determination of heavy metals (Cu^{2+}, Pb^{2+}, Cd^{2+}), anions (ClO_3^-, NO_2^-), gases (NO, NO_2, DMMP), CWAs (mustard agent) microbial pathogens (*E. coli*), or other biotoxics (antibiotics or pesticides).

In the clinical field, e-PADs have been utilized for the determination of a wide variety of biomolecules including miRNAs, hormones (17β-estradiol, FSH), viruses (HPV, HIV), proteomic biomarkers of relevance in cancer and cardiovascular diseases (CRP, BNP, IFN-γ, CA125, cTnI, AFP, BChE activity, MMP9, and CEA), and other clinically relevant analytes (3-NT, glutathione, glucose, H_2O_2, serotonin, L-Tyr, Phe, acetylcholine and Cl^-). Paper-based electrochemical (bio)sensors have been employed to determine target analytes in highly variable matrices—soils, exhaust gases, waters and industrial sewage, cellular extracted DNA, blood, plasma, serum, urine, sweat, exhaled breath, and pharmaceutical capsules. Remarkable achievements include the development of disposable gPADs and the combination of SPEs with simple paper-based microfluidics (μ-PEI) with great interest in clinical diagnostics for prototyping new POCTs.

Apart from the interesting features derived from the use of paper as substrate such as porosity, capillarity, high surface area/volume ratio, disposability, lightness, flexibility, eco-friendliness, and low-cost, the filtration properties of this particular substrate have also been exploited. Moreover, some of these (bio)devices exhibit antifouling properties that are highly pursued to ensure the proper functioning of the devices in real world matrices and their use to determine low levels of analytes directly in such matrices involving simple and straightforward protocols.

Focusing on printed wearable devices devoted to electrochemical biosensing, new generation of printed wearable devices include soft, biocompatible, stretchable, and anatomically compliant devices with multifunctional characteristics that enable efficient bio-integration and withstand high tensile stress associated with on-body applications. Over the past two years, their applications have been geared mostly toward healthcare and environmental fields. To date, they have demonstrated preliminary potential for individual or multiplexed electrochemical determination in or near real time of relevant analytes in the agro-food, clinical and environmental areas (ethanol, drugs of abuse, lactate, glucose, vitamins, TYR, OP nerve agents) in different matrixes such as biofluids (sweat, saliva, tears, and ISF), vapor phases, and skin. So far, printed temporary tattoo and bandage sensors and printed sensors mounted in mouthguards, eyeglasses, or rings have been proposed for healthcare applications. Regarding environmental applications sensors have been printed on textile and gloves. Particularly noteworthy is the use of this type of sensors to screen biomarkers related with cancer in skin, the pioneering coupling with a parallel iontophoretic mechanism (extraction and delivery) for simultaneous sampling of different biofluids at separate locations, the enclosing within a microfluidic chamber for continuous monitoring and the development of a wearable taste-sensing robotic technology to discriminate between different flavors in liquid and solid food samples and an epidermal enzymatic biosensor for noninvasive nutrition status assessments.

However, it should be noted that, at present, the remarkable capabilities for taking preventive intervention of health and environmental risks, have been proved using proof-of-concept prototypes and for a limited number of samples, analytes, and biosensing approaches. Therefore, a thorough validation with large population studies and coordination and collaboration with medical practitioners to correctly interpret the data, and a better understanding of the correlations between analyte concentrations in the blood and noninvasive biofluids are required to underpin clinical acceptance. In addition, their extension to bioaffinity assays, particularly challenging since they require complex, multistep and usually not reversible protocols, is highly demanded to make other important biomarkers (proteins, DNAs, and RNAs) accessible to monitoring. Indeed, despite the great customization potential of these printed wearable electrochemical devices, important efforts are required for advancing them from prototypes to field devices and for their widespread commercial exploitation. Additional efforts are required to improve durability and robustness of wearables' batteries. Solutions are required to decrease the power consumption of devices for extended monitoring periods, which include battery consumption minimization, and developing of replaceable and flexible power source with continuous and long-time output or self-powering wearable devices.

It is evident that the development of wearable printed electrochemical devices, which so far has only scratched the surface of their tremendous potential, is poised to grow very rapidly over the next decade, bringing a considerable advance to the field of wearable devices. Although many challenges that impede the widespread adaptation of this field for commercial applications have been addressed due to the recent years endeavors in material science, microfluidics, nanotechnology, and biotechnology, together with the work in unison of researchers from diverse fields, there are still some outstanding issues before their full potential will be realized and exploited in our real lives. However, there is clearly room for them within many applications (forensic, food assessment, healthcare, security), where rapid screening and timely chemical information is critical.

The exciting new developments anticipated to come in the foreseeable future in both paper-based and wearable electrochemical (bio)sensors will certainly change and improve our daily lives providing eco-friendly, affordable and efficient solutions for smart healthcare (preventive medicine, precision medication and management of chronic diseases) and wellness moving the lab to our body (skin, mouth, and eyes), minimizing risks of exposure to chemical threats and drug impaired driving concerns. Taking into account the demands for ordinary users facing other issues such as the diversity of analytical targets in the practical applications and the selectivity, stability and recyclability of these screen-printing biodevices, additional future research can be predicted in this field. Moreover, continuous investment in material preparation and fabrication process perfection (including ingenious structural designs) and in achieving higher integration between the multifunction sensing units and auxiliary components (power supply, communication and even signal processing and displaying) will play a significant role in constructing cost-effective and consistent (bio)sensors, easily adopted by current society.

Indeed, today, it is timely to stress that the adaptation of mask filtration systems to include transducers for aerosol/viral detection, although very difficult, would represent a disruptive technology for the detection of pandemics such as the SARS-CoV-2 coronavirus we are experiencing.

Author Contributions: Writing—review and editing, funding acquisition: S.C., P.Y.-S., and J.M.P. All authors have read and agreed to the published version of the manuscript.

Funding: This research was funded by Spanish Ministerio de Ciencia, Innovación y Universidades, research project RTI2018-096135-B-I00; Spanish Ministerio de Ciencia e Innovación, research project PID2019-103899RB-I00; Comunidad de Madrid TRANSNANOAVANSENS-CM Program, Grant S2018/NMT-4349.

Acknowledgments: The financial support of the RTI2018-096135-B-I00 (Spanish Ministerio de Ciencia, Innovación y Universidades) and PID2019-103899RB-I00 (Spanish Ministerio de Ciencia e Innovación) Research Projects and the TRANSNANOAVANSENS-CM Program from the Comunidad de Madrid (Grant S2018/NMT-4349) are gratefully acknowledged.

Conflicts of Interest: The authors declare no conflict of interest.

References

1. Lamas-Ardisana, P.J.; Casuso, P.; Fernandez-Gauna, I.; Martínez-Paredes, G.; Jubete, E.; Añorga, L.; Cabañero, G.; Grande, H.J. Disposable electrochemical paper-based devices fully fabricated by screen-printing technique. *Electrochem. Commun.* **2020**, *75*, 25–28. [CrossRef]
2. Dungchai, W.; Chailapakul, O.; Henry, C.S. Electrochemical detection for paper-based microfluidics. *Anal. Chem.* **2009**, *81*, 5821–5826. [CrossRef]
3. Dungchai, W.; Chailapakul, O.; Henry, C.S. A low-cost, simple, and rapid fabrication method for paper-based microfluidics using wax screen-printing. *Analyst* **2011**, *136*, 77–82. [CrossRef]
4. Sameenoi, Y.; Nongkai, P.N.; Nouanthavong, S.; Henry, C.S.; Nacapricha, D. One-step polymer screen-printing for microfluidic paper-based analytical device (μPAD) fabrication. *Analyst* **2014**, *139*, 6580–6588. [CrossRef] [PubMed]
5. Cinti, S.; Mazzaracchio, V.; Cacciotti, I.; Moscone, D.; Arduini, F. Carbon black-modified electrodes screen-printed onto paper towel, waxed paper, and parafilm M®. *Sensors* **2017**, *17*, 2267. [CrossRef] [PubMed]
6. Yong, H.; Yan, W.; Jian-Zhong, F.; Wen-Bin, W. Fabrication of paper-based microfluidic analysis devices: A review. *RSC Adv.* **2015**, *5*, 78109–78127.
7. Metters, P.; Houssein, S.M.; Kampouris, D.K.; Banks, C.E. Paper-based electroanalytical sensing platforms. *Anal. Methods* **2013**, *5*, 103–110. [CrossRef]
8. Noviana, E.; McCord, C.P.; Clark, K.M.; Jang, I.; Henry, C.S. Elctrochemical paper-based devices: Sensing approaches and progress toward practical applications. *Lab Chip* **2020**, *20*, 9–34. [CrossRef]
9. Akyazi, T.; Basabe-Desmonts, L.; Benito-Lopez, F. Review on microfluidic paper-based analytical devices towards commercialisation. *Anal. Chim. Acta* **2018**, *1001*, 1–17. [CrossRef]
10. Shriver-Lake, L.C.; Zabetakis, D.; Dressick, W.J.; Stengerand, D.A.; Trammell, S.A. Paper-based electrochemical detection of chlorate. *Sensors* **2018**, *18*, 328. [CrossRef]
11. Colozza, N.; Kehe, K.; Dionisi, G.; Popp, T.; Tsoutsoulopoulos, A.; Steinritz, D.; Moscone, D.; Arduini, F. A wearable origami-like paper-based electrochemical biosensor for sulfur mustard detection. *Biosens. Bioelectron.* **2019**, *129*, 15–23. [CrossRef] [PubMed]
12. Figueredo, F.; González-Pabón, M.J.; Cortón, E. Low cost layer by layer construction of CNT/chitosan flexible paper-based electrodes: A versatile electrochemical platform for point of care and point of need testing. *Electroanalysis* **2018**, *30*, 497–508. [CrossRef]
13. Wang, P.; Wang, M.; Zhou, F.; Yang, G.; Qu, L.; Miao, X. Development of a paper-based, inexpensive, and disposable electrochemical sensing platform for nitrite detection. *Electrochem. Commun.* **2017**, *81*, 74–78. [CrossRef]
14. Pungjunun, K.; Chaiyo, S.; Praphairaksit, N.; Siangproh, W.; Ortner, A.; Kalcher, K.; Chailapakul, O.; Mehmeti, E. Electrochemical detection of NOx gas based on disposable paper-based analytical device using a copper nanoparticles-modified screen-printed graphene electrode. *Biosens. Bioelectron.* **2019**, *143*, 111606. [CrossRef] [PubMed]
15. Yu, H.; Han, H.; Jang, J.; Cho, S. Fabrication and optimization of conductive paper based on screen printed polyaniline/graphene patterns for nerve agent detection. *ACS Omega* **2019**, *4*, 5586–5594. [CrossRef]
16. Rengaraj, S.; Cruz-Izquierdo, Á.; Scott, J.L.; Di Lorenzo, M. Impedimetric paper-based biosensor for the detection of bacterial contamination in water. *Sens. Actuators B* **2018**, *265*, 50–58. [CrossRef]
17. Channon, R.B.; Yang, Y.; Feibelman, K.M.; Geiss, B.J.; Dandy, D.S.; Henry, C.S. Development of an electrochemical paper-based analytical device for trace detection of virus particles. *Anal. Chem.* **2018**, *90*, 7777–7783. [CrossRef]

18. Alatraktchi, F.A.; Noori, J.S.; Tanev, G.P.; Mortensen, J.; Dimaki, M.; Johansen, H.K.; Madsen, J.; Molin, S.; Svendsen, W.E. Paper-based sensors for rapid detection of virulence factor produced by *Pseudomonas aeruginosa*. *PLoS ONE* **2018**, *13*, e0194157. [CrossRef]
19. Adkins, J.A.; Boehle, K.; Friend, C.; Chamberlain, B.; Bisha, B.; Henry, C.S. Colorimetric and electrochemical bacteria detection using printed paper- and transparency-based analytic devices. *Anal. Chem.* **2017**, *89*, 3613–3621. [CrossRef]
20. Zhang, J.; Yang, Z.; Liu, Q.; Liang, H. Electrochemical biotoxicity detection on a microfluidic paper-based analytical device via cellular respiratory inhibition. *Talanta* **2019**, *202*, 384–391. [CrossRef]
21. Arduini, F.; Cinti, S.; Caratelli, V.; Amendola, L.; Palleschi, G.; Moscone, D. Origami multiple paper-based electrochemical biosensors for pesticide detection. *Biosens. Bioelectron.* **2019**, *126*, 346–354. [CrossRef] [PubMed]
22. Wang, H.; Wang, J.; Liu, G.; Zhang, Z.; Hou, X. Electrochemical sensing of Pb(II) and Cd(II) in decorative material of wood panel using nano-cellulose paper-based electrode modified using graphene/multi-walled carbon nanotubes/bismuth film. *Int. J. Electrochem. Sci.* **2019**, *14*, 11253–11266. [CrossRef]
23. Cinti, S.; Minotti, C.; Moscone, D.; Palleschi, G.; Arduini, F. Fully integrated ready-to-use paper-based electrochemical biosensor to detect nerve agents. *Biosens. Bioelectron.* **2017**, *93*, 46–51. [CrossRef] [PubMed]
24. Sánchez-Calvo, A.; Fernández-Abedul, M.T.; Blanco-López, M.C.; Costa-García, A. Paper-based electrochemical transducer modified with nanomaterials for mercury determination in environmental waters. *Sens. Actuators B* **2019**, *290*, 87–92. [CrossRef]
25. Scala-Benuzzi, M.L.; Raba, J.; Soler-Illia, G.J.A.A.; Schneider, R.J.; Messina, G.A. Novel electrochemical paper-based immunocapture assay for the quantitative determination of ethinyl estradiol in water samples. *Anal. Chem.* **2018**, *90*, 77–82. [CrossRef] [PubMed]
26. Jemmeli, D.; Marcoccio, E.; Moscone, D.; Dridi, C.; Arduini, F. Highly sensitive paper-based electrochemical sensor for reagent free detection of bisphenol A. *Talanta* **2020**, *216*, 120924. [CrossRef]
27. Martins, G.V.; Marques, A.C.; Fortunato, E.; Sales, M.G.F. Wax-printed paper-based device for direct electrochemical detection of 3-nitrotyrosine. *Electrochim. Acta* **2018**, *284*, 60–68. [CrossRef]
28. Tomei, M.R.; Cinti, S.; Interino, N.; Manovella, V.; Moscone, D.; Arduini, F. Paper-based electroanalytical strip for user-friendly blood glutathione detection. *Sens. Actuators B* **2019**, *294*, 291–297. [CrossRef]
29. Maier, D.; Laubender, E.; Basavanna, A.; Schumann, S.; Güder, F.; Urban, G.A.; Dincer, C. Toward continuous monitoring of breath biochemistry: A paper based wearable sensor for real-time hydrogen peroxide measurement in simulated breath. *ACS Sens.* **2019**, *4*, 2945–2951. [CrossRef]
30. Amatatongchai, M.; Sitanurak, J.; Sroysee, W.; Sodanat, S.; Chairam, S.; Jarujamrus, P.; Nacapricha, D.; Lieberzeit, P.A. Highly sensitive and selective electrochemical paper-based device using a graphite screen-printed electrode modified with molecularly imprinted polymers coated Fe_3O_4@Au@SiO_2 for serotonin determination. *Anal. Chim. Acta* **2019**, *1077*, 255–265. [CrossRef]
31. Pinyorospathum, C.; Chaiyo, S.; Sae-Ung, P.; Hoven, V.P.; Damsongsang, P.; Siangproh, W.; Chailapakul, O. Disposable paper-based electrochemical sensor using thiol-terminated poly(2-methacryloyloxyethyl phosphorylcholine) for the label-free detection of C-reactive protein. *Microchim. Acta* **2019**, *186*, 472. [CrossRef] [PubMed]
32. Thompson, D.; Pepys, M.M.; Wood, S.P. The physiological structure of human C-reactive protein and its complex with phosphocholine. *Structure* **1999**, *7*, 169–177. [CrossRef]
33. Moreira, C.M.; Pereira, S.V.; Raba, J.; Bertolino, F.A.; Messina, G.A. Paper-based enzymatic platform coupled to screen printed graphene modified electrode for the fast neonatal screening of phenylketonuria. *Clin. Chim. Acta* **2018**, *486*, 59–65. [CrossRef] [PubMed]
34. Panraksa, Y.; Siangproh, W.; Khampieng, T.; Chailapakul, O.; Apilux, A. Paper-based amperometric sensor for determination of acetylcholinesterase using screen-printed graphene electrode. *Talanta* **2018**, *178*, 1017–1023. [CrossRef] [PubMed]
35. Cao, L.; Fang, C.; Zeng, R.; Zhao, X.; Zhao, F.; Jiang, Y.; Chen, Z. A disposable paper-based microfluidic immunosensor based on reduced graphene oxide-tetraethylene pentamine/Au nanocomposite decorated carbon screen-printed electrodes. *Sens. Actuators B* **2017**, *252*, 44–54. [CrossRef]
36. Bird, T.G.; Dimitropoulou, P.; Turner, R.M.; Jenks, S.J.; Cusack, P.; Hey, S.; Blunsum, A.; Kelly, S.; Sturgeon, C.; Hayes, P.C.; et al. Alpha-fetoprotein detection of hepatocellular carcinoma leads to a standardized analysis of dynamic AFP to improve screening based detection. *PLoS ONE* **2016**, *11*, e0156801. [CrossRef]

37. Moazeni, M.; Karimzadeh, F.; Kermanpura, A. Peptide modified paper based impedimetric immunoassay with nanocomposite electrodes as a point-of-care testing of Alpha-fetoprotein in human serum. *Biosens. Bioelectron.* **2018**, *117*, 748–757. [CrossRef] [PubMed]
38. Liu, X.; Li, X.; Gao, X.; Ge, L.; Sun, X.; Li, F. A universal paper-based electrochemical sensor for zero-background assay of diverse biomarkers. *ACS Appl. Mater. Interfaces* **2019**, *11*, 15381–15388. [CrossRef]
39. Rahimi-Mohseni, M.; Raoof, J.B.; Ojani, R.; Aghajanzadeh, T.A.; Hashkavayi, A.B. Development of a new paper based nano-biosensor using the co-catalytic effect of tyrosinase from banana peel tissue (*Musa Cavendish*) and functionalized silica nanoparticles for voltammetric determination of L-tyrosine. *Int. J. Biol. Macromol.* **2018**, *113*, 648–654. [CrossRef] [PubMed]
40. Wang, Y.; Luo, J.; Liu, J.; Li, X.; Kong, Z.; Jin, H.; Cai, X. Electrochemical integrated paper-based immunosensor modified with multiwalled carbon nanotubes nanocomposites for point-of-care testing of 17β-estradiol. *Biosens. Bioelectron.* **2018**, *107*, 47–53. [CrossRef]
41. Zheng, Y.; Huang, Z.; Zhang, J. Paper-based microfluidic immunoassay for electrochemical detection of B-type natriuretic peptide. *Int. J. Electrochem. Sci.* **2018**, *13*, 7246–7254. [CrossRef]
42. Luo, J.; Kong, Z.; Wang, Y.; Xie, J.; Liu, J.; Jin, H.; Cai, X. Label-free paper-based immunosensor with graphene nanocomposites for electrochemical detection of follicle-stimulating hormone. *Conf. Proc. IEEE. Eng. Med. Biol. Soc.* **2018**, *2018*, 2901–2904. [CrossRef]
43. Boonkaew, S.; Chaiyo, S.; Jampasa, S.; Rengpipat, S.; Siangproh, W.; Chailapakul, O. An origami paper-based electrochemical immunoassay for the C-reactive protein using a screen-printed carbon electrode modified with graphene and gold nanoparticles. *Microchim. Acta* **2019**, *186*, 153. [CrossRef]
44. Pavithra, M.; Muruganand, S.; Parthiban, C. Development of novel paper based electrochemical immunosensor with self-made gold nanoparticle ink and quinone derivate for highly sensitive carcinoembryonic antigen. *Sens. Actuators B* **2018**, *257*, 496–503. [CrossRef]
45. Ruecha, N.; Shin, K.; Chailapakul, O.; Rodthongkum, N. Label-free paper-based electrochemical impedance immunosensor for human interferon gamma detection. *Sens. Actuators B* **2019**, *279*, 298–304. [CrossRef]
46. Fan, Y.; Shi, S.; Ma, J.; Guo, Y. A paper-based electrochemical immunosensor with reduced graphene oxide/thionine/gold nanoparticles nanocomposites modification for the detection of cancer antigen 125. *Biosens. Bioelectron.* **2019**, *135*, 1–7. [CrossRef] [PubMed]
47. Vasantham, S.; Alhans, R.; Singhal, C.; Nagabooshanam, S.; Nissar, S.; Basu, T.; Ray, S.C.; Wadhwa, S.; Narang, J.; Mathur, A. Paper based point of care immunosensor for the impedimetric detection of cardiac troponin I biomarker. *Biomed. Microdevices* **2020**, *22*, 6. [CrossRef] [PubMed]
48. Sánchez-Tirado, E.; González-Cortés, A.; Yáñez-Sedeño, P.; Pingarrón, J.M. Electrochemical immunosensor for the determination of the cytokine interferon gamma (IFN-γ) in saliva. *Talanta* **2020**, *211*, 120761. [CrossRef]
49. Cinti, S.; Fiore, L.; Massoud, R.; Cortese, C.; Moscone, D.; Palleschi, G.; Arduini, F. Low-cost and reagent-free paper-based device to detect chloride ions in serum and sweat. *Talanta* **2018**, *179*, 186–192. [CrossRef]
50. Chaiyo, S.; Mehmeti, E.; Siangproh, W.; Long Hoang, T.; Phong Nguyen, H.; Chailapakula, O.; Kalcher, K. Non-enzymatic electrochemical detection of glucose with a disposable paper-based sensor using a cobalt phthalocyanine–ionic liquid–graphene composite. *Biosens. Bioelectron.* **2018**, *102*, 113–120. [CrossRef]
51. Núnez-Bajo, E.; Blanco-López, M.C.; Costa-García, A.; Fernández-Abedul, M.T. In situ gold-nanoparticle electrogeneration on gold films deposited on paper for non-enzymatic electrochemical determination of glucose. *Talanta* **2018**, *178*, 160–165. [CrossRef] [PubMed]
52. Scordo, G.; Moscone, D.; Palleschi, G.; Arduini, F. A reagent-free paper-based sensor embedded in a 3D printing device for cholinesterase activity measurement in serum. *Sens. Actuators B* **2018**, *258*, 1015–1021. [CrossRef]
53. Ruiz-Vega, G.; Garcia-Berrocoso, T.; Montaner, J.; Baldrich, E. Paper microfluidics on screen-printed electrodes for simple electrochemical magneto-immunosensor performance. *Sens. Actuators B* **2019**, *298*, 126897. [CrossRef]
54. Teengam, P.; Siangproh, W.; Tuantranont, A.; Henry, C.S.; Vilaivan, T.; Chailapakul, O. Electrochemical paper-based peptide nucleic acid biosensor for detecting human papillomavirus. *Anal. Chim. Acta* **2017**, *952*, 32–40. [CrossRef] [PubMed]
55. Cinti, S.; Proietti, E.; Casotto, F.; Moscone, D.; Arduini, F. Paper-based strips for the electrochemical detection of single and double stranded DNA. *Anal. Chem.* **2018**, *90*, 13680–13686. [CrossRef] [PubMed]

56. Boobphahom, S.; Ruecha, N.; Rodthongkum, N.; Chailapakul, O.; Remcho, V.T. A copper oxide-ionic liquid/reduced graphene oxide composite sensor enabled by digital dispensing: Non-enzymatic paper-based microfluidic determination of creatinine in human blood serum. *Anal. Chim. Acta* **2019**, *1083*, 110–118. [CrossRef] [PubMed]
57. Cao, L.; Han, G.C.; Xiao, H.; Chen, Z.; Fang, C. A novel 3D paper-based microfluidic electrochemical glucose biosensor based on rGO-TEPA/PB sensitive film. *Anal. Chim. Acta* **2020**, *1096*, 34–43. [CrossRef]
58. Income, K.; Ratnarathorn, N.; Khamchaiyo, N.; Srisuvo, C.; Ruckthong, L.; Dungchai, W. Disposable nonenzymatic uric acid and creatinine sensors using μPAD coupled with screen-printed reduced graphene oxide-gold nanocomposites. *Int. J. Anal. Chem.* **2019**, 3457247. [CrossRef]
59. Wang, Y.; Luo, J.; Liu, J.; Sun, S.; Xiong, Y.; Ma, Y.; Yan, S.; Yang, Y.; Yin, H.; Cai, X. Label-free microfluidic paper-based electrochemical aptasensor for ultrasensitive and simultaneous multiplexed detection of cancer biomarkers. *Biosens. Bioelectron.* **2019**, *136*, 84–90. [CrossRef]
60. Cao, Q.; Liang, B.; Tu, T.; Wei, J.; Fang, L.; Ye, X. Three-dimensional paper-based microfluidic electrochemical integrated devices (3D-PMED) for wearable electrochemical glucose detection. *RSC Adv.* **2019**, *9*, 5674–5681. [CrossRef]
61. Bandodkar, A.J.; Jeerapan, I.; Wang, J. Wearable chemical sensors: Present challenges and future prospects. *ACS Sens.* **2016**, *1*, 464–482. [CrossRef]
62. Wang, S.; Chinnasamy, T.; Lifson, M.A.; Inci, F.; Demirci, U. Flexible substrate-based devices for point-of-care diagnostics. *Trends Biotechnol.* **2016**, *34*, 909–921. [CrossRef]
63. Kim, J.; Kumar, R.; Bandodkar, A.J.; Wang, J. Advanced materials for printed wearable electrochemical devices: A review. *Adv. Electron. Mater.* **2017**, *3*, 1600260. [CrossRef]
64. Kim, J.; Jeerapan, I.; Sempionatto, J.R.; Barfidokht, A.; Mishra, R.K.; Campbell, A.S.; Hubble, L.J.; Wang, J. Wearable Bioelectronics: Enzyme-Based Body-Worn Electronic Devices. *Acc. Chem. Res.* **2018**, *51*, 2820–2828. [CrossRef] [PubMed]
65. Hubble, L.J.; Wang, J. Sensing at Your Fingertips: Glove-based wearable chemical sensors. *Electroanalysis* **2019**, *31*, 428–436. [CrossRef]
66. Khan, S.; Ali, S.; Bermak, A. Recent developments in printing flexible and wearable sensing electronics for healthcare applications. *Sensors* **2019**, *19*, 1230. [CrossRef] [PubMed]
67. Tu, J.; Torrente-Rodríguez, R.M.; Wang, M.; Gao, W. The era of digital health: A review of portable and wearable affinity biosensors. *Adv. Funct. Mater.* **2019**, 1906713. [CrossRef]
68. Kim, J.; Campbell, A.S.; Esteban-Fernández de Ávila, B.; Wang, J. Wearable biosensors for healthcare monitoring. *Nat. Biotechnol.* **2019**, *37*, 389–406. [CrossRef]
69. Sempionatto, J.R.; Jeerapan, I.; Krishnan, S.; Wang, J. Wearable chemical sensors: Emerging systems for on-body analytical chemistry. *Anal. Chem.* **2020**, *92*, 378–396. [CrossRef]
70. Kim, J.; Campbell, A.S.; Wang, J. Wearable non-invasive epidermal glucose sensors: A review. *Talanta* **2018**, *177*, 163–170. [CrossRef]
71. Campbell, A.S.; Kim, J.; Wang, J. Wearable electrochemical alcohol biosensors. *Curr. Opin. Electrochem.* **2018**, *10*, 126–135. [CrossRef] [PubMed]
72. Ciui, B.; Martin, A.; Mishra, R.K.; Nakagawa, T.; Dawkins, T.J.; Lyu, M.; Cristea, C.; Sandulescu, R.; Wang, J. Chemical sensing at the robot fingertips: Toward automated taste discrimination in food samples. *ACS Sens.* **2018**, *3*, 2375–2384. [CrossRef] [PubMed]
73. Zhao, S.; Li, J.; Cao, D.; Zhang, G.; Li, J.; Li, K.; Yang, Y.; Wang, W.; Jin, Y.; Sun, R.; et al. Recent advancements in flexible and stretchable electrodes for electromechanical sensors: Strategies, materials, and features. *ACS Appl. Mater. Interfaces* **2017**, *9*, 12147–12164. [CrossRef] [PubMed]
74. Heikenfeld, J.; Jajack, A.; Rogers, J.; Gutruf, P.; Tian, L.; Pan, T.; Li, R.; Khine, M.; Kim, J.; Wang, J.; et al. Wearable sensors: Modalities, challenges, and prospects. *Lab Chip* **2018**, *18*, 217. [CrossRef] [PubMed]
75. Gao, W.; Ota, H.; Kiriya, D.; Takei, K.; Javey, A. Flexible electronics toward wearable sensing. *Acc. Chem. Res.* **2019**, *52*, 523–533. [CrossRef]
76. Song, Y.; Min, J.; Gao, W. Wearable and implantable electronics: Moving toward precision therapy. *ACS Nano* **2019**, *13*, 12280–12286. [CrossRef]
77. Stoppa, M.; Chiolerio, A. Wearable electronics and smart textiles: A critical review. *Sensors* **2014**, *14*, 11957–11992. [CrossRef]

78. Tang, Y.; Li, X.; Lv, H.; Wang, W.; Zhi, C.; Li, H. Integration designs toward new-generation wearable energy supply-sensor systems for real-time health monitoring: A minireview. *InfoMat* **2020**, in press. [CrossRef]
79. Liu, Y.; Wang, H.; Zhao, W.; Zhang, M.; Qin, H.; Xie, Y. Screen-printed electrodes: Promising paper and wearable transducers for (bio)sensing. *Sensors* **2018**, *18*, 645. [CrossRef]
80. Guk, K.; Han, G.; Lim, J.; Jeong, K.; Kang, T.; Lim, E.-K.; Jung, J. Evolution of wearable devices with real-time disease monitoring for personalized healthcare. *Nanomaterials* **2019**, *9*, 813. [CrossRef]
81. Ciui, B.; Martin, A.; Mishra, R.K.; Brunetti, B.; Nakagawa, T.; Dawkins, T.J.; Lyu, M.; Cristea, C.; Sandulescu, R.; Wang, J. Wearable wireless tyrosinase bandage and microneedle sensors: Toward melanoma screening. *Adv. Healthcare Mater.* **2018**, *7*, 1701264. [CrossRef] [PubMed]
82. Mishra, R.K.; Martín, A.; Nakagawa, T.; Barfidokht, A.; Lu, X.; Sempionatto, J.R.; Lyu, K.M.; Karajic, A.; Musameh, M.M.; Kyratzis, I.L.; et al. Detection of vapor-phase organophosphate threats using wearable conformable integrated epidermal and textile wireless biosensor systems. *Biosens. Bioelectron.* **2018**, *101*, 227–234. [CrossRef] [PubMed]
83. Kim, J.; Sempionatto, J.R.; Imani, S.; Hartel, M.C.; Barfidokht, A.; Tang, G.; Campbell, A.S.; Mercier, P.P.; Wang, J. Simultaneous monitoring of sweat and interstitial fluid using a single wearable biosensor platform. *Adv. Sci.* **2018**, *5*, 1800880. [CrossRef] [PubMed]
84. Mannoor, M.S.; Tao, H.; Clayton, J.D.; Sengupta, A.; Kaplan, D.L.; Naik, R.R.; Verma, N.; Omenetto, F.G.; McAlpine, M.C. Graphene-based wireless bacteria detection on tooth enamel. *Nat. Commun.* **2012**, *3*, 763. [CrossRef]
85. Sempionatto, J.R.; Brazaca, L.C.; García-Carmona, L.; Bolat, G.; Campbell, A.S.; Martin, A.; Tang, G.; Shah, R.; Mishra, R.K.; Kim, J.; et al. Eyeglasses-based tear biosensing system: Non-invasive detection of alcohol, vitamins and glucose. *Biosens. Bioelectron.* **2019**, *137*, 161–170. [CrossRef]
86. Mishra, R.K.; Sempionatto, J.R.; Li, Z.; Brown, C.; Galdino, N.M.; Shah, R.; Liu, S.; Hubble, L.J.; Bagot, K.; Tapert, S.; et al. Simultaneous detection of salivary Δ9-tetrahydrocannabinol and alcohol using a wearable electrochemical ring sensor. *Talanta* **2020**, *211*, 120757. [CrossRef] [PubMed]
87. Payne, M.E.; Zamarayeva, A.; Pister, V.I.; Yamamoto, N.A.D.; Arias, A.C. Printed, flexible lactate sensors: Design considerations before performing on-body measurements. *Sci. Rep.* **2019**, *9*, 13720. [CrossRef] [PubMed]
88. Bandodkar, A.J.; O'Mahony, A.M.; Ramírez, J.; Samek, I.A.; Anderson, S.M.; Windmiller, J.R.; Wang, J. Solid-state Forensic Finger sensor for integrated sampling and detection of gunshot residue and explosives: Towards 'Lab-on-a-finger'. *Analyst* **2013**, *138*, 5288. [CrossRef] [PubMed]
89. Sempionatto, J.R.; Khorshed, A.A.; Ahmed, A.; De Loyola e Silva, A.N.; Barfidokht, A.; Yin, L.; Goud, K.Y.; Mohamed, M.A.; Bailey, E.; May, J.; et al. Epidermal enzymatic biosensors for sweat vitamin C: Toward personalized nutrition. *ACS Sens.* **2020**, *5*, 1804–1813. [CrossRef]
90. Mishra, R.K.; Barfidokht, A.; Karajic, A.; Sempionatto, J.R.; Wang, J.; Wang, J. Wearable potentiometric tattoo biosensor for on-body detection of G-type nerve agents simulants. *Sens. Actuator B Chem.* **2018**, *273*, 966–972. [CrossRef]

biosensors

MDPI

Review

Paper-Based Screen-Printed Electrodes: A New Generation of Low-Cost Electroanalytical Platforms †

Estefanía Costa-Rama and María Teresa Fernández-Abedul *

Department of Physical and Analytical Chemistry, University of Oviedo, Av. Julián Clavería 8, 33006 Oviedo, Spain; costaestefania@uniovi.es

* Correspondence: mtfernandeza@uniovi.es; Tel.: +34-985-102-968

† In memoriam of Prof. Agustín Costa García who liked to walk on the surface of the electrodes to observe the electron transfer.

Abstract: Screen-printed technology has helped considerably to the development of portable electrochemical sensors since it provides miniaturized but robust and user-friendly electrodes. Moreover, this technology allows to obtain very versatile transducers, not only regarding their design, but also their ease of modification. Therefore, in the last decades, the use of screen-printed electrodes (SPEs) has exponentially increased, with ceramic as the main substrate. However, with the growing interest in the use of cheap and widely available materials as the basis of analytical devices, paper or other low-cost flat materials have become common substrates for SPEs. Thus, in this revision, a comprehensive overview on paper-based SPEs used for analytical proposes is provided. A great variety of designs is reported, together with several examples to illustrate the main applications.

Keywords: microfluidics; electroanalysis; paper-based devices (μPADs); paper-based electroanalytical devices (ePADs); screen-printed electrodes

Citation: Costa-Rama, E.; Fernández-Abedul, M.T. Paper-Based Screen-Printed Electrodes: A New Generation of Low-Cost Electroanalytical Platforms . *Biosensors* **2021**, *11*, 51. https://doi.org/10.3390/bios11020051

Received: 16 December 2020
Accepted: 1 February 2021
Published: 16 February 2021

1. Introduction

In the last decades, the great advances in (micro)electronics, (nano)technology, and material science have led to an easy availability and management of increasing information. In turn, society requires real-time information to deliver an immediate feedback if necessary. In this knowledgeable society, it is obvious the interest in developing low-cost, miniaturized and easy-to-use analytical devices that provide on-site quantitative information in a fast and easy way. The high potential of these devices makes them useful in very assorted fields: from clinical and biomedical applications [1–5] to food analysis and quality control [6–8], as well as environmental monitoring [9–12].

As it is well-known, screen-printed electrodes (SPEs) have contributed enormously to the great development of electroanalytical devices. Conventional cells employed in amperometric/voltammetric measurements usually consisted of pen-like electrodes; namely, working (WE), reference (RE), and counter (CE) electrodes in a potentiostatic system of three electrodes. If only two electrodes were employed, apart from the WE, an auxiliary electrode acts as both, RE to apply a stable potential and CE to close the electrical circuit. Before the spread of solid electrodes, a mercury electrode delivered renewable drops from a glass capillary, following also a pen-like format. All the electrodes were introduced in a glass container of ca. 10 or 20 mL of volume (Figure 1A).

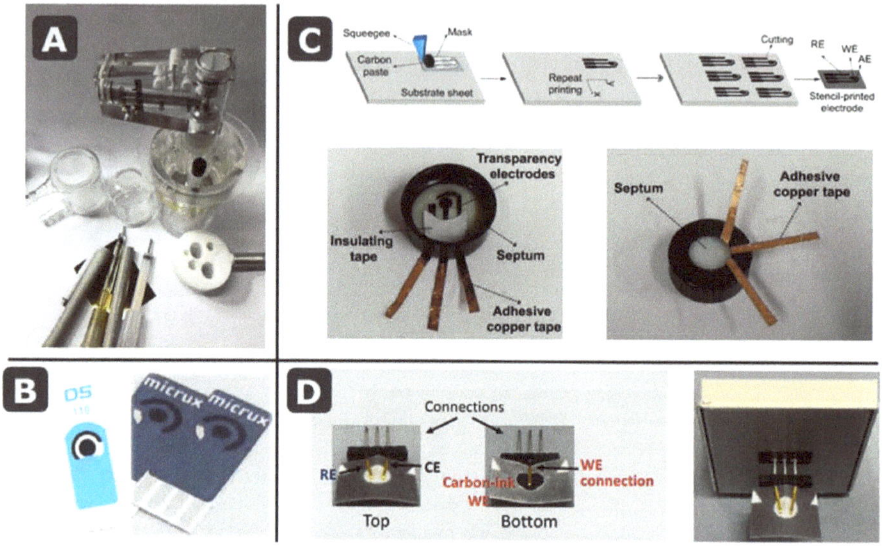

Figure 1. (**A**) Picture of conventional electrochemical cells with pen-like electrodes. (**B**) Examples of commercial screen-printed electrode cards from DropSens (ceramic substrate [13]) and MicruX (polymeric substrate [14]). (**C**) Schematic drawing of the stencil-printing process to fabricate several low-cost electrodes on a transparency sheet, with pictures showing the integration of the transparency electrode onto the cap of a sample vial for on-site water analysis. (**D**) Pictures of a paper-based electrochemical cell containing the three electrodes (top and bottom views), also inserted in a commercial interface that connects electrodes to the potentiostat. (**C**) and (**D**) are reprinted from [15] (Chapters 4 and 25), Copyright (2020), with permission from Elsevier.

This configuration, that resulted appropriate for centralized labs, was associated to an instrumentation (potentiostat) that was also a bench equipment, with clips and cables that usually connected to the top of each pen-like electrode. A stand with mechanized holes was used for adding the solutions and also for nitrogen purging in case deoxygenation was required. Different cell designs were available, even thermostatized. Mass transport was easily controlled by locating the cell over a magnetic stirrer and introducing a stir bar in the cell. Alternatively, rotating rods and electrodes were also applicable. In this way, measurements could be done under forced convection or pure diffusion regimes. Apart from this, flow cells were also accessible, with the WE located in a flow cell with wall-jet or thin-layer configurations as the most common, with RE/CE placed downstream, in a specific container leading to the waste.

The use of thick-film technology (producing layers of thickness in the micrometer scale) employed by electronic engineers (as e.g., screen-printing) using conductive inks allowed the development of flat platforms that included all the three electrodes of the electrochemical cell in the same plane (Figure 1B). This takes advantage of the interfacial nature of electrochemical measurements. Since only a surface able to transfer electrons in one way or another (oxidation or reduction) is required, the thickness is not an important variable. This simplifies enormously the design of the electrochemical cell and expands the possibilities. Then, the 3D-electrode/cell ensemble is converted into a 2D electrochemical cell, where the electrolyte can be deposited in such a way that drop analysis becomes possible. Although these flat cells could also be introduced in a conventional glass container, measurement-on-drop procedures have led the field. This change supposed a very important contribution to Green (Analytical) Chemistry (GAC) [16]. Regarding e.g., pollution detection, this was historically conducted with field sampling protocols that require extensive effort to be brought to a laboratory where extensive work-up generated large volumes of solvents and waste. Therefore, analysis of an environmental problem often contributed to other environmental problems. With the use of real-time in-field analysis, the necessary measurements can be taken without wasting time, material and energy. Actually, most of the 12 principles of GAC [17] are followed by miniaturized electroanalytical devices.

Then, screen-printing technology allows to easily obtain small-size flat electrochemical cells which are robust, cheap, and mass-produced. Alternatively, in many cases, a stencil is used instead of a screen, especially in handmade devices. Therefore, the term stencil-printed electrodes, closely related, is also found in the bibliography. The disposability of platforms avoids polishing/cleaning/activating treatments required by other solid electrodes [18] such as carbon paste [19], glassy carbon [20,21], or noble metal [22] electrodes. Moreover, (bio)assays could be performed in parallel. Although the most common commercial screen-printed electrode (SPE) card contains the three electrodes (WE, RE and CE) of the electrochemical cell, different configurations are possible due to the high versatility in design, one of the most important advantages of SPEs [2,6]. Currently, platforms with more than one electrochemical cell and cells with more than one WE can be found in the market [13,14,23]. Multianalyte determination (with e.g., spatial separation), as well as recording simultaneous measurements of both redox or non-redox active analytes, becomes available, and constitutes a field of enormous interest [24].

There is a great variety of inks that can be used for printing the electrochemical cells. Carbon continues to be the most common because of its good characteristics for electroanalytical applications, although metallic inks based on e.g., gold or platinum are also used [13,25,26]. (Bio/nano)modifiers can also be added, before or after the screen-printing process [2,27,28]. For the RE, silver-based inks are common, resulting in pseudo-reference (literally "false" reference) or quasi-reference ("almost" or "essentially" reference) electrodes [29]. The main difference between true reference and pseudo-reference electrodes is the lack of thermodynamic equilibrium in the latter case, since there is no common component in the two adjacent phases. Apart from their simplicity, and because they are immersed directly into the electrolyte used in the cell, the ohmic drop is small, no liquid junction appears and, usually, there is no contamination of the test solution by molecules/ions that a conventional reference electrode might transfer. Although there are also some drawbacks (e.g., they are not ideally nonpolarizable and work over a limited range of conditions such as pH or temperature), under selected conditions, the potential (although unknown) might be surprisingly constant during experiments. Direct exposure to the test environment could limit its applicability in complex sample matrices, but their short conditioning time allows performing measurements with insignificant potential drift. Moreover, since paper-based devices are often conceived for single-use measurements, the required operational stability of the RE is limited to one unique measurement and not to a large number, as in the case of conventional REs. However, where uninterrupted gathering of sensor data over longer periods of time is required, as in the case of inaccessible or remote locations where sensor replacement is difficult, potential stability has to be thoroughly studied [30].

The increasing use of multimodal detection systems that merge electrochemical and optical techniques such as spectroelectrochemistry, has led to the commercialization of electrochemical cells that combine screen-printed RE and CE with optically transparent WEs made of e.g., indium tin oxide, carbon (made of carbon nanomaterials) or gold (obtained by sputtering) [13]. Regarding the substrate, the electrochemical cell was usually printed on rigid materials such as ceramic [13]; however, nowadays the use of SPEs on flexible polymers are increasingly common, either commercial [14] or homemade [31] (Figure 1B,C, respectively).

Taking into account all the advantageous characteristics of the SPEs (mainly their disposability combined with precision), their success as transducers in electroanalytical devices is understandable. While the use of SPEs on ceramic and polymeric materials became widespread, in 2007, paper was demonstrated as innovative substrate for developing promising microfluidic analytical devices (μPADs) [32]. When compared with conventional microfluidic analytical devices, commonly based on glass or polymers, μPADs result simpler, cheaper, and then disposable. Moreover, paper is a lightweight and flexible material with the ability of transporting liquids without the need of external forces [33,34]. The first μPADs were colorimetric but this detection principle provided just qualitative or semiquantitative information and poor sensitivity [35,36]. However, due to the above mentioned advantages of screen-printing technology for the fabrication of electrodes, it is not surprising that the first microfluidic paper-based electroanalytical device (ePAD) was based on SPEs [37] (Figure 5A). After this pioneering design, electrochemical detection was continuously and easily integrated with paper-based devices because of its ability for miniaturization, low-cost and simplicity of the required instrumentation [38]. In the last years, the development of paper-based electroanalytical devices has experienced an enormous increase and, although other kind of electrodes (e.g., metallic film and wires [39–41]) have been also integrated in these devices, paper-based SPEs continue to be among the most reported. SPEs printed on or integrated with paper-based devices have been developed for the construction of different types of biosensors (e.g., enzymatic, immunosensors, DNA/aptasensors), with many different designs and for a wide variety of applications (clinical, food or environmental analysis) [5,11,12,34–36,38]. Alternatively, paper and ink can be combined by simple deposition of a conductive carbon dispersion (WE), very useful when both paper faces are employed in the design of the electrochemical cell (Figure 1D) [15,42].

Considering that the field of paper-based analytical devices is currently a very active research area and that SPEs are the most common transducers in biosensors and point-of-need devices, the purpose of this work is to review their integration to produce interesting miniaturized devices with different innovative designs and promising applications. The main approaches and trends in the development of these paper/SPE-based devices will be here discussed.

2. Paper as Substrate in (Electro)analytical Platforms

2.1. Paper as Material: Some Properties

A rapid development of calligraphy by archaic Chinese scholars, their spontaneous adoption of a camel-hair brush and fluid pigment, together with the urgent need of a writing substance cheaper and more practical than those already used, e.g., woven textile, inspired Ts'ai Lun in the year 105 to invent true paper [43]. It was defined as a thin, felted material formed on flat, porous molds from macerated vegetable fiber. This process separated each individual filament as a unit, and after adding water, fibers were filtered and dried leaving a sheet upon. Paper sheets were very advantageous compared to previous substrates because e.g., wooden strips were difficult to write upon and difficult to store to preserve records, as they had to be tied into bundles consuming much space. Nowadays, this flat configuration is still one of the main advantages for the design of electrochemical cells.

The process of paper manufacturing involves the mechanical or (bio)chemical conversion of a fibrous raw material (commonly wood) into pulp (free fibers separated to the unusable fraction) and later, bleaching and further treatment (with mineral fillers, polymeric additives ...) depends on the type and grade of paper that is to be produced [44]. In the paper factory, the pulp is dried and pressed to produce paper sheets.

2.1.1. Paper Source

Although vegetable (and then cellulosic) materials are mainly considered when referring to paper, other materials such as e.g., glass or polymers that also form fibers (Figure 2) and can be pressed to form sheets, could also be included. Then, a first classification (Figure 3) would distinguish among cellulosic (based on cellulose, a polymer of ß-linked D-glucose units) and non-cellulosic paper, depending on its source. Both are very interesting, although cellulosic materials exceed, by far, the rest of paper materials. In each of the classes, subclasses could be made according to the different vegetable species (also bacteria) in the first or the different materials in the second one. Bacterial cellulose, an extracellular polymer produced by some microorganisms, chemically pure and with high water-holding capacity and mechanical stability, has been proposed to develop enzymatically active paper including lipase [45] and also lactate oxidase, in this case for electrochemical detection of lactate in sweat [46].

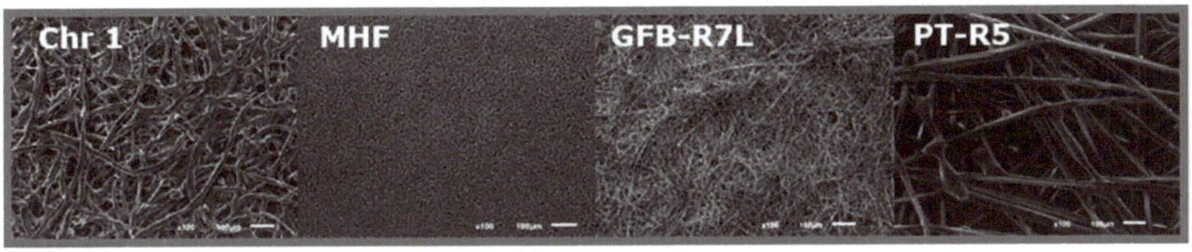

Figure 2. SEM images (zoom ×100) of different paper substrates: Chr 1: Whatman Grade 1; MHF: Millipore Hi-Flow nitrocellulose membrane; GFB-R7L: Mdi glass fiber; and PT-R5: Mdi polyester. Reprinted (adapted) from [40], Copyright (2017), with permission from Elsevier.

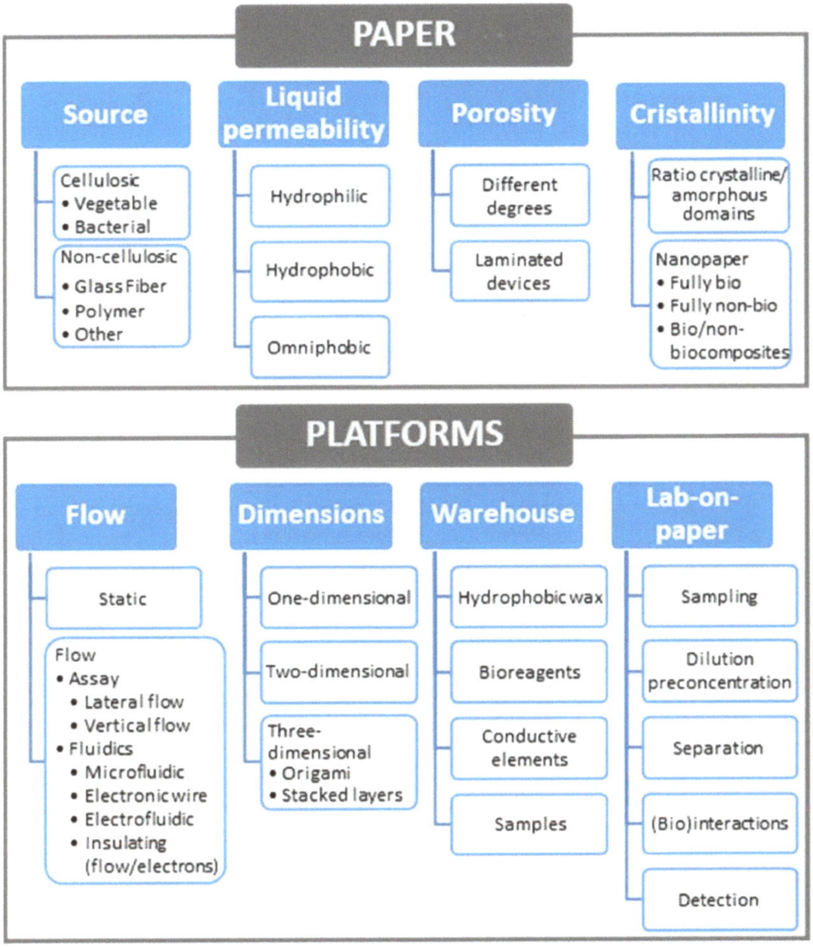

Figure 3. Classification of paper-based platforms according to different criteria and properties.

Cellulose can be also modified for changing the properties of the final product, being cellulose acetate or nitrocellulose the most common outputs. Cellulose acetate [47] is one of the most important esters of cellulose, obtained by reaction of cellulose with acetic anhydride and acetic acid in the presence of sulfuric acid. Nitrocellulose is the nitrate ester, obtained by nitrating cellulose with nitric and sulfuric acids. The analytical applications are numerous in both cases, as e.g., electrophoresis on cellulose acetate [48] or lateral flow immunoassays employing nitrocellulose membranes [49]. There are other different cellulose-based materials that can be employed for analytical purposes, such as the case of cellophane, brand name for cellulose films that are manufactured by regenerating cellulose from cellulose xanthate (viscose). It has been proposed for use in microfluidics either coated (with a micron thick coating of nitrocellulose or polyvinylidene chloride) or uncoated [50].

2.1.2. Hydrophilicity

Cellulose fibers, the building blocks of paper, are hollow tubes (*ca.* 1.5 mm long, 20-µm wide, with a wall thickness of ca. 2 µm) that form a porous, hydrophilic material that take up more than their own mass of water [51]. However, although hydrophilicity is one of the main characteristics that defines paper, in some cases is not desirable. In the fabrication process, "sizing" agents can be employed to lower surface energy, increasing water-contact angle and lowering rates of water penetration [52]. Alternatively, cellophane or also pre-treated cellulose-based paper, as in the case of this silanized with decyltrichlorosilane [53] to render it hydrophobic, or with fluoroalkyl trichlorosilanes to obtain omniphobic paper [54], could be employed. In this last case, the paper is both hydrophobic and oleophobic, repelling water and aqueous solutions containing ionic and non-ionic surfactants but also organic liquids or other complex liquids such as blood.

2.1.3. Porosity

Surface chemistry and porosity influence wet properties of paper, very important especially in the development of devices for flow assays. In non-porous paper, only the macroscopic external surface is accessible. The availability of the inner surface and then, the specific surface area that is accessible, increases with porosity. Related to this, there are two correlated macroscopic properties of paper that have to be considered: thickness (in m) and basis weight (g.m^{-2}), which is the mass of dry paper per square meter [51]. Using both parameters, the total pore volume can be estimated. In the case of the most commonly used paper in analytical devices (Whatman No. 1 filter paper, 180-µm thick with a basis weight of 87 g.m^{-2}, and then a density of 483 kg.m^{-3}), considering a density of the solid component of wood fiber of 1540 kg.m^{-3}, the pore/volume ratio results approximately 0.69 [51]. This porosity arises from spaces between the fibers, uncollapsed fiber lumens, and the intrinsic porosity of the fiber walls.

Porosity changes depending upon tree species, pulping type and drying after pulping, which can cause some of the pores to collapse. Alternatively, the porosity of paper-based platforms can be mechanically tuned, as is the case of the laminated paper-based analytical devices [55], enclosed between plastic sheets using a roll laminator. Apart from providing mechanical strength to the paper device, compression changes the structure since lamination reduces the thickness of the paper and the effective pore size, and in turn, the flow rate. The Washburn's equation [56] that describes the capillary flow in a bundle of parallel cylindrical tubes, can be extended to a paper strip: $L^2 = \gamma Dt/4\eta$, where L is the length of the strip, γ the effective surface tension, D the pore diameter, t the time, and η the dynamic viscosity of the fluid, can be rearranged as $L/t = \gamma D/4\eta L$, where L/t is the flow rate. Its linear dependence upon the pore diameter (D) is clear. Although it could seem an inconvenience, decreasing the flow rate increases the residence time what could be interesting when using paper as a (bio)reactor. This useful equation indicates variables that alter linear velocity. Thus, to increase velocity, a high surface tension, low viscosity or low density (if the liquid runs upward) liquid could be employed [57]. On the other hand, the surface energy of paper can be lowered to increase the contact angle of a liquid, what can be achieved by depositing material into the pores or chemically bonding groups to hydroxyl groups on cellulose. As commented before, treatment with organosilanes in gas phase [54] is a way to increase the contact angle ($\theta_{H_2O} >> 140^{\circ}$).

2.1.4. Cristallinity

Another parameter that can be correlated to paper hydrophilicity is the degree of crystallinity of cellulosic fibers (usually 50%). Cellulose is made up of bundles of fibrils called microfibrils, each individual consisting of crystalline and amorphous regions. Crystalline domains do not swell with water but by contrast, amorphous cellulose swells in water and is more susceptible to chemical reactions. Depending on the applications, paper surfaces can be saturated with water that forms slightly anionic water swollen hydrogel of amorphous cellulose and hemicellulose. Then, negatively charged particles, polymers or DNA have little tendency to adsorb [51]. In this context, cellulose nanofibers and nanocrystals (rod-shaped crystalline part that remains after removal of the amorphous domains) can be extracted from vegetable cellulose to produce nanopaper with different properties [58]. Apart from the crystallinity, the size of cellulose microfibrils in a paper is in the micrometre scale and consequently their specific surface area and hydrogen bond intensity are much less than that of nano-sized cellulose fibres [59]. As happens with the definition of paper, nanopaper can also come from different sources. Being a thin sheet mainly made of tightly packed nanostructures, those based on cellulose, chitin or chitosan nanofibers, are considered as fully bionanopapers. At the other side of the classification are the fully non-bionanopapers, based on nanographene (oxide), silicon carbide, carbon nanotubes or even synthetic polymers (polyurethane, polypropylene, polyvinylfluoride, etc.). In between, mixed bio/non-bionanocomposites including a main matrix component with other secondary components embedded are possible. In all the cases, interesting applications are being developed, including also the incorporation of screen-printed electrodes [60].

The classification of paper-based platforms according to different criteria, including properties of the material, is summarized in the schematics of Figure 3.

2.2. Paper in (Electro)analytical Platforms

The main use of paper in analytical (bio)platforms dates back to the 1930s with the development of paper chromatography [61], although estimation of pH with litmus paper and employ of paper as filter are even older (see Figure 4, [57]). Home-based pregnancy tests were based on the use of porous materials and pioneered analysis decentralization in the 1980s [62]. An impressive renaissance was done by Whitesides' group who patterned microfluidic channels in two dimensional platforms for multianalyte clinical analysis [32]. They were devised in a time when the need for cheap diagnostics in low-resource settings was urged [63]. The developing world does not have access to many of the best medical diagnostic technologies; these require air-conditioned laboratories, refrigerated storage of chemicals, a constant supply of calibrators and reagents, stable electrical power, highly trained personnel, and rapid transportation of samples. In remote zones, running water and electricity may or not be available, power is at best intermittent with wide fluctuations in voltage. The ambient temperature can range from 10 °C to more than 40 °C. Dust, wind and contaminating pathogens are very common. Potentially high-risk human samples are routinely handled with few precautions. Maintaining and calibrating even moderately complex instruments still presents a challenge. Actually, microfluidic paper-based analytical devices (μPADs) were conceived for useful application to the diagnostics in the developing world [33]. The scheme in Figure 4 shows the milestones in the history of paper science, with special focus on analytical devices. It includes the paper spray ionization, where the sample is deposited onto a sharp tip cut out of paper, aligned in front of a mass spectrometer [64] and should be completed with the introduction of the electrochemical detection by Henry's group in 2009 [37] to generate electrochemical paper-based analytical devices (ePADs) (Figure 5A). This detection, initially based on the use of screen-printed electrodes, the topic of this revision, fits perfectly with the concept of decentralization analysis.

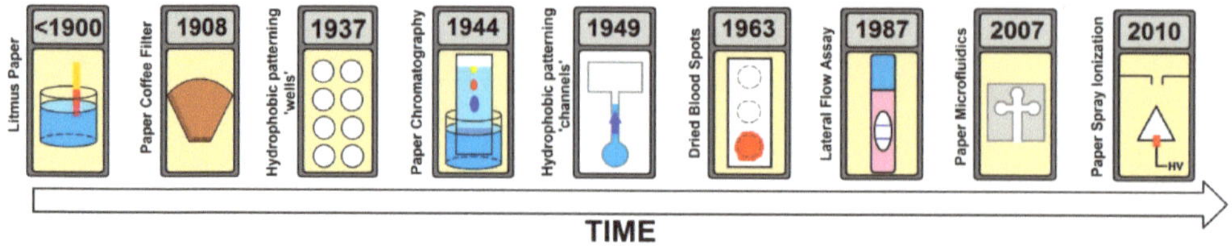

Figure 4. A history of paper in science, highlighting important milestones. Reprinted from [57] (open access, https://pubs.acs.org/doi/10.1021/acs.analchem.8b04825 accessed on 16 February 2021, further permissions should be directed to the ACS).

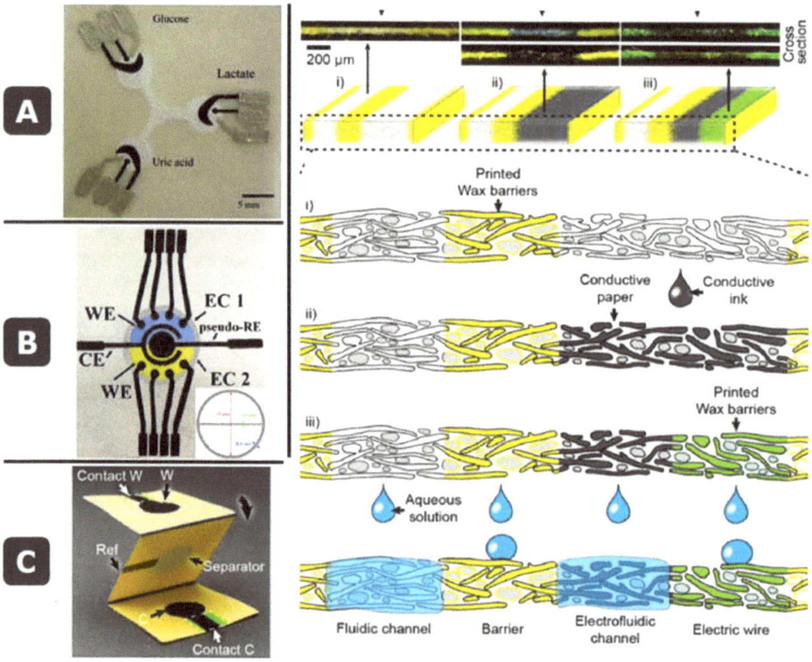

Figure 5. (**A**) Pioneer work on ePADs with screen-printed electrodes (SPEs) for multianalyte enzymatic detection of glucose, uric acid, and lactate. Reprinted with permission from [37]. Copyright (2009) American Chemical Society. (**B**) Picture of the paper-based electrochemical device developed by de Oliveira et al. with two electrochemical cells, with 4 working (WE) each and sharing reference (RE) and counter (CE). Reprinted from [65], Copyright (2019), with permission from Elsevier. (**C**) Integration of microfluidic and electronics, exemplified in an origami-based electroanalytical cell and visual explanation of the concept of electrofluidics. Reprinted with permission from [66], Copyright (2016) Wiley-VCH Verlag GmbH and Co.

2.2.1. Configuration: Static or Flow Assays

Hydrophilicity and porosity, commented in Section 2.1, are the main causes of capillary forces, in such a way that a liquid can easily flow through the gaps in porous media without using any extra energy due to the inherent capillary action. Then, paper allows passive transport of solutions and also can be used to absorb liquids. Then, both static and flow analytical devices are possible, as shown in the classification of Figure 3. In the first case, a volume of liquid can be deposited over a delimited hydrophilic area [42] or aspirated through a sampler [67] to perform the analysis. In the case of flow assays, lateral or vertical flow platforms [68] have been designed. Lateral flow platforms have been the pillar of rapid point-of-care diagnostics due to its simplicity, rapid process, and low cost. However, vertical (or flow-through) assays have emerged as an interesting alternative due to its rapidity and design possibilities.

2.2.2. Dimensions

Additionally, a basic classification of analytical platforms can be made according to their dimensions. One-dimensional channels can be made from paper sheets in two ways: by cutting two-dimensional paper to obtain physical boundaries or by forming hydrophobic boundaries through impregnation of agents. In both cases the methodology can be simple and cheap or more complicate and expensive. Precision depends also on the procedure (handmade or automatized). Thus, scissors and razor blades could be used as well as craft-cutting machines or lasers for delimiting the paper-based platform. In this case, due to the fibrous structure, paper tearing can occur and due to the fiber orientation, different profiles are obtained when cutting is made in parallel to the y- or x-axis, which have to be previously considered since fluid transport may depend upon the angle at which the paper strip is cut. Additionally, the z-axis (thickness) has to be taken into account, since the mass distribution in a paper sheet is usually not constant, with the maximum density in the center and decreasing near the surfaces [51]. One of the best examples of one-dimensional platforms are those employed for already introduced lateral flow immunoassays [69], where after assembling all the components required for the assay on a two-dimensional adhesive backing (i.e., sample pad, conjugate pad with detection antibodies, nitrocellulose membrane with test and control lines of immobilized capture antibodies, and absorbent pad) different units are cut. Being one-dimensional is one of the main advantages of other low-cost platforms, e.g., thread-based devices, allowing not to delimit channels in colorimetric [70] or electrochemical, either amperometric [71] or potentiometric [72] platforms. However, when delimiting a hydrophilic working area (by photolithography, wax printing, etc.) is required in paper-based devices, a two-dimensional structure usually is employed, including in most of the cases several one-dimensional channels. Apart from this, the flat structure of paper allows easy generation of 3D platforms, by stacking layers of similar or different composition (e.g., adhesive tape, polymeric films, etc.). Separate layers can be piled up or, alternatively, taking advantage of the foldability of paper, origami-based devices [50,73] can be employed, as well as other "pop-up" electrochemical platforms [74]. In this last case, a single sheet of paper is folded into a three-dimensional device that changes the shape and integrates fluidic and electrical connectivity by simply folding and unfolding the structure. The reconfigurable 3D structure provides mechanical support and allows control timing. Similarly, a sliding strip microfluidic device enables perform colorimetric ELISAs on paper [75]. The possibilities of multilayer devices are enormous. An interesting example is the use of both paper and hollow channels in the same analytical platform [76,77]. Removing the cellulose matrix from within a predefined channel and leaving the bottom of the channel hydrophilic, the flow rate of solution in the channel can be enhanced without requiring external equipment. In this way, analysis time is reduced, and larger fluidic networks can be employed. Non-specific adsorption is decreased, and micrometer-sized objects could flow freely. Colorimetric [77] and electrochemical [76] detection have been demonstrated, in the last case for voltammetry and amperometry, under flow and no flow conditions.

2.2.3. Functions: Lab-on-Paper Devices

These multilayer devices (made of a folded continuous layer or several piled up) allow the integration of different functions, approaching the concept of lab-on-paper. Although these could be included in single-layer devices, the use of several layers brings more possibilities. The analytical process is that performed with the aim of obtaining information that allows solving social, economic, scientific, or technical problems. The analysis can be qualitative (with a binary yes/no answer obtained) or (semi)quantitative. Steps go from sampling to data treatment, including detection and several operations carried out with the aim of obtaining sensitive and selective methodologies (here most separation steps are included as well as (bio)interactions for recognition purposes or to obtain detectable species). Technological improvements, especially in sample processing, fluid flow control, signal amplification and component integration increased the applicability of PADs to real-world problems [78]. Several strategies can be employed for controlled fluid actuation and manipulation and thus, slipping, channeling, delaying delivery of reagents, switching to initiate flow on demand, or fluid mixing (considering that the flow is generally laminar) is possible. Moreover, inclusion of nano- and microsized objects as well as filtration [79] or separation [80], and also dilution or preconcentration [81], have been demonstrated.

One of the reasons why paper is excellent substrate for the development of analytical platforms is its versatility. Many different systems and applications can be converted into paper-based formats, taking advantage of its tunable properties. As an example, regarding the detection, multiplexed devices including several electrochemical cells with multiple electrodes can be easily designed [65] (Figure 1B).

2.2.4. Paper as a Warehouse

In a more general way, paper can be considered as a huge warehouse where different elements can be stored using weak or strong bindings, mainly: (i) hydrophobic materials, already commented, to delimit microfluidic areas in two-dimensional platforms, (ii) biological materials, as active reagents for the development of bioassays, (iii) samples that could be even dried for further analysis, allowing transportation from remote settings [82], and (iv) conductive materials, very interesting for electroanalytical purposes. Combination of some of them introduces the concept of electrofluidics [66], by monolithic integration of microfluidics and electronics (Figure 5C). Thus, in the same paper device a hydrophobic barrier (insulated non-microfluidic electrical conductor), and microfluidic channels with and without electronic conductivity could coexist by modification with wax and carbon nanotubes (or conductive polymers). In this case, since an aqueous dispersion of conductive elements is employed, the flow is not impeded, what could happen when dense inks are employed. The same idea could be exploited using micropatterned conductive structures where the conductors (carbon nanotubes or silver nanowires) are monolithically integrated with nanocellulose-based paper [83].

Introducing electrode integration and coming back to the history of papermaking, even when paper dates back from year 105, printing was long retarded, with the first text printed upon paper finally completed in the year 770 by Empress Shotoku of Japan, more than 665 years later [43]. The use of a soft and pliable paper made possible to make an impression from a wood block spread with pigment. After putting the paper over the block, a ball was rubbed by hand until a definite impression was made upon the paper. In Europe, fibers from macerated linen and cotton cloth formed sheets that were impregnated with gelatin when Gutenberg established his printing office in Mainz using metal blocks. Similar to what happens nowadays, the type of paper influenced the process and readout. Thin and transparent Oriental paper allowed writing on one side, meanwhile thick and opaque European paper allowed writing or printing on both sides. Obviously, the process has evolved and improved with time and nowadays, screen-printed thick-film technology is one of the more successful in the development of electroanalytical devices, also in paper-based devices, as commented in more detail in the following section. Actually, the first electrochemical cell for paper devices was made by a screen-printing process of conductive inks [37]. Since fibers are pressed and acquire a flat configuration, paper is very appropriate for integration of electrochemical cells, in a similar way that is done for the rest of the substrates. However, similarly to what happens at the beginning of the process, as paper is a porous substrate, if an electrolyte is employed, ionic conduction exists between the two faces. Then, electrodes do not need to "share" the same surface, and screen-printed technology, that commonly uses a pattern to locate all the electrodes required for electrochemical readout in the same face, is not required. Simple deposition of ink in a wax-delimited area, without the need of screens or stencils, can be employed. Two more pins are used as electrodes that connect the electrolyte by the other side (Figure 1D, [42,67]).

However, since this technology is commonly employed with non-porous substrates (as ceramics or polymers), the translation into paper was done performing the same procedure, considering paper as one more flat surface with unconnected faces. However, this is a very differential advantage when compared to the rest of substrates employed in SPEs. Moreover, printing and coating technologies are based on the application of almost any fluid onto dry paper. Aqueous or organic solutions could be employed. Usually, aqueous solutions are particularly easy because capillary forces and the hydrophilic nature of cellulose promote rapid sorption, but in SPE technology conductive elements are dispersed in organic solvents and curing is necessary. Moreover, biomolecules can be also spotted or printed onto dry paper without denaturation.

3. Paper-Based Screen-Printed Electrodes

As mentioned before, besides being a flat substrate, paper can have other utilities such as storage of reagents or sample [84,85], support for (biological) reactions [42,86–89], or platform for taking [67,90] or treating the sample (e.g., preconcentration [81,91] or separation [79,92,93]). Taking this into account, paper-based electroanalytical devices integrating SPEs can be designed in different formats: (i) combining paper with a SPE card fabricated on ceramic or polymeric materials [79,84–89,91,92,94–102]; (ii) combining one electrode of the electrochemical cell (e.g., WE) made on paper with other electrodes (e.g., RE and CE) of a SPE card printed on a conventional material [42,103–105]; and (iii) printing the SPE directly on paper [5,106–108]. In the last case, 2D devices are the most basic but, by stacking and/or folding the paper along the vertical axis, devices with 3D formats (multilayer and origami) can be easily constructed. Obviously, the fabrication of 2D devices is much simpler but 3D platforms can improve analytical characteristics (mainly reproducibility and sensitivity) and reduce the steps of the analytical procedure as well as time analysis [108]. Here, electrodes could be included in the same or different layers.

3.1. SPE Cards Combined with Paper Elements

In the case of the lowest degree of integration (paper and SPE ceramic card), the paper is mainly used as support for biological components (enzymes [85,86,95,97–99], but also antibodies [87] or nucleic acids [88]) where reactions with components of the sample take place. Paper was also used as a medium for cell culture. In this case, the SPE platform allowed the on-line evaluation of cell viability by monitoring dopamine release from cell damage models [109]. Alternatively, the paper was also combined with a SPE card for sampling, as in the tear sampling system developed by Honikel et al. using Whatman No 41 ashless filter paper [90].

In a following step of the analytical process, the assembling of layers of different types of paper has been also used for the treatment of the sample. A clear example of this application is the analysis of whole blood as in the glucose sensor developed by Noiphung et al. [79] in which Whatman No. 1 and blood separation paper, VF1 and VF2, were combined with a Prussian Blue screen-printed electrode. A wax-dipping technique was used to design microfluidic patterns that consisted of two separation zones and one more for detection, where the enzyme glucose oxidase was immobilized (Figure 6A). Paper can be employed for separating analytes before their detection as in the microfluidic platform developed by Primpray et al. [92]. This device used Whatman SG81 paper, which is an ion exchange paper that combines cellulose with large pore silica gel, for the separation of dexamethasone or prednisolone steroids (Figure 6B). Moreover, and taking advantage of the white color of paper and of the intense blue color produced by the interaction of tetrazolium blue with steroids, this reagent was added to the channels with the aim of visualizing the exact position of each steroid, facilitating the subsequent electrochemical detection [92]. Lateral flow immunodevices are also composed of different layers overlapping one another. Then, screen-printed electrodes can be included just below the test line (this containing the antibody to capture the analyte) of the membrane for detection, in this case taking advantage of the vertical flow [110], or even on the same membrane [111,112]. Alternatively, another possibility is to include the capture antibody directly over the screen-printed working electrode in a mixed biosensing approach [113]. Apart from these examples in which a lateral flow is produced, the combination of paper-based working electrodes with electrodes from screen-printed cards have been used in static systems with the aim of reducing the cost by reusing the RE and CE of the same SPE for different paper WEs [42,103], but also for the in situ electrogeneration of nanoparticles on paper-based carbon [104] or gold [105] electrodes, the last one employed for the chronoamperometric determination of arsenic in white wines (Figure 6C).

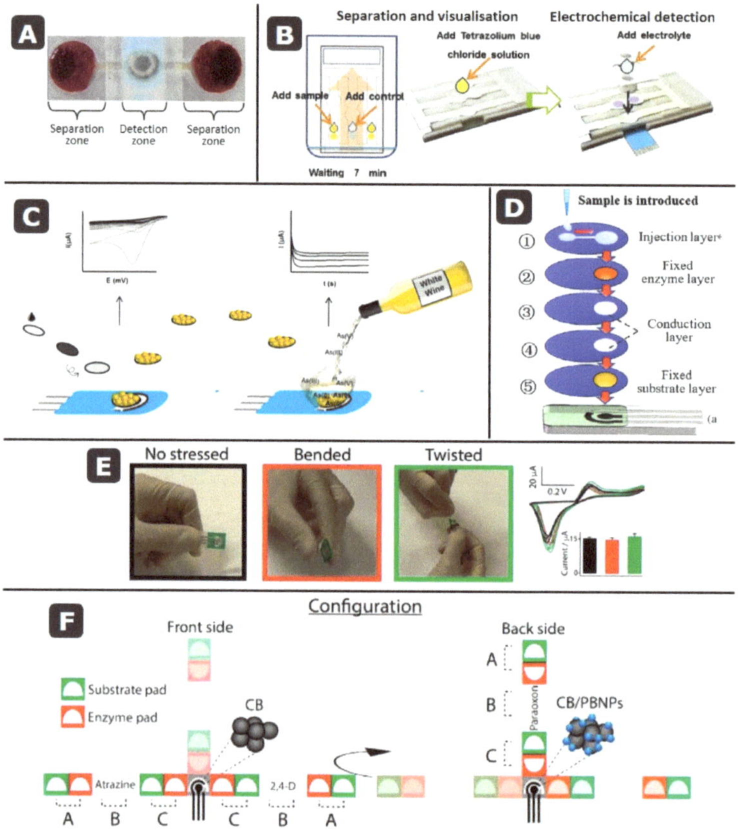

Figure 6. SPE cards combined with paper elements: (**A**) Picture of the paper-based device developed by Noiphung et al. for whole blood separation and glucose determination. Reprinted from [79], Copyright (2013), with permission from Elsevier. (**B**) Schematic representation of the separation, visualization and measurement steps in the paper-based devices developed by Primpray et al. for the determination of steroids. Reprinted from [92], Copyright (2019), with permission from Elsevier. (**C**) Schematic representation of the paper-based system for in-situ generation of gold nanoparticles and further analysis of arsenic developed by Nuñez-Bajo et al. Reprinted with permission from [105]. Copyright (2017) American Chemical Society. (**D**) Schematic diagram of the multilayer platform developed by Yang et al. for pesticide detection. Reprinted with permission from [95], Copyright (2020) Wiley Periodicals LLC. Fully-integrated paper-based electrochemical devices: (**E**) Evaluation of the robustness of the paper-based device developed by Cinti et al. for the detection of chloride ions, after 100 repeated bending (red) and twisting (green) tests. Reprinted from [114], Copyright (2017), with permission from Elsevier. (**F**) Schematic diagrams of the configuration of the origami device developed by Arduini et al. for the detection of pesticides. Reprinted from [115], copyright (2018), with permission from Elsevier.

On the other hand, several layers of paper can be added in vertical flow configurations, with each layer for a different function as in this multiplex platform developed by Yang et al. for the detection of organophosphorus pesticides [95]. The detection and identification of the pesticides were performed recording impedance time-sequence spectra after inhibition of the enzymatic activity of acetylcholinesterase by pesticides. This platform consisted of five layers for: (1) injection, (2) enzyme immobilization, (3 and 4) sample transport, and (5) substrate (indophenol acetate) immobilization (Figure 6D). The layers 3 and 4 were hollow and their function was to allow interaction between sample and enzyme for a long enough time. In a similar way, stacking different layers over a SPE has been employed for electrochemical detection of specific sequences of DNA by combining recombinase polymerase amplification with an electroactive mediator [116]. In this case, a disposable paper strip that incorporates a carbon SPE card is included in a handheld device that accomplishes thermoregulation and enables electrochemical detection of *Mycobacterium smegmatis* and *M. tuberculosis*. On the other hand, origami paper devices (folded multilayer platforms) can be also combined with SPEs printed on ceramics or polymers [85,117] as in the case of the device developed by Pinyorospathum et al. [117] for the determination of human C-reactive protein (CRP), which is an important biomarker for different cardiovascular diseases. This platform consisted of a SPE printed on a PVC substrate, that was wrapped in between the two paper folded layers containing reagents for detection.

3.2. Fully-Integrated Devices: Electrochemical Cells Printed on Paper

Although there are numerous different designs that combine SPEs (in ceramic/polymer substrates) with paper devices, the most integrated approach is achieved by printing electrodes directly onto the paper creating 2D and 3D devices, with all the three electrodes in the same or different layers [78,108]. In two-dimensional devices, the sample is deposited in the same layer in which the electrochemical cell is printed or flows toward it by capillarity. As commented before, platforms are commonly fabricated by patterning a single piece of paper with a hydrophobic material (e.g., wax or photoresist) that delimits hydrophilic microfluidic channels, areas or reservoirs [118–136]. Other materials such as adhesive tape can be used either as an additional stencil for printing the electrodes [121,137] or to delimit the area of the working electrode once the ink has been spread on a specific piece of paper [138]. In this last case, screen or stencil printing is not necessary since the ink is directly deposited on a piece of paper. Delimiting the area of the working electrode with hydrophobic wax would avoid the need of an additional tape layer [42,67]. Different specific (bio)reagents (enzymes [129,130], antibodies [139], nucleic acids [120,131,140]), or nanomaterials [124,134,141]), which react with the sample, can be immobilized on delimited areas. On the other hand, multiplexed devices can be easily designed [65]. The combination of paper and screen-printed technology allows to obtain flexible devices, able to be bended or even twisted without loss of analytical signal as in the case of this designed by Cinti et al. [114] (Figure 6E).

Taking advantage of how easy it is to fold, stack, or cut the paper, there was a subsequent development of 3D multilayer and origami paper-based devices [73,93,117,142–156], including integration of electrodes directly on paper. These seem to require more laborious procedures of construction but the use of software for transferring wax and ink designs to paper makes this an easy task, especially useful when more sophisticated operations are enabled: e.g., sampling [143,157,158], sample delivery [146,159], or sample treatment [93]. As mentioned before, in many of these devices, the electrodes of the electrochemical cell are printed on different layers of the devices with the aim of improving the contact of the electrodes with the sample or decreasing the size of the device [73,142,144,152,160–166]. Moreover, vertical microfluidics, in comparison with lateral flow, helps to reduce the consumption or reagents and the loss of sensitivity due to the diffusion of the analyte. An example of these origami devices is the one developed by Arduini et al. for the multiplexed detection of three pesticides (paraoxon, 2,4-dichlorophenoxyacetic acid and atrazine) of different classes exploiting their capability to inhibit the activity of enzymes butyrylcholinesterase, alkaline phosphatase and tyrosinase [115]. The configuration of this device can be seen in Figure 6F. It consisted of two office paper-based SPEs cells, printed on the front and the backside of the device. Different filter paper strips were combined with these electrodes using adhesive tape, with the aim of determining the initial and residual enzymatic activity and therefore, the concentration of the pesticides. Each filter paper strip (from a total of three) was constituted of two pairs of different pads (A and C in the figure), one for containing the enzyme, and the other the substrate, separated by an empty paper. To determine the initial enzymatic activity, enzyme- and substrate-pads were contacted, and distilled water was added to perform the measurement. After cutting the first used pads, the sample was analyzed putting in contact the other enzyme- and substrate pads of the following strip and letting the sample incubate. After addition of water, the analytical signal was recorded. This device performs 6 measurements for recording the signals corresponding to the three pesticides [115].

Origami devices have been widely developed for clinical and biological applications [5,106,107]. For many of them, biological reagents such as enzymes or antibodies are used but those based on molecularly imprinted polymers (MIPs) are also useful approaches [121,137,155]. Amatatongchai et al. developed a MIP-based origami paper using alkyl ketene dymer (AKD)-inkjet printing to create circular hydrophobic areas [155]. The graphite working electrode of the screen-printed electrochemical cell was modified with a Fe_3O_4@Au@SiO_2-MIP nanocomposite, showing electrocatalytic activity toward the oxidation of serotonin, determined by linear sweep voltammetry in pharmaceutical capsules and urine samples [155].

Foldability is also highly useful for the creation of wearable devices for on-site monitoring of important parameters in non-invasive samples. One example is the origami paper-based platform developed by Cao et al. for the determination of glucose in sweat [158]. This 3D-microfluidic device was fabricated patterning the paper with wax and folding it to form five layers with different functions: sweat collector, vertical and transverse channels, electrode layer, and sweat evaporator (Figure 7A). In this way, sweat was absorbed by the collector and flowed toward the electrochemical cell by capillary forces. Sweat evaporation on the evaporator allowed continuous flow keeping fresh sample flowing across the electrochemical cell and avoiding sweat accumulation [158]. An on-body test was carried out to validate the device, but still the stability has to be evaluated and the size reduced. On the other hand, using 3D or origami approaches, paper-based SPEs can be directly integrated in wearable devices as in the case of the mask developed by Maier et al. for the continuous real-time monitoring of exhaled hydrogen peroxide (H_2O_2), an important biomarker in respiratory diseases [167]. The electrochemical cell was printed on a wax-patterned paper in which a solution of electrolyte was evaporated to form a solid electrolyte. Two carbon working electrodes were printed: one with Prussian Blue as mediator, for the H_2O_2 detection, and the other without modification, for subtracting the background signals (Figure 7B). This screen-printed electrochemical cell was placed inside a respiratory mask so the user breath directly onto it. This sensing mask was tested in simulated breath.

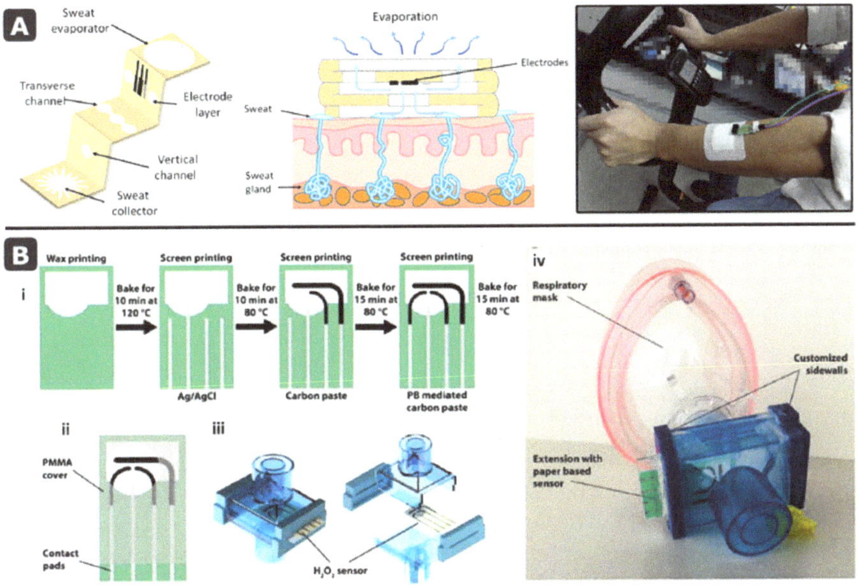

Figure 7. (**A**) Schematic diagram of the 3D microfluidic device developed by Cao et al. for sweat analysis. Schematic representation of the device applied on skin; a channel was formed by folding the origami device through which the sweat flow from the skin to the electrochemical cell. Picture of the device applied on the forearm of a user. Reproduced from [158] with permission from The Royal Society of Chemistry. (**B**) Schematics of the procedure of fabrication and the final design of the screen-printed electrochemical cell inserted in the mask developed by Maier et al. (**i** and **ii**). Model of the customized filter extension in which the screen-printed electrochemical cell is inserted (**iii**). Photograph of the respiratory mask with the filter extension containing the screen-printed electrode (**iv**) Reproduced from [167] (open access).

In a similar way, the high versatility of screen-printing technology, which allows to produce miniaturized electrodes, is also applicable not only to paper and other two-dimensional materials such as transparency sheets [31,168–170], but also other such as textiles or plastics, leading to a wide variety of wearable devices [106,171–173] such as bandages [174–176], gloves [177], rings [178,179], or pacifiers [180].

4. Conclusions and Perspectives

Paper has demonstrated to be an excellent substrate for the design of extremely innovative analytical platforms. The thick-film technology of screen-printing fits perfectly with this material that, in turn, is very appropriate for electroanalysis. Paper is an excellent container for both electrodes and electrolytes required for interfacial techniques. The possibilities are enormous, and the classification includes many varieties according to paper properties (porosity, crystallinity, hydrophilicity, etc.) and platform characteristics (dimensions, flow, etc.). Most of them have been applied to the clinical field and a challenge is to expand their utilization to other areas. Actually, paper-based electroanalytical devices would be very useful to the agri-food sector, to evaluate the presence of certain molecules and/or the freshness of a beverage/foodstuff [6–8]. In this way, these devices could reduce the gap between complex laboratory analysis and simple point-of-need testing, lowering costs, simplifying procedures, and reducing waste generation.

As reported in the bibliography as recommendation for future paper-based research by Verpoorte's group [57], if these devices are to provide society with the tools to perform on-site analyses, one have to learn from previous successes (e.g., regarding LFAs, avoiding the use of pipets for sample application (as e.g., in [67]) or of complicated readout systems) and look beyond the piece of paper to increase the functionality, by including various elements or by integrating electronics (as e.g., in [66]). Paper is so versatile that true lab-on-paper platforms are possible by incorporating the different steps of the analytical process. Moreover, combination with portable electrochemical readers approaches the analysis to the point where is required. It is possible to find nowadays potentiostats in the "do-it-yourself" format [181–183], with wireless control from smartphones [184,185], or wired in case of older mobile generations [186]. Apart from the instrumentation, other components required such as the energy source could also be incorporated in paper format, as e.g., paper-based fuel cells [187]. Paper potentiostats have also been suggested for integrated biosensing [188], even including a screen-printed manganese dioxide battery and an electrochromic display, which approaches the concept of "use-and-throw" instruments in what can be considered as promising paradigm-changing new products [189].

Notwithstanding the above, the commercialization of these devices continues to be a challenge since the knowledge transfer from laboratories to the society is still complicated. Although general printed electrochemical devices represent a unique opportunity to enable low-cost, fast, non-invasive and/or continuous monitoring of analytes, metrological aspects such as sensitivity, repeatability and stability represent very challenging aspects [190]. It is important to note that, in the laboratory, many of the fabrication and assembling steps are handmade, which affects the precision. Therefore, automatization and mass production could help to improve it. Although operational stability is not a concern when single-use platforms are aimed, an important issue is storage stability, especially relevant when biological reagents are involved. However, despite these yet unmet challenges, the potential of paper-based printed devices is clear as well as the relevance they will have in the future [191]. In an increasingly knowledgeable society with growing globalized problems, as is the current pandemic, simple tools that provide fast, reliable, and on-site responses will be undoubtedly, and increasingly required.

Author Contributions: E.C.-R. and M.T.F.-A. are equally responsible for the ideation, the literature search and also for drafting and revising the work, All authors have read and agreed to the published version of the manuscript.

Funding: This research was funded by the Spanish Ministry of Science and Innovation by the Project PID2019-107838RB-100 and also Banco Santander, CRUE and CSIC by the Project LIFE (Fondo Supera COVID-19).

Institutional Review Board Statement: Not applicable.

Informed Consent Statement: Not applicable.

Data Availability Statement: Not applicable.

Conflicts of Interest: The authors declare no conflict of interest.

References

1. Carneiro, P.; Morais, S.; do Carmo Pereira, M. Biosensors on the road to early diagnostic and surveillance of Alzheimer's disease. *Talanta* **2020**, *211*, 120700. [CrossRef]
2. Couto, R.A.S.; Lima, J.L.F.C.; Quinaz, M.B. Recent developments, characteristics and potential applications of screen-printed electrodes in pharmaceutical and biological analysis. *Talanta* **2016**, *146*, 801–814. [CrossRef]
3. Rama, E.C.; Costa-García, A. Screen-printed Electrochemical Immunosensors for the Detection of Cancer and Cardiovascular Biomarkers. *Electroanalysis* **2016**, *28*, 1700–1715. [CrossRef]
4. Toyos-Rodríguez, C.; García-Alonso, F.J.; de la Escosura-Muñiz, A. Electrochemical biosensors based on nanomaterials for early detection of alzheimer's disease. *Sensors* **2020**, *20*, 4748. [CrossRef]
5. Gutiérrez-Capitán, M.; Baldi, A.; Fernández-Sánchez, C. Electrochemical paper-based biosensor devices for rapid detection of biomarkers. *Sensors* **2020**, *20*, 967. [CrossRef] [PubMed]
6. Torre, R.; Costa-Rama, E.; Nouws, H.P.A.; Delerue-Matos, C. Screen-Printed Electrode-Based Sensors for Food Spoilage Control: Bacteria and Biogenic Amines Detection. *Biosensors* **2020**, *10*, 139. [CrossRef]
7. Smart, A.; Crew, A.; Pemberton, R.; Hughes, G.; Doran, O.; Hart, J.P. Screen-printed carbon based biosensors and their applications in agri-food safety. *Trac Trends Anal. Chem.* **2020**, *127*, 115898. [CrossRef]
8. Griesche, C.; Baeumner, A.J. Biosensors to support sustainable agriculture and food safety. *Trac Trends Anal. Chem.* **2020**, *128*, 115906. [CrossRef]

9. Pérez-Fernández, B.; Costa-García, A.; De La Escosura- Muñiz, A. Electrochemical (bio)sensors for pesticides detection using screen-printed electrodes. *Biosensors* **2020**, *10*, 32. [CrossRef]
10. Rebelo, P.; Costa-Rama, E.; Seguro, I.; Pacheco, J.G.; Nouws, H.P.A.; Cordeiro, M.N.D.S.; Delerue-Matos, C. Molecularly imprinted polymer-based electrochemical sensors for environmental analysis. *Biosens. Bioelectron.* **2020**, *172*, 112719. [CrossRef]
11. Almeida, M.I.G.S.; Jayawardane, B.M.; Kolev, S.D.; McKelvie, I.D. Developments of microfluidic paper-based analytical devices (μPADs) for water analysis: A review. *Talanta* **2018**, *177*, 176–190. [CrossRef]
12. Kung, C.T.; Hou, C.Y.; Wang, Y.N.; Fu, L.M. Microfluidic paper-based analytical devices for environmental analysis of soil, air, ecology and river water. *Sens. Actuators B Chem.* **2019**, *301*, 126855. [CrossRef]
13. Metrohm DropSens. Available online: http://www.dropsens.com/ (accessed on 31 December 2020).
14. MicruX Technologies. Available online: http://www.micruxfluidic.com/ (accessed on 31 December 2020).
15. Fernández-Abedul, M.T. (Ed.) *Laboratory Methods in Dynamic Electroanalysis*; Elsevier: Amsterdam, The Netherlands, 2020.
16. Anastas, P.T.; Zimmerman, J.B. The periodic table of the elements of green and sustainable chemistry. *Green Chem.* **2019**, *21*, 6545–6566. [CrossRef]
17. Gałuszka, A.; Migaszewski, Z.; Namieśnik, J. The 12 principles of green analytical chemistry and the SIGNIFICANCE mnemonic of green analytical practices. *Trac Trends Anal. Chem.* **2013**, *50*, 78–84. [CrossRef]
18. Swain, G.M. Solid Electrode Materials: Pretreatment and Activation. In *Handbook of Electrochemistry*; Zoski, C.G., Ed.; Elsevier: Amsterdan, The Netherlands, 2007; pp. 111–153. ISBN 9780444519580.
19. Abedul, M.T.F.; Rodríguez, J.R.B.; García, A.C.; Blanco, P.T. Voltammetric determination of cocaine in confiscated samples. *Electroanalysis* **1991**, *3*, 409–412. [CrossRef]
20. Nagaoka, T.; Yoshino, T. Surface properties of electrochemically pretreated glassy carbon. *Anal. Chem.* **1986**, *58*, 1037–1042. [CrossRef]
21. Thornton, D.C.; Corby, K.T.; Spendel, V.A.; Jordan, J.; Robbat, A.; Rutstrom, D.J.; Gross, M.; Ritzier, G. Pretreatment and Validation Procedure for Glassy Carbon Voltammetric Indicator Electrodes. *Anal. Chem.* **1985**, *57*, 150–155. [CrossRef]
22. Lamas-Ardisana, P.J.; Fanjul-Bolado, P.; Costa-García, A. Hydrogen evolution: Electrochemical pretreatment for voltammetric analysis with gold electrodes. *Electroanalysis* **2015**, *27*, 1073–1077. [CrossRef]
23. PalmSens. Available online: https://www.palmsens.com/ (accessed on 31 December 2020).
24. Noviana, E.; Henry, C.S. Simultaneous electrochemical detection in paper-based analytical devices. *Curr. Opin. Electrochem.* **2020**, *23*, 1–6. [CrossRef]
25. Cano-Raya, C.; Denchev, Z.Z.; Cruz, S.F.; Viana, J.C. Chemistry of solid metal-based inks and pastes for printed electronics—A review. *Appl. Mater. Today* **2019**, *15*, 416–430. [CrossRef]
26. Sanati, A.; Jalali, M.; Raeissi, K.; Karimzadeh, F.; Kharaziha, M.; Mahshid, S.S.; Mahshid, S. A review on recent advancements in electrochemical biosensing using carbonaceous nanomaterials. *Microchim. Acta* **2019**, *186*, 773. [CrossRef]
27. Chu, Z.; Peng, J.; Jin, W. Advanced nanomaterial inks for screen-printed chemical sensors. *Sens. Actuators B Chem.* **2017**, *243*, 919–926. [CrossRef]
28. Trojanowicz, M. Impact of nanotechnology on design of advanced screen-printed electrodes for different analytical applications. *Trac Trends Anal. Chem.* **2016**, *84*, 22–47. [CrossRef]
29. Inzelt, G. Pseudo-reference Electrodes. In *Handbook of Reference Electrodes*; Inzelt, G., Lewenstam, A., Scholz, F., Eds.; Springer: Berlin/Heidelberg, Germany, 2013; pp. 331–332. ISBN 9783642361883.
30. Søpstad, S.; Johannessen, E.A.; Seland, F.; Imenes, K. Long-term stability of screen-printed pseudo-reference electrodes for electrochemical biosensors. *Electrochim. Acta* **2018**, *287*, 29–36. [CrossRef]
31. Martín-Yerga, D.; Álvarez-Martos, I.; Blanco-López, M.C.; Henry, C.S.; Fernández-Abedul, M.T. Point-of-need simultaneous electrochemical detection of lead and cadmium using low-cost stencil-printed transparency electrodes. *Anal. Chim. Acta* **2017**, *981*, 24–33. [CrossRef]
32. Martinez, A.W.; Phillips, S.T.; Butte, M.J.; Whitesides, G.M. Patterned paper as a platform for inexpensive, low-volume, portable bioassays. *Angew. Chem. Int. Ed.* **2007**, *46*, 1318–1320. [CrossRef]
33. Martinez, A.W.; Phillips, S.T.; Whitesides, G.M.; Carrilho, E. Diagnostics for the developing world: Microfluidic paper-based analytical devices. *Anal. Chem.* **2010**, *82*, 3–10. [CrossRef] [PubMed]
34. Noviana, E.; McCord, C.P.; Clark, K.M.; Jang, I.; Henry, C.S. Electrochemical paper-based devices: Sensing approaches and progress toward practical applications. *Lab Chip* **2020**, *20*, 9–34. [CrossRef] [PubMed]
35. Cate, D.M.; Adkins, J.A.; Mettakoonpitak, J.; Henry, C.S. Recent Developments in Paper-Based Microfluidic Devices. *Anal. Chem.* **2015**, *87*, 19–41. [CrossRef] [PubMed]
36. Yang, Y.; Noviana, E.; Nguyen, M.P.; Geiss, B.J.; Dandy, D.S.; Henry, C.S. Paper-Based Microfluidic Devices: Emerging Themes and Applications. *Anal. Chem.* **2017**, *89*, 71–91. [CrossRef]
37. Dungchai, W.; Chailapakul, O.; Henry, C.S. Electrochemical detection for paper-based microfluidics. *Anal. Chem.* **2009**, *81*, 5821–5826. [CrossRef]
38. Mettakoonpitak, J.; Boehle, K.; Nantaphol, S.; Teengam, P.; Adkins, J.A.; Srisa-Art, M.; Henry, C.S. Electrochemistry on paper-based analytical devices: A Review. *Electroanalysis* **2016**, *28*, 1420–1436. [CrossRef]
39. Fosdick, S.E.; Anderson, M.J.; Renault, C.; Degregory, P.R.; Loussaert, J.A.; Crooks, R.M. Wire, Mesh, and Fiber Electrodes for Paper-Based Electroanalytical Devices. *Anal. Chem.* **2014**, *86*, 3659–3666. [CrossRef]

40. Núnez-Bajo, E.; Blanco-López, M.C.; Costa-García, A.; Fernández-Abedul, M.T.; Carmen Blanco-López, M.; Costa-García, A.; Teresa Fernández-Abedul, M. Integration of gold-sputtered electrofluidic paper on wire-included analytical platforms for glucose biosensing. *Biosens. Bioelectron.* **2017**, *91*, 824–832. [CrossRef]
41. Adkins, J.A.; Henry, C.S. Electrochemical detection in paper-based analytical devices using microwire electrodes. *Anal. Chim. Acta* **2015**, *891*, 247–254. [CrossRef] [PubMed]
42. Amor-Gutiérrez, O.; Costa Rama, E.; Costa-García, A.; Fernández-Abedul, M.T. Paper-based maskless enzymatic sensor for glucose determination combining ink and wire electrodes. *Biosens. Bioelectron.* **2017**, *93*, 40–45. [CrossRef] [PubMed]
43. Hunter, D. Before Paper: The Writing Substances of the Ancients. In *Papermaking: The History and Technique of an Ancient Craft*; Dover Publications: New York, NY, USA, 1978.
44. Bajpai, P. *Brief Description of the Pulp and Papermaking Process*, 2nd ed.; Springer: Berlin/Heidelberg, Germany, 2018.
45. Buruaga-Ramiro, C.; Valenzuela, S.V.; Valls, C.; Roncero, M.B.; Pastor, F.I.J.; Díaz, P.; Martínez, J. Bacterial cellulose matrices to develop enzymatically active paper. *Cellulose* **2020**, *27*, 3413–3426. [CrossRef]
46. Gomes, N.O.; Carrilho, E.; Machado, S.A.S.; Sgobbi, L.F. Bacterial cellulose-based electrochemical sensing platform: A smart material for miniaturized biosensors. *Electrochim. Acta* **2020**, *349*, 136341. [CrossRef]
47. Fischer, S.; Thümmler, K.; Volkert, B.; Hettrich, K.; Schmidt, I.; Fischer, K. Properties and applications of cellulose acetate. *Macromol. Symp.* **2008**, *262*, 89–96. [CrossRef]
48. Kumar, R.; Derbigny, W.A. Cellulose Acetate Electrophoresis of Hemoglobin. In *Electrophoretic Separation of Proteins. Methods in Molecular Biology*; Humana Press: New York, NY, USA, 2019; Volume 1855.
49. Schenk, F.; Weber, P.; Vogler, J.; Hecht, L.; Dietzel, A.; Gauglitz, G. Development of a paper-based lateral flow immunoassay for simultaneous detection of lipopolysaccharides of Salmonella serovars. *Anal. Bioanal. Chem.* **2018**, *410*, 863–868. [CrossRef]
50. Hamedi, M.M.; Ünal, B.; Kerr, E.; Glavan, A.C.; Fernandez-Abedul, M.T.; Whitesides, G.M. Coated and uncoated cellophane as materials for microplates and open-channel microfluidics devices. *Lab Chip* **2016**, *16*, 3885–3897. [CrossRef] [PubMed]
51. Pelton, R. Bioactive paper provides a low-cost platform for diagnostics. *Trac Trends Anal. Chem.* **2009**, *28*, 925–942. [CrossRef]
52. Hubbe, M.A.; Rojas, O.J.; Sulic, N.; Sezaki, T. Unique behaviour of polyamholytes as dry-strength additives. *Appita J.* **2007**, *60*, 106–111.
53. Glavan, A.C.; Christodouleas, D.C.; Mosadegh, B.; Yu, H.D.; Smith, B.S.; Lessing, J.; Fernández-Abedul, M.T.; Whitesides, G.M. Folding Analytical Devices for Electrochemical ELISA in Hydrophobic R H Paper. *Anal. Chem.* **2014**, *86*, 11999–12007. [CrossRef]
54. Glavan, A.C.; Martinez, R.V.; Subramaniam, A.B.; Yoon, H.J.; Nunes, R.M.D.; Lange, H.; Thuo, M.M.; Whitesides, G.M. Omniphobic "rF paper" produced by silanization of paper with fluoroalkyltrichlorosilanes. *Adv. Funct. Mater.* **2014**, *24*, 60–70. [CrossRef]
55. Cassano, C.L.; Fan, Z.H. Laminated paper-based analytical devices (LPAD): Fabrication, characterization, and assays. *Microfluid. Nanofluidics* **2013**, *15*, 173–181. [CrossRef]
56. Washburn, E.W. The dynamics of capillary flow. *Phys. Rev.* **1921**, *17*, 273–283. [CrossRef]
57. Salentijn, G.I.J.; Grajewski, M.; Verpoorte, E. Reinventing (Bio)chemical Analysis with Paper. *Anal. Chem.* **2018**, *90*, 13815–13825. [CrossRef]
58. Mtibe, A.; Linganiso, L.Z.; Mathew, A.P.; Oksman, K.; John, M.J.; Anandjiwala, R.D. A comparative study on properties of micro and nanopapers produced from cellulose and cellulose nanofibres. *Carbohydr. Polym.* **2015**, *118*, 1–8. [CrossRef] [PubMed]
59. Naghdi, T.; Yousefi, H.; Sharifi, A.R.; Golmohammadi, H. Nanopaper-based sensors. In *Comprehensive Analytical Chemistry*; Barceló, D., Ed.; *Paper-based Sensors* (Merkoçi, A., Volume Ed.); Elsevier: Amsterdam, The Netherlands, 2020; pp. 257–312.
60. Eynaki, H.; Kiani, M.A.; Golmohammadi, H. Nanopaper-based screen-printed electrodes: A hybrid sensing bioplatform for dual opto-electrochemical sensing applications. *Nanoscale* **2020**, *12*, 18409–18417. [CrossRef]
61. Yagoda, H. Applications of confined spot tests in analytical chemistry: Preliminary paper. *Ind. Eng. Chem. Anal. Ed.* **1937**, *9*, 79–82. [CrossRef]
62. Banik, U.K.; Hirsch, M.A.; Irvine, D.S.; Krupey, J.; Hurwitz, A.; Singh, K.; Wetzel, J.; Givner, M.L. A simple and sensitive nonradioactive method for the detection of urinary human chorionic gonadotropin and diagnosis of early human pregnancy. II. Single-unit test. *Fertil. Steril.* **1979**, *32*, 426–432. [CrossRef]
63. Yager, P.; Edwards, T.; Fu, E.; Helton, K.; Nelson, K.; Tam, M.R.; Weigl, B.H. Microfluidic diagnostic technologies for global public health. *Nature* **2006**, *442*, 412–418. [CrossRef] [PubMed]
64. Wang, H.; Liu, J.; Graham Cooks, R.; Ouyang, Z. Paper spray for direct analysis of complex mixtures using mass spectrometry. *Angew. Chem. Int. Ed.* **2010**, *49*, 877–880. [CrossRef] [PubMed]
65. de Oliveira, T.R.; Fonseca, W.T.; de Oliveira Setti, G.; Faria, R.C. Fast and flexible strategy to produce electrochemical paper-based analytical devices using a craft cutter printer to create wax barrier and screen-printed electrodes. *Talanta* **2019**, *195*, 480–489. [CrossRef] [PubMed]
66. Hamedi, M.M.; Ainla, A.; Güder, F.; Christodouleas, D.C.; Fernández-Abedul, M.T.; Whitesides, G.M. Integrating Electronics and Microfluidics on Paper. *Adv. Mater.* **2016**, *28*, 5054–5063. [CrossRef]
67. Amor-Gutiérrez, O.; Costa-Rama, E.; Fernández-Abedul, M.T. Sampling and multiplexing in lab-on-paper bioelectroanalytical devices for glucose determination. *Biosens. Bioelectron.* **2019**, *135*, 64–70. [CrossRef] [PubMed]
68. Jiang, N.; Ahmed, R.; Damayantharan, M.; Ünal, B.; Butt, H.; Yetisen, A.K. Lateral and Vertical Flow Assays for Point-of-Care Diagnostics. *Adv. Healthc. Mater.* **2019**, *8*, 1–19. [CrossRef]

69. Wang, K.; Qin, W.; Hou, Y.; Xiao, K.; Yan, W. The application of lateral flow immunoassay in point of care testing: A review. *Nano Biomed. Eng.* **2016**, *8*, 172–183. [CrossRef]
70. Reches, M.; Mirica, K.A.; Dasgupta, R.; Dickey, M.D.; Butte, M.J.; Whitesides, G.M. Thread as a Matrix for Biomedical Assays. *ACS Appl. Mater. Interfaces* **2010**, *2*, 1722–1728. [CrossRef]
71. Glavan, A.C.; Ainla, A.; Hamedi, M.M.; Fernández-Abedul, M.T.; Whitesides, G.M. Electroanalytical devices with pins and thread. *Lab Chip* **2016**, *16*, 112–119. [CrossRef]
72. Mousavi, M.P.S.; Ainla, A.; Tan, E.K.W.; Abd El-Rahman, M.; Yoshida, Y.; Yuan, L.; Sigurslid, H.H.; Arkan, N.; Yip, M.C.; Abrahamsson, C.K.; et al. Ion sensing with thread-based potentiometric electrodes. *Lab Chip* **2018**, *18*, 2279–2290. [CrossRef]
73. Boonkaew, S.; Chaiyo, S.; Jampasa, S.; Rengpipat, S.; Siangproh, W.; Chailapakul, O. An origami paper-based electrochemical immunoassay for the C-reactive protein using a screen-printed carbon electrode modified with graphene and gold nanoparticles. *Microchim. Acta* **2019**, *186*. [CrossRef]
74. Wang, C.C.; Hennek, J.W.; Ainla, A.; Kumar, A.A.; Lan, W.J.; Im, J.; Smith, B.S.; Zhao, M.; Whitesides, G.M. A Paper-Based Pop-up Electrochemical Device for Analysis of Beta-Hydroxybutyrate. *Anal. Chem.* **2016**, *88*, 6326–6333. [CrossRef]
75. Verma, M.S.; Tsaloglou, M.-N.; Sisley, T.; Christodouleas, D.; Chen, A.; Milette, J.; Whitesides, G.M. Sliding-strip microfluidic device enables ELISA on paper. *Biosens. Bioelectron.* **2018**, *99*, 77–84. [CrossRef]
76. Renault, C.; Anderson, M.J.; Crooks, R.M. Electrochemistry in hollow-channel paper analytical devices. *J. Am. Chem. Soc.* **2014**, *136*, 4616–4623. [CrossRef] [PubMed]
77. Renault, C.; Li, X.; Fosdick, S.E.; Crooks, R.M. Hollow-channel paper analytical devices. *Anal. Chem.* **2013**, *85*, 7976–7979. [CrossRef]
78. Cunningham, J.C.; DeGregory, P.R.; Crooks, R.M. New Functionalities for Paper-Based Sensors Lead to Simplified User Operation, Lower Limits of Detection, and New Applications. *Annu. Rev. Anal. Chem.* **2016**, *9*, 183–202. [CrossRef]
79. Noiphung, J.; Songjaroen, T.; Dungchai, W.; Henry, C.S.; Chailapakul, O.; Laiwattanapaisal, W. Electrochemical detection of glucose from whole blood using paper-based microfluidic devices. *Anal. Chim. Acta* **2013**, *788*, 39–45. [CrossRef] [PubMed]
80. González-López, A.; García-Manrique, P.; Blanco-López, M.C.; Fernández-Abedul, M.T. Integrated Electrophoresis Separation and Electrochemical Detection in a Paper-based Device. *Procedia Technol.* **2017**, *27*, 21–22. [CrossRef]
81. Costa-Rama, E.; Nouws, H.P.A.; Delerue-Matos, C.; Blanco-López, M.C.; Fernández-Abedul, M.T. Preconcentration and sensitive determination of the anti-inflammatory drug diclofenac on a paper-based electroanalytical platform. *Anal. Chim. Acta* **2019**, *1074*, 89–97. [CrossRef] [PubMed]
82. Edelbroek, P.M.; Heijden, J.V.D.; Stolk, L.M.L. Dried blood spot methods in therapeutic drug monitoring: Methods, assays, and pitfalls. *Ther. Drug Monit.* **2009**, *31*, 327–336. [CrossRef]
83. Hajian, A.; Wang, Z.; Berglund, L.A.; Hamedi, M.M. Cellulose Nanopaper with Monolithically Integrated Conductive Micropatterns. *Adv. Electron. Mater.* **2019**, *5*, 1800924. [CrossRef]
84. Kong, F.Y.; Gu, S.X.; Li, W.W.; Chen, T.T.; Xu, Q.; Wang, W. A paper disk equipped with graphene/polyaniline/Au nanoparticles/glucose oxidase biocomposite modified screen-printed electrode: Toward whole blood glucose determination. *Biosens. Bioelectron.* **2014**, *56*, 77–82. [CrossRef] [PubMed]
85. Cao, Q.; Liang, B.; Yu, C.; Fang, L.; Tu, T.; Wei, J.; Ye, X. High accuracy determination of multi metabolite by an origami-based coulometric electrochemical biosensor. *J. Electroanal. Chem.* **2020**, *873*, 114358. [CrossRef]
86. Kuretake, T.; Kawahara, S.; Motooka, M.; Uno, S. An electrochemical gas biosensor based on enzymes immobilized on chromatography paper for ethanol vapor detection. *Sensors* **2017**, *17*, 281. [CrossRef]
87. Scala-Benuzzi, M.L.; Raba, J.; Soler-Illia, G.J.A.A.; Schneider, R.J.; Messina, G.A. Novel Electrochemical Paper-Based Immunocapture Assay for the Quantitative Determination of Ethinylestradiol in Water Samples. *Anal. Chem.* **2018**, *90*, 4104–4111. [CrossRef]
88. Liu, X.; Li, X.; Gao, X.; Ge, L.; Sun, X.; Li, F. A Universal Paper-Based Electrochemical Sensor for Zero-Background Assay of Diverse Biomarkers. *ACS Appl. Mater. Interfaces* **2019**, *11*, 15381–15388. [CrossRef]
89. Kuek Lawrence, C.S.; Tan, S.N.; Floresca, C.Z. A "green" cellulose paper based glucose amperometric biosensor. *Sens. Actuators B Chem.* **2014**, *193*, 536–541. [CrossRef]
90. Honikel, M.M.; Lin, C.E.; Cardinell, B.A.; LaBelle, J.T.; Penman, A.D. Direct Measurement of a Biomarker's Native Optimal Frequency with Physical Adsorption Based Immobilization. *ACS Sens.* **2018**, *3*, 823–831. [CrossRef]
91. Rattanarat, P.; Dungchai, W.; Siangproh, W.; Chailapakul, O.; Henry, C.S. Sodium dodecyl sulfate-modified electrochemical paper-based analytical device for determination of dopamine levels in biological samples. *Anal. Chim. Acta* **2012**, *744*, 1–7. [CrossRef] [PubMed]
92. Primpray, V.; Chailapakul, O.; Tokeshi, M.; Rojanarata, T.; Laiwattanapaisal, W. A paper-based analytical device coupled with electrochemical detection for the determination of dexamethasone and prednisolone in adulterated traditional medicines. *Anal. Chim. Acta* **2019**, *1078*, 16–23. [CrossRef] [PubMed]
93. Zhang, J.; Yang, Z.; Liu, Q.; Liang, H. Electrochemical biotoxicity detection on a microfluidic paper-based analytical device via cellular respiratory inhibition. *Talanta* **2019**, *202*, 384–391. [CrossRef]
94. Sekar, N.C.; Ge, L.; Mousavi Shaegh, S.A.; Ng, S.H.; Tan, S.N. A mediated turnip tissue paper-based amperometric hydrogen peroxide biosensor. *Sens. Actuators B Chem.* **2015**, *210*, 336–342. [CrossRef]

95. Yang, N.; Zhou, X.; Yu, D.; Jiao, S.; Han, X.; Zhang, S.; Yin, H.; Mao, H. Pesticide residues identification by impedance time-sequence spectrum of enzyme inhibition on multilayer paper-based microfluidic chip. *J. Food Process Eng.* **2020**, *43*, e13544. [CrossRef]
96. Shriver-Lake, L.C.; Zabetakis, D.; Dressick, W.J.; Stenger, D.A.; Trammell, S.A. Paper-based electrochemical detection of chlorate. *Sensors* **2018**, *18*, 328. [CrossRef]
97. Moreira, C.M.; Pereira, S.V.; Raba, J.; Bertolino, F.A.; Messina, G.A. Paper-based enzymatic platform coupled to screen printed graphene-modified electrode for the fast neonatal screening of phenylketonuria. *Clin. Chim. Acta* **2018**, *486*, 59–65. [CrossRef] [PubMed]
98. Rahimi-Mohseni, M.; Raoof, J.B.; Ojani, R.; Aghajanzadeh, T.A.; Bagheri Hashkavayi, A. Development of a new paper based nano-biosensor using the co-catalytic effect of tyrosinase from banana peel tissue (Musa Cavendish) and functionalized silica nanoparticles for voltammetric determination of L-tyrosine. *Int. J. Biol. Macromol.* **2018**, *113*, 648–654. [CrossRef]
99. Rahimi-Mohseni, M.; Raoof, J.B.; Aghajanzadeh, T.A.; Ojani, R. Rapid Determination of Phenolic Compounds in Water Samples: Development of a Paper-based Nanobiosensor Modified with Functionalized Silica Nanoparticles and Potato Tissue. *Electroanalysis* **2019**, *31*, 2311–2318. [CrossRef]
100. Delaney, J.L.; Hogan, C.F.; Tian, J.; Shen, W. Electrogenerated chemiluminescence detection in paper-based microfluidic sensors. *Anal. Chem.* **2011**, *83*, 1300–1306. [CrossRef]
101. Cao, L.; Fang, C.; Zeng, R.; Zhao, X.; Zhao, F.; Jiang, Y.; Chen, Z. A disposable paper-based microfluidic immunosensor based on reduced graphene oxide-tetraethylene pentamine/Au nanocomposite decorated carbon screen-printed electrodes. *Sens. Actuators B Chem.* **2017**, *252*, 44–54. [CrossRef]
102. Tan, S.N.; Ge, L.; Tan, H.Y.; Loke, W.K.; Gao, J.; Wang, W. Paper-based enzyme immobilization for flow injection electrochemical biosensor integrated with reagent-loaded cartridge toward portable modular device. *Anal. Chem.* **2012**, *84*, 10071–10076. [CrossRef]
103. Sánchez-Calvo, A.; Blanco-López, M.C.; Costa-García, A. Paper-based working electrodes coated with mercury or bismuth films for heavy metals determination. *Biosensors* **2020**, *10*, 52. [CrossRef]
104. Núnez-Bajo, E.; Blanco-López, M.C.; Costa-García, A.; Fernández-Abedul, M.T. In situ gold-nanoparticle electrogeneration on gold films deposited on paper for non-enzymatic electrochemical determination of glucose. *Talanta* **2018**, *178*, 160–165. [CrossRef]
105. Nunez-Bajo, E.; Blanco-López, M.C.; Costa-García, A.; Fernández-Abedul, M.T. Electrogeneration of Gold Nanoparticles on Porous-Carbon Paper-Based Electrodes and Application to Inorganic Arsenic Analysis in White Wines by Chronoamperometric Stripping. *Anal. Chem.* **2017**, *89*, 6415–6423. [CrossRef]
106. Yáñez-Sedeño, P.; Campuzano, S.; Pingarrón, J.M. Screen-printed electrodes: Promising paper and wearable transducers for (bio)sensing. *Biosensors* **2020**, *10*, 76. [CrossRef]
107. Chinnadayyala, S.R.; Park, J.; Le, H.T.N.; Santhosh, M.; Kadam, A.N.; Cho, S. Recent advances in microfluidic paper-based electrochemiluminescence analytical devices for point-of-care testing applications. *Biosens. Bioelectron.* **2019**, *126*, 68–81. [CrossRef] [PubMed]
108. Ozer, T.; McMahon, C.; Henry, C.S. Advances in Paper-Based Analytical Devices. *Annu. Rev. Anal. Chem.* **2020**, *13*, 85–109. [CrossRef]
109. Liu, M.M.; Guo, Z.Z.; Liu, H.; Li, S.H.; Chen, Y.; Zhong, Y.; Lei, Y.; Lin, X.H.; Liu, A.L. Paper-based 3D culture device integrated with electrochemical sensor for the on-line cell viability evaluation of amyloid-beta peptide induced damage in PC12 cells. *Biosens. Bioelectron.* **2019**, *144*, 111686. [CrossRef] [PubMed]
110. Aller Pellitero, M.; Kitsara, M.; Eibensteiner, F.; Del Campo, F.J. Rapid prototyping of electrochemical lateral flow devices: Stencilled electrodes. *Analyst* **2016**, *141*, 2515–2522. [CrossRef] [PubMed]
111. Ruiz-Vega, G.; Garcia-Berrocoso, T.; Montaner, J.; Baldrich, E. Paper microfluidics on screen-printed electrodes for simple electrochemical magneto-immunosensor performance. *Sens. Actuators B Chem.* **2019**, *298*, 126897. [CrossRef]
112. Ruiz-Vega, G.; Kitsara, M.; Pellitero, M.A.; Baldrich, E.; del Campo, F.J. Electrochemical Lateral Flow Devices: Towards Rapid Immunomagnetic Assays. *ChemElectroChem* **2017**, *4*, 880–889. [CrossRef]
113. Sinawang, P.D.; Rai, V.; Ionescu, R.E.; Marks, R.S. Electrochemical lateral flow immunosensor for detection and quantification of dengue NS1 protein. *Biosens. Bioelectron.* **2016**, *77*, 400–408. [CrossRef] [PubMed]
114. Cinti, S.; Fiore, L.; Massoud, R.; Cortese, C.; Moscone, D.; Palleschi, G.; Arduini, F. Low-cost and reagent-free paper-based device to detect chloride ions in serum and sweat. *Talanta* **2018**, *179*, 186–192. [CrossRef]
115. Arduini, F.; Cinti, S.; Caratelli, V.; Amendola, L.; Palleschi, G.; Moscone, D. Origami multiple paper-based electrochemical biosensors for pesticide detection. *Biosens. Bioelectron.* **2019**, *126*, 346–354. [CrossRef] [PubMed]
116. Tsaloglou, M.N.; Nemiroski, A.; Camci-Unal, G.; Christodouleas, D.C.; Murray, L.P.; Connelly, J.T.; Whitesides, G.M. Handheld isothermal amplification and electrochemical detection of DNA in resource-limited settings. *Anal. Biochem.* **2018**, *543*, 116–121. [CrossRef] [PubMed]
117. Pinyorospathum, C.; Chaiyo, S.; Sae-ung, P.; Hoven, V.P.; Damsongsang, P.; Siangproh, W.; Chailapakul, O. Disposable paper-based electrochemical sensor using thiol-terminated poly(2-methacryloyloxyethyl phosphorylcholine) for the label-free detection of C-reactive protein. *Microchim. Acta* **2019**, *186*, 472. [CrossRef]
118. Lou, B.; Chen, C.; Zhou, Z.; Zhang, L.; Wang, E.; Dong, S. A novel electrochemical sensing platform for anions based on conducting polymer film modified electrodes integrated on paper-based chips. *Talanta* **2013**, *105*, 40–45. [CrossRef] [PubMed]

119. Kit-Anan, W.; Olarnwanich, A.; Sriprachuabwong, C.; Karuwan, C.; Tuantranont, A.; Wisitsoraat, A.; Srituravanich, W.; Pimpin, A. Disposable paper-based electrochemical sensor utilizing inkjet-printed Polyaniline modified screen-printed carbon electrode for Ascorbic acid detection. *J. Electroanal. Chem.* **2012**, *685*, 72–78. [CrossRef]
120. Teengam, P.; Siangproh, W.; Tuantranont, A.; Henry, C.S.; Vilaivan, T.; Chailapakul, O. Electrochemical paper-based peptide nucleic acid biosensor for detecting human papillomavirus. *Anal. Chim. Acta* **2017**, *952*, 32–40. [CrossRef]
121. Martins, G.V.; Marques, A.C.; Fortunato, E.; Sales, M.G.F. Wax-printed paper-based device for direct electrochemical detection of 3-nitrotyrosine. *Electrochim. Acta* **2018**, *284*, 60–68. [CrossRef]
122. Pungjunun, K.; Nantaphol, S.; Praphairaksit, N.; Siangproh, W.; Chaiyo, S.; Chailapakul, O. Enhanced sensitivity and separation for simultaneous determination of tin and lead using paper-based sensors combined with a portable potentiostat. *Sens. Actuators B Chem.* **2020**, *318*, 128241. [CrossRef]
123. Cinti, S.; Talarico, D.; Palleschi, G.; Moscone, D.; Arduini, F. Novel reagentless paper-based screen-printed electrochemical sensor to detect phosphate. *Anal. Chim. Acta* **2016**, *919*, 78–84. [CrossRef]
124. Chaiyo, S.; Mehmeti, E.; Siangproh, W.; Hoang, T.L.; Nguyen, H.P.; Chailapakul, O.; Kalcher, K. Non-enzymatic electrochemical detection of glucose with a disposable paper-based sensor using a cobalt phthalocyanine–ionic liquid–graphene composite. *Biosens. Bioelectron.* **2018**, *102*, 113–120. [CrossRef] [PubMed]
125. Colozza, N.; Kehe, K.; Popp, T.; Steinritz, D.; Moscone, D.; Arduini, F. Paper-based electrochemical sensor for on-site detection of the sulphur mustard. *Env. Sci. Pollut. Res.* **2018**, *80*, 6928–6934. [CrossRef]
126. Cinti, S.; De Lellis, B.; Moscone, D.; Arduini, F. Sustainable monitoring of Zn(II) in biological fluids using office paper. *Sens. Actuators B Chem.* **2017**, *253*, 1199–1206. [CrossRef]
127. Tomei, M.R.; Cinti, S.; Interino, N.; Manovella, V.; Moscone, D.; Arduini, F. Paper-based electroanalytical strip for user-friendly blood glutathione detection. *Sens. Actuators B Chem.* **2019**, *294*, 291–297. [CrossRef]
128. Jemmeli, D.; Marcoccio, E.; Moscone, D.; Dridi, C.; Arduini, F. Highly sensitive paper-based electrochemical sensor for reagent free detection of bisphenol A. *Talanta* **2020**, *216*, 120924. [CrossRef] [PubMed]
129. Lamas-Ardisana, P.J.; Martínez-Paredes, G.; Añorga, L.; Grande, H.J. Glucose biosensor based on disposable electrochemical paper-based transducers fully fabricated by screen-printing. *Biosens. Bioelectron.* **2018**, *109*, 8–12. [CrossRef]
130. Suresh, V.; Qunya, O.; Kanta, B.L.; Yuh, L.Y.; Chong, K.S.L. Non-invasive paper-based microfluidic device for ultra-low detection of urea through enzyme catalysis. *R. Soc. Open Sci.* **2018**, *5*, 171980. [CrossRef] [PubMed]
131. Wei, B.; Mao, K.; Liu, N.; Zhang, M.; Yang, Z. Graphene nanocomposites modified electrochemical aptamer sensor for rapid and highly sensitive detection of prostate specific antigen. *Biosens. Bioelectron.* **2018**, *121*, 41–46. [CrossRef]
132. Nantaphol, S.; Jesadabundit, W.; Chailapakul, O.; Siangproh, W. A new electrochemical paper platform for detection of 8-hydroxyquinoline in cosmetics using a cobalt phthalocyanine-modified screen-printed carbon electrode. *J. Electroanal. Chem.* **2019**, *832*, 480–485. [CrossRef]
133. Scordo, G.; Moscone, D.; Palleschi, G.; Arduini, F. A reagent-free paper-based sensor embedded in a 3D printing device for cholinesterase activity measurement in serum. *Sens. Actuators B Chem.* **2018**, *258*, 1015–1021. [CrossRef]
134. Boobphahom, S.; Ruecha, N.; Rodthongkum, N.; Chailapakul, O.; Remcho, V.T. A copper oxide-ionic liquid/reduced graphene oxide composite sensor enabled by digital dispensing: Non-enzymatic paper-based microfluidic determination of creatinine in human blood serum. *Anal. Chim. Acta* **2019**, *1083*, 110–118. [CrossRef]
135. Lamas-Ardisana, P.J.; Casuso, P.; Fernandez-Gauna, I.; Martínez-Paredes, G.; Jubete, E.; Añorga, L.; Cabañero, G.; Grande, H.J. Disposable electrochemical paper-based devices fully fabricated by screen-printing technique. *Electrochem. Commun.* **2017**, *75*, 25–28. [CrossRef]
136. Deroco, P.B.; Fatibello-Filho, O.; Arduini, F.; Moscone, D. Electrochemical determination of capsaicin in pepper samples using sustainable paper-based screen-printed bulk modified with carbon black. *Electrochim. Acta* **2020**, *354*, 136628. [CrossRef]
137. Martins, G.V.; Marques, A.C.; Fortunato, E.; Sales, M.G.F. Paper-based (bio)sensor for label-free detection of 3-nitrotyrosine in human urine samples using molecular imprinted polymer. *Sens. Bio-Sens. Res.* **2020**, *28*, 100333. [CrossRef]
138. Pradela-Filho, L.A.; Andreotti, I.A.A.; Carvalho, J.H.S.; Araújo, D.A.G.; Orzari, L.O.; Gatti, A.; Takeuchi, R.M.; Santos, A.L.; Janegitz, B.C. Glass varnish-based carbon conductive ink: A new way to produce disposable electrochemical sensors. *Sens. Actuators B Chem.* **2020**, *305*, 127433. [CrossRef]
139. Pavithra, M.; Muruganand, S.; Parthiban, C. Development of novel paper based electrochemical immunosensor with self-made gold nanoparticle ink and quinone derivate for highly sensitive carcinoembryonic antigen. *Sens. Actuators B Chem.* **2018**, *257*, 496–503. [CrossRef]
140. Cinti, S.; Proietti, E.; Casotto, F.; Moscone, D.; Arduini, F. Paper-Based Strips for the Electrochemical Detection of Single and Double Stranded DNA. *Anal. Chem.* **2018**, *90*, 13680–13686. [CrossRef] [PubMed]
141. Cinti, S.; Mazzaracchio, V.; Cacciotti, I.; Moscone, D.; Arduini, F. Carbon black-modified electrodes screen-printed onto paper towel, waxed paper and parafilm M®. *Sensors* **2017**, *17*, 2267. [CrossRef]
142. Wang, Y.; Sun, S.; Luo, J.; Xiong, Y.; Ming, T.; Liu, J.; Ma, Y.; Yan, S.; Yang, Y.; Yang, Z.; et al. Low sample volume origami-paper-based graphene-modified aptasensors for label-free electrochemical detection of cancer biomarker-EGFR. *Microsyst. Nanoeng.* **2020**, *6*, 32. [CrossRef]

143. Pungjunun, K.; Chaiyo, S.; Praphairaksit, N.; Siangproh, W.; Ortner, A.; Kalcher, K.; Chailapakul, O.; Mehmeti, E. Electrochemical detection of NOx gas based on disposable paper-based analytical device using a copper nanoparticles-modified screen-printed graphene electrode. *Biosens. Bioelectron.* **2019**, *143*, 111606. [CrossRef] [PubMed]
144. Ruecha, N.; Shin, K.; Chailapakul, O.; Rodthongkum, N. Label-free paper-based electrochemical impedance immunosensor for human interferon gamma detection. *Sens. Actuators B Chem.* **2019**, *279*, 298–304. [CrossRef]
145. Cao, L.; Han, G.C.; Xiao, H.; Chen, Z.; Fang, C. A novel 3D paper-based microfluidic electrochemical glucose biosensor based on rGO-TEPA/PB sensitive film. *Anal. Chim. Acta* **2020**, *1096*, 34–43. [CrossRef] [PubMed]
146. Wang, Y.; Luo, J.; Liu, J.; Li, X.; Kong, Z.; Jin, H.; Cai, X. Electrochemical integrated paper-based immunosensor modified with multi-walled carbon nanotubes nanocomposites for point-of-care testing of 17β-estradiol. *Biosens. Bioelectron.* **2018**, *107*, 47–53. [CrossRef]
147. Boonyasit, Y.; Chailapakul, O.; Laiwattanapaisal, W. A folding affinity paper-based electrochemical impedance device for cardiovascular risk assessment. *Biosens. Bioelectron.* **2019**, *130*, 389–396. [CrossRef]
148. Yakoh, A.; Siangproh, W.; Chailapakul, O.; Ngamrojanavanich, N. Optical Bioelectronic Device Based on a Screen-Printed Electroluminescent Transducer. *ACS Appl. Mater. Interfaces* **2020**, *12*, 22543–22551. [CrossRef]
149. Cao, L.; Fang, C.; Zeng, R.; Zhao, X.; Jiang, Y.; Chen, Z. Paper-based microfluidic devices for electrochemical immunofiltration analysis of human chorionic gonadotropin. *Biosens. Bioelectron.* **2017**, *92*, 87–94. [CrossRef]
150. Punjiya, M.; Moon, C.H.; Matharu, Z.; Rezaei Nejad, H.; Sonkusale, S. A three-dimensional electrochemical paper-based analytical device for low-cost diagnostics. *Analyst* **2018**, *143*, 1059–1064. [CrossRef]
151. Wang, P.; Ge, L.; Yan, M.; Song, X.; Ge, S.; Yu, J. Paper-based three-dimensional electrochemical immunodevice based on multi-walled carbon nanotubes functionalized paper for sensitive point-of-care testing. *Biosens. Bioelectron.* **2012**, *32*, 238–243. [CrossRef] [PubMed]
152. Li, L.; Zhang, Y.; Zhang, L.; Ge, S.; Yan, M.; Yu, J. Steric paper based ratio-type electrochemical biosensor with hollow-channel for sensitive detection of Zn2+. *Sci. Bull.* **2017**, *62*, 1114–1121. [CrossRef]
153. Wu, Y.; Xue, P.; Hui, K.M.; Kang, Y. A paper-based microfluidic electrochemical immunodevice integrated with amplification-by-polymerization for the ultrasensitive multiplexed detection of cancer biomarkers. *Biosens. Bioelectron.* **2014**, *52*, 180–187. [CrossRef]
154. Panraksa, Y.; Siangproh, W.; Khampieng, T.; Chailapakul, O.; Apilux, A. Paper-based amperometric sensor for determination of acetylcholinesterase using screen-printed graphene electrode. *Talanta* **2018**, *178*, 1017–1023. [CrossRef] [PubMed]
155. Amatatongchai, M.; Sitanurak, J.; Sroysee, W.; Sodanat, S.; Chairam, S.; Jarujamrus, P.; Nacapricha, D.; Lieberzeit, P.A. Highly sensitive and selective electrochemical paper-based device using a graphite screen-printed electrode modified with molecularly imprinted polymers coated Fe3O4@Au@SiO2 for serotonin determination. *Anal. Chim. Acta* **2019**, *1077*, 255–265. [CrossRef]
156. Colozza, N.; Kehe, K.; Dionisi, G.; Popp, T.; Tsoutsoulopoulos, A.; Steinritz, D.; Moscone, D.; Arduini, F. A wearable origami-like paper-based electrochemical biosensor for sulfur mustard detection. *Biosens. Bioelectron.* **2019**, *129*, 15–23. [CrossRef]
157. Cinti, S.; Minotti, C.; Moscone, D.; Palleschi, G.; Arduini, F. Fully integrated ready-to-use paper-based electrochemical biosensor to detect nerve agents. *Biosens. Bioelectron.* **2017**, *93*, 46–51. [CrossRef]
158. Cao, Q.; Liang, B.; Tu, T.; Wei, J.; Fang, L.; Ye, X. Three-dimensional paper-based microfluidic electrochemical integrated devices (3D-PMED) for wearable electrochemical glucose detection. *RSC Adv.* **2019**, *9*, 5674–5681. [CrossRef]
159. Wang, Y.; Luo, J.; Liu, J.; Sun, S.; Xiong, Y.; Ma, Y.; Yan, S.; Yang, Y.; Yin, H.; Cai, X. Label-free microfluidic paper-based electrochemical aptasensor for ultrasensitive and simultaneous multiplexed detection of cancer biomarkers. *Biosens. Bioelectron.* **2019**, *136*, 84–90. [CrossRef]
160. Wang, Y.; Zang, D.; Ge, S.; Ge, L.; Yu, J.; Yan, M. A novel microfluidic origami photoelectrochemical sensor based on CdTe quantum dots modified molecularly imprinted polymer and its highly selective detection of S-fenvalerate. *Electrochim. Acta* **2013**, *107*, 147–154. [CrossRef]
161. Fan, Y.; Shi, S.; Ma, J.; Guo, Y. A paper-based electrochemical immunosensor with reduced graphene oxide/thionine/gold nanoparticles nanocomposites modification for the detection of cancer antigen 125. *Biosens. Bioelectron.* **2019**, *135*, 1–7. [CrossRef] [PubMed]
162. Yan, J.; Yan, M.; Ge, L.; Ge, S.; Yu, J. An origami electrochemiluminescence immunosensor based on gold/graphene for specific, sensitive point-of-care testing of carcinoembryonic antigen. *Sens. Actuators B Chem.* **2014**, *193*, 247–254. [CrossRef]
163. Su, Y.; Liang, Y.; Wu, H.; Jiang, J.; Lai, W.; Zhang, C. A three-dimensional cloth-based microfluidic label-free proximity hybridization-electrochemiluminescence biosensor for ultrasensitive detection of K-ras gene. *Sens. Actuators B Chem.* **2019**, *296*, 126654. [CrossRef]
164. Li, L.; Li, W.; Ma, C.; Yang, H.; Ge, S.; Yu, J. Paper-based electrochemiluminescence immunodevice for carcinoembryonic antigen using nanoporous gold-chitosan hybrids and graphene quantum dots functionalized Au@Pt. *Sens. Actuators B Chem.* **2014**, *202*, 314–322. [CrossRef]
165. Li, L.; Xu, J.; Zheng, X.; Ma, C.; Song, X.; Ge, S.; Yu, J.; Yan, M. Growth of gold-manganese oxide nanostructures on a 3D origami device for glucose-oxidase label based electrochemical immunosensor. *Biosens. Bioelectron.* **2014**, *61*, 76–82. [CrossRef]
166. Lu, J.; Ge, S.; Ge, L.; Yan, M.; Yu, J. Electrochemical DNA sensor based on three-dimensional folding paper device for specific and sensitive point-of-care testing. *Electrochim. Acta* **2012**, *80*, 334–341. [CrossRef]

167. Maier, D.; Laubender, E.; Basavanna, A.; Schumann, S.; Güder, F.; Urban, G.A.; Dincer, C. Toward Continuous Monitoring of Breath Biochemistry: A Paper-Based Wearable Sensor for Real-Time Hydrogen Peroxide Measurement in Simulated Breath. *ACS Sens.* **2019**, *4*, 2945–2951. [CrossRef]
168. Shkodra, B.; Abera, B.D.; Cantarella, G.; Douaki, A.; Avancini, E.; Petti, L.; Lugli, P. Flexible and printed electrochemical immunosensor coated with oxygen plasma treated SWCNTs for histamine detection. *Biosensors* **2020**, *10*, 35. [CrossRef]
169. Uliana, C.V.; Peverari, C.R.; Afonso, A.S.; Cominetti, M.R.; Faria, R.C. Fully disposable microfluidic electrochemical device for detection of estrogen receptor alpha breast cancer biomarker. *Biosens. Bioelectron.* **2018**, *99*, 156–162. [CrossRef]
170. Adkins, J.A.; Boehle, K.; Friend, C.; Chamberlain, B.; Bisha, B.; Henry, C.S. Colorimetric and Electrochemical Bacteria Detection Using Printed Paper- and Transparency-Based Analytic Devices. *Anal. Chem.* **2017**, *89*, 3613–3621. [CrossRef]
171. Khan, S.; Ali, S.; Bermak, A. Recent developments in printing flexible and wearable sensing electronics for healthcare applications. *Sensors* **2019**, *19*, 1230. [CrossRef]
172. Sempionatto, J.R.; Jeerapan, I.; Krishnan, S.; Wang, J. Wearable Chemical Sensors: Emerging Systems for On-Body Analytical Chemistry. *Anal. Chem.* **2019**, *92*, 378–396. [CrossRef] [PubMed]
173. Kim, J.; Campbell, A.S.; de Ávila, B.E.F.; Wang, J. Wearable biosensors for healthcare monitoring. *Nat. Biotechnol.* **2019**, *37*, 389–406. [CrossRef] [PubMed]
174. Ciui, B.; Martin, A.; Mishra, R.K.; Brunetti, B.; Nakagawa, T.; Dawkins, T.J.; Lyu, M.; Cristea, C.; Sandulescu, R.; Wang, J. Wearable Wireless Tyrosinase Bandage and Microneedle Sensors: Toward Melanoma Screening. *Adv. Healthc. Mater.* **2018**, *7*, e1701264. [CrossRef]
175. Guinovart, T.; Valdés-Ramírez, G.; Windmiller, J.R.; Andrade, F.J.; Wang, J. Bandage-Based Wearable Potentiometric Sensor for Monitoring Wound pH. *Electroanalysis* **2014**, *26*, 1345–1353. [CrossRef]
176. Kassal, P.; Kim, J.; Kumar, R.; De Araujo, W.R.; Steinberg, I.M.; Steinberg, M.D.; Wang, J. Smart bandage with wireless connectivity for uric acid biosensing as an indicator of wound status. *Electrochem. Commun.* **2015**, *56*, 6–10. [CrossRef]
177. Hubble, L.J.; Wang, J. Sensing at Your Fingertips: Glove-based Wearable Chemical Sensors. *Electroanalysis* **2019**, *31*, 428–436. [CrossRef]
178. Sempionatto, J.R.; Mishra, R.K.; Martín, A.; Tang, G.; Nakagawa, T.; Lu, X.; Campbell, A.S.; Lyu, K.M.; Wang, J. Wearable Ring-Based Sensing Platform for Detecting Chemical Threats. *ACS Sens.* **2017**, *2*, 1531–1538. [CrossRef] [PubMed]
179. Mishra, R.K.; Sempionatto, J.R.; Li, Z.; Brown, C.; Galdino, N.M.; Shah, R.; Liu, S.; Hubble, L.J.; Bagot, K.; Tapert, S.; et al. Simultaneous detection of salivary Δ9-tetrahydrocannabinol and alcohol using a Wearable Electrochemical Ring Sensor. *Talanta* **2020**, *211*, 120757. [CrossRef]
180. García-Carmona, L.; Martín, A.; Sempionatto, J.R.; Moreto, J.R.; González, M.C.; Wang, J.; Escarpa, A. Pacifier Biosensor: Toward Noninvasive Saliva Biomarker Monitoring. *Anal. Chem.* **2019**, *91*, 13883–13891. [CrossRef] [PubMed]
181. Rowe, A.A.; Bonham, A.J.; White, R.J.; Zimmer, M.P.; Yadgar, R.J.; Hobza, T.M.; Honea, J.W.; Ben-Yaacov, I.; Plaxco, K.W. Cheapstat: An open-source, "do-it-yourself" potentiostat for analytical and educational applications. *PLoS ONE* **2011**, *6*, e23783. [CrossRef] [PubMed]
182. Dryden, M.D.M.; Wheeler, A.R. DStat: A versatile, open-source potentiostat for electroanalysis and integration. *PLoS ONE* **2015**, *10*, e0140349. [CrossRef] [PubMed]
183. Glasscott, M.W.; Verber, M.D.; Hall, J.R.; Pendergast, A.D.; McKinney, C.J.; Dick, J.E. SweepStat: A Build-It-Yourself, Two-Electrode Potentiostat for Macroelectrode and Ultramicroelectrode Studies. *J. Chem. Educ.* **2020**, *97*, 265–270. [CrossRef]
184. Ainla, A.; Mousavi, M.P.S.; Tsaloglou, M.N.; Redston, J.; Bell, J.G.; Fernández-Abedul, M.T.; Whitesides, G.M. Open-Source Potentiostat for Wireless Electrochemical Detection with Smartphones. *Anal. Chem.* **2018**, *90*, 6240–6246. [CrossRef]
185. Giordano, G.F.; Vicentini, M.B.R.; Murer, R.C.; Augusto, F.; Ferrão, M.F.; Helfer, G.A.; da Costa, A.B.; Gobbi, A.L.; Hantao, L.W.; Lima, R.S. Point-of-use electroanalytical platform based on homemade potentiostat and smartphone for multivariate data processing. *Electrochim. Acta* **2016**, *219*, 170–177. [CrossRef]
186. Nemiroski, A.; Christodouleas, D.C.; Hennek, J.W.; Kumar, A.A.; Maxwell, E.J.; Fernandez-Abedul, M.T.; Whitesides, G.M. Universal mobile electrochemical detector designed for use in resource-limited applications. *Proc. Natl. Acad. Sci. Usa* **2014**, *111*, 11984–11989. [CrossRef]
187. Esquivel, J.P.; Buser, J.R.; Lim, C.W.; Domínguez, C.; Rojas, S.; Yager, P.; Sabaté, N. Single-use paper-based hydrogen fuel cells for point-of-care diagnostic applications. *J. Power Sources* **2017**, *342*, 442–451. [CrossRef]
188. Bezuidenhout, P.; Smith, S.; Joubert, T.H. A low-cost inkjet-printed paper-based potentiostat. *Appl. Sci.* **2018**, *8*, 968. [CrossRef]
189. Turner, A. The Paper Potentiostat. In Proceedings of the 4th International Conference on Bio-Sensing Technology, Lisbon, Portugal, 10–13 May 2015; Available online: https://www.diva-portal.org/smash/record.jsf?pid=diva2%3A813164&dswid=1323 (accessed on 31 December 2020).
190. Sardini, E.; Serpelloni, M.; Tonello, S. Printed electrochemical biosensors: Opportunities and metrological challenges. *Biosensors* **2020**, *10*, 166. [CrossRef] [PubMed]
191. Kuswandi, B.; Ensafi, A.A. Perspective—Paper-based biosensors: Trending topic in clinical diagnostics developments and commercialization. *J. Electrochem. Soc.* **2020**, *167*, 037509. [CrossRef]

biosensors

MDPI

Article

Graphene Oxide Bulk-Modified Screen-Printed Electrodes Provide Beneficial Electroanalytical Sensing Capabilities

Samuel J. Rowley-Neale [1], Dale A. C. Brownson [1], Graham Smith [2] and Craig E. Banks [1,*]

[1] Faculty of Science and Engineering, Manchester Metropolitan University, Chester Street, Manchester M1 5GD, UK; S.Rowley-Neale@mmu.ac.uk (S.J.R.-N.); d.brownson@mmu.ac.uk (D.A.C.B.)
[2] Department of Natural Sciences, Faculty of Science and Engineering, University of Chester, Thornton Science Park, Pool Lane, Ince, Chester CH2 4NU, UK; graham.smith@chester.ac.uk
* Correspondence: c.banks@mmu.ac.uk; Tel.: +(0)-161-2471-196; Fax: +(0)-161-2476-831

Received: 27 February 2020; Accepted: 12 March 2020; Published: 19 March 2020

check for updates

Abstract: We demonstrate a facile methodology for the mass production of graphene oxide (GO) bulk-modified screen-printed electrodes (GO-SPEs) that are economical, highly reproducible and provide analytically useful outputs. Through fabricating GO-SPEs with varying percentage mass incorporations (2.5%, 5%, 7.5% and 10%) of GO, an electrocatalytic effect towards the chosen electroanalytical probes is observed, which increases with greater GO incorporated compared to bare/graphite SPEs. The optimum mass ratio of 10% GO to 90% carbon ink produces an electroanalytical signal towards dopamine (DA) and uric acid (UA) which is ca. ×10 greater in magnitude than that achievable at a bare/unmodified graphite SPE. Furthermore, 10% GO-SPEs exhibit a competitively low limit of detection (3σ) towards DA at ca. 81 nM, which is superior to that of a bare/unmodified graphite SPE at ca. 780 nM. The improved analytical response is attributed to the large number of oxygenated species inhabiting the edge and defect sites of the GO nanosheets, which are able to exhibit electrocatalytic responses towards inner-sphere electrochemical analytes. Our reported methodology is simple, scalable, and cost effective for the fabrication of GO-SPEs that display highly competitive LODs and are of significant interest for use in commercial and medicinal applications.

Keywords: graphene oxide; electroanalytical sensing; dopamine; uric acid; screen-printed electrodes

1. Introduction

Graphene oxide (GO), a two-dimensional oxygenated carbon nanosheet, previously considered by many researchers as solely a precursor for the synthesis of graphene, possesses a number of unique chemical properties that make it a beneficial material in its own right [1–3]. Whilst researchers have found niche applications for GO in an array of technologies, such as hydrogen storage [4], supercapacitors [5], and biosensors [6], GO is often overlooked due to its limited application in electrically active devices/materials. This is a result of its reported high electrical resistance that stems from carboxyl, hydroxyl, and epoxy groups located on the periphery of the GO sheet [7]. It is however, these hydrophilic oxygenated functional groups which assist in biorecognition during biosensing by promoting favourable interactions with specific analytes [1,8–10], allowing GO to be used as the underlying electrode material for a biosensor towards a number of biological/organic molecules, such as DNA [11,12] and peptides [13]. In many cases where GO is utilised for sensing applications, it is as a component/supporting framework within a more complex catalyst [14,15]. GO's ability to act singularly as a (bio)sensor has yet to be observed within the literature. A study by Brownson et al. [16]

demonstrated that GO, when immobilised upon the surface of graphitic electrodes, exhibited intriguing electrochemical responses, with the redox probes studied giving rise to electrochemical responses dependent upon the C/O content [17]. This suggests that GO could be beneficially utilised as an electrochemical platform where oxygenated electrocatalytic reactions are involved.

In this paper, we take this prior work one-step further [17] by fabricating GO bulk-modified SPEs and explore their performance towards a range of electroanalytically interesting analytes, namely dopamine (DA) and uric acid (UA). The preferred method of detection is via electrochemical techniques, as they offer rapid, portable and low cost analysis. It is evident that the literature focuses (See Table 1) on graphene rather than GO as an electrochemical sensing platform, where the chosen nanomaterial is drop casted upon a supporting carbon electrode, allowing it to be electrochemically wired. The use of drop casting as a method to modify a supporting electrode has several drawbacks, such as the supporting electrode has to be prepared for each measurement, which can be time consuming, and the drop-casting process results in an uncontrollable distribution of the nanomaterial upon the electrode's surface, that in turn results in poor reproducibility [18,19]. In order to overcome these issues, screen-printed electrodes (SPEs) have proven to be mass-producible electrochemical sensing platforms that offer versatility in electrode design and repeatability in the signal output [20]. The screen-printing technique can produce a vast number of SPEs that exhibit uniform heterogeneous electron transfer kinetics, thereby enabling separate electrodes to be used for independent measurements and give consistent/reliable responses. SPEs can also be readily adapted with respect to the composition of the ink utilised in their production, allowing for the incorporation of materials that alter the electrocatalytic behaviour displayed by the SPE [18].

Table 1. Comparison of current literature reporting the use of graphene and related electrocatalytic materials explored towards the electroanalytical sensing of dopamine (DA) and uric acid (UA).

Electrocatalyst	Electrode Material	Deposition Technique	Dopamine LOD (M)	Uric Acid LOD (M)	Electrochemical Method	Reference
GO-MWCNT/ MnO_2AuNP	GC	Drop Cast	1.7×10^{-7}	–	CV	[14]
pCu_2O NS-rGO	GC	Drop Cast	1.5×10^{-8}	1.1×10^{-7}	DPV	[21]
G-SnO_2	GC	Drop Cast	1.0×10^{-6}	–	DPV	[22]
DA-ERG/PMB	GC	Drop Cast	1.0×10^{-7}	–	DPV	[23]
GSCR-MIPs	GC	Drop Cast	1.0×10^{-7}	–	LSV	[24]
NG	GC	Drop Cast	2.5×10^{-7}	4.5×10^{-8}	DPV	[25]
Bare/unmodified	SPE	Screen Printed	7.8×10^{-7}	2.3×10^{-6}	CV	This Work
2.5% GO-ink	SPE	Screen Printed	2.9×10^{-7}	1.6×10^{-6}	CV	This Work
5% GO-ink	SPE	Screen Printed	1.3×10^{-7}	1.0×10^{-6}	CV	This Work
7.5% GO-ink	SPE	Screen Printed	1.0×10^{-7}	9.6×10^{-7}	CV	This Work
10% GO-ink	SPE	Screen Printed	8.1×10^{-8}	6.1×10^{-7}	CV	This Work

GC, glassy carbon; GO-MWCNT/MnO_2AuNP, graphene oxide multi-walled carbon nanotubes with manganese dioxide, poly(diallyldimethylammonium chloride) and gold nanoparticles; –, value unknown or not applicable; CV, cycling voltammetry; pCu_2O NS-rGO, porous cuprous oxide nanospheres on reduced graphene oxide; DPV, differential pulse voltammetry; G-SnO_2, graphene-tin oxide; DA-ERG/PMB, dopamine-grafted reduced graphene oxide/poly(methylene blue); GSCR-MIPs, graphene sheets/Congo red molecular imprinted polymers; LSV, linear sweep voltammetry; NG, nitrogen doped graphene; SPE, screen-printed electrode.

In order to explore this principle, this paper reports the bulk modification of SPEs, with varying percentage mass incorporations of GO and electrochemically exploring the capabilities of GO bulk-modified screen-printed electrodes (GO-SPEs), in comparison to bare/unmodified SPEs, as potential electroanalytical sensing platforms towards DA and UA (separately) for the first time.

2. Experimental Section

All chemicals used were of analytical grade and were used as received from Sigma-Aldrich without any further purification. All solutions were prepared with deionised water of resistivity no less than 18.2 MΩ cm^{-1} and were vigorously degassed prior to electrochemical measurements with high purity, oxygen free nitrogen. The GO powder utilised was commercially purchased from Graphene Supermarket [26].

Electrochemical measurements were performed using an Ivium CompactstatTM (Eindhoven, The Netherlands) potentiostat. Measurements were carried out using a typical three-electrode system, with a Pt wire counter electrode and a saturated calomel electrode (SCE) reference. The working electrodes were screen-printed graphite electrodes (SPEs), which have a 3.1 mm diameter working electrode. The SPEs were fabricated in house, the methodology of which is outlined in the electronic supporting information (ESI). Following production of the standard SPE, modification/production of the GO variation was achieved as follows: the GO powder was incorporated into the bulk graphitic ink on the basis of the weight percentage of MP to MI, where MP is the mass of particulate (in this case the GO) and MI is the mass of the ink formulation used in the printing process, i.e., % = (MP/MI) × 100. The weight percentage of MP to MI varied from 2.5%, 5%, 7.5% to 10%, resulting in 4 separate GO bespoke inks that are then screen printed upon the working area of bare SPEs; see the ESI for further details. Note, the maximum amount of GO that can be incorporated into the graphitic ink was found to correspond to 10% with any further percentage incorporation resulting in an increase in the resultant ink viscosity to where it is not screen printable via the technique used within this manuscript.

Physicochemical characterisation was performed utilising Raman spectroscopy, transmission electron microscopy (TEM), X-ray diffraction (XRD) and X-ray photoelectron spectroscopy (XPS). Details of the instrumentation utilised are reported in the ESI.

3. Results and Discussion

Initially, it was essential to perform a full physicochemical characterisation of the commercially purchased GO powder in order to ascertain its quality/properties prior to being incorporated into the SPEs (as reported in the experimental section). Raman spectroscopy, SEM, TEM, XPS and XRD analysis were all conducted. Figure 1A displays a TEM of the GO nano-platelets indicating that they exhibit a particle size (lateral width) of between 300 and 600 nm, which strongly agrees with the size stated by the commercial manufacturer, of ca. 500 nm [26].

Next, Raman spectroscopy was utilised to confirm the presence of GO by structural characterisation. The obtained spectra can be viewed in Figure 1B and displays the D and G vibrational band peaks at ca. 1350 and 1590 cm^{-1}, respectively; which are typically characteristic of GO [27,28]. Additionally, the composition of the GO sample is confirmed via XRD in Figure 1C, in which a characteristic 'sharp' peak is evident at $2\theta = 11.5°$, corresponding to the (001) diffraction peak of disordered GO [29]. Lastly, XPS analysis was performed to determine the GO's elemental composition, with Figure 1D showing the gathered survey spectra and Figure S1 displaying the individual spectra for the C and O regions. The GO was observed to contain 66.8% carbon and 28.6% oxygen, with trace amounts of nitrogen, sulphur and chlorine, which are likely mere contaminants. The combination of surface and physicochemical analysis presented above and expanded upon within the ESI confirm that the commercially sourced GO herein utilised is of high quality/purity.

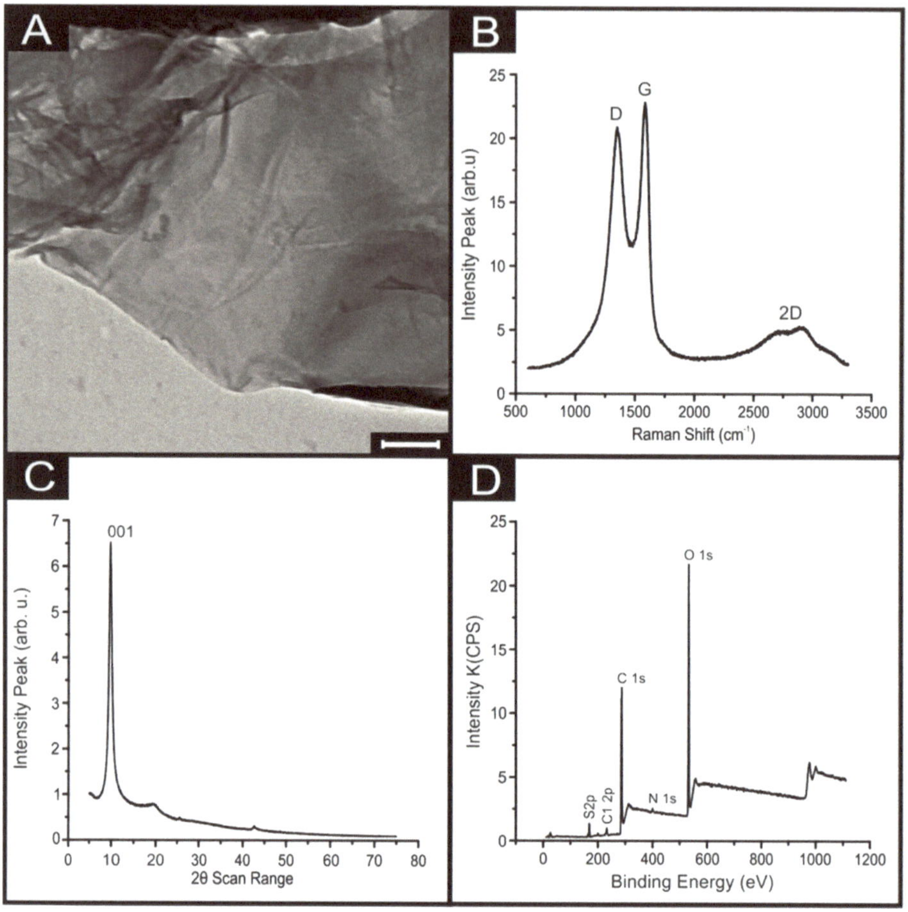

Figure 1. Characterisation of the commercially sourced GO; (**A**) image of the GO nanosheet (Scale bar: 100 nm), (**B**) Raman spectra of GO deposited onto a silicon wafer between 100 and 3400 cm, (**C**) X-ray diffraction (XRD) spectra between 5 and 75 2θ, and (**D**) high-resolution XPS survey spectra.

The GO-SPEs (the design and fabrication of which are outlined within the ESI) were electrochemically evaluated using the near ideal 'outer-sphere' redox probe 1 mM $[Ru(NH_3)_6]^{3+/2+}$ in 0.1 M KCl [30]. SEM was utilised to image the surface of a bare SPE and a 10% GO-SPE. However, the obtained images were indistinguishable due to the GO nanosheets having a very similar appearance to graphitic nanoplatelets found within the SPE bulk ink (see Figure S2). Whilst the bare/unmodified SPEs and the GO-SPEs were visually indistinguishable at the microscale, the incorporation of the GO into the SPEs bulk ink significantly altered their electrochemical performance, as described below. Utilising a 10% GO-SPE as a representative example, the observed voltammetric profiles are presented in Figure S3. Note that the electrochemical reduction peak current increased from 3.6 to 32 μA on the bare SPE compared to the 10% GO-SPE, respectively. Note, however, that the 10% GO-SPE displayed a smaller oxidation peak than the bare SPE. This alteration in the obtained cyclic voltammetric (CV) response is characteristic of an EC′ type reaction as described previously by Brownson et al. [16], who explored the electrochemistry of GO towards select redox probes by drop casting it onto an edge plane pyrolytic graphite (EPPG) support electrode. Such a response suggests that, as the amount of GO incorporation into the GO-SPEs is increased, so too is the proportion of oxygenated species present, resulting in a larger amount of oxygenated species available to catalyse the chemical reaction. Note that the electrochemical response of "graphene" towards $[Ru(NH_3)_6]^{3+/2+}$ does not display the catalytic behaviour herein observed at GO [31]. This inference could allow for an electrochemical test to differentiate the presence of "true" graphene and GO, as they have unique CV signal responses. The proposition that it is the C-O groups that produce such a response is as pointed out by Brownson et al. [17], who observed similar electrochemical signatures [16], making GO a much more promising electrocatalyst for sensing applications than graphene—especially when the amount and coverage of GO is highly controlled, as is the case with the GO-SPEs produced herein.

Next, the electroanalytical efficacy of the GO-SPEs was explored towards the sensing of dopamine (DA). DA is a neurotransmitter essential for bodily functions, such as memory and emotional regulation [32,33], where the detection of DA within body fluids is widely studied, as its concentration within bodily systems is linked to numerous neurological disorders [16].

Additions of DA were made into to a phosphate buffer (pH 7) solution, incrementing the DA concentration from 5 to 50 μM. The obtained CVs and calibration plots are presented within Figure 2. Using the 10% GO-SPEs as a representative example of all the GO-SPEs, Figure 2A shows that the oxidation peak current at a 5 μM DA concentration was 1.21 μA, which subsequently increased to 15.24 μA by 50 μM. There was a corresponding anodic shift in the onset potential from + 0.212 to + 0.316 V (all values are deduced from an average of N = 3). Of note is the large capacitive effect observed when GO is incorporated into the bulk of the SPEs (see Figures 2 and 3). This is to be expected, as previous literature has noted GO's capacitive nature [34]. The bare/unmodified SPEs do not display this capacitive effect (see Figures S4 and S5). It is clearly observable from Figure 2B that in agreement with the 10% GO-SPE, all the GO-SPEs display a greater anodic peak current than the bare SPE (see Figure S4). This can be associated with the oxygenated species present on GO facilitating the oxygenated electrocatalytic reactions. This is further supported by the observation that the greater percentage incorporation of GO into the GO-SPE the larger the observed anodic peak current (see Figure 2B). However, as the percentage of GO within the electrode increases from 0 to 10%, the activation potential for DA oxidation increases. A similar trend was observed when UA was utilised in the exact manner as above rather than DA (see Figure 3 and Figure S5), with a 10% GO-SPE displaying a ca. ×10 increase in the achievable peak current density when compared to a bare SPE. For a full description, see the ESI.

In terms of the analytical utility of the GO-SPE towards DA and UA sensing, there is a clear correlation between the percentage mass incorporation of GO and the electrode's limit of detection. Of note is the appearance of two linear ranges within a number of the trend lines for the separate electrodes in Figures 2 and 3. In these cases, the initial linear range was utilised as the slope for LOD calculations. As shown in Table 1, a bare/unmodified SPE displays an analytical useful limit of detection (LOD, based on 3σ) for DA and UA at 0.78 and 2.3 μM respectively. The 10% GO-SPE exhibited the lowest limit of detection of 81 nM and 0.61 μM for DA and UA respectively. The LOD values for the GO-SPE are highly competitive to those found within the current literature. They are also within the medically relevant range, given that the baseline concentration of DA within the striatum is ca. 10–20 nM, with unusual activity (i.e., burst firing) associated with neurological disorders exhibited by high DA concentrations in the hundreds of μM range [35]. The above observations suggest that the synergy between GO and the SPE offers huge beneficial electrocatalytic responses towards DA.

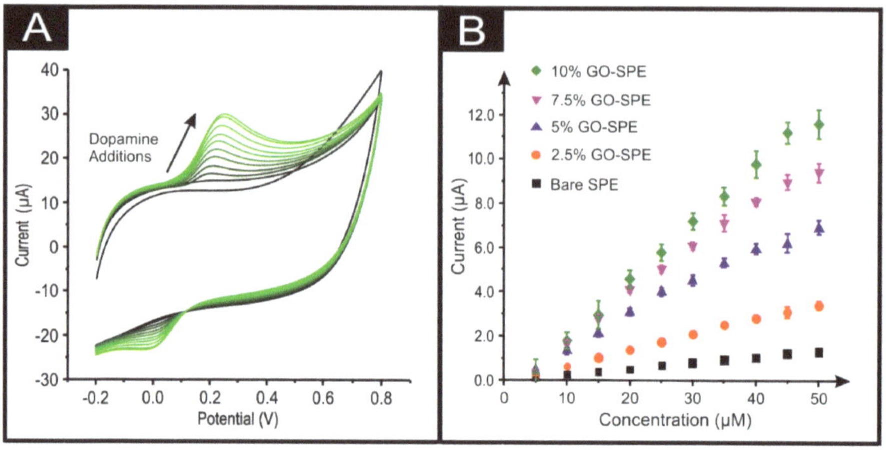

Figure 2. (**A**) Typical cyclic voltammetric response obtained utilising 10% GO-SPEs by sequentially adding aliquots of DA into pH 7.4 PBS, from 5 to 50 μM. (**B**) Calibration plot of the anodic peak current associated with the electroanalytical oxidation of DA over the concentration range for a bare SPE (black square), a 2.5% GO-SPE (orange circle), a 5% GO-SPE (blue triangle), a 7.5% GO-SPE (purple inverted triangle), and a 10% GO-SPE (green star). Error bars are on the data points and represent the average standard deviation (N = 3). Scan rate utilised: 100 mVs^{-1} (vs. SCE).

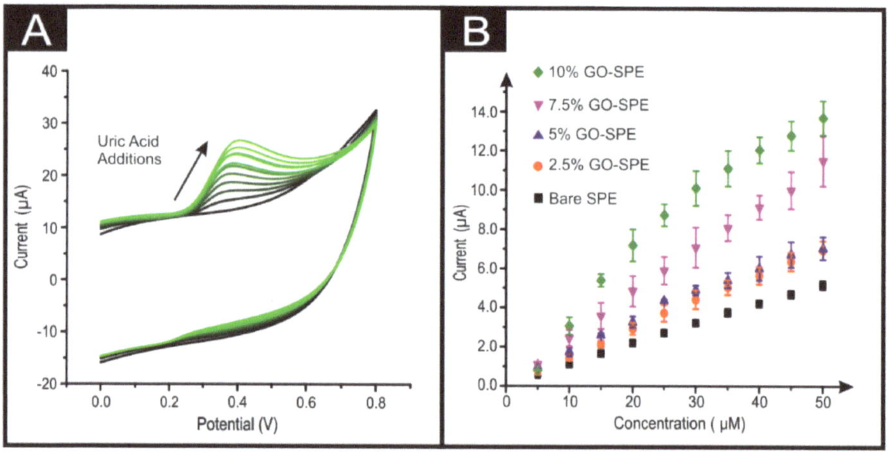

Figure 3. (**A**) Typical cyclic voltammetric response obtained utilising 10% GO-SPEs by sequentially adding aliquots of UA to pH 7.4 PBS, from 20 to 200 μM. (**B**) Calibration plot of the anodic peak current associated with the oxidation of UA over the concentration range for a bare SPE (black square), a 2.5% GO-SPE (orange circle), a 5% GO-SPE (blue triangle), a 7.5% GO-SPE (purple inverted triangle), and a 10% GO-SPE (green star). Error bars are on the data points and represent the average standard deviation (N = 3). Scan rate utilised: 100 mVs^{-1} (vs. SCE).

The intra-repeatability of the GO-SPEs was tested (N = 3). The percentage relative standard deviation (%RSD) for the observed peak current observed at the bare/unmodified SPE, 2.5%, 5%, 7.5%, and 10% GO-SPEs is shown via error bars in Figures 2B and 3B. With respect to the observed oxidation peak current, there is clearly a trend of increasing %RSD corresponding to an increase in the percentage of GO within the GO-SPEs. We postulate that this is due to a greater percentage of GO present, leading to a larger number of variations within the orientation of the modified GO structure, whereby there will be a greater chance for a different proportional of the GO oxygenated species to be present on the electrodes surface. The %RSDs at 50 μM for the bare/unmodified SPE, 2.5%, 5%, 7.5%, and 10% GO-SPEs are 1.7, 2.2, 3.4, 5.1 and 5.8 percent, respectively. These low %RSD values for the anodic oxidation peak attest to the high/favourable reproducibility of the screen-printing technique utilised herein to produce the GO-SPEs.

4. Conclusions

We have designed, fabricated and evaluated GO bulk-modified SPEs, which demonstrate electrocatalytic capabilities towards the sensing of DA and UA. The application of GO in this manner takes advantage of the oxygenated surface species inhabiting the edge and defect sites of the GO nanosheets to create a cheap, mass producible and tailorable sensing platform for applications requiring oxygenated electrocatalysis. Through increasing the amount of GO present (to a maximum of 10%), we observe a correlation between the number of oxygenated species and the magnitude of DA and UA electroanalytical signals.

Supplementary Materials: The following are available online at http://www.mdpi.com/2079-6374/10/3/27/s1. Includes the following sections: Electrode production, Experimental details on physicochemical characterization, Scan rate study, Dopamine electrochemistry, and Uric acid electrochemistry. Supporting Information figures: Figure S1. High-resolution XPS spectra of C and O regions of the GO utilised herein (A and B respectively). Figure S2. SEM images of the graphite and GO electrode surfaces in the supercapacitor device show little variation in the surface morphology of the surfaces with variation in GO content. Given this, it is apparent that the dominating influence of the morphology of the electrodes is in fact the carbon ink. This indicates that the improvement in the performance is a result of physicochemical properties of the graphene oxide, and not a result of any morphological differences induced by the addition of the GO. Figure S3. Typical cyclic voltammetric response of a bare SPE and a 10% GO-SPE recorded 1 mM $[Ru(NH_3)_6]^{3+/2+}$ in 0.1 M KCl solution. Scan rate utilised: 5 mVs^{-1} (vs. SCE). Figure S4. Typical cyclic voltammetric response obtained utilising a Bare/unmodified SPE by sequentially adding aliquots of 0.5 mM DA to pH 7.4 PBS, additions from 5 to 50 μM. Figure S5. Typical cyclic voltammetric response obtained utilising a bare/unmodified SPE by sequentially adding aliquots of 2 mM UA to pH 7.4 PBS, altering the bulk solution from 20 to 200 μM.

Author Contributions: C.E.B., conceptualization, experiment design and manuscript writing; D.A.C.B., experiment design and manuscript writing; G.S., XPS data acquisition and analysis; S.J.R.-N., data acquisition, manuscript writing and graphical visualization. All authors have read and agreed to the published version of the manuscript.

Funding: This research was funded British Council Institutional Link grant (No. 172726574).

Conflicts of Interest: The authors declare no conflict of interest.

References

1. Lee, J.; Kim, J.; Kim, S.; Min, D.-H. Biosensors based on graphene oxide and its biomedical application. *Adv. Drug Deliv. Rev.* **2016**, *105 Pt B*, 275–287. [CrossRef]
2. Ferrari, A.G.-M.; Brownson, D.; Banks, C. Investigating the Integrity of Graphene towards the Electrochemical Oxygen Evolution Reaction. *ChemElectroChem* **2019**, *6*, 5446–5453.
3. Rowley-Neale, S.J.; Randviir, E.P.; Dena, A.S.A. Banks, An overview of recent applications of reduced graphene oxide as a basis of electroanalytical sensing platforms. *C.E. Appl. Mater. Today* **2018**, *10*, 218–226. [CrossRef]
4. Cho, E.S.; Ruminski, A.M.; Aloni, S.; Liu, Y.S.; Guo, S.; Urban, J.J. Graphene oxide/metal nanocrystal multilaminates as the atomic limit for safe and selective hydrogen storage. *Nat. Commun.* **2016**, *7*, 10804.
5. Fan, T.; Zeng, W.; Niu, Q.; Tong, S.; Cai, K.; Liu, Y.; Huang, W.; Min, Y.; Epstein, A.J. Fabrication of high-quality graphene oxide nanoscrolls and application in supercapacitor. *Nanoscale Res. Lett.* **2015**, *10*, 192. [CrossRef] [PubMed]
6. Liu, Y.; Yu, D.; Zeng, C.; Miao, Z.; Dai, L. Biocompatible Graphene Oxide-Based Glucose Biosensors. *Langmuir* **2010**, *26*, 6158–6160. [CrossRef]
7. Mkhoyan, K.A.; Contryman, A.W.; Silcox, J.; Stewart, D.A.; Eda, G.; Mattevi, C.; Miller, S.; Chhowalla, M. Atomic and Electronic Structure of Graphene-Oxide. *Nano Lett.* **2009**, *9*, 1058–1063. [CrossRef]
8. Pumera, M. Graphene in biosensing. *Mater. Today* **2011**, *14*, 308–315.

9. Sharma, D.; Kanchi, S.; Sabela, M.I.; Bisetty, K. Insight into the biosensing of graphene oxide: Present and future prospects. *Arab. J. Chem.* **2016**, *9*, 238–261. [CrossRef]
10. Zhu, Y.; Murali, S.; Cai, W.; Li, X.; Suk, J.W.; Potts, J.R.; Ruoff, R.S. Graphene and Graphene Oxide: Synthesis, Properties, and Applications. *Adv. Mater.* **2010**, *22*, 3906–3924. [CrossRef]
11. Alonso-Cristobal, P.; Vilela, P.; El-Sagheer, A.; Lopez-Cabarcos, E.; Brown, T.; Muskens, O.L.; Rubio-Retama, J.; Kanaras, A.G. Highly Sensitive DNA Sensor Based on Upconversion Nanoparticles and Graphene Oxide. *ACS Appl. Mater. Interfaces* **2015**, *7*, 12422–12429. [CrossRef] [PubMed]
12. Gao, L.; Lian, C.; Zhou, Y.; Yan, L.; Li, Q.; Zhang, C.; Chen, L.; Chen, K. Graphene oxide–DNA based sensors. *Biosens. Bioelectron.* **2014**, *60*, 22–29. [CrossRef]
13. Zhang, Y.; Wu, C.; Guo, S.; Zhang, J. Interactions of graphene and graphene oxide with proteins and peptides. *Nanotechnol. Rev.* **2013**, *2*, 27. [CrossRef]
14. Rao, D.; Zhang, X.; Sheng, Q.; Zheng, J. Highly improved sensing of dopamine by using glassy carbon electrode modified with MnO2, graphene oxide, carbon nanotubes and gold nanoparticles. *Microchim. Acta* **2016**, *183*, 2597–2604. [CrossRef]
15. Bahrami, S.; Abbasi, A.R.; Roushani, M.; Derikvand, Z.; Azadbakht, A. An electrochemical dopamine aptasensor incorporating silver nanoparticle, functionalized carbon nanotubes and graphene oxide for signal amplification. *Talanta* **2016**, *159*, 307–316. [CrossRef] [PubMed]
16. Brownson, D.A.C.; Lacombe, A.C.; Gomez-Mingot, M.; Banks, C.E. Graphene oxide gives rise to unique and intriguing voltammetry. *RSC Adv.* **2012**, *2*, 665–668. [CrossRef]
17. Brownson, D.A.C.; Smith, G.C.; Banks, C.R. Graphene oxide electrochemistry: The electrochemistry of graphene oxide modified electrodes reveals coverage dependent beneficial electrocatalysis. *Soc. Open Sci.* **2017**, *4*, 171128. [CrossRef] [PubMed]
18. Rowley-Neale, S.J.; Foster, C.W.; Smith, G.C.; Brownson, D.A.C.; Banks, C.E. Mass-producible 2D-MoSe2 bulk modified screen-printed electrodes provide significant electrocatalytic performances towards the hydrogen evolution reaction. *Sustain. Energy Fuels* **2017**, *1*, 74–83. [CrossRef]
19. Randviir, E.P.; Brownson, D.A.C.; Metters, J.P.; Kadara, R.O.; Banks, C.E. The fabrication, characterisation and electrochemical investigation of screen-printed graphene electrodes. *Phys. Chem. Chem. Phys.* **2014**, *16*, 4598–4611. [CrossRef]
20. Baccarin, M.; Rowley-Neale, S.J.; Cavalheiro, É.T.G.; Smith, G.C.; Banks, C.E. Nanodiamond based surface modified screen-printed electrodes for the simultaneous voltammetric determination of dopamine and uric acid. *Microchim. Acta* **2019**, *186*, 200. [CrossRef]
21. Mei, L.-P.; Feng, J.-J.; Wu, L.; Chen, J.-R.; Shen, L.; Xie, Y.; Wang, A.-J. A glassy carbon electrode modified with porous Cu_2O nanospheres on reduced graphene oxide support for simultaneous sensing of uric acid and dopamine with high selectivity over ascorbic acid. *Microchim. Acta* **2016**, *183*, 2039–2046. [CrossRef]
22. Nurzulaikha, R.; Lim, H.N.; Harrison, I.; Lim, S.S.; Pandikumar, A.; Huang, N.M.; Lim, S.P.; Thien, G.S.H.; Yusoff, N.; Ibrahim, I. Graphene/SnO_2 nanocomposite-modified electrode for electrochemical detection of dopamine. *Sens. Bio-Sens. Res.* **2015**, *5*, 42–49. [CrossRef]
23. Gorle, D.B.; Kulandainathan, M.A. Electrochemical sensing of dopamine at the surface of a dopamine grafted graphene oxide/poly(methylene blue) composite modified electrode. *RSC Adv.* **2016**, *6*, 19982–19991. [CrossRef]
24. Mao, Y.; Bao, Y.; Gan, S.; Li, F.; Niu, L. Electrochemical sensor for dopamine based on a novel graphene-molecular imprinted polymers composite recognition element. *Biosens. Bioelectron.* **2011**, *28*, 291–297. [CrossRef]
25. Sheng, Z.-H.; Zheng, X.-Q.; Xu, J.-Y.; Bao, W.-J.; Wang, F.-B.; Xia, X.-H. Electrochemical sensor based on nitrogen doped graphene: Simultaneous determination of ascorbic acid, dopamine and uric acid. *Biosens. Bioelectron.* **2012**, *34*, 125–131. [CrossRef]
26. Graphene Supermarket. Available online: https://graphene-supermarket.com/Single-Layer-Graphene-Oxide-small-flakes-1g.html (accessed on 31 February 2016).
27. King, A.A.K.; Davies, B.R.; Noorbehesht, N.; Newman, P.; Church, T.L.; Harris, A.T.; Razal, J.M.; Minett, A.I. A New Raman Metric for the Characterisation of Graphene oxide and its Derivatives. *Sci. Rep.* **2016**, *6*, 19491. [CrossRef]
28. Kudin, K.N.; Ozbas, B.; Schniepp, H.C.; Prud'homme, R.K.; Aksay, I.A.; Car, R. Raman Spectra of Graphite Oxide and Functionalized Graphene Sheets. *Nano Lett.* **2008**, *8*, 36–41. [CrossRef]

29. Chowdhuri, A.R.; Tripathy, S.; Chandra, S.; Roy, S.; Sahu, S.K. A ZnO decorated chitosan-graphene oxide nanocomposite shows significantly enhanced antimicrobial activity with ROS generation. *RSC Adv.* **2015**, *5*, 49420–49428. [CrossRef]
30. Ji, X.; Banks, C.E.; Crossley, A.; Compton, R.G. Oxygenated Edge Plane Sites Slow the Electron Transfer of the Ferro-/Ferricyanide Redox Couple at Graphite Electrodes. *ChemPhysChem* **2006**, *7*, 1337–1344. [CrossRef]
31. Brownson, D.A.C.; Munro, L.J.; Kampouris, D.K.; Banks, C.E. Electrochemistry of graphene: Not such a beneficial electrode material? *RSC Adv.* **2011**, *1*, 978–988. [CrossRef]
32. Nichkova, M.; Wynveen, P.M.; Marc, D.T.; Huisman, H.; Kellermann, G.H.J. Validation of an ELISA for urinary dopamine: Applications in monitoring treatment of dopamine-related disorders. *Neurochem* **2013**, *125*, 724–735. [CrossRef]
33. Silva, L.I.B.; Ferreira, F.D.P.; Freitas, A.C.; Rocha-Santos, T.A.P.; Duarte, A.C. Optical fiber biosensor coupled to chromatographic separation for screening of dopamine, norepinephrine and epinephrine in human urine and plasma. *Talanta* **2009**, *80*, 853–857. [CrossRef]
34. Zhang, J.; Zhao, X.S.J. Conducting Polymers Directly Coated on Reduced Graphene Oxide Sheets as High-Performance Supercapacitor Electrodes. *Phys. Chem. C* **2012**, *116*, 5420–5426. [CrossRef]
35. Goto, Y.; Otani, S.; Grace, A.A. The Yin and Yang of dopamine release: A new perspective. *Neuropharmacology* **2007**, *53*, 583–587. [CrossRef]

biosensors

MDPI

Review

Printed Electrodes in Microfluidic Arrays for Cancer Biomarker Protein Detection

Lasangi Dhanapala [1], Colleen E. Krause [2], Abby L. Jones [1] and James F. Rusling [1,3,4,5,*]

1 Department of Chemistry, University of Connecticut, 55 North Eagleville Road, Storrs, CT 06269, USA; lasangi.dhanapala@uconn.edu (L.D.); abby.jones@uconn.edu (A.L.J.)
2 Department of Chemistry, University of Hartford, Bloomfield Ave., West Hartford, CT 06117, USA; ckrause@hartford.edu
3 Department of Surgery and Neag Cancer Center, UConn Health, Farmington, CT 06232, USA
4 Institute of Materials Science, University of Connecticut, 97 N. Eagleville Road, Storrs, CT 06269, USA
5 School of Chemistry, National University of Ireland Galway, University Road, H91 TK33 Galway, Ireland
* Correspondence: james.rusling@uconn.edu; Tel.: +1-(860)-486-4909

Received: 31 July 2020; Accepted: 1 September 2020; Published: 7 September 2020

check for updates

Abstract: Medical diagnostics is trending towards a more personalized future approach in which multiple tests can be digitized into patient records. In cancer diagnostics, patients can be tested for individual protein and genomic biomarkers that detect cancers at very early stages and also be used to monitor cancer progression or remission during therapy. These data can then be incorporated into patient records that could be easily accessed on a cell phone by a health care professional or the patients themselves on demand. Data on protein biomarkers have a large potential to be measured in point-of-care devices, particularly diagnostic panels that could provide a continually updated, personalized record of a disease like cancer. Electrochemical immunoassays have been popular among protein detection methods due to their inherent high sensitivity and ease of coupling with screen-printed and inkjet-printed electrodes. Integrated chips featuring these kinds of electrodes can be built at low cost and designed for ease of automation. Enzyme-linked immunosorbent assay (ELISA) features are adopted in most of these ultrasensitive detection systems, with microfluidics allowing easy manipulation and good fluid dynamics to deliver reagents and detect the desired proteins. Several of these ultrasensitive systems have detected biomarker panels ranging from four to eight proteins, which in many cases when a specific cancer is suspected may be sufficient. However, a grand challenge lies in engineering microfluidic-printed electrode devices for the simultaneous detection of larger protein panels (e.g., 50–100) that could be used to test for many types of cancers, as well as other diseases for truly personalized care.

Keywords: screen-printed electrodes; inkjet-printed electrodes; immunoassays; cancer; biomarkers; protein

1. Introduction

Growing rates of incidence and mortality have made cancer a global pandemic, a key impediment for increasing life expectancy and quality of life [1,2]. Most of the current cancer diagnostic tests rely on tissue biopsies and imaging techniques. First of all, these techniques require a physical tumor to be detectable. Biopsies analyze tissue morphology and cellular arrangement, but the precise location of the tumor is needed, they are highly invasive and can easily miss tumor cells. Imaging techniques such as mammograms or colonoscopies only allow tumor detection and do not allow early diagnosis before the onset of tumor development [3,4]. These techniques are relatively expensive and require expert technical knowledge for the test and its interpretation (especially for mammograms), limiting their widespread accessibility. Relying only on these tests, cancer cells can spread through the body before being diagnosed, drastically reducing survival rates [1,5,6]. Thus, a new strategy has been developing for personalized cancer diagnosis which will enable patients to access health or disease conditions early. This approach will rely on fast, quick, sensitive, and accurate assays with no or minimum invasion, utilizing samples such as blood, urine, and saliva. Full implementation of this approach, which has thus far been very slow to take hold in the clinical realm, has the potential to greatly improve rates of survival and quality of life of millions of cancer patients across the globe.

Cancer progression is marked by abnormal cell division, followed by the expression of specific molecules (biomarkers) that are otherwise absent or typically expressed in small amounts in healthy cells [7–10]. A biomarker is an indicator of a person's health, disease condition, or responses to therapy and collective of a broad range of biomolecules from proteins, glycoproteins, nucleic acids to a variety of other small molecules found in bodily fluids [3,4,6,8–11]. Nucleic acids such as DNA and RNA mainly act as prognostic biomarkers providing a risk assessment of cancer [4,9], whereas protein biomarkers serve as diagnostic/predictive biomarkers of cancer, to provide a quick picture of the patient's health [9,12,13]. The most accurate way to use these tools is to measure multiple biomarkers for each type of risk, or in the case type of cancer (Figure 1). Thus, the majority of cancer biomarker research focuses on protein detection for early diagnosis, post-surgery reoccurrence, and cancer staging [5,9].

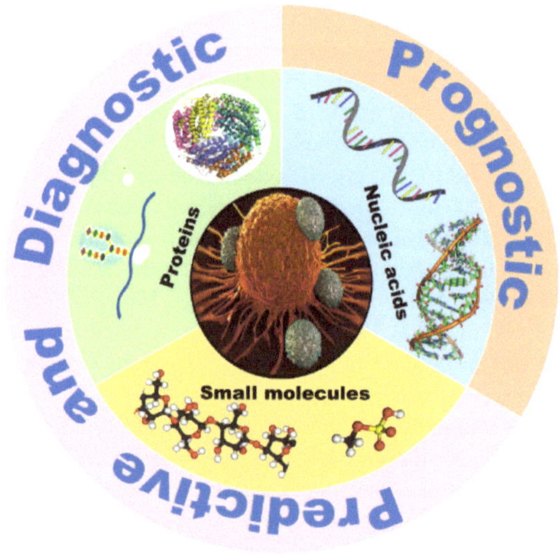

Figure 1. A simple overview of cancer biomarkers and their clinical importance.

Despite significant research gains in this area, only a handful of protein cancer biomarkers and no protein panels are approved by the US Food and Drug Administration (FDA) [3,5,14,15] (Table 1) for clinical practice. This discrepancy between research and clinical practice is due to three main factors: (a) Lag in development of point-of-care (POC) devices for detection of panels of biomarkers [5,9,13] (b) lack of sufficient National Institutes of Health (NIH) funding for translational research in development and validation of effective biomarker panels, and (c) inability to detect rare biomarkers due to poor sensitivity for low concentration biomarkers coupled in a panel with more abundant proteins [16–19]. Detection of a panel of biomarkers enables better diagnostic accuracy than single proteins. Current commercial assays, including multiplexed enzyme-linked immunosorbent assay (ELISA), LC-MS/MS [5,9,19], mesoscale electrochemiluminescence (ECL) [20], luminex [19,21], and single-molecule counting system Simoa-HD [22], struggle to truly answer the above challenges, slowing translation of protein biomarker research into clinical practice.

Table 1. FDA approved protein biomarkers [3,5,14,15].

Biomarker		Sample	Cancer Type	Clinical Use	Assay
Abbreviation	Name				
Free PSA/fPSa	Free Prostate Specific Antigen	Serum	Prostate	Benign Hyperplasia vs. Cancer diagnosis	Immunoassay
tPSA	Total PSA	Serum	Prostate	S, M	Immunoassay
cPSA	Complex PSA	Serum	Prostate	S, M	Immunoassay
p63	Transformation-related protein 63	FFPE tissue†	Prostate	S, M	Immunohistochemistry
TG	Thyroglobulin	Serum	Thyroid	S, M	Immunoassay
EGFR	Epidermal growth factor receptor	Colon tissue	Colon	Pre	Immunoassay
CEA	Carcinoembryonic antigen	Serum	Colon	M	Immunoassay
MW CEA	High molecular weight CEA	Urine	Bladder	M	Immunofluorescence
FDP (AMDL-ELISA DR-70)	Fibrin/fibrinogen degradation products	Urine/Serum	Bladder	M	Immunoassay
NMP/22	Nuclear matrix protein 22	Urine	Bladder	S, M	Immunoassay
BTA	Bladder tumor antigen	Urine	Bladder	M	Immunoassay
HER2	Human epidermal growth factor receptor	Serum	Breast	M	Immunohistochemistry
CA15-3 *	Carbohydrate antigen 15-3	Serum, plasma	Breast	M	Immunoassay
CA27-29 *	Carbohydrate antigen 27-29	Serum	Breast	M	Immunoassay
HER/NEU	Human epidermal growth factor receptor 2	FFPE tissue†	Breast	P, Pre	Immunohistochemistry
ER	Estrogen factor	FFPE tissue†	Breast	P, Pre	Immunohistochemistry
PR	Progesterone factor	FFPE tissue†	Breast	P, Pre	Immunohistochemistry
AFP*	α-fetoprotein	Serum, plasma, amniotic fluid	Testicular	St	Immunoassay
β-hGC *	Human chorionic gonadotropin-β	Serum	Testicular	St	Immunoassay
AFP-L3%	α-fetoprotein L3% isoform	Serum	Hepatocellular	P	HPLC, microfluidic capillary electrophoresis
KIT	Receptor Tyrosine Kinase	FFPE tissue†	Gastrointestinal stromal tumors	Pre	Immunohistochemistry
CA 19-9 *	Carbohydrate antigen 19-9	Serum	Pancreatic	M	Immunoassay
CA 125 *	Carbohydrate antigen 125	Serum	Ovarian	M	Immunoassay
HE4	Human epididymis protein 4	Serum	Ovarian	M	Immunoassay
OVA1 (Multiprotein test	CA125, Apolipoprotein A1, Beta-2 microglobulin, Transferrin, Pre-albumin	Ovarian	Serum	P	Immunoassay

M—monitoring, S—screening, P—prognosis, Pre—prediction of therapy, St—staging, *—Glycoproteins, FFPE tissue†—formalin-fixed paraffin-embedded (FFPE) tissue slides [23].

The inherent sensitivity and ease of multiplexing of electrochemical immunosensors has placed them among the most utilized methods of biosensors. Coupled with simplicity of instrumentation, low cost, the capability of miniaturization, automation, and ease of integration to microfluidic devices has made electrochemical immunosensors an excellent platform for fabrication of POC devices [24,25]. The detection ability of each electrode-sensor is based on the desired function of a specific electrode, the electrode material, surface modification, and dimensions [26]. Over the past several decades miniaturization and the production cost of electrodes that make up the electrochemical sensor platform has improved with the advent of a variety of thin-film technologies. Common methods in developing these thin-film electronic devices include chemical vapor deposition (CVD) [27], photolithography [28], stencil printing [29], screen printing [30], and inkjet printing [31,32]. Inkjet printing has also been used to make flexible thin-film transistors that can be used as sensors [33]. All these methods allow for well-defined electrode fabrication and insulation. However, in terms of mass production, screen-printing and inkjet printing have emerged to become the favored approach to produce disposable electrochemical sensing platforms and arrays [9,34–36].

In this review, we cover some of the key cancer biomarker research done using printed electrodes focusing on the evolution of immunosensors invented by our group, as well as examples from recent literature landmarks of printed electrode biosensors. We focus mainly on screen-printed electrodes (SPEs) with a few illustrations of inkjet-printed electrode sensors. The primary focus on SPEs is driven by the numerous applications of SPEs to POC devices, including the poster-child of biosensors—the glucose biosensor [25].

2. Immunoassay Techniques for Printed Electrodes

2.1. Immunoassay Protocol

Heineman and coworkers pioneered the introduction of ELISA-type techniques into electrochemical biosensors, allowing a floodgate of research that led to modern electrochemical immunoassays [37]. Many immunoassays on printed electrodes adopted the sandwich ELISA platform, where the capture or primary antibodies (Ab_1) are bound onto the sensor electrode surface and capture the protein antigens (Ag), which then entrap a detection or secondary antibody (Ab_2), providing, in many cases, higher sensitivity and selectivity than classical optical detection ELISA and related techniques [3] (Scheme 1H). When coupled with microfluidics, electrochemical immunosensors can surpass the traditional drawbacks associated with ELISA to build a diagnosis platform at a lower sample volume, low reagent consumption, more efficient mixing, faster response times, and continuous monitoring [38–41].

Layer-by-layer (LBL) techniques have gained popularity in biosensor fabrication, following its introduction by Lvov and Decher in the early 1990s, and have been used frequently for the attachment of Ab_1 on printed electrodes [42–45] (Scheme 1B). Carbon electrode surfaces carry a net negative charge in neutral solutions, and metal electrodes can be derivatized with anionic thiols, like mercapto-propionic acid, to give them a negative charge at $pH < 6$ [43]. Hence these negative surfaces can be layered with adsorbed polycations such as polyethyleneimine (PEI), poly(diallyldimethylammonium chloride) (PDDA), or poly(allylamine hydrochloride) (PAH) [44] (Scheme 1B). Then, the negatively charged nanomaterials can be adsorbed as the next layer (see Section 3.1 for use of nanomaterials in surface modification of electrodes) (Scheme 1C), followed with Ab_1 immobilization, commonly through amidization [46] (Scheme 1D,E). Sharafeldin et al. further discussed different strategies of Ab_1 immobilization on screen-printed carbon electrodes (SPCEs), such as (a) electrochemical adsorption of glutathione-coated gold nanoparticles (GSH-AuNP); (b) electrochemical deposition of graphene; (c) passive adsorption of antibodies directly to the carbon surface through the Fc region, with Fab region orientated towards the antigen-binding; (d) passive adsorption with random orientation; (e) chitosan films; and (f) covalent bonding of antibodies to GSH-AuNP. According to their study, sensitivity of the sensor mainly depends on the active area and antibody coverage enhanced by nanostructures (discussed later in detail), whereas a greater degree of orientation will extend the linear dynamic range to higher concentrations but not lower detection limits [47] (Figure 2).

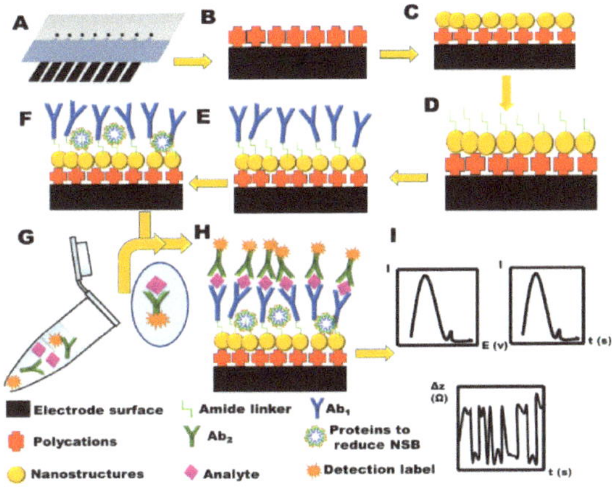

Scheme 1. Representation of the general protocol followed (**A–I**) during an electrochemical immunoassay. (**A**) A typical screen-printed carbon electrode, (**B**) Surface modification by LBL where polycations are coated on the electrode surface, (**C**) Equipping the surface with nanostructures, (**D**) Amidization reaction with surface functional group via linkers such as EDC 1-(3-(Dimethylamino)propyl)-3-ethyl carbodiimide hydrochloride (EDC) and N-hydroxysulfosuccinimide (NHSS) [17], (**E**) Ab_1 immobilization followed by incubation, (**F**) Reduction of NSB by washing with detergents and BSA or casein proteins, (**G**) Introduction of a signal generating labeled Ab_2 for detection of analyte, (**H**) Final arrangement of Ab_1, Ag and Ab_2 to give a sandwich type ELISA, (**I**) Electrochemical detection through voltammetry, amperometry, or impedance.

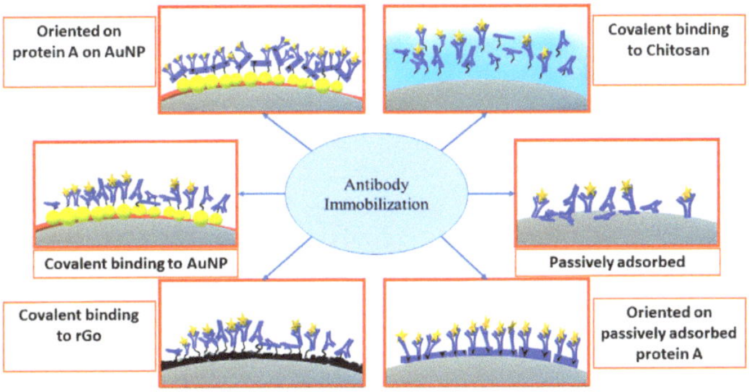

Figure 2. Different Ab_1 immobilization strategies on SPCE. Reprinted with permission from [47], copyright 2019 The Royal Society of Chemistry.

Following Ab_1 immobilization, analyte protein is introduced into a solution contacting the sensor, which then selectively captures Ab_2, which is often accompanied by a detection label [9–11] (Figure 3). Labels can be anything that can produce a signal, including conductive polymers, nanoparticles, electroactive metal ions and complexes, and liposomes with electroactive species. Very often enzyme labels are used for detection [6,9–11,48]. Commonly, enzymes such as horseradish peroxidase (HRP), glucose oxidase, or alkaline phosphate (ALP) are used as labels to provide a high, reproducible, stable signal amplification [45]. A considerable amount of recent research focuses on the use of Ab_2-HRP conjugates to immobilize the labels on the electrodes to avoid diffusional cross-talk between sensor elements [6]. Multiple HPR labels associated with one Ab_2 can also enhance sensitivity tremendously, and, in some cases, nanoparticles have been used to bring Ab_2 and HRP or other labels together [9,10]. Ab_2-HRP can be achieved by streptavidin–biotin association with most of the Ab_2 now commercially available in biotinylated form, or can easily be biotinylated using commercial kits [49–51]. Other similar associations with biotin include avidin and neutravidin [52,53]. Avidin tends to increase non-specific binding (NSB) due to its basic isoelectric pH (10.5), making it highly positively charged in physiological pH, hence, interacting with negatively charged surfaces [54]. Neutravidin is a promising alternative over both streptavidin and avidin due to its higher association with biotin and lower NSB [49,55].

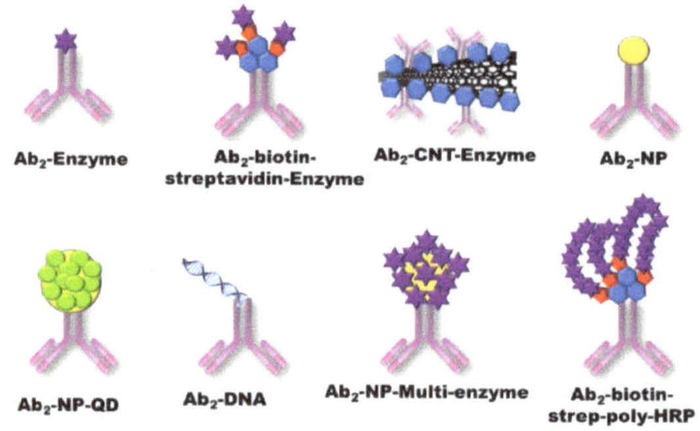

Figure 3. Different labels conjugated with the secondary antibody (Ab_2) during electro immunoassays performed on printed electrodes.

Typically, after the immobilization of Ab_1 and the analytes captured, the electrodes are washed by detergent and/or an NSB blocking protein, such as bovine serum albumin (BSA) or casein, to reduce the NSB that greatly decreases the signal/noise (S/N) ratio [9–11,17,24]. (Scheme 1F) NSB arises when the Ab_2 molecules bind to non-antigen sites of the electrodes, providing a signal which is not proportionate to the analyte concentration, increasing the LOD and damaging the sensitivity. Casein and BSA are used to prevent NSB by sterically blocking the non-analyte sites, whereas detergent such as phosphate-buffered saline-Tween 20 (PBS-T20) or Tris buffer solution (TBS), with sodium dodecyl sulfate (SDS) or Triton 100-X, help to wash away weakly-bound Ab_2 on non-analyte sites [17,56]. Another approach on reduction of NSB is through chemical attachments of the electrode surface to reduce protein adsorption through (a) polymerization strategies (polyethylene glycol (PEG), conducting polymers), (b) modification of allotropic carbons (carbon nanotubes), (c) sol-gel modification, (d) surface modification by diazonium salts, (e) metal nanoparticles (magnetic beads, gold and silver nanoparticles), (f) self-assembled monolayers (SAMs) [52]. In our hands, these chemical modifications are not so efficient in reducing NSB, and detergent and BSA or casein are still needed [6,9,10].

The signal generated by Ab_1-Ag-Ab_2 conjugation on the sensor may be detected through a variety of electrochemical techniques, including amperometry, voltammetry, square wave voltammetry (SWV), differential pulse voltammetry (DPV), stripping voltammetry, cyclic voltammetry, and impedance [11] (Scheme 1I).

2.2. Ultrasensitive Detection

The signal amplification provided through the enzyme label alone is not sufficient to reach the ultra-sensitivities required for many clinical cancer biomarker diagnoses. Thus, additional amplification strategies have to be adopted on the printed electrode systems [3,6,9–11]. Approaches include (a) redox cycling by using reversible redox couples regenerated electrochemically, chemically, or enzymatically (discussed in detail by Yang et al.) [48]; (b) multiple labeling such as multi-enzyme nanoparticles or polymers that provide a manifold of electrochemical events per bound analyte [48]. Other approaches include secondary antibody tags with dissolvable nanoparticles and multi-enzyme Ab_2 conjugates (Figure 3) [9–11,17,48].

2.2.1. Dissolvable Nanoparticles

Dissolvable nanoparticle (NP) labeling has been used in biosensor development due to ease of mass-scale synthesis, chemical stability, high surface: volume ratio, low cost, and higher stability over enzyme labels, increasing the shelf life of the sensor [57–59]. Replacement of enzymes by NPs has opened a new chapter of research coined "nanozymes". Examples include metal NPs, quantum dots (QDs), and most recently, carbon-based nanoparticles [57,60–62].

Pioneering work on metal NPs on SPEs was done by Dequaire et al. using colloidal gold particles tagged with Ab_2, following acid digestion and detection of the produced Au (III) particles via anodic stripping voltammetry (ASV) detecting immunoglobulin G (IgG) at μg mL^{-1} LODs [63]. Following, Joe Wang and his team did ground-breaking research on metal NPs for protein detection during the 2000s, including cyclic accumulation of gold nanoparticles to catalyze the precipitation of silver [64–67]. AuNPs were used by Lai at el. with a less destructive mode for multiplexed detection of α-fetoprotein (AFP) and carcinoembryonic antigen (CEA) with 3.9 and 3.5 pg mL^{-1} LODs, respectively. After immobilization of Ab_1 on the electrode surface using chitosan linking, electrodes were incubated with AFP and CEA, followed by secondary antibodies anti-AFP and anti-CEA conjugated with AuNP tags. Next, silver nanoparticles (AgNPs) solution was deposited on the SPE and the Ag reduction was detected by linear sweep voltammetry (LVS) from −0.15 to 0.25 V at 50 mV s^{-1} in 1.0 M KCl solution [68] (Figure 4). Following this, a variety of NPs were used for protein biomarker detection, including silver NPS [68–70], platinum NPs [71–74], palladium NPs [57,75], copper NPs [76], and iridium NPs [57,77].

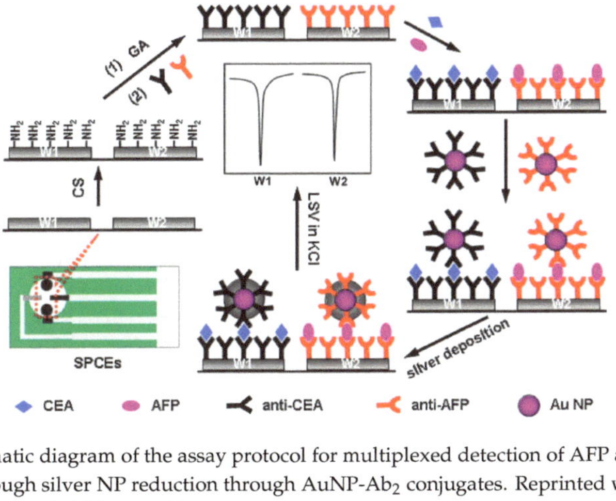

Figure 4. Schematic diagram of the assay protocol for multiplexed detection of AFP and CEA tumor biomarkers through silver NP reduction through AuNP-Ab_2 conjugates. Reprinted with permission from [68], copyright 2012 Elsevier B.V.

Additionally, multi-metal NP tags and hybrids of different metals are being reported [57,75]. Kalyoncu and coworkers used three different hybrid nanotags CuAu@γFe_2O_3, ZnAu@γFe_2O_3, and PbAu@γFe_2O_3 for multiplexed detection of vascular endothelial growth factor (VEGF), AFP, and CEA by binding Ab_2 molecules with the respective nanotags. DPV was used for the detection of the oxidation peaks of Cu, Zn, and Pb ions obtained by acid dissolution of the hybrid NPS at +0.8, +1.3, and +0.1 V, respectively [78] A similar study was done by Putnin et al. with polyethylenimine-coated gold nanoparticles (PEI-AuNPs) conjugated with four different electroactive metal ions—Cd (II), Ag (I), Pb (II), and Cu (II)—as labels. They also used the same PEI-AuNPs for surface modification of the SPEs and detected four tumor biomarkers—PSA, AFP, CEA, and interleukin-8 (IL-8)—with LODs from 0.9 to 1.7 fg mL^{-1} [79].

QDs are attractive for designing bar-coded labels for multiplexed detection of biomarkers by differential tuning of the potentials specific for each QD [57,75,80,81]. In principle, such barcodes could be used for the simultaneous detection of thousands of biomarkers using computational deconvolution. The most widespread practice in detecting QDs is through ASV proceeded by acid digestion [82,83]. Merkoci et al. were among the first to demonstrate a less aggressive strategy by direct detection of CdS through the redox cycling of Cd (II). Water-soluble, glutathione-modified CdS QDs were introduced onto the electrode surface, followed by a deposition potential of −1.1 V for 120 s to reduce the Cd(II) in QDs to Cd(0), and SWV detection with a potential ranging from −1.1 to −0.7 V to oxidize the accumulated Cd(0) [84].

Carbon-based nanoparticles became a "hot topic" after the discovery of graphene that worked its way into immunosensors. Graphene, fullerenes, and carbon nanotubes (CNTs) are among the most used carbon-based NPs [75]. Wang and coworkers pioneered the use of CNTs in immunosensors for the detection of DNA and proteins [85] and upright high surface area carbon nanotube forest sensors, with Ab_1 attached at their ends, were developed by our research team [86–89] for detection of tumor biomarkers on pyrolytic graphite electrodes, and can be easily adapted to printed electrodes. More recently, we used graphene oxide (GO) nanosheets on SPEs to detect prostate-specific antigen (PSA) and prostate-specific membrane antigen (PSMA) with LODs of 5–15 fg mL^{-1}. Mediator-free, Ab_2_Fe_3O_4@GO particles were used as labels, and Fe_3O_4 shows peroxidase activity mimicking HRP, and reduces H_2O_2 to generate amperometric current at −0.3 V vs. Ag/AgCl in a microfluidic device. GO nanosheets are bound with multiple Fe_3O_4 nanoparticles to obtain paramagnetic Fe_3O_4@GO, to which Ab_2 are immobilized using EDC/NHSS amidation to obtain Ab_2_Fe_3O_4@GO, which first captures the biomarkers. This system is then introduced to a microfluidic device housing screen-printed carbon (Kanichi) arrays decorated with electrochemically-reduced graphene oxide (ERGO) immobilized with Ab_1. The dynamic range was tunable based on the concentration of Fe_3O_4@GO used [90] (Figure 5). Pingarrón's group designed a nanotag with graphene quantum dots (GQDs) and multiwalled carbon nanotubes (MWCNTs) hybridization for the detection of two metastatic proteins, IL-13 receptor-α2 (IL-13Rα2) and cadherin-17 (CDH-17), with sub-ng mL^{-1} detection using SPCE modified with p-aminobenzoic acid for covalent immobilization of Ab_1. Assays were performed for 3 h with 0.5 μg of sample [91].

2.2.2. Multi-Enzyme Conjugates

HRP and ALP are frequently used for multi-enzyme labels in conjugates, such as Ab_2-magnetic beads, carbon-based multi-enzymes (Carbon nanotubes, graphene, carbon spheres), silica nanoparticles, metal nanoparticles (Au, Ag), and polymerized enzymes. Generally, these multi-enzyme tags require a mediator for their effective functionality as the bulky particles provide a steric and distance barrier for direct electron transfer between the enzyme and the sensor electrode surface. Hydroquinone is a frequently used mediator coupled with H_2O_2 that is being used on SPE systems [17,48,82,84].

Magnetic beads (MBs) have drawn attention to the conjugation of multiple enzymes due to advantages over other nanoparticles such as ease of separation and ease of bioconjugation through the surrounding polymer layer [92]. Our research team exploited MB-based nanotags for the detection of various cancer biomarkers proteins. In our first paper using MBs in 2009 with Mani et al., we developed a sensitive assay for PSA with 7500 HRP molecules per MB to attain LODs of 0.5 pg mL^{-1} [93]. Later, collaborations with Munge et al. synthesized MBs with about 50,000 HRP enzymes, pushing LODs down to 1 fg mL^{-1} for interleukin 8 (IL-8) detection [94]. These studies were then extended for ultrasensitive single and multiplexed detection of cancer biomarker proteins on both SPEs and inkjet-printed electrodes coupled in microfluidic devices [31,95–103] (Figure 6).

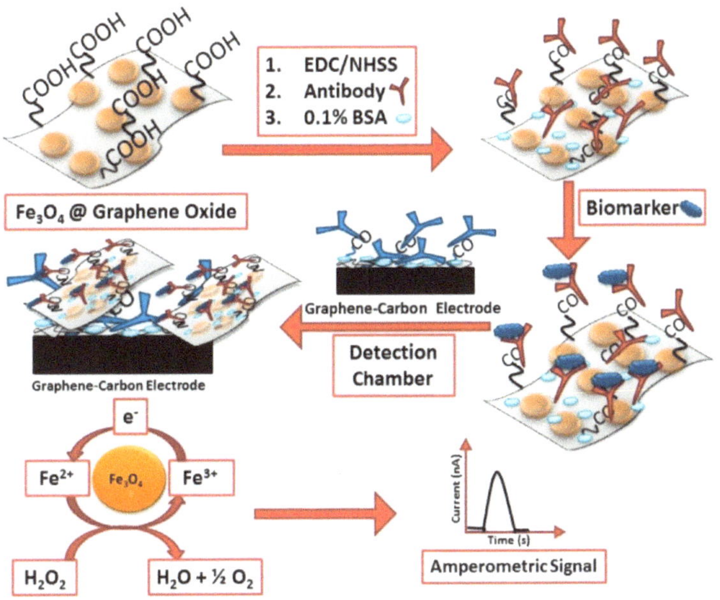

Figure 5. Schematic illustration of protein capture and detection by Fe_3O_4@GO sheets. After the analyte is captured by the GO sheets, it is captured on the SPCE surface coated with Ab_2 and an amperometric signal is generated using H_2O_2. Reprinted with permission from [90], copyright 2016 Elsevier B.V.

Clinical diagnosis of oral cancer through the multiplexed detection of IL-6, IL-8, VEGF-C, and VEGF with LODs from 5 to 50 fg mL^{-1} [96] and prediction of mucositis reaction from oral cancer treatment through detection of cytokines; tumor necrosis factor α (TNF-α), IL-6, Interleukin-1β (IL-1β), and c-reactive protein (CPR) at 10–50 fg mL^{-1} LODs [101] are good examples of clinical applications of MB-based systems. Furthermore, Krause et al. developed a sandwich immunoassay on inkjet-printed electrodes with greatly reduced assay times to achieve clinically-relevant 5 pg mL^{-1} LODs for IL-6 and IL-8, trading off sensitivity for speed, to achieve a clinically promising system for quick diagnosis [98]. Otieno et al. then modified this microfluidic device with an online protein capture chamber, where the analytes are separated from the serum in an isolated compartment with Mb-Ab_2, before being introduced into the detection channel with the electrodes [99]. Otieno et al. further detected parathyroid hormone-related peptide (PTHrP) fragments which act as biomarkers for bone metastasis of breast and prostate cancer obtaining LODs of 3 fg mL^{-1},1000-fold less than the traditionally used immunoradiometric assay (IRMA) [100].

Extending the application of MB-enzyme conjugates further, Uliana et al. developed an elegant and inexpensive method to produce a fully disposable microfluidic electrochemical device (μFED) using SPE arrays for detection of breast cancer biomarker estrogen receptor alpha (ERα). Silhouette software was used for the designing of the electrodes and the patterned electrodes were transferred on to the vinyl sheets, followed by a vinyl mask and carbon ink from Henkel, which was poured on to the surface and then allowed to cure. Ag/AgCl ink was then deposited on the electrodes to obtain a μFED by sandwiching the SPEs using a double adhesive polystyrene card. Using this method, authors note that dozens of inexpensive ($0.20) devices could be constructed in just a few hours. Working electrodes were modified with DNA sequences known as estrogen response elements that are specific for Erα. A bioconjugate was developed using magnetic particles heavily labeled with ERα antibodies and enzyme labels of HRP were used to capture ERα in solution. The bioconjugate was injected into the μFED. Limits of detection of 10.0 fg mL^{-1} were achieved through amperometric detection with reduced incubation times [104].

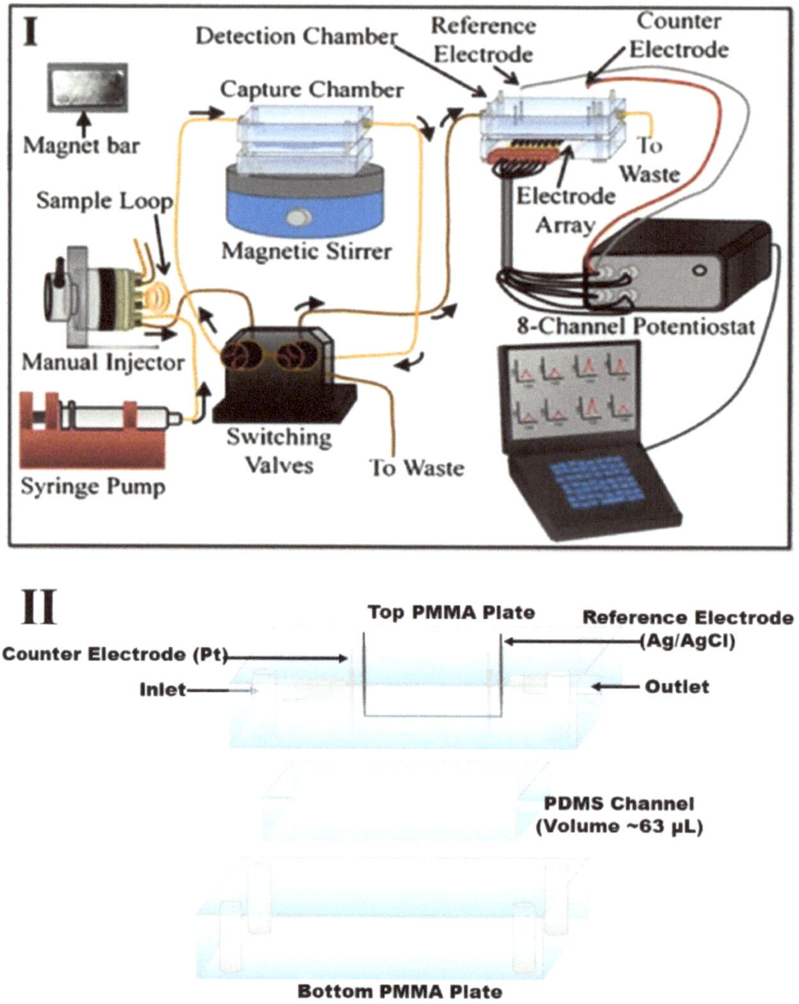

Figure 6. The microfluidic set-up used by our group in electroimmunosensors. (**I**) Schematic diagram of the pump (used for the flow of wash buffers and hydroquinone mediator into the detection chamber), HPLC injector (used for the injection of the sample into the detection chamber), switching valve (for alternation of the direction of fluids into the detection chamber, capture chamber or the waste), microfluidic chamber (houses the electrodes and the detection chamber), and the eight-channel multipotentionstat for amperometric detection of the analytes. (**II**) The microfluidic chamber with a poly (dimethoxy) silane channel (1.5 mm × 2.8 cm) sandwiched between two hard poly (methylmetacrylate) (PMMA) plates. The upper PMMA plate houses the platinum counter electrode and the Ag/AgCl reference electrode. Reprinted with permission from [101], copyright 2015 Springer Nature.

While MBs are attractive labels to reach ultrasensitive recognition of proteins, the synthesis of MBs is a cumbersome process with issues in reproducibility, bioconjugation, and characterization. Moreover, when integrated into microfluidic devices, the bulky MBs are often found to block the tubing, requiring additional cleaning or periodic replacement of the tubes [92].

Recent developments in producing multi-enzyme tags for detection of printed electrodes include the use of the polymerized form of enzymes, particularly HRP (poly-HRP) [105,106]. These polymers are conjugated with streptavidin molecules to bind with biotin-Ab_2 via streptavidin–biotin interactions. Commercially, streptavidin-poly-HRP conjugates, such as strep-poly-HRP20, strep-poly-HRP40, and strep-poly-HRP80, are readily available. These conjugates contain five homopolymers, each carrying either 20, 40, or 80 HRP molecules covalently coupled with multiple streptavidin molecules [107]. Our group used strep-poly-HRP on a disposable, patterned array made from gold compact discs (CDs) for the detection of IL-6 with LOD 10 fg mL^{-1}, at a cost of $0.2 per array [108]. Pingarrón et al. developed a magnetoimmunosensor on SPCE for detection of IL-6 coupled with carboxyl-functionalized magnetic microparticles for Ab_1, but only reached a LOD of 0.4 pg mL^{-1} using poly-HRP80 [109]. The same strategy was later used for transforming growth factor-β (TGF-β) detection with LOD of 10 pg mL^{-1} [110]. They further modified the array with 4-carboxyphenyl-functionalized double-walled carbon nanotubes to improve the detection for multiplex proteins, TNF-α and IL-1β, with LODs 0.85 and 0.38 pg mL^{-1}, respectively [111]. Meanwhile, our group recently published a multiplexed detection assay of four prostate cancer biomarkers, PSA, ETS-related gene protein (ERG), insulin-like growth factor-1 (IGF-1), and VEGF-D, using a system with strep-poly-HRP80 coupled with SPCE from Kanichi Ltd. modified with the glutathione-AuNP layers from Mani et al. [93]. Signal enhancement from high HRP/Ab_2 ratio and the AuNP surface modification generated unprecedented sub-zeptomole LODs [17], representing the most sensitive multiplexed protein assay thus far reported (Figure 7). PSA, ERG, IGF-1, and VEGF-D LODs were 0.13, 0.063, 0.013, and 0.088 fg mL^{-1}, respectively, accounting for less than 100 protein molecules (Figure 8).

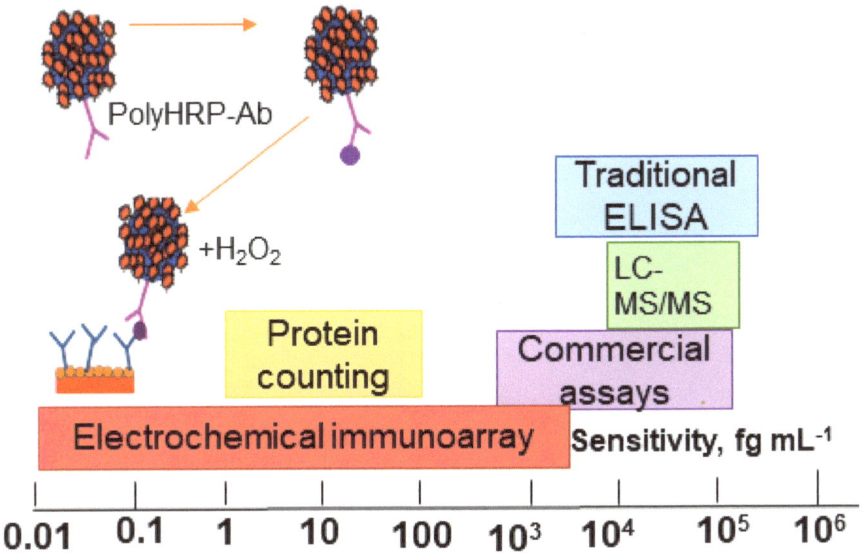

Figure 7. Comparison of the sensitivity of the new electrochemical immunoassay with poly-HRP and AuNP-modified carbon electrodes developed by Dhanapala et al. with other protein detection methods currently available. Reprinted with permission from [17], copyright 2020 American Chemical Society.

2.3. Label-Free Detection

In the past decade, with the advancement of label-free methods on printed electrodes, advantages of label vs. label-free methods have been debated. While label-free methods offer ease of fabrication, less incubation time, and ease of automation, they still trail well behind labeled assays in sensitivity [112,113]. Label free methods use potentiometry, amperometry, voltammetry, conductometry, field-effect transistor (FET), and electrochemical impedance spectroscopy (EIS) as modes of detection [113,114]. Among these, EIS is the most frequently used mode due to its innate capability to sensitively measure any changes in the charging capacity, conductivity, or resistivity of an electrochemical surface when modified with biomolecules [113,114]. Davis et al. developed strategies for detecting C-reactive protein (CRP) on gold electrodes using aptamers, reaching LODs down to 300 pM using EIS measurements [115]. Aptamers are oligonucleotide or protein recognition probes that are selected for binding a target molecule from a large random sequence pool. Davis' group has done interesting work on EIS-based label-free immunoassays that have catalyzed progress in unlabeled immunoassays and can be adapted to printed electrodes [116–121].

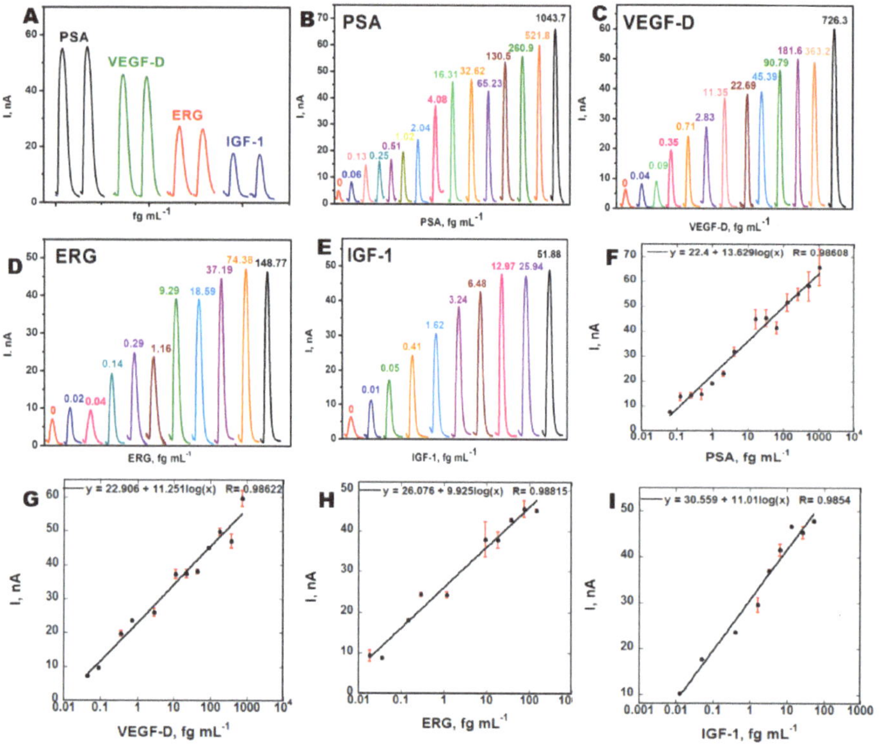

Figure 8. Calibration plots obtained with (**A**) duplicate amperometric peaks of PSA, VEGF-D, ERG, and IGF-1 showing reproducibility; (**B–E**) show amperometric peaks obtained for each biomarker standard; and (**F–I**) show the calibration plots with dynamic ranges from 0.013 to 1000 fg mL^{-1}. Reprinted with permission from [17], copyright 2020 American Chemical Society.

Zhao et al. developed a system with multiple working electrodes on a SPCE, each for specific target analytes coated with four different hydrogels activated by sodium alginate-Au nanoparticle (SA-AuNP) composites. They detected four proteins—CA-125, neuron-specific enolase (NSE), fragment antigen 21-1 (Cyfra21-1), and squamous cell carcinoma antigen (SCCA) at 0.0054 U mL^{-1}, 2.3 pg mL^{-1}, 5.5 pg mL^{-1}, and 4.8 pg mL^{-1}, respectively, using amperometric detection [122] (Figure 9). Many other works, in label-free protein detection, have attained clinically-relevant detection limits, but they have only detected single biomarkers and improvements are required for multiplexing [123–126].

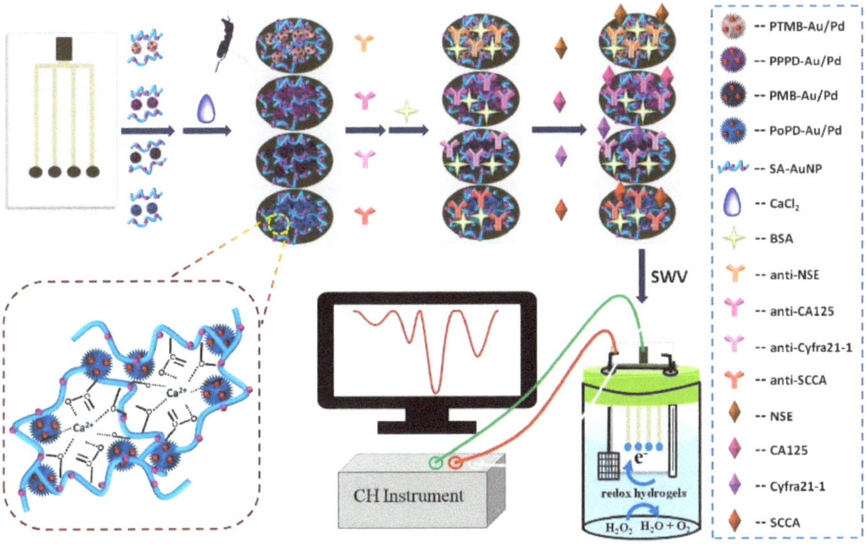

Figure 9. Label-free amperometric detection of four tumor biomarkers using specific hydrogel activation for each biomarker, with multiple working electrodes on a SPCE. Reprinted with permission from [122], copyright 2017 Elsevier B.V.

3. Screen Printing

Screen printing employs a viscous ink or paste being passed over a screened mesh using a blade (also known as a squeegee) onto a substrate service [127]. This technique is also used for printing designs and labels on T-shirts. Several commercially-available inks are available from suppliers such as Dupont [128], Sun Chemicals [129], Acheson [130], and Tekra/Henkel [131]. Typically, for SPEs, the ink contains powdered silver gold, platinum, copper, or carbon along with adhesives and additives including resin, cellulose acetate, cyclohexanone, or ethylene glycol [30]. However, the formulation of the conductive pastes is typically not released due to its commercial value [127]. Substrates for thin-film electronics include plastics (polyester films, polyvinylchloride, polycarbonate, etc.), alumina, glass, ceramic, and paper. Several recent reviews outline the advantages and disadvantages of these inks and substrates [30,127]. Ready-to-use SPEs can also be purchased from commercial suppliers including Metrohm [132], Pine Research, [133] PalmSens [134], BASi [135], and Kanichi Research Services Ltd. [136] (Figure 10B–F).

First-generation SPEs included working and reference electrodes. Second-generation included the three-electrode configuration of working, reference, and auxiliary electrodes [30,34] (Figure 10A). Using these configurations, several groups have developed single analyte detection platforms. Without prior surface treatment of the SPE, Tallapragada et al. recently developed an immunosensor for the detection of breast cancer biomarker human epidermal growth factor receptor-2 (HER-2) [137]. Full sandwich immunoassays were developed on their previously designed SPE [138], which consisted of carbon paste working and auxiliary electrodes, and an Ag/AgCl paste reference electrode. Incubation times were consistent with ELISA and there was no improvement to total assay time. The redox reaction between the streptavidin-conjugated HRP label and 3,3′5,5′-tetramethylbenzidine was measured using cyclic voltammetry at a scan rate of 50 mV s^{-1} to quantify HER-2 in solution. The authors noted that it was the biotin–avidin chemistry that facilitated detection of HER-2 into the low nanogram levels of the detection limit of 4 ng mL^{-1} [137].

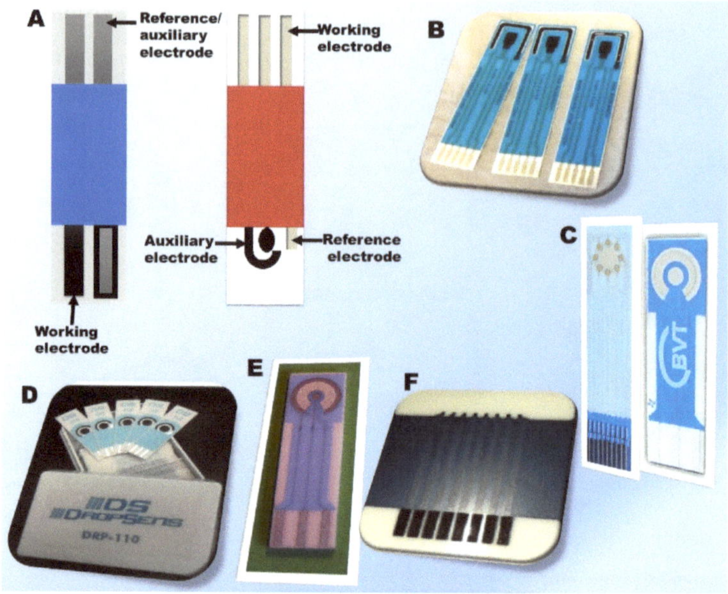

Figure 10. Screen-printed electrodes that are commercially available. (**A**) schematic diagram of generation one (left) and two (right) SPEs. (**B**) SPE from Pine Research including a carbon working electrode, a carbon counter electrode, and an Ag/AgCl reference electrode [133]. (**C**) left—AC9C Sensor array with graphite working electrode and Ag/AgCl reference electrode, right—AC1 Sensor with graphite working and auxiliary electrodes and Ag/AgCl Reference Electrode from BASi [135]. (**D**) metrohm SPE based on carbon, gold, platinum, silver, or carbon nanotube inks [132]. (**E**) PalmSens electrodes with same material working and counter electrodes (C, Pot, Ag, Au or Au-Pt alloy) and reference electrode made from Ag or Ag/AgCl [134] and (**F**) SPCE from Kanichi Research Services Ltd. housing eight working electrodes [136].

3.1. Nanoparticle Surface Coatings of SPEs

Enhanced performance in terms of sensitivity and LOD has been achieved on coated SPEs with nanoparticles [25]. These nanoparticles increase the surface roughness greatly, enhance the electrochemical surface areas, and allow for attachment of large quantities of capture antibodies for target analytes. Shana Kelly's group has done a lot of interesting research on nanostructured modifications of electrode surfaces to show how nanostructured surfaces could be tuned to achieve high sensitivity. They detected carbohydrate antigen 125 (CA-125) down to 0.1 U mL^{-1} by optimizing the size of the chip sensors without using the conventional sandwich platform or labels, and this is a classic example of how much the surface modifications could enhance sensitivity even without further amplification strategies. Their findings indicate larger surface areas hinder reaching ultrasensitive measurements and suppression of the background signal can easily be achieved by minimal surface area [139]. We have already presented many examples of signal amplification by surface modification of SPE's by nanoparticles [17,79,90,91,93–103,109–111]. A few more examples are included in this section.

Electrochemical deposition by Chan et al. generated a graphene-gold nanocomposite on the surface of Dropsens SPE for the detection of cancer biomarker, CEA [140]. A sandwich immunoassay was constructed on this nanocomposite surface using HRP as the redox label and H_2O_2 as an activator, and cyclic voltammetry was used to determine the concentration of CEA in the sample. The LOD was 0.28 ng mL^{-1} [140]. Incubation steps here were also consistent with conventional ELISA. Suresh et al. modified the surface of a SPE using chitosan and gold nanoparticles (AuNPs) for the detection of

PSA [141]. A sandwich immunoassay protocol was followed with HRP as the enzyme, but they amplified the signal using methylene blue as a redox mediator. Electrochemical detection of PSA was monitored by cyclic voltammetry and square wave voltammetry. A dynamic range of 1–18 ng mL^{-1} was observed with LOD 1 pg mL^{-1} [141].

Giannetto et al. demonstrated a competitive immunosensor for determination of bladder cancer biomarker p53 protein on a DropSens SPE modified with a carbon nanotubes/AuNP composite [142]. Protein p53 was immobilized on the modified SPE followed by a single anti-p53 mouse monoclonal antibody that recognizes both wild-type and mutant p53. After immunocompetition in the sample, an alkaline phosphatase-conjugated reading antibody was added for electrochemical detection. In urine, this nano-modified SPE exhibited a wide linear range, from 20 pM to 10 nM, with LOD of 14 pM for p53 protein [142]. DropSens SPE was also used by Marques et al. for the simultaneous detection of breast cancer biomarkers CA 15-3 and HER-2 [143]. The SPE was coated with AuNP. Then the authors enzymatically deposited metallic silver using alkaline phosphatase [144]. Voltammograms were recorded to observe the electrochemical oxidation current of the enzymatically-deposited silver for LODs of 5.0 U mL^{-1} for CA 15-3 and 2.9 ng mL^{-1} for HER-2 [143].

3.2. SPEs and Molecular Imprinting

SPEs have also served as the base for a molecular imprinted sensor [145]. Molecular imprinting involves developing affinity polymers for target analytes. The template or target analyte interacts during polymerization on the surface to form a polymerized binding site on the surface of SPE. Once the template is removed, it leaves the binding sites free for the subsequent capture of the imprinted analyte [145]. Bozal-Palabiyik et al. developed a molecularly-imprinted polymer on the surface of a SPE for cancer biomarker VEGF [146]. They used label-free impedance to detect VEGF with a dynamic range from 20 to 200 pg mL^{-1} and a limit of detection of 0.08 pg mL^{-1} [146] (Figure 11). Gomes et al. developed an electropolymerized SPE with amine-substituted benzene rings as monomers and specifically charged monomers for creating protein imprinting materials for breast cancer biomarker CA 15-3 [147]. The commercial Au SPE from DropSens consists of gold working and auxiliary electrodes, and reference and electrical contacts made of silver. Using this sensor, Gomes et al. demonstrated a linear dynamic range of 0.25 to 20.00 U mL^{-1}, with detection limit of 0.05 U mL^{-1}, for CA 15-3, from square wave voltammograms in diluted serum, with only 15 min incubation between sensor and solution of CA-15-3 [147].

3.3. SPE in μPADs

Microfluidic electrochemical immunosensing was pioneered by Heineman in the early 2000s [148,149], and paper microfluidic devices (μPADs) were first reported by Whiteside's group to improve the ease of operation and ability to work without external pumps [150]. SPEs on the surface of the paper were explored by several groups [151]. Fan et al. fabricated a SPE on the surface of paper modified with graphene oxide/thionine/gold nanoparticles (rGo/Thi/AuNPs) for detection of CA 125 [152]. Using custom-designed screens, electrodes were patterned onto wax-printed paper. Carbon ink was used for both working and counter electrodes, and Ag/AgCl ink was used for a reference. The patterned paper was folded with double-sided tape for use. The (rGo/Thi/AuNPs) were used for not only antibody immobilization but also signal amplification. The detection principle followed that the CA 125 antigen could decrease the current response of thionine, which corresponded to the concentration of CA 125 [152]. Using differential pulse voltammetry, the immunosensor displayed a linear range from 0.1 to 200 U mL^{-1} with LOD of 0.01 U mL^{-1} [152] (Figure 12).

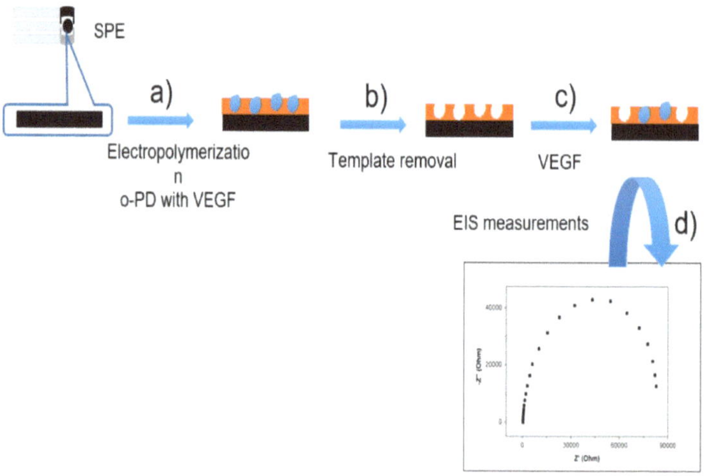

Figure 11. Schematic representation for the process of molecular imprinting (Bozal-Palabiyik et al.) through the electropolymerization of o-phenylenediamine (o-PD) for label-free detection. Reprinted with permission from [146], copyright 2019 Wiley-VCH Verlag GmbH and Co. KGaA, Weinheim.

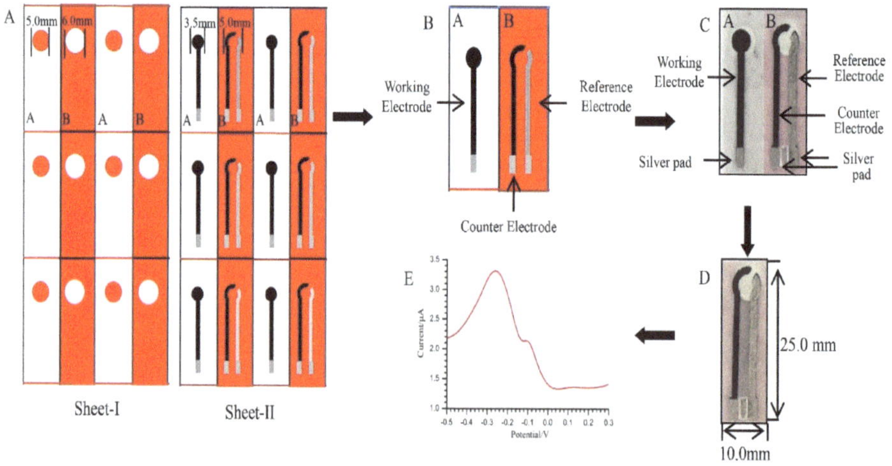

Figure 12. The fabrication of the paper-based electrodes by Fan et al. (**A**) Sheet I and II indicate the screen-printed wax patterns. Paper A and B of Sheet I contain the sample zone (5.00 mm) and the auxiliary zone (6.00 mm), which were designed for the carbon counter, working, and Ag/AgCl reference electrodes. Sheet II contains Paper A and B with patterns for the above three electrodes. (**B**) Represent a full screen-printed electrode system with all the three electrodes: counter, working, and reference. (**C**) A photograph of the produced electrode system with a silver pad. (**D**) Folded electrode. (**E**) A typical signal from the DPV used for the detection of the immunosensor. Reprinted with permission from [152], copyright 2019 Elsevier B.V.

4. Inkjet Printing

In efforts to decrease the cost of electrochemical arrays, our team has developed inkjet-printed arrays [31,98]. Inkjet printing offers several advantages over screen printing in that it is a contactless method for fabrication that does not require a stencil or template [32,35,153]. Design patterns can be developed using digital software patterns that can be sent to the relatively simple materials inkjet printer in a similar manner to using an office inkjet printer. Several groups developed simple techniques using a regular office printer with slight modifications [154,155]. However, for further control and precision, more sophisticated material printers are available at a reasonable cost [31,98]. Commercially available inks exist from the same suppliers. Several groups have synthesized their own custom inks. Recent reviews on inkjet-printed platforms provide additional details on the printing process and inks available [31,98] (Figure 13).

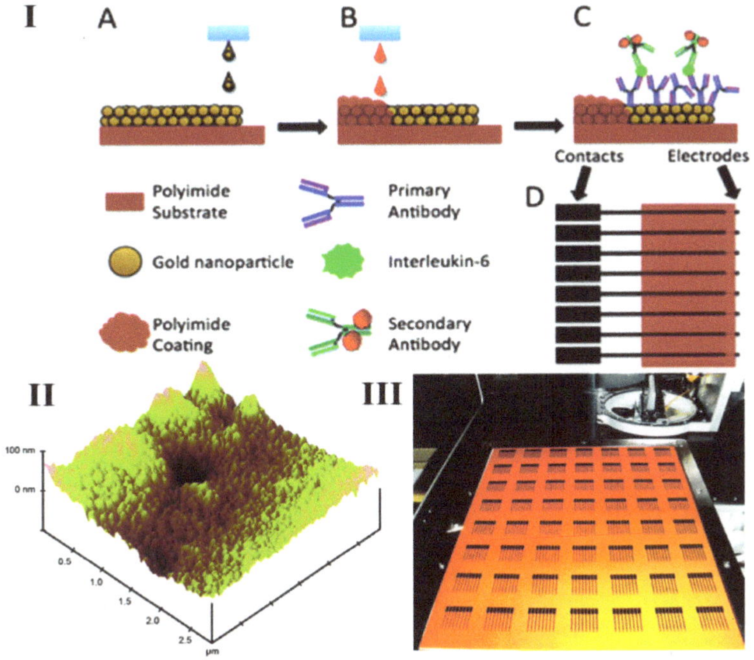

Figure 13. The inkjet-printed electrode immunoassay. (**I**) Schematic diagram of the immunoassay procedure indicating (**A**) AuNP inkjet printing on to the Kapton polyimide substrate, (**B**) insulation with poly(amic) acid, (**C**) immunoassay build-up, and (**D**) the electrode surface with AuNP patterns in black and insulating poly(amic acid) patterns in orange. (**II**) Tapping mode Atomic Force Microscopy (AFM) image of the AuNP working electrode surface. (**III**) A total of 56 eight-electrode arrays printed on a Kapton surface by Dimatix Materials Printer. Reprinted with permission from [31], copyright 2012 The Royal Society of Chemistry.

Using a custom-fabricated ink made from dodecane thiol-protected gold nanoparticle in toluene, we printed arrays of gold electrodes on the surface of a heat-resistant polyimide Kapton plastic sheet as a platform for cancer detection [31,98]. The inkjet-printed array was printed using a FUJIFILM Dimatix Materials Printer. With less than 2 mL of ink, dozens of arrays can be printed. Electrode arrays were insulated with a printed overcoating of poly(amic) acid, a precursor that converts to Kapton when heated. The inkjet-printed working electrodes had reproducible surface areas with relative standard deviation (RSD) < 3% [56,57]. The material cost was \$0.2 [31,98]. Using this process, the methods could easily be scaled up using industrial-sized inkjet printers [31,98]. For immunoassay applications, the electrode arrays were heated at 200 °C for 30 min, cleaned in sulfuric acid, and sensors coated with a self-assembled monolayer (SAM) of mercaptopropionic acid (MPA) to introduce carboxyl groups for cross-linking capture antibodies to the working electrodes. Arrays were used in microfluidic platforms with external counter and reference electrodes. Using this strategy, we reported the first ultrasensitive multiplex peptide assay to measure intact parathyroid hormone-related peptide PTHrP 1-173, as well as circulating N-terminal and C-terminal peptide fragments [100].

Fully inkjet-printed electrochemical sensors with integrated counter and reference electrodes have also been developed [156,157]. We recently reported a fully disposable inkjet-printed electrochemical platform for HER-2 [158]. The electrochemical sensor platform consisted of an inkjet-printed gold working eight-electrode array (WEA) and counter electrode, along with an inkjet-printed silver electrode that was chlorinated with bleach to produce an Ag/AgCl reference electrode. The array sensors were treated with MPA to allow for conjugation of capture antibodies. Once decorated with capture antibodies, the arrays were used in a microfluidic channel to implement a sandwich immunoassay on the surface of the WEA, following a simultaneous injection of target protein, biotinylated antibody, and polyHRP label [158]. Clinically-relevant LODs of 12 pg mL^{-1} were achieved with an assay time of only 15 min [158].

5. Challenges for Printed Electrodes in Microfluidic Assays

Microfluidics coupled with printed electrodes have been in the forefront of emerging electrochemical microsystems, synergistically enhancing inherent features of each other in lab-on-chip platforms for POC. Despite its success in production of wearables and self-powered devices [25,159,160], these approaches need to fill gaps in technical and clinical aspects to be used in commercialized cancer detection platforms.

Adaptation of microfluidics with printed electrodes into clinical platforms is relatively new as it requires a multi-disciplinary foundation of chemistry, physics, engineering, medicine, and biology [161]. Fabrication of microfluidic devices involve a multistep process including designing, material selection, processing and surface treatment with each step having their own technical issues [162]. Fabrication protocols such as photolithography, polymer molding, lase ablation, lamination, soft lithography, and, most recently, 3D printing [163] are available for chip production, but bottlenecks at the closing stages of the microstructure which may introduce defects in design, and is also a time-consuming process [163,164]. Integration of printed electrodes into microfluids also has its own set of technical difficulties, such as complex initial designs where all aspects of the final platform must be optimized such as mixing, binding, separation, washing, and detection. Incorporation of electrodes on microfluidic chips must avoid channel deformations and electrode damage. New materials and fabrication protocols are being introduced to solve these technical issues. Replacement of the silicon- and glass-based chips with polydimethylsiloxane (PDMS) provided several advantages, including mechanical flexibility, biocompatibility, electric insulation, optical transparency, ease of handling and manipulation, and low cost [28,164–166]. PDMS chips have challenges such as leaching, channel deformation, evaporation, sample adsorption, low acid-base resistivity, and hydrophobic recovery, providing a challenge to use in routine clinical practice [165]. Novel fabrication protocols, such as 3D printing and nanofabrication, are now emerging as solutions to overcome some of the traditional complications involved in microfluidics and to extend it to a more commercialized platform [159,163,165,167].

Reliable future diagnostics will require rapid detection of a panel of biomarkers at low cost, in clinically-relevant ranges. We discussed above many examples where such systems have been fabricated, especially for multiplexed detection of two to four biomarkers. Multiplexing beyond four biomarkers for biomedical samples remain a challenge, but not impossible. These issues will need to be addressed through creative engineering to produce ideal POC platforms for protein panels.

6. Conclusions

We have summarized above many different strategies of fabricating printed electrodes for the detection of cancer biomarker proteins in clinically-relevant ranges. Our focus was on inkjet-printed and SPEs, with the majority of examples covering the latter due to its widespread use in immunosensor development. Advantages include low cost, speed, and ease of fabrication in sensor designs tailored specifically for the microfluidic system being developed. Disadvantages include lack of reproducibility of sensor areas in arrays in some cases, but this can often be overcome with surface area measurements. Electrochemical ELISA-based approaches used in most cases can lead to low cost, ultrasensitive multiplexing of immunoassays, with excellent potential for future clinical applications. One particular electrochemical immunoassay holds the current record for the world's lowest protein LODs in serum, in the sub-fg mL^{-1} range [17]. Electrochemical detection has some limitations for POC applications; however, since individually addressed electrode sensors and a multiplexed electrochemical work station are needed, as opposed to detection by ECL and chemiluminescence, which do not require sophisticated electronic measurements and can be measured with a CCD camera [164]. Automation is also easier with the ECL and CL detection.

Most of the immunosensors discussed have used a variety of labels to achieve ultrasensitive multiplexed detection, with some new research trending towards label-free detection methods. Though unlabeled strategies have improved in sensitivities over the years, they still trail behind labeled biomarkers that have reached record LODs down to 0.02 fg mL^{-1} and below, enabling detection of fewer than 100 protein molecules per sample [17]. Multi-enzyme labeling coupled with nanostructured electrode surfaces are key factors for electrochemical signal enhancement and higher sensitivity. One challenge in SPE immunoassays is a fully automated system, preferably with every component integrated into a single chip. Up to now, only semi-automated units have been exploited with washing, sample preparation, and reagent addition being stumbling blocks in the path to automation.

Despite the delay in automation, immunoassays based on printed electrodes have a huge potential in cancer diagnostics, especially with the ability to be adopted into multiplex detection platforms, which is essential for accurate cancer detection and staging. Unfortunately, current research advances in immunoarrays have not reached active clinical practice for the detection and monitoring of cancer. The transition to clinical use will most likely require commercial development. We hope that the printed electrode immunoassays will overcome this barrier in the near future and be used in future widespread clinical cancer diagnostics.

Funding: Preparation of this article and the authors' work described herein was supported by Grant no. EB016707 from the National Institute of Biomedical Imaging and Bioengineering (NIBIB), NIH, and by an Academic Plan Grant from the University of Connecticut.

Acknowledgments: J.F.R. conceived the idea for the paper and supervised organization and final revisions. L.D., C.E.K., and A.L.J. researched the literature and, wrote the initial drafts, and participated in revisions. The authors thank their many collaborators for excellent contributions to our papers referenced in this article.

Conflicts of Interest: The authors declare no conflict of interest.

References

1. Bray, F.; Ferlay, J.; Soerjomataram, I.; Siegel, R.L.; Torre, L.A.; Jemal, A. Global Cancer Statistics 2018: GLOBOCAN Estimates of Incidence and Mortality Worldwide for 36 Cancers in 185 Cuntries. *CA Cancer J. Clin.* **2018**, *68*, 394–424. [CrossRef] [PubMed]
2. Siegel, R.L.; Miller, K.D.; Jemal, A. Cancer Statistics, 2020. *CA Cancer J. Clin.* **2020**, *70*, 7–30. [CrossRef]
3. Dixit, C.K.; Kadimisetty, K.; Otieno, B.A.; Tang, C.; Malla, S.; Krause, C.E.; Rusling, J.F. Electrochemistry-Based Approaches to Low Cost, High Sensitivity, Automated, Multiplexed Protein Immunoassays for Cancer Diagnostics. *Analyst* **2016**, *141*, 536–547. [CrossRef] [PubMed]
4. Kalia, M. Biomarkers for Personalized Oncology: Recent Advances and Future Challenges. *Metabolism* **2015**, *64*, S16–S21. [CrossRef] [PubMed]
5. Borrebaeck, C.A.K. Precision Diagnostics: Moving towards Protein Biomarker Signatures of Clinical Utility in Cancer. *Nat. Rev. Cancer* **2017**, *17*, 199–204. [CrossRef] [PubMed]
6. Rusling, J.F.; Kumar, C.V.; Gutkind, J.S.; Patel, V. Measurement of Biomarker Proteins for Point-of-Care Early Detection and Monitoring of Cancer. *Analyst* **2010**, *135*, 2496–2511. [CrossRef]
7. Hanahan, D.; Weinberg, R.A. Hallmarks of Cancer: The next Generation. *Cell* **2011**, *144*, 646–674. [CrossRef]
8. Hassan, E.M.; DeRosa, M.C. Recent Advances in Cancer Early Detection and Diagnosis: Role of Nucleic Acid Based Aptasensors. *TrAC—Trends Anal. Chem.* **2020**, *124*, 115806. [CrossRef]

9. Rusling, J.F. Multiplexed Electrochemical Protein Detection and Translation to Personalized Cancer Diagnostics. *Anal. Chem.* **2013**, *85*, 5304–5310. [CrossRef]
10. Rusling, J.F.; Bishop, G.W.; Doan, N.M.; Papadimitrakopoulos, F. Nanomaterials and Biomaterials in Electrochemical Arrays for Protein Detection. *J. Mater. Chem. B* **2014**, *2*, 12–30. [CrossRef]
11. Munge, B.S.; Stracensky, T.; Gamez, K.; DiBiase, D.; Rusling, J.F. Multiplex Immunosensor Arrays for Electrochemical Detection of Cancer Biomarker Proteins. *Electroanalysis* **2016**, *28*, 2644–2658. [CrossRef] [PubMed]
12. Ludwig, J.A.; Weinstein, J.N. Biomarkers in Cancer Staging, Prognosis and Treatment Selection. *Nat. Rev. Cancer* **2005**, *5*, 845–856. [CrossRef] [PubMed]
13. Jayanthi, V.S.P.K.S.A.; Das, A.B.; Saxena, U. Recent Advances in Biosensor Development for the Detection of Cancer Biomarkers. *Biosens. Bioelectron.* **2017**, *91*, 15–23. [CrossRef] [PubMed]
14. Füzéry, A.K.; Levin, J.; Chan, M.M.; Chan, D.W. Translation of Proteomic Biomarkers into FDA Approved Cancer Diagnostics: Issues and Challenges. *Clin. Proteom.* **2013**, *10*, 13. [CrossRef] [PubMed]
15. Bertok, T.; Lorencova, L.; Chocholova, E.; Jane, E.; Vikartovska, A.; Kasak, P.; Tkac, J. Electrochemical Impedance Spectroscopy Based Biosensors: Mechanistic Principles, Analytical Examples and Challenges towards Commercialization for Assays of Protein Cancer Biomarkers. *ChemElectroChem* **2019**, *6*, 989–1003. [CrossRef]
16. Duffy, D. Short Keynote Paper: Single Molecule Detection of Protein Biomarkers to Define the Continuum from Health to Disease. *IEEE J. Biomed. Health Inform.* **2020**, *24*, 1864–1868. [CrossRef]
17. Dhanapala, L.; Jones, A.L.; Czarnecki, P.; Rusling, J.F. Sub-Zeptomole Detection of Biomarker Proteins Using a Microfluidic Immunoarray with Nanostructured Sensors. *Anal. Chem* **2020**, *92*, 8021–8025. [CrossRef]
18. Duffy, D.C.; Walt, D.R. Protein Detection by Counting Molecules. *Clin. Chem.* **2019**, *65*, 809–810. [CrossRef]
19. Wu, C.; Maley, A.M.; Walt, D.R. Single-Molecule Measurements in Microwells for Clinical Applications. *Crit. Rev. Clin. Lab. Sci.* **2020**, *57*, 270–290. [CrossRef]
20. MSD Assays on Meso Scale Discovery Platform (MSD-E). Available online: https://www.bioagilytix.com/meso-scale-discovery-electrochemiluminescence (accessed on 22 June 2020).
21. Luminex Assays | Thermo Fisher Scientific US. Available online: https://www.thermofisher.com/us/en/home/life/science/antibodies/immunoassays/procartaplex-assays-luminex.html (accessed on 22 June 2020).
22. Akkilic, N.; Geschwindner, S.; Höök, F. Single-Molecule Biosensors: Recent Advances and Applications. *Biosens. Bioelectron.* **2020**, *151*, 111944. [CrossRef]
23. Jin, W.; Tang, Q.; Wan, M.; Cui, K.; Zhang, Y.; Ren, G.; Ni, B.; Sklar, J.; Przytycka, T.M.; Childs, R.; et al. Genome-Wide Detection of DNase i Hypersensitive Sites in Single Cells and FFPE Tissue Samples. *Nature* **2015**, *528*, 142–146. [CrossRef] [PubMed]
24. Jones, A.; Dhanapala, L.; Kankanamage, R.N.T.; Kumar, C.V.; Rusling, J.F. Multiplexed Immunosensors and Immunoarrays. *Anal. Chem.* **2020**, *92*, 345–362. [CrossRef] [PubMed]
25. Arduini, F.; Micheli, L.; Moscone, D.; Palleschi, G.; Piermarini, S.; Ricci, F.; Volpe, G. Electrochemical Biosensors Based on Nanomodified Screen-Printed Electrodes: Recent Applications in Clinical Analysis. *TrAC—Trends in Anal. Chem.* **2016**, *79*, 114–126. [CrossRef]
26. Dorothee Grieshaber, R.M.J.V.E.R. Electrochemical Biosensors—Sensor Principles and Architectures. *Sensors* **2008**, *8*, 1400–1458. [CrossRef]
27. Wang, X.; Gao, D.; Li, M.; Li, H.; Li, C.; Wu, X.; Yang, B. CVD graphene as an electrochemical sensing platform for simultaneous detection of biomolecules. *Sci. Rep.* **2017**, *7*, 1–9. [CrossRef]
28. Lu, L.; Gunasekaran, S. Dual-channel ITO-microfluidic electrochemical immunosensor for simultaneous detection of two mycotoxins. *Talanta* **2019**, *194*, 709–716. [CrossRef] [PubMed]
29. Stromberg, L.R.; Hondred, J.A.; Sanborn, D.; Mendivelso-Perez, D.; Ramesh, S.; Rivero, I.V.; Kogot, J.; Smith, E.; Gomes, C.; Claussen, J.C. Stamped multilayer graphene laminates for disposable in-field electrodes: Application to electrochemical sensing of hydrogen peroxide and glucose. *Microchim. Acta* **2019**, *186*, 1–13. [CrossRef]
30. Rama, E.C.; Costa-García, A. Screen-printed Electrochemical Immunosensors for the Detection of Cancer and Cardiovascular Biomarkers. *Electroanalysis* **2016**, *28*, 1700–1715. [CrossRef]
31. Jensen, G.C.; Krause, C.E.; Sotzing, G.A.; Rusling, J.F. Inkjet-printed gold nanoparticle electrochemical arrays on plastic. Application to immunodetection of a cancer biomarker protein. *Phys. Chem. Chem. Phys.* **2011**, *13*, 4888–4894. [CrossRef]

32. Sui, Y.; Zorman, C.A. Review—Inkjet Printing of Metal Structures for Electrochemical Sensor Applications. *J. Electrochem. Soc.* **2020**, *167*, 037571. [CrossRef]
33. Kim, D.; Lee, S.H.; Jeong, S.H.; Moon, J. All-Ink-Jet Printed Flexible Organic Thin-Film Transistors on Plastic Substrates. *Electrochem. Solid State Lett.* **2009**, *12*, H195–H197. [CrossRef]
34. Taleat, Z.; Khoshroo, A.; Mazloum-Ardakani, M. Screen-printed electrodes for biosensing: A review (2008–2013). *Microchim. Acta* **2014**, *181*, 865–891. [CrossRef]
35. Tao, R.; Ning, H.; Chen, J.; Zou, J.; Fang, Z.; Yang, C.; Zhou, Y.; Zhang, J.; Yao, R.; Peng, J. Inkjet Printed Electrodes in Thin Film Transistors. *IEEE J. Electron. Devices Soc.* **2018**, *6*, 774–790. [CrossRef]
36. Beitollahi, H.; Mohammadi, S.Z.; Safaei, M.; Tajik, S. Applications of Electrochemical Sensors and Biosensors Based on Modified Screen-Printed Electrodes: A Review. *Anal. Methods* **2020**, *12*, 1547–1560. [CrossRef]
37. Heineman, W.R.; Halsall, H.B. Strategies for electrochemical immunoassay. *Anal. Chem.* **1985**, *75*, 1321A–1331A.
38. Lopez, G.A.; Estevez, M.C.; Soler, M.; Lechuga, L.M. Recent Advances in Nanoplasmonic Biosensors: Applications and Lab-on-a-Chip Integration. *Nanophotonics* **2017**, *6*, 123–136. [CrossRef]
39. Song, Y.; Lin, B.; Tian, T.; Xu, X.; Wang, W.; Ruan, Q.; Guo, J.; Zhu, Z.; Yang, C. Recent Progress in Microfluidics-Based Biosensing. *Anal. Chem.* **2019**, *91*, 388–404. [CrossRef]
40. Reverté, L.; Prieto-Simón, B.; Campàs, M. New Advances in Electrochemical Biosensors for the Detection of Toxins: Nanomaterials, Magnetic Beads and Microfluidics Systems. A Review. *Anal. Chim. Acta* **2016**, *908*, 8–21. [CrossRef]
41. Pandey, C.M.; Augustine, S.; Kumar, S.; Kumar, S.; Nara, S.; Srivastava, S.; Malhotra, B.D. Microfluidics Based Point-of-Care Diagnostics. *Biotechnol. J.* **2018**, *13*, 1700047. [CrossRef]
42. Lvov, Y.M.; Lu, Z.; Schenkman, J.B.; Rusling, J.F. Direct Electrochemistry of Myoglobin and Cytochrome P450cam in Alternate Polyion Layer-by-Layer Films with DNA and other polyions. *J. Am. Chem. Soc.* **1998**, *120*, 4073–4080. [CrossRef]
43. Rusling, J.F. Electroactive and Enzyme-Active Protein-Polyion Films Assembled Layer-by-Layer. In *Protein Architecture: Interfacing Molecular Assemblies and Immobilization Biotechnology*; Marcel Dekker: New York, NY, USA, 2000; pp. 337–354.
44. Tang, Z.; Wang, Y.; Podsiadlo, P.; Kotov, N.A. Biomedical Applications of Layer-by-Layer Assembly: From Biomimetics to Tissue Engineering. *Adv. Mater.* **2006**, *18*, 3203–3224. [CrossRef]
45. Zhao, W.; Xu, J.J.; Chen, H.Y. Electrochemical Biosensors Based on Layer-by-Layer Assemblies. *Electroanalysis* **2006**, *18*, 1737–1748. [CrossRef]
46. Koniev, O.; Wagner, A. Developments and Recent Advancements in the Field of Endogenous Amino Acid Selective Bond Forming Reactions for Bioconjugation. *Chem. Soc. Rev.* **2015**, *44*, 5495–5551. [CrossRef] [PubMed]
47. Sharafeldin, M.; McCaffrey, K.; Rusling, J.F. Influence of Antibody Immobilization Strategy on Carbon Electrode Immunoarrays. *Analyst* **2019**, *144*, 5108–5116. [CrossRef]
48. Yang, H. Enzyme-Based Ultrasensitive Electrochemical Biosensors. *Curr. Opin. Chem. Biol.* **2012**, *16*, 422–428. [CrossRef]
49. Nguyen, T.T.; Sly, K.L.; Conboy, J.C. Comparison of the Energetics of Avidin, Streptavidin, NeutrAvidin, and Anti-Biotin Antibody Binding to Biotinylated Lipid Bilayer Examined by Second-Harmonic Generation. *Anal. Chem.* **2012**, *84*, 201–208. [CrossRef]
50. Pierce, T.M. Antibody Biotinylation Kit for IP. Available online: https://www.thermofisher.com/order/catalog/product/90407#/90407 (accessed on 30 June 2020).
51. Udeshi, N.D.; Pedram, K.; Svinkina, T.; Fereshetian, S.; Myers, S.A.; Aygun, O.; Krug, K.; Clauser, K.; Ryan, D.; Ast, T.; et al. Antibodies to Biotin Enable Large-Scale Detection of Biotinylation Sites on Proteins. *Nat. Methods* **2017**, *14*, 1167–1170. [CrossRef]
52. Jain, A.; Cheng, K. The principles and applications of avidin-based nanoparticles in drug delivery and diagnosis. *J. Control. Release* **2017**, *245*, 27–40. [CrossRef]
53. Jain, A.; Barve, A.; Zhao, Z.; Jin, W.; Cheng, K. Comparison of avidin, neutravidin, and streptavidin as nanocarriers for efficient siRNA Delivery. *Mol. Pharm.* **2017**, *14*, 1517–1527. [CrossRef]

54. Avidin and Streptavidin Conjugates—Section 7.6 | Thermo Fisher Scientific—US. Available online: https://www.thermofisher.com/us/en/home/references/molecular-probes-the-handbook/antibodies-avidins-lectins-and-related-products/avidin-streptavidin-neutravidin-and-captavidin-biotin-binding-proteins-and-affinity-matrices.html (accessed on 30 June 2020).
55. Orelma, H.; Johansson, L.S.; Filpponen, I.; Rojas, O.J.; Laine, J. Generic Method for Attaching Biomolecules via Avidin—Biotin Complexes Immobilized on Films of Regenerated and Nanofibrillar Cellulose. *Biomacromolecules* **2012**, *13*, 2802–2810. [CrossRef]
56. Contreras-Naranjo, J.E.; Aguilar, O. Suppressing Non-Specific Binding of Proteins onto Electrode Surfaces in the Development of Electrochemical Immunosensors. *Biosensors* **2019**, *9*, 15. [CrossRef] [PubMed]
57. Iglesias-Mayor, A.; Amor-Gutiérrez, O.; Costa-García, A.; de la Escosura-Muñiz, A. Nanoparticles as Emerging Labels in Electrochemical Immunosensors. *Sensors* **2019**, *19*, 5137. [CrossRef] [PubMed]
58. Zhou, Y.; Liu, B.; Yang, R.; Liu, J. Filling in the Gaps between Nanozymes and Enzymes: Challenges and Opportunities. *Bioconjug. Chem.* **2017**, *28*, 2903–2909. [CrossRef]
59. Wu, J.; Wang, X.; Wang, Q.; Lou, Z.; Li, S.; Zhu, Y.; Qin, L.; Wei, H. Nanomaterials with Enzyme-like Characteristics (Nanozymes): Next-Generation Artificial Enzymes (II). *Chem. Soc. Rev.* **2019**, *48*, 1004–1076. [CrossRef] [PubMed]
60. Pirsaheb, M.; Mohammadi, S.; Salimi, A. Current advances of carbon dots based biosensors for tumor marker detection, cancer cells analysis and bioimaging. *TrAC—Trends Anal. Chem.* **2019**, *115*, 83–99. [CrossRef]
61. Campuzano, S.; Yáñez-Sedeño, P.; Pingarrón, J.M. Carbon dots and graphene quantum dots in electrochemical biosensing. *Nanomaterials* **2019**, *9*, 634. [CrossRef] [PubMed]
62. Karimi-Maleh, H. Electrochemical Sensors Based on Metal Nanoparticles, Carbon Based Nanomaterials or Ionic Liquids. *Curr. Anal. Chem.* **2017**, *13*, 4. [CrossRef]
63. Dequaire, M.; Degrand, C.; Limoges, B. An Electrochemical Metalloimmunoassay Based on a Colloidal Gold Label. *Anal. Chem.* **2000**, *72*, 5521–5528. [CrossRef]
64. Wang, J. Nanoparticle-Based Electrochemical Bioassays of Proteins. *Electroanalysis* **2007**, *19*, 769–776. [CrossRef]
65. Liu, G.; Wang, J.; Kim, J.; Jan, M.R.; Collins, G.E. Electrochemical Coding for Multiplexed Immunoassays of Proteins. *Anal. Chem.* **2005**, *76*, 7126–7130. [CrossRef]
66. Wang, J.; Liu, G.; Munge, B.; Lin, L.; Zhu, Q. DNA-Based Amplified Bioelectronic Detection and Coding of Proteins. *Angew. Chem.* **2004**, *116*, 2210–2213. [CrossRef]
67. Chumbimuni-Torres, K.Y.; Dai, Z.; Rubinova, N.; Xiang, Y.; Pretsch, E.; Wang, J.; Bakker, E. Potentiometric Biosensing of Proteins with Ultrasensitive Ion-Selective Microelectrodes and Nanoparticle Labels. *J. Am. Chem. Soc.* **2006**, *128*, 13676–13677. [CrossRef] [PubMed]
68. Zhang, C.; Zhang, S.; Jia, Y.; Li, Y.; Wang, P.; Liu, Q.; Xu, Z.; Li, X. Sandwich-type electrochemical immunosensor for sensitive detection of CEA based on the enhanced effects of Ag NPs@CS spaced Hemin/rGO. *Biosens. Bioelectron.* **2019**, *126*, 785–791. [CrossRef] [PubMed]
69. Zhang, H.; Ma, L.; Li, P.; Zheng, J. A novel electrochemical immunosensor based on nonenzymatic Ag@Au-Fe3O4 nanoelectrocatalyst for protein biomarker detection. *Biosens. Bioelectron.* **2016**, *85*, 343–350. [CrossRef]
70. Lai, G.; Wang, L.; Wu, J.; Ju, H.; Yan, F. Electrochemical Stripping Analysis of Nanogold Label-Induced Silver Deposition for Ultrasensitive Multiplexed Detection of Tumor Markers. *Anal. Chim. Acta* **2012**, *721*, 1–6. [CrossRef]
71. Li, F.; Li, Y.; Feng, J.; Gao, Z.; Lv, H.; Ren, X.; Wei, Q. Facile synthesis of MoS2@Cu2O-Pt nanohybrid as enzyme-mimetic label for the detection of the Hepatitis B surface antigen. *Biosens. Bioelectron.* **2018**, *100*, 512–518. [CrossRef]
72. Feng, J.; Li, Y.; Li, M.; Li, F.; Han, J.; Dong, Y.; Chen, Z. A novel sandwich-type electrochemical immunosensor for PSA detection based on PtCu bimetallic hybrid (2D/2D) rGO/g-C3N4. *Biosens. Bioelectron.* **2017**, *91*, 441–448. [CrossRef]
73. Liu, L.; Tian, L.; Zhao, G.; Huang, Y.; Wei, Q.; Cao, W. Ultrasensitive electrochemical immunosensor for alpha fetoprotein detection based on platinum nanoparticles anchored on cobalt oxide/graphene nanosheets for signal amplification. *Anal. Chim. Acta* **2017**, *986*, 138–144. [CrossRef]
74. Charoenkitamorn, K.; Tue, P.; Kawai, K.; Chailapakul, O.; Takamura, Y. Electrochemical Immunoassay Using Open Circuit Potential Detection Labeled by Platinum Nanoparticles. *Sensors* **2018**, *18*, 444. [CrossRef]

75. Zhang, Z.; Cong, Y.; Huang, Y.; Du, X. Nanomaterials-Based Electrochemical Immunosensors. *Micromachines* **2019**, *10*, 397. [CrossRef]
76. Zhu, Y.; Wang, H.; Wang, L.; Zhu, J.; Jiang, W. Cascade Signal Amplification Based on Copper Nanoparticle-Reported Rolling Circle Amplification for Ultrasensitive Electrochemical Detection of the Prostate Cancer Biomarker. *ACS Appl. Mater. Interfaces* **2016**, *8*, 2573–2581. [CrossRef] [PubMed]
77. Rivas, L.; de la Escosura-Muñiz, A.; Pons, J.; Merkoçi, A. Alzheimer Disease Biomarker Detection through Electrocatalytic Water Oxidation Induced by Iridium Oxide Nanoparticles. *Electroanalysis* **2014**, *26*, 1287–1294. [CrossRef]
78. Kalyoncu, D.; Buyuksunetci, Y.T.; Anık, Ü. Development of a Sandwich Immunosensor for concurrent detection of carcinoembryonic antigen (CEA), vascular endothelial growth factor (VEGF) and α-fetoprotein (AFP) biomarkers. *Mater. Sci. Eng. C* **2019**, *101*, 88–91. [CrossRef] [PubMed]
79. Putnin, T.; Ngamaroonchote, A.; Wiriyakun, N.; Ounnunkad, K.; Laocharoensuk, R. Dually functional polyethylenimine-coated gold nanoparticles: A versatile material for electrode modification and highly sensitive simultaneous determination of four tumor markers. *Microchim. Acta* **2019**, *186*, 305–317. [CrossRef]
80. Tang, D.; Hou, L.; Niessner, R.; Xu, M.; Gao, Z.; Knopp, D. Multiplexed Electrochemical Immunoassay of Biomarkers Using Metal Sulfide Quantum Dot Nanolabels and Trifunctionalized Magnetic Beads. *Biosens. Bioelectron.* **2013**, *46*, 37–43. [CrossRef]
81. Zhang, B.; Tang, D.; Goryacheva, I.Y.; Niessner, R.; Knopp, D. Anodic-Stripping Voltammetric Immunoassay for Ultrasensitive Detection of Low-Abundance Proteins Using Quantum Dot Aggregated Hollow Microspheres. *Chemistry* **2013**, *19*, 2496–2503. [CrossRef]
82. Martín-Yerga, D.; González-García, M.B.; Costa-García, A. Electrochemical immunosensor for anti-tissue transglutaminase antibodies based on the in situ detection of quantum dots. *Talanta* **2014**, *130*, 598–602. [CrossRef]
83. Martín-Yerga, D.; Costa-García, A. Towards a blocking-free electrochemical immunosensing strategy for anti-transglutaminase antibodies using screen-printed electrodes. *Bioelectrochemistry* **2015**, *105*, 88–94. [CrossRef]
84. Merkoçi, A.; Marcolino-Junior, L.H.; Marín, S.; Fatibello-Filho, O.; Alegret, S. Detection of cadmium sulphide nanoparticles by using screen-printed electrodes and a handheld device. *Nanotechnology* **2007**, *18*, 035502–035507. [CrossRef]
85. Wang, J.; Liu, G.; Jan, M.R. Ultrasensitive Electrical Biosensing of Proteins and DNA: Carbon-Nanotube Derived Amplification of the Recognition and Transduction Events. *J. Am. Chem. Soc.* **2004**, *126*, 3010–3011. [CrossRef]
86. Malhotra, R.; Patel, V.; Vaqué, J.P.; Gutkind, J.S.; Rusling, J.F. Ultrasensitive Electrochemical Immunosensor for Oral Cancer Biomarker IL-6 Using Carbon Nanotube Forest Electrodes and Multilabel Amplification. *Anal. Chem.* **2010**, *82*, 3118–3123. [CrossRef] [PubMed]
87. Munge, B.S.; Fisher, J.; Millord, L.N.; Krause, C.E.; Dowd, R.S.; Rusling, J.F. Sensitive Electrochemical Immunosensor for Matrix Metalloproteinase-3 Based on Single-Wall Carbon Nanotubes. *Analyst* **2010**, *135*, 1345–1350. [CrossRef] [PubMed]
88. Yu, X.; Munge, B.; Patel, V.; Jensen, G.; Bhirde, A.; Gong, J.D.; Kim, S.N.; Gillespie, J.; Gutkind, J.S.; Papadimitrakopoulos, F.; et al. Carbon Nanotube Amplification Strategies for Highly Sensitive Immunodetection of Cancer Biomarkers. *J. Am. Chem. Soc.* **2006**, *128*, 11199–11205. [CrossRef] [PubMed]
89. Kim, S.N.; Rusling, J.F.; Papadimitrakopoulos, F. Carbon Nanotubes for Electronic and Electrochemical Detection of Biomolecules. *Adv. Mater.* **2007**, *19*, 3214–3228. [CrossRef]
90. Sharafeldin, M.; Bishop, G.W.; Bhakta, S.; El-Sawy, A.; Suib, S.L.; Rusling, J.F. Fe_3O_4 Nanoparticles on Graphene Oxide Sheets for Isolation and Ultrasensitive Amperometric Detection of Cancer Biomarker Proteins. *Biosens. Bioelectron.* **2017**, *91*, 359–366. [CrossRef]
91. Serafín, V.; Valverde, A.; Garranzo-Asensio, M.; Barderas, R.; Campuzano, S.; Yáñez-Sedeño, P.; Pingarrón, J.M. Simultaneous amperometric immunosensing of the metastasis-related biomarkers IL-13Rα2 and CDH-17 by using grafted screen-printed electrodes and a composite prepared from quantum dots and carbon nanotubes for signal amplification. *Microchim. Acta* **2019**, *186*, 411. [CrossRef]
92. Otieno, B.A.; Krause, C.E.; Rusling, J.F. Bioconjugation of Antibodies and Enzyme Labels onto Magnetic Beads. *Methods Enzymol.* **2016**, *571*, 135–150.

93. Mani, V.; Chikkaveeraiah, B.V.; Patel, V.; Gutkind, J.S.; Rusling, J.F. Ultrasensitive Immunosensor for Cancer Biomarker Proteins Using Gold Nanoparticle Film Electrodes and Multienzyme-Particle Amplification. *ACS Nano* **2009**, *3*, 585–594. [CrossRef]
94. Munge, B.S.; Coffey, A.L.; Doucette, J.M.; Somba, B.K.; Malhotra, R.; Patel, V.; Gutkind, J.S.; Rusling, J.F. Nanostructured Immunosensor for Attomolar Detection of Cancer Biomarker Interleukin-8 Using Massively Labeled Superparamagnetic Particles. *Angew. Chem.* **2011**, *123*, 8061–8064. [CrossRef]
95. Chikkaveeraiah, B.V.; Mani, V.; Patel, V.; Gutkind, J.S.; Rusling, J.F. Microfluidic Electrochemical Immunoarray for Ultrasensitive Detection of Two Cancer Biomarker Proteins in Serum. *Biosens. Bioelectron.* **2011**, *26*, 4477–4483. [CrossRef]
96. Malhotra, R.; Patel, V.; Chikkaveeraiah, B.V.; Munge, B.S.; Cheong, S.C.; Zain, R.B.; Abraham, M.T.; Dey, D.K.; Gutkind, J.S.; Rusling, J.F. Ultrasensitive Detection of Cancer Biomarkers in the Clinic by Use of a Nanostructured Microfluidic Array. *Anal. Chem.* **2012**, *84*, 6249–6255. [CrossRef] [PubMed]
97. Patel, V.; Martin, D.; Malhotra, R.; Marsh, C.A.; Doçi, C.L.; Veenstra, T.D.; Nathan, C.A.O.; Sinha, U.K.; Singh, B.; Molinolo, A.A.; et al. DSG3 as a Biomarker for the Ultrasensitive Detection of Occult Lymph Node Metastasis in Oral Cancer Using Nanostructured Immunoarrays. *Oral Oncol.* **2013**, *49*, 93–101. [CrossRef] [PubMed]
98. Krause, C.E.; Otieno, B.A.; Latus, A.; Faria, R.C.; Patel, V.; Gutkind, J.S.; Rusling, J.F. Rapid Microfluidic Immunoassays of Cancer Biomarker Proteins Using Disposable Inkjet-Printed Gold Nanoparticle Arrays. *ChemistryOpen* **2013**, *2*, 141–145. [CrossRef] [PubMed]
99. Otieno, B.A.; Krause, C.E.; Latus, A.; Chikkaveeraiah, B.V.; Faria, R.C.; Rusling, J.F. On-Line Protein Capture on Magnetic Beads for Ultrasensitive Microfluidic Immunoassays of Cancer Biomarkers. *Biosens. Bioelectron.* **2014**, *53*, 268–274. [CrossRef]
100. Otieno, B.A.; Krause, C.E.; Jones, A.L.; Kremer, R.B.; Rusling, J.F. Cancer Diagnostics via Ultrasensitive Multiplexed Detection of Parathyroid Hormone-Related Peptides with a Microfluidic Immunoarray. *Anal. Chem.* **2016**, *88*, 9269–9275. [CrossRef]
101. Krause, C.E.; Otieno, B.A.; Bishop, G.W.; Phadke, G.; Choquette, L.; Lalla, R.V.; Peterson, D.E.; Rusling, J.F. Ultrasensitive Microfluidic Array for Serum Pro-Inflammatory Cytokines and C-Reactive Protein to Assess Oral Mucositis Risk in Cancer Patients. *Anal. Bioanal. Chem.* **2017**, 7239–7243. [CrossRef]
102. Phadke, G.S.; Satterwhite-Warden, J.E.; Choudhary, D.; Taylor, J.A.; Rusling, J.F. A Novel and Accurate Microfluidic Assay of CD62L in Bladder Cancer Serum Samples. *Analyst* **2018**, *143*, 5505–5511. [CrossRef]
103. Mercer, C.; Jones, A.; Rusling, J.F.; Leech, D. Multiplexed electrochemical cancer diagnostics with automated microfluidics. *Electroanalysis* **2019**, *31*, 208–211. [CrossRef]
104. Uliana, C.V.; Peverari, C.R.; Afonso, A.S.; Cominetti, M.R.; Faria, R.C. Fully disposable microfluidic electrochemical device for detection of estrogen receptor alpha breast cancer biomarker. *Biosens. Bioelectron.* **2018**, *99*, 156–162. [CrossRef]
105. Fukuda, T.; Tani, Y.; Kobayashi, T.; Hirayama, Y.; Hino, O. A New Western Blotting Method Using Polymer Immunocomplexes: Detection of Tsc1 and Tsc2 Expression in Various Cultured Cell Lines. *Anal. Biochem.* **2000**, *285*, 274–276. [CrossRef]
106. Dhawan, S. Design and Construction of Novel Molecular Conjugates for Signal Amplification (I): Conjugation of Multiple Horseradish Peroxidase Molecules to Immunoglobulin via Primary Amines on Lysine Peptide Chains. *Peptides* **2002**, *23*, 2091–2098. [CrossRef]
107. Mishra, M.; Tiwari, S.; Gunaseelan, A.; Li, D.; Hammock, B.D.; Gomes, A.V. Improving the Sensitivity of Traditional Western Blotting via Streptavidin Containing Poly-horseradish Peroxidase (PolyHRP). *Electrophoresis* **2019**, *40*, 12–13. [CrossRef] [PubMed]
108. Tang, C.K.; Vaze, A.; Rusling, J.F. Fabrication of Immunosensor Microwell Arrays from Gold Compact Discs for Detection of Cancer Biomarker Proteins. *Lab. Chip* **2012**, *12*, 281–286. [CrossRef] [PubMed]
109. Ojeda, I.; Moreno-Guzmán, M.; González-Cortés, A.; Yáñez-Sedeño, P.; Pingarrón, J.M. Electrochemical Magnetoimmunosensor for the Ultrasensitive Determination of Interleukin-6 in Saliva and Urine Using Poly-HRP Streptavidin Conjugates as Labels for Signal Amplification. *Anal. Bioanal. Chem.* **2014**, *406*, 6363–6371. [CrossRef]
110. Sánchez-Tirado, E.; Martínez-García, G.; González-Cortés, A.; Yáñez-Sedeño, P.; Pingarrón, J.M. Electrochemical Immunosensor for Sensitive Determination of Transforming Growth Factor (TGF)—B1 in Urine. *Biosens. Bioelectron.* **2017**, *88*, 9–14. [CrossRef]

111. Sánchez-Tirado, E.; Salvo, C.; González-Cortés, A.; Yáñez-Sedeño, P.; Langa, F.; Pingarrón, J.M. Electrochemical immunosensor for simultaneous determination of interleukin-1 beta and tumor necrosis factor alpha in serum and saliva using dual screen printed electrodes modified with functionalized double–walled carbon nanotubes. *Anal. Chim. Acta* **2017**, *959*, 66–73. [CrossRef]
112. Pastucha, M.; Farka, Z.; Lacina, K.; Mikušová, Z.; Skládal, P. Magnetic Nanoparticles for Smart Electrochemical Immunoassays: A Review on Recent Developments. *Microchim. Acta* **2019**, *186*, 1–26. [CrossRef]
113. Kanyong, P.; Patil, A.V.; Davis, J.J. Functional Molecular Interfaces for Impedance-Based Diagnostics. *Annu. Rev. Anal. Chem.* **2020**, *13*, 183–200. [CrossRef]
114. Luo, X.; Davis, J.J. Electrical Biosensors and the Label Free Detection of Protein Disease Biomarkers. *Chem. Soc. Rev.* **2013**, *42*, 5944–5962. [CrossRef]
115. Johnson, A.; Song, Q.; Ko Ferrigno, P.; Bueno, P.R.; Davis, J.J. Sensitive Affimer and Antibody Based Impedimetric Label-Free Assays for C-Reactive Protein. *Anal. Chem.* **2012**, *84*, 6553–6560. [CrossRef]
116. Fernandes, F.C.B.; Góes, M.S.; Davis, J.J.; Bueno, P.R. Label Free Redox Capacitive Biosensing. *Biosens. Bioelectron.* **2013**, *50*, 437–440. [CrossRef] [PubMed]
117. Luo, X.; Xu, M.; Freeman, C.; James, T.; Davis, J.J. Ultrasensitive Label Free Electrical Detection of Insulin in Neat Blood Serum. *Anal. Chem.* **2013**, *85*, 4129–4134. [CrossRef] [PubMed]
118. Wang, W.; Fan, X.; Xu, S.; Davis, J.J.; Luo, X. Low Fouling Label-Free DNA Sensor Based on Polyethylene Glycols Decorated with Gold Nanoparticles for the Detection of Breast Cancer Biomarkers. *Biosens. Bioelectron.* **2015**, *71*, 51–56. [CrossRef] [PubMed]
119. Santos, A.; Bueno, P.R.; Davis, J.J. A Dual Marker Label Free Electrochemical Assay for Flavivirus Dengue Diagnosis. *Biosens. Bioelectron.* **2018**, *100*, 519–525. [CrossRef] [PubMed]
120. Kanyong, P.; Davis, J.J. Homogeneous Functional Self-Assembled Monolayers: Faradaic Impedance Baseline Signal Drift Suppression for High-Sensitivity Immunosensing of C-Reactive Protein. *J. Electroanal. Chem.* **2020**, *856*, 113675. [CrossRef]
121. Kanyong, P.; Catli, C.; Davis, J.J. Ultrasensitive Impedimetric Immunosensor for the Detection of C-Reactive Protein in Blood at Surface-Initiated-Reversible Addition–Fragmentation Chain Transfer Generated Poly (2-Hydroxyethyl Methacrylate) Brushes. *Anal. Chem.* **2020**, *92*, 4707–4710. [CrossRef]
122. Zhao, L.; Han, H.; Ma, Z. Improved Screen-Printed Carbon Electrode for Multiplexed Label-Free Amperometric Immuniosensor: Addressing Its Conductivity and Reproducibility Challenges. *Biosens. Bioelectron.* **2018**, *101*, 304–310. [CrossRef]
123. Khoshroo, A.; Mazloum-Ardakani, M.; Forat-Yazdi, M. Enhanced Performance of Label-Free Electrochemical Immunosensor for Carbohydrate Antigen 15-3 Based on Catalytic Activity of Cobalt Sulfide/Graphene Nanocomposite. *Sens. Actuators B Chem.* **2018**, *255*, 580–587. [CrossRef]
124. Shafaat, A.; Faridbod, F.; Ganjali, M.R. Label-Free Detection of Cytochrome: C by a Conducting Polymer-Based Impedimetric Screen-Printed Aptasensor. *New J. Chem.* **2018**, *42*, 6034–6039. [CrossRef]
125. Srivastava, M.; Nirala, N.R.; Srivastava, S.K.; Prakash, R. A Comparative Study of Aptasensor vs. Immunosensor for Label-Free PSA Cancer Detection on GQDs-AuNRs Modified Screen-Printed Electrodes. *Sci. Rep.* **2018**, *8*, 1–11. [CrossRef]
126. Mollarasouli, F.; Serafín, V.; Campuzano, S.; Yáñez-Sedeño, P.; Pingarrón, J.M.; Asadpour-Zeynali, K. Ultrasensitive Determination of Receptor Tyrosine Kinase with a Label-Free Electrochemical Immunosensor Using Graphene Quantum Dots-Modified Screen-Printed Electrodes. *Anal. Chim. Acta* **2018**, *1011*, 28–34. [CrossRef] [PubMed]
127. Ahmed, M.U.; Hossain, M.M.; Safavieh, M.; Wong, Y.L.; Rahman, I.A.; Zourob, M.; Tamiya, E. Toward the development of smart and low cost point-of-care biosensors based on screen printed electrodes. *Crit. Rev. Biotechnol.* **2016**, *36*, 495–505. [CrossRef] [PubMed]
128. DuPont Conductive Inks for Digital Printing. Available online: https://www.dupont.com/products/inkjet-silver-conductor-inks.html (accessed on 15 May 2020).
129. A Range of Conductive Inks That Are Optimized for Various Applications on Multiple Substrates. Available online: https://www.sunchemical.com/product/conductive-inks/ (accessed on 15 May 2020).
130. Acheson | Electrodag | Lumidag | Minico | Flexible Circuit | PCB |. Available online: https://www.materialtech.biz/index.php/brands/acheson (accessed on 15 May 2020).
131. Conductive Silver Inks | Conductive Carbon Inks | Tekra, LLC. Available online: https://www.tekra.com/products/conductive-inks (accessed on 15 May 2020).

132. Metrohm DropSens Screen-Printed Electrodes. Available online: http://www.dropsens.com/en/screen_printed_electrodes_pag.html (accessed on 15 May 2020).
133. Carbon Screen-Printed Electrodes. Available online: https://pineresearch.com/shop/products/electrodes/screen-printed-electrodes/carbon-spes (accessed on 15 May 2020).
134. Screen Printed Electrodes. Available online: https://www.palmsens.com/products/sensors/screen-printed-electrodes (accessed on 15 May 2020).
135. Screen Printed Electrodes. Available online: https://www.basinc.com/products/ec/screen-printed-electrodes (accessed on 15 May 2020).
136. Kanichi Research Services. Available online: http://www.gmclifesciencesfund.com/Medical-Devices/kanichi-research-services/30980 (accessed on 15 May 2020).
137. Tallapragada, S.D.; Layek, K.; Mukherjee, R.; Mistry, K.K.; Ghosh, M. Development of screen-printed electrode based immunosensor for the detection of HER2 antigen in human serum samples. *Bioelectrochemistry* **2017**, *118*, 25–30. [CrossRef] [PubMed]
138. Mistry, K.K.; Layek, K.; Chell, T.N.; Chaudhuri, C.R.; Saha, H. Design and development of an amperometric immunosensor based on screen-printed electrodes. *Anal. Methods* **2016**, *8*, 3096–3101. [CrossRef]
139. Das, J.; Kelley, S.O. Protein detection using arrayed microsensor chips: Tuning sensor footprint to achieve ultrasensitive readout of CA-125 in serum and whole blood. *Anal. Chem.* **2011**, *83*, 1167–1172. [CrossRef] [PubMed]
140. Chan, K.F.; Lim, H.N.; Shams, N.; Jayabal, S.; Pandikumar, A.; Huang, N.M. Fabrication of graphene/gold-modified screen-printed electrode for detection of carcinoembryonic antigen. *Mater. Sci. Eng. C* **2016**, *58*, 666–674. [CrossRef]
141. Suresh, L.; Brahman, P.K.; Reddy, K.R.; Bondili, J.S. Development of an electrochemical immunosensor based on gold nanoparticles incorporated chitosan biopolymer nanocomposite film for the detection of prostate cancer using PSA as biomarker. *Enzyme Microb. Technol.* **2018**, *112*, 43–51. [CrossRef]
142. Giannetto, M.; Bianchi, M.V.; Mattarozzi, M.; Careri, M. Competitive amperometric immunosensor for determination of p53 protein in urine with carbon nanotubes/gold nanoparticles screen-printed electrodes: A potential rapid and noninvasive screening tool for early diagnosis of urinary tract carcinoma. *Anal. Chim. Acta* **2017**, *991*, 133–141. [CrossRef]
143. Marques, R.C.B.; Costa-Rama, E.; Viswanathan, S.; Nouws, H.P.A.; Costa-García, A.; Delerue-Matos, C.; González-García, M.B. Voltammetric immunosensor for the simultaneous analysis of the breast cancer biomarkers CA 15-3 and HER2-ECD. *Sens. Actuators B Chem.* **2018**, *255*, 918–925. [CrossRef]
144. Fanjul-Bolado, P.; Hernández-Santos, D.; González-García, M.B.; Costa-García, A. Alkaline phosphatase-catalyzed silver deposition for electrochemical detection. *Anal. Chem.* **2007**, *79*, 5272–5277. [CrossRef]
145. Selvolini, G.; Marrazza, G. MIP-based sensors: Promising new tools for cancer biomarker determination. *Sensors* **2017**, *17*, 718. [CrossRef] [PubMed]
146. Bozal-Palabiyik, B.; Lettieri, M.; Uslu, B.; Marrazza, G. Electrochemical Detection of Vascular Endothelial Growth Factor by Molecularly Imprinted Polymer. *Electroanalysis* **2019**, *31*, 1458–1464. [CrossRef]
147. Gomes, R.S.; Moreira, F.T.C.; Fernandes, R.; Sales, M.G.F. Sensing CA 15-3 in point-of-care by electropolymerizing O-phenylenediamine (oPDA) on Au-screen printed electrodes. *PLoS ONE* **2018**, *13*, e0196656. [CrossRef] [PubMed]
148. Ronkainen-Matsuno, N.J.; Thomas, J.H.; Halsall, H.B.; Heineman, W.R. Electrochemical immunoassay moving into the fast lane. *TrAC—Trends Anal. Chem.* **2002**, *21*, 213–225. [CrossRef]
149. Bange, A.; Halsall, H.B.; Heineman, W.R. Microfluidic immunosensor systems. *Biosens. Bioelectron.* **2005**, *20*, 2488–2503. [CrossRef] [PubMed]
150. Martinez, A.W.; Phillips, S.T.; Butte, M.J.; Whitesides, G.M. Patterned Paper as a Platform for Inexpensive, Low-Volume, Portable Bioassays. *Angew. Chem. Int. Ed.* **2007**, *46*, 1318–1320. [CrossRef]
151. Gutiérrez-Capitán, M.; Baldi, A.; Fernández-Sánchez, C. Electrochemical Paper-Based Biosensor Devices for Rapid Detection of Biomarkers. *Sensors* **2020**, *20*, 967. [CrossRef]
152. Fan, Y.; Shi, S.; Ma, J.; Guo, Y. A paper-based electrochemical immunosensor with reduced graphene oxide/thionine/gold nanoparticles nanocomposites modification for the detection of cancer antigen 125. *Biosens. Bioelectron.* **2019**, *135*, 1–7. [CrossRef]
153. Tortorich, R.; Shamkhalichenar, H.; Choi, J.-W. Inkjet-Printed and Paper-Based Electrochemical Sensors. *Appl. Sci.* **2018**, *8*, 288. [CrossRef]

154. Huther Da Costa, T.; Song, E. A Paper-Based Electrochemical Sensor Using Inkjet-Printed Carbon Nanotube Electrodes. *ECS J. Solid State Sci. Technol.* **2015**, *4*, S3044–S3047. [CrossRef]
155. Tortorich, R.P.; Song, E.; Choi, J.-W. Inkjet-Printed Carbon Nanotube Electrodes with Low Sheet Resistance for Electrochemical Sensor Applications. *J. Electrochem. Soc.* **2014**, *161*, B3044–B3048. [CrossRef]
156. Määttänen, A.; Vanamo, U.; Ihalainen, P.; Pulkkinen, P.; Tenhu, H.; Bobacka, J.; Peltonen, J. A low-cost paper-based inkjet-printed platform for electrochemical analyses. *Sens. Actuators B Chem.* **2013**, *177*, 153–162. [CrossRef]
157. Shamkhalichenar, H.; Choi, J.-W. An Inkjet-Printed Non-Enzymatic Hydrogen Peroxide Sensor on Paper. *J. Electrochem. Soc.* **2017**, *164*, B3101–B3106. [CrossRef]
158. Carvajal, S.; Fera, S.N.; Jones, A.L.; Baldo, T.A.; Mosa, I.M.; Rusling, J.F.; Krause, C.E. Disposable inkjet-printed electrochemical platform for detection of clinically relevant HER-2 breast cancer biomarker. *Biosens. Bioelectron.* **2018**, *104*, 158–162. [CrossRef] [PubMed]
159. Fernández-la-Villa, A.; Pozo-Ayuso, D.F.; Castaño-Álvarez, M. Microfluidics and Electrochemistry: An Emerging Tandem for next-Generation Analytical Microsystems. *Curr. Opin. Electrochem.* **2019**, *15*, 175–185. [CrossRef]
160. Sanjay, S.T.; Zhou, W.; Dou, M.; Tavakoli, H.; Ma, L.; Xu, F.; Li, X.J. Recent Advances of Controlled Drug Delivery Using Microfluidic Platforms. *Adv. Drug Deliv. Rev.* **2018**, *128*, 3–28. [CrossRef]
161. Shi, H.; Nie, K.; Dong, B.; Long, M.; Xu, H.; Liu, Z. Recent Progress of Microfluidic Reactors for Biomedical Applications. *Chem. Eng. J.* **2019**, *361*, 635–650. [CrossRef]
162. van Heeren, H.; Tantra, R.; Salomon, P. Microfluidic Devices: A Road Forward by Standardization of Interconnects and Classification. *Microfluidics Nanofluidics* **2015**, *19*, 1203–1207. [CrossRef]
163. Sharafeldin, M.; Kadimisetty, K.; Bhalero, K.R.; Chen, T.; Rusling, J.F. 3D-printed Immunosensor arrays for cancer diagnostics. *Sensors* **2020**, *20*, 4514. [CrossRef]
164. Iliescu, C.; Taylor, H.; Avram, M.; Miao, J.; Franssila, S. A Practical Guide for the Fabrication of Microfluidic Devices Using Glass and Silicon. *Biomicrofluidics* **2012**, *6*, 16505. [CrossRef]
165. Gale, B.K.; Jafek, A.R.; Lambert, C.J.; Goenner, B.L.; Moghimifam, H.; Nze, U.C.; Kamarapu, S.K.A. Review of Current Methods in Microfluidic Device Fabrication and Future Commercialization Prospects. *Inventions* **2018**, *3*, 60. [CrossRef]
166. Lee, G.H.; Lee, J.K.; Kim, J.H.; Choi, H.S.; Kim, J.; Lee, S.H.; Lee, H.Y. Single Microfluidic Electrochemical Sensor System for Simultaneous Multi-Pulmonary Hypertension Biomarker Analyses. *Sci. Rep.* **2017**, *7*, 1–8. [CrossRef] [PubMed]
167. De Oliveira, R.A.G.; Materon, E.M.; Melendez, M.E.; Carvalho, A.L.; Faria, R.C. Disposable Microfluidic Immunoarray Device for Sensitive Breast Cancer Biomarker Detection. *ACS Appl. Mater. Interfaces* **2017**, *9*, 27433–27440. [CrossRef] [PubMed]

biosensors

MDPI

Article

On the Electrochemical Detection of Alpha-Fetoprotein Using Aptamers: DNA Isothermal Amplification Strategies to Improve the Performance of Weak Aptamers

Ramón Lorenzo-Gómez [1,2], Daniel González-Robles [1], Rebeca Miranda-Castro [1,2], Noemí de-los-Santos-Álvarez [1,2] and María Jesús Lobo-Castañón [1,2,*]

[1] Departamento de Química Física y Analítica, Universidad de Oviedo, Av. Julián Clavería 8, 33006 Oviedo, Spain; lorenzoramon@uniovi.es (R.L.-G.); dgonrob88@gmail.com (D.G.-R.); mirandarebeca@uniovi.es (R.M.-C.); santosnoemi@uniovi.es (N.d.-l.-S.-Á.)

[2] Instituto de Investigación Sanitaria del Principado de Asturias, Avenida de Roma, 33011 Oviedo, Spain

* Correspondence: mjlc@uniovi.es; Tel.: +34-98-510-6235

Received: 9 April 2020; Accepted: 28 April 2020; Published: 30 April 2020

check for updates

Abstract: Affinity characterization is essential to develop reliable aptamers for tumor biomarker detection. For alpha-fetoprotein (AFP), a biomarker of hepatocellular carcinoma (HCC), two DNA aptamers were described with very different affinity. In this work, we estimate the dissociation constant of both of them by means of a direct assay on magnetic beads modified with AFP and electrochemical detection on carbon screen-printed electrodes (SPCE). Unlike previous works, both aptamers showed similar dissociation constant (K_d) values, in the subμM range. In order to improve the performance of these aptamers, we proposed the isothermal amplification of the aptamers by both terminal deoxynucleotidyl transferase (TdT) and rolling circle amplification (RCA). Both DNA amplifications improved the sensitivity and also the apparent binding constants from 713 nM to 189 nM for the short aptamer and from 526 nM to 32 nM for the long aptamer. This improvement depends on the true affinity of the binding pair, which ultimately limits the analytical usefulness.

Keywords: aptamer; alpha-fetoprotein; dissociation constant; rolling circle amplification; terminal deoxynucleotidyl transferase

1. Introduction

Cancer management is a multifactorial decision-making task that relies on a combination of clinical evidences, biochemical parameters and tumor biomarker values. Ideally, a tumor biomarker is any biological characteristic that is detected only in the presence of cancer. However, the perfect biomarker does not exist, and most approved for clinical usage present variable degrees of false negative (limited sensitivity) or false positive (limited specificity) results. In fact, the cut-off values and recommended use are under continuous scrutiny and could be modified over the years.

This is the case of alpha-fetoprotein (AFP), one of the two biomarkers suitable for large-scale screening in addition to the controversial prostate specific antigen [1]. AFP serum levels rapidly decrease after birth and remain below 10 ng mL^{-1} under healthy conditions except in pregnant women [2,3]. Since the expression of AFP is highly specific for the liver, it was found elevated in hepatocellular carcinoma (HCC) patients and proposed as a tumor biomarker. However, AFP alone is not enough for HCC diagnosis, screening or surveillance [4]. Its suboptimal performance, that is, increased levels in benign liver pathologies and teratomas or low levels in small HCC [5], has prompted that US and European guidelines allow HCC surveillance with or without AFP control, in contrast to Asian guidelines that recommend its use in combination with imaging techniques or even other biomarkers [2]. Nonetheless, it remains as a supplemental test when only low-quality or inconclusive images are available. AFP has also found applicability as a prognostic (disease outcome in the absence of therapy), monitoring of therapy and recurrence biomarker and to identify the best candidates for liver transplantation. The optimal cut-off level for each purpose has not been definitively adopted [4,6]. This means that AFP remains widely employed in the clinical setting in spite of its limitations.

As with other protein biomarkers, in large centralized facilities, the methods of choice for AFP detection are immunoassays performed in automated instruments [7]. To reduce the ever-increasing health expenditures and decentralize the analysis, the extreme success of glucometer for diabetes patients, an electrochemical biosensor, is a model to imitate. To this end, stable, cost-effective and reproducible reagents are needed. Antibodies do not fully meet all these criteria. In contrast, aptamers considered as the chemical antibodies of non-protein nature have demonstrated their better chemical and thermal stability, inherent reproducibility among batches due to chemical synthesis and lower manufacturing costs than antibodies.

The last decade has witnessed the selection of three aptamers for AFP; an RNA aptamer with anti-proliferative activity on HCC cells via the downregulation of AFP expression [8] and two DNA aptamers for diagnostic applications [9,10]. The LOD achieved ranged between 6.97 ag mL^{-1} [11] and 259 ng mL^{-1} [12], for the 72 nt-long aptamer selected by microfluidics [9], and 0.23 pg mL^{-1} [13] to 500 ng mL^{-1} [14], for the 75 nt-long aptamer selected by capillary electrophoresis (CE) [10]. This means that the LODs span over 10 and 6 orders of magnitude, respectively. The lowest values exceed by several orders of magnitude the LOD expected for binding pairs having K_d in the nM or subμM range in the absence of any target amplification scheme [15,16]. Nonetheless, some of these reports use other strategies recently reviewed by Gooding et al. to reach ultralow limits of detection [16], mainly through the use of nanostructures to obtain surfaces with high density of receptors. This way, the equilibrium is shifted toward the complex aptamer-target due to the increased chance of rebinding. Additional strategies comprise the restriction of the volume to the nanoscale, the implementation of amplification schemes or the use of magnetic beads to improve the target mass transport. To the best of our knowledge, DNA amplification schemes have not been employed to improve the performance of anti-AFP aptamers.

DNA isothermal amplification methods can be easily implemented in aptamer-based assays because of the nucleic acid nature of the receptor [17,18]. Their aim is to push down the LOD of DNA-based sensors (both aptasensors and genosensors) and to facilitate the full integration of all analytical steps on a single biosensor platform, without requiring additional instrumentation as in PCR-based amplification strategies [19–21]. Several isothermal methods have been developed in the last two decades [22]. We have selected two of them: terminal deoxynucleotidyl transferase (TdT) and rolling circle amplification (RCA). TdT is a non-templated elongation of DNA by its 3′ end that allows the incorporation of labeled nucleotides in a tunable ratio [23,24]. RCA benefits from the quickness and high fidelity of phi29 polymerase to copy thousands of times the sequence of a pre-ligated circular DNA (known as padlock), thus providing multiple binding sites for reporting probes [25]. While TdT DNA elongation of an aptamer is directly performed, RCA requires tagging the aptamer with an RCA primer, a highly specific site for circularizing the padlock and thus triggering the elongation.

In this work, we have studied and compared the binding affinity of the two DNA aptamers evolved against AFP. We used magnetic beads to anchor the protein and aptamers with several tags for conventional enzymatic amplification. Our results support the outperformance of the 72 nt-long aptamer. Then, TdT and RCA were tested in combination with these aptamers. TdT was more efficient in shifting the apparent binding constant toward lower values, while RCA showed superior amplification power.

2. Materials and Methods

All reagents and instrumentations are listed in the Supplementary Information. The following protocols were used in this work.

2.1. Modification of Tosylactivated Magnetic Particles with AFP

Dynabeads® M-280 Tosylactivated were modified according to manufacturer recommendations. Briefly, 5 mg of magnetic beads (MBs) were washed twice with 0.1 M Na-phosphate buffer pH 7.4. Then, 250 μL of 0.1 M Na-phosphate buffer containing 200 μg mL^{-1} human AFP and 1.2 M ammonium sulfate was added to the MBs and incubated for 12–18 h at 37 °C with continuous stirring at 1300 rpm. After removing the supernatant, the unreacted tosyl groups were blocked with a 0.5% BSA solution in 1× PBS incubated at 37 °C for 1 h, under stirring. Then, two washing steps with 1× PBS—0.1% BSA were performed and finally the MBs were resuspended in this BSA-containing buffer to a storage concentration of 20 mg mL^{-1}. The modified beads were kept at 4 °C while not in use.

The supernatant from the MBs modification was analyzed by the Bradford assay, applying the protocol recommended by the manufacturer of the Bradford reagent. The amount of immobilized protein was calculated as the subtraction of the amount of AFP found in the supernatant from the amount of AFP used for immobilization.

2.2. Binding Assays on AFP-Modified Magnetic Particles

First, 60 μg of AFP-modified MBs were incubated with 250 μL of increasing concentrations of anti-AFP aptamers prepared in 1× PBS, for 30 min at 25 °C, with continuous stirring at 1300 rpm. Then, the supernatant was discarded, and the MBs were washed twice with 1× PBS—0.01% Tween 20 (washing buffer).

When no nucleic acid amplification was used for signal enhancement, the step following the incubation with anti-AFP aptamers (Table S1) (AFP-S-biotin, AFP-S-FITC or AFP-L-FITC; Scheme 1A, step 1) was the binding of an enzymatic conjugate. A volume of 250 μL of anti-fluorescein-Fab-fragment-peroxidase conjugate (antiF-POD, 0.5 U mL^{-1} in 1× PBS—0.5% casein) or streptavidin-peroxidase (SA-POD, 2.5 μg mL^{-1} in 1× PBS—0.01% Tween 20) was added, depending on the label of the aptamer, and incubated for 30 min at 25 °C, with continuous shaking (Scheme 1A, step 2). After two washing steps with washing buffer and one with 1× PBS, the MBs were resuspended in 30 μL of this latter buffer. For the electrochemical measurement, 10 μL of MBs were dropped on the working electrode of the SPCE with a 4 mm diameter and circular Neodymium magnet placed exactly under it. In this way, MBs are distributed onto the surface and entrapped magnetically for 1 min (Scheme 1A, step 3). Finally, 40 μL of a ready-to-use TMB solution was added, and the enzymatic reaction proceeded for 60 s. Chronoamperometry was carried out immediately after at 0 V for 60 s.

When TdT amplification was used, the incubation with anti-AFP aptamers (AFP-L-TdT or AFP-S-TdT; Scheme 1B, step 1) was followed by the amplification reaction. After two washings with washing buffer, the MBs were resuspended in 50 μL of TdT mix, containing 10 U TdT, 0.25 mM $CoCl_2$, 500 μM dNTPs, and 25 μM biotin-dATP prepared in 1× reaction buffer. The reaction proceeded for 1 h at 37 °C, under continuous stirring (Scheme 1B, step 2). Next, the supernatant was discarded, and the MBs were washed twice. The enzymatic labeling with SA-POD (Scheme 1B, step 3), enzymatic reaction of TMB and chronoamperometric measurement (Scheme 1B, step 4) were performed as stated above.

In the case of RCA, the incubation with anti-AFP aptamers (AFP-S-RCA; Scheme 1C, step 1) was followed by the annealing and ligation of the circularizable RCA template (padlock). To this aim, the washed MBs were resuspended in 30 μL of ligation mix, containing 3.75 Weiss T4 DNA ligase and 10 nM padlock in 1× ligase buffer, and incubated for 30 min at 25 °C, under continuous stirring (Scheme 1C, step 2). After two washings, the MBs were then resuspended in 30 μL of RCA mix, containing 7.5 U phi29 DNA polymerase and 500 μM dNTPs in 1× polymerase buffer, and incubated for 15 min at 37 °C, under continuous stirring (Scheme 1C, step 3). Next, the MBs were washed twice and incubated with 250 µL of 100 nM 6-FAM-labeled reporter probe in 1× PBS for 30 min at 25 °C (Scheme 1C, step 4). The enzymatic labeling with antiF-POD (Scheme 1C, step 5), enzymatic reaction of TMB and chronoamperometric measurement (Scheme 1C, step 6) were performed as mentioned above.

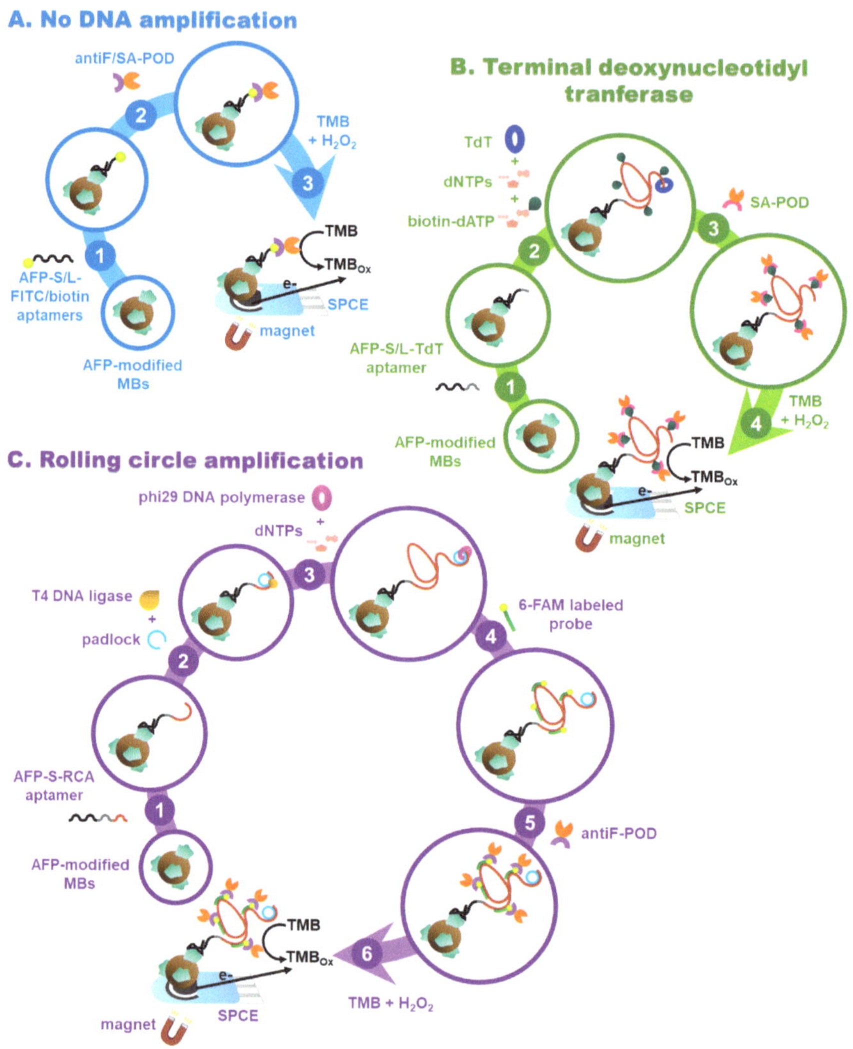

Scheme 1. Depiction of the methodology employed in the binding curves on alpha-fetoprotein (AFP)-modified magnetic beads (MBs). (**A**) Without DNA amplification: 1. Aptamer binding; 2. Enzyme labeling with streptavidin-peroxidase (POD) or anti-fluorescein Fab fragment-POD conjugates; 3. MBs magnetic entrapment on screen-printed carbon electrodes (SPCEs), substrate addition, enzymatic reaction and chronoamperometric measurement. (**B**) With terminal deoxynucleotidyl transferase (TdT) amplification: 1. Aptamer binding; 2. TdT elongation; 3. Enzyme labeling with the streptavidin-POD conjugate; 4. MBs magnetic entrapment on SPCEs, substrate addition, enzymatic reaction and chronoamperometric measurement. (**C**) With rolling circle amplification (RCA): 1. Aptamer binding; 2. Padlock annealing and ligation; 3. RCA elongation; 4. Reporter probe hybridization; 5. Enzyme labeling with antiF-POD conjugate; 6. MBs magnetic entrapment on SPCEs, substrate addition, enzymatic reaction and chronoamperometric measurement.

2.3. TdT Amplification Study by Gel Electrophoresis

TdT amplification was performed in solution, incubating 20 μL of the reaction mix (1.5 μM aptamer, 30 U TdT, 500 μM dATP, and 0–50 μM biotin-dATP prepared in 1× reaction buffer, for 1 h at 37 °C under gentle stirring. After that, 5 μL of 10 μM T40 probe was added, and the mixture was heated at 70 °C for 10 min to deactivate the enzyme. Next, the solution was cooled at room temperature to allow the hybridization between the poliA-tailed aptamer and the T40 probe, resulting in a double stranded DNA. A volume of 5 μL of this product was mixed with 1 μL of 6× loading buffer and loaded in a 2% agarose gel (prepared in 1× TBE containing SimplySafe DNA staining dye). The electrophoresis was run for 45 min under a potential difference of 80 V. DNA bands were revealed under an UV lamp, and the size of the elongated aptamer was estimated by comparison of its displacement on the gel to that of a DNA ladder of 20 bp.

3. Results

3.1. Comparison of Binding Affinity in the Absence of DNA Amplification

Two DNA aptamers selected against AFP were used in this study: a 72-nt aptamer and the truncated version of a 75-nt aptamer that has been previously tested to keep its affinity without the flanking primer sites [26,27]. For brevity, in this work, we will denote those aptamers as "long" (AFP-L) and "short" (AFP-S), respectively. First, the short aptamer tagged with a biotin (AFP-S-biotin, Table S1) or with a fluorescein (AFP-S-FITC, Table S1) was tested on AFP-modified magnetic beads using increasing concentrations of the aptamer (0.25–2 μM). Direct binding of the aptamer on the beads is revealed through the labeling with a peroxidase conjugate (SA-POD or antiF-POD) and further addition of the TMB substrate. After 1 min, the current of the reduction of the TMB oxidized enzymatically is measured by chronoamperometry.

Figure 1 shows the corresponding binding curves. As expected, the multivalence of streptavidin-POD conjugate reduces the current measured by chronoamperometry in comparison to the monovalent anti-fluorescein Fab fragment-POD, so the former was discarded as the reporter conjugate. Comparatively, the fluorescein-tagged long aptamer showed much higher currents, which indicates that the enzyme activity immobilized on the magnetic beads is higher, and more aptamers bind AFP at each specific concentration. Fitting the binding curves to the Hill equation yields dissociation constant (K_d) values of 713 ± 63 nM and 526 ± 101 nM, for the short and long aptamers, respectively.

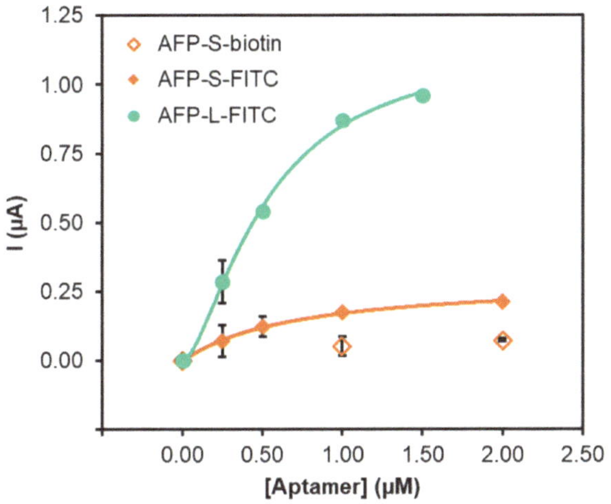

Figure 1. Binding curves obtained with AFP-modified MBs and the biotinylated or fluorescein-tagged short aptamer and the fluorescein-tagged long aptamer.

Interestingly, the long aptamer shows certain cooperativity ($n = 1.6 \pm 0.4$) while the short one fits very well to a 1:1 stoichiometry (Langmuir model). The K_d for the short aptamer is close to the reported for the full sequence. The long aptamer, however, seems to be a much poorer binder than expected. The K_d is more than two orders of magnitude higher. It is important to bear in mind that this equilibrium model assumes that target binding does not change the concentration of the ligand in solution (the aptamer here). This may not be true when there is a high packing density of the probes (the protein here) or when working with low sample volumes. In such cases, the binding signal, which is a measurement of the probe occupancy, is no longer related to the ligand concentration in solution but to the relative amount of ligand in solution and probes on the surface. Under those conditions, the midpoint of the binding curve represents half the effective probe concentration [15]. To check whether this regime is operating in the present case, the effective AFP concentration was estimated to be about 34 nM. This value is 20-fold higher than the reported K_d. This means that depletion of the ligand due to rapid binding might occur, theoretically resulting in an apparent K_d of 17 nM (34/2). However, the midpoint of our experimental binding curve is well above this value, which excludes the operation under the ligand-depletion regimen. The reported K_D was measured by SPR in a reverse set up where the aptamer was anchored at an unknown density, and the protein was in solution. Accordingly, the discrepancy could arise from limited access of the aptamer to the protein binding site. However, the aptamer was selected on AFP-modified epoxy-MBs, which binds to primary amines and sulfhydryl groups as the tosylactivated MBs herein used, ruling out this explanation. Together, everything points to a "true" K_d higher than the reported one.

3.2. Amplification by Terminal Deoxynucleotidyl Transferase Elongation

Isothermal amplification of the aptamer is an effective way of enhancing the sensitivity and shifting the apparent K_d to lower ligand concentrations. TdT elongates any ss-DNA provided that a free 3′-OH end is available. In addition to the four natural nucleotides, it incorporates a wide variety of unnatural analogues including ribonucleotides, biotinylated or fluorescent-labeled nucleotides [28]. The rate of incorporation is nucleotide-, ion- and label-dependent [29]. We selected the biotinylated dATP (biotin-dATP) as the labeled nucleotide for subsequent enzyme conjugation using SA-POD. First, we studied the appropriate ratio of biotin-dATP to dATP in order to obtain the longest tail with the optimum number of biotin labels. To this aim, the TdT was first performed in solution and the products visualized on agarose gel. Figure 2a shows the results obtained with the AFP-S-TdT. Note that although a certain length distribution appeared in the elongated products as a consequence of the enzyme random kinetics, their mean size depends on the ratio of biotin-dATP. The higher the biotin-dATP:dATP ratio, the shorter the products are, confirming the preference of the enzyme for the natural nucleotide. About 600 dATP nucleotides can be incorporated in 60 min (Figure 2a, lane 3) while only half this value when 10% of the nucleotides are biotinylated (Figure 2a, lane 4).

In the electrochemical assay, TdT was carried out on AFP-modified MBs after the interaction with 1 μM of the short aptamer. Then the SA-POD conjugate was added, and the reduction of the enzymatic product was measured by chronoamperometry. The number of biotinylated dATP determines the amount of enzymes carried by each aptamer-AFP complex and thus the amplification of the analytical signal. The current increased when the biotin-dATP:dATP ratio was varied between 1:20 and 1:1 (5–50% of biotin-dATP). This increase is related with the higher concentration of labeled nucleotides that can be incorporated to the rising strand (Figure 2b) when the proportion of biotin-dATP increases. The net current multiplies by 3 for a 4-fold increase in biotin-dATP concentration (from 25 to 100 μM), while it dramatically decreases when the concentration further increases 5-fold (from 100 to 500 μM) due to a strong increase in the blank current (no aptamer present). Unspecific binding to the MBs of biotin-dATP might account for this increase of the blank signal. A "trade off" between increasing the reaction rate (by increasing the amount of unlabeled substrates) and the amount of labeled nucleotides needed for sensing with proper reproducibility provides the optimum biotin-dATP:dATP ratio when 5% (1:20) of the nucleotides are biotinylated This ratio was used in subsequent experiments.

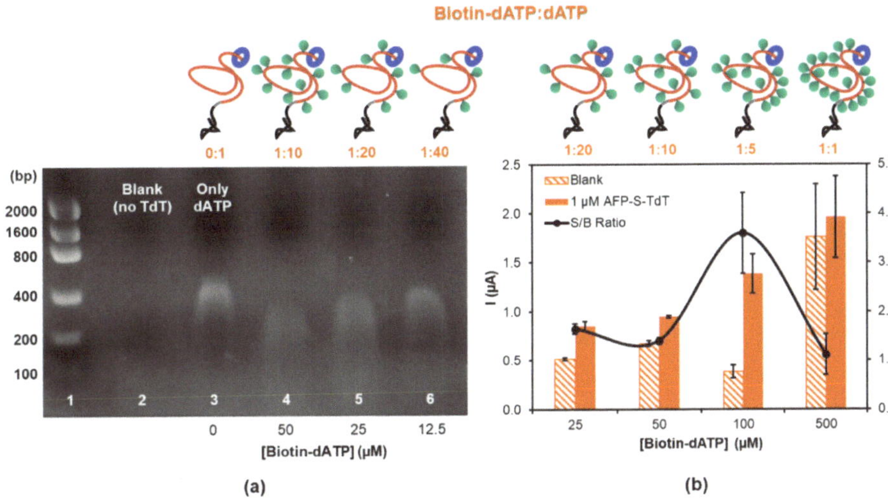

Figure 2. (**a**) Agarose gel electrophoresis of TdT elongation of short aptamer. Lane 1: DNA ladder. Lane 2: TdT blank. TdT elongated product obtained with Lane 3: 100% dATP. Lane 4: 10% biotin-dATP. Lane 5: 5% biotin-dATP. Lane 6: 2.5% biotin-dATP. (**b**) Current obtained in the electrochemical assay with different biotin-dATP:dATP ratios in the absence (blank, striped bars) and in the presence of the short aptamer at 1 μM (solid bars) and the corresponding signal-to-blank (S/B) ratio for each condition.

Since the elongation rate depends on the nucleotide used, we examine the influence of adding just dATP or an equimolar mixture of the four nucleotides (dNTP), maintaining the total concentration. Figure 3 shows that the current dramatically increases when the mixture of dNTPs is employed, suggesting that the TdT is somehow impeded when only dATP is used as a source of unlabeled nucleotides. In this experiment the long aptamer with an 8-thymine tail at the 3′ end was used in order to avoid steric hindrance between the binding event and the enzyme elongation. It has been demonstrated that the formation of duplexes or G-quartets can halt the TdT elongation [30]. In our case, the thymine spacer and the adenine-based nucleotides can hybridize slowing down the reaction as it is schematized in Figure 3. Of note, in the absence of TdT, the current is much lower due to the presence of a single enzyme-labeling site. The amplification power of TdT, in terms of current, by incorporation of multiple enzymes per recognition event is 3.3 and 9-fold for the dATP and dNTP elongations, respectively.

The effect of TdT elongation on the apparent binding affinity was studied by obtaining the corresponding binding curve on AFP-modified MBs under the optimized conditions with electrochemical detection. Figure 4a compares the TdT-amplified and the non-TdT amplified binding curves obtained with the long aptamer. The dramatic increase in the magnitude of the analytical signal and the displacement of the TdT amplified curve toward smaller concentrations of aptamer are apparent. This is reflected in the apparent K_d value estimated from the fitting to the Langmuir equation, 32 ± 11 nM, more than an order of magnitude lower than the above calculation for the non-TdT amplified assay. This value is still an order of magnitude higher than the reported by SPR.

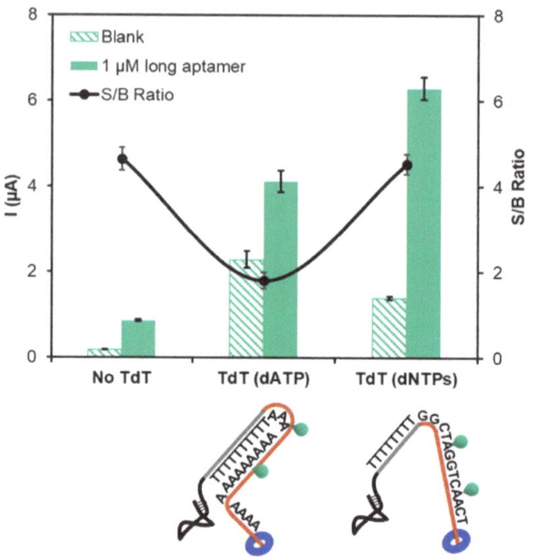

Figure 3. Chronoamperometric currents obtained without and with TdT amplification using only dATP or a mixture of dNTPs at 500 µM concentration in the absence (striped bars) and in the presence (solid bars) of the long aptamer at 1 µM and the corresponding signal-to-blank ratio at each condition.

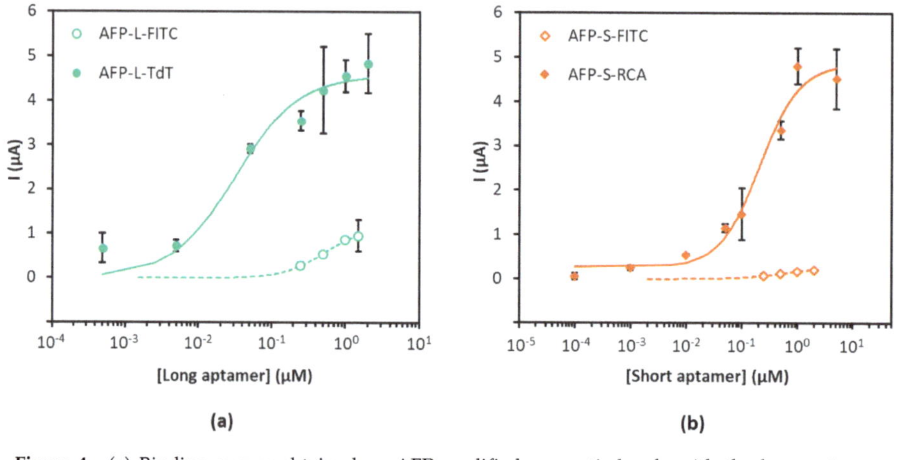

Figure 4. (**a**) Binding curves obtained on AFP-modified magnetic beads with the long aptamer without (open circles) and with (solid circles) TdT amplification. (**b**) Binding curves obtained on AFP-modified magnetic beads with the short aptamer without (open diamonds) and with RCA amplification (solid diamonds).

3.3. Rolling Circle Amplification

Previously we were able to observe an increase in the apparent affinity of a candidate tumor marker, NGAL, of three orders of magnitude using RCA as isothermal amplification technique [31]. In order to understand whether this improvement is general for any binding pair or it depends on the true affinity, we combined RCA with the short aptamer. RCA requires the design of an appropriate primer that is added to the 3′ end of the aptamer. This primer acts as a complementary strand for the hybridization of the circularizable padlock that triggers the amplification. It is recommended to add a spacer between the aptamer and the primer to secure that the target binding and padlock hybridization occur simultaneously. Unfortunately, this makes the synthetic strand quite long, precluding the use of the long aptamer. A fluorescein-tagged short reporting strand, whose sequence corresponds to a region of the padlock complementary to the nascent strand, allows the incorporation of thousands of enzymes per aptamer, and thus the signal amplification.

The binding curve obtained with the short aptamer elongated with the RCA primer is shown in Figure 4b and compared with the non-RCA amplified one. Again, a dramatic increase in the current is observed, which is ascribed to the multilabeling of the elongated aptamer. Fitting to the Langmuir equation yields an apparent K_d of 188 ± 33 nM. The improvement in affinity is modest (about 6-fold) supporting that the affinity gain depends on the aptamer-target pair.

4. Discussion

The affinity of the aptamer for its ligand is a crucial parameter to evaluate its potential utility in diagnostic applications. However, characterization of the binding ability must be carefully conducted in order not to overestimate its performance. In this sense, the use of several techniques makes more robust and reliable the final value, usually in terms of dissociation constant. In this work, we have examined two DNA aptamers derived against the tumor biomarker AFP. Their affinity differs in more than two orders of magnitude. In our hands and using electrochemical detection after aptamer binding on AFP-modified MBs, both aptamers show modest affinity, in the sub-micromolar range. The K_d for the short aptamer agrees well with the reported by CE, that is, in solution. However, Dong et al. estimated a much lower value, around 19 nM, when using a sandwich assay. In that experiment, the aptamer was used as a capture element and an antibody as a reporter one. In an analogue experiment, the long aptamer was tested obtaining a K_d of 17 nM, almost one order of magnitude higher than the original value, 2.37 nM [10]. The observation of similar K_d with two different aptamers and an identical antibody might point to the non-negligible influence of the latter in the value measured, which could result in an overestimation of the K_d for the weaker aptamer. If the K_d value estimated by the sandwich assay for the short aptamer was true, this would mean that the heterogeneous K_d would be one and a half orders of magnitude smaller than the homogeneous one. This is not usually true because of the lack of steric hindrance in solution. Nonetheless, when surfaces with a high load of receptor are used, avidity might occur. For this reason, to estimate the true affinity high loading is not recommended [32].

When estimating the K_d by using the Hill model, it is important to verify that the assumptions implicit in the model are met. Of special relevance is to be sure that the concentration of the partner in solution is not affected by the binding; that is, it can be considered constant. If this is not the case, the true concentration of the partner in solution should be precisely calculated. From our experiments, depletion of the aptamer in solution is not observed, so the Hill model can be applied.

The analytical sensitivity can be improved by coupling isothermal DNA amplification strategies. We have demonstrated that both TdT and RCA can enhance the analytical signal. For example, the current improvement for the long aptamer at 500 nM was 7.8-fold when TdT was used. The RCA showed much higher amplification, about 24-fold for the same aptamer concentration. Interestingly, the improvement in the apparent dissociation constant was higher with TdT amplification, 16-fold versus the 6-fold for RCA. We attribute this fact to the different aptamers used. The shift of the binding curves seems to be affinity dependent. The better the affinity, the larger the shift is. In any case, even with the TdT amplification, it was not possible to obtain K_d values as low as those previously reported for the aptamer that shows the highest affinity for AFP.

5. Conclusions

In this work, we have studied the affinity of two anti-AFP aptamers previously reported to have very different affinities. Using a heterogeneous direct binding assay with electrochemical detection, different from the techniques used to evaluate the affinity in the original work, similar binding constants were estimated for both receptors. However, they are rather poor for analytical applications. By taking advantage of the nucleic acid character of the aptamers, we demonstrated that DNA amplification of the aptamer enables an improvement in sensitivity and in the apparent dissociation constant. DNA amplification strategies can enhance the performance of weak binding aptamers but their magnitude depends on the true affinity. These results are of general significance, as a number of electrochemical biosensors relying on the recognition by poor aptamers could benefit from these results.

Supplementary Materials: The following are available online at http://www.mdpi.com/2079-6374/10/5/46/s1. The reagents and instrumentation and Table S1: Sequences of the aptamers and other DNA probes used in this work.

Author Contributions: Conceptualization, M.J.L.-C., N.d.-l.-S.-Á. and R.M.-C.; performed the experiments, R.L.-G. and D.G.-R.; analyzed the data, M.J.L.-C., N.-S.-Á., R.M.-C. and R.L.-G.; writing—original, review and editing, M.J.L.-C., N.d.-l.-S.-Á., R.M.C. and R.L.G. All authors have read and agreed to the published version of the manuscript.

Funding: This research was funded by the Spanish Government (project RTI-2018-095756-B-I00) and Principado de Asturias Government (IDI2018-000217), co-financed by FEDER funds. R.L.-G. thanks the Spanish Government for a predoctoral FPU fellowship (FPU16/05670).

Acknowledgments: The Unit of Biomedical and Biotechnological Tests of the Scientific and Technological Resources from Universidad de Oviedo is gratefully acknowledged.

Conflicts of Interest: The authors declare no conflict of interest. The funders had no role in the design of the study; in the collection, analyses, or interpretation of data; in the writing of the manuscript or in the decision to publish the results.

References

1. Diamandis, E.P. Cancer Biomarkers: Can We Turn Recent Failures into Success? *J. Natl. Cancer Inst.* **2010**, *102*, 1462–1467. [CrossRef] [PubMed]
2. Trevisani, F.; Garuti, F.; Neri, A. Alpha-fetoprotein for Diagnosis, Prognosis, and Transplant Selection. *Semin. Liver Dis.* **2019**, *39*, 163–177. [CrossRef] [PubMed]
3. Ruoslahti, E.; Seppälä, M. α-Foetoprotein in Normal Human Serum. *Nature* **1972**, *235*, 161–162. [CrossRef] [PubMed]
4. Galle, P.R.; Foerster, F.; Kudo, M.; Chan, S.L.; Llovet, J.M.; Qin, S.; Schelman, W.R.; Chintharlapalli, S.; Abada, P.B.; Sherman, M.; et al. Biology and Significance of alpha-Fetoprotein in Hepatocellular Carcinoma. *Liver Int.* **2019**, *39*, 2214–2229. [CrossRef]
5. Wang, W.; Wei, C. Advances in the Early Diagnosis of Hepatocellular Carcinoma. *Genes Dis.* **2020**. [CrossRef]
6. Zhang, J.; Chen, G.; Zhang, P.; Zhang, J.; Li, X.; Gan, D.; Cao, X.; Han, M.; Du, H.; Ye, Y. The Threshold of alpha-Fetoprotein (AFP) for the Diagnosis of Hepatocellular Carcinoma: A Systematic Review and Meta-analysis. *PLoS ONE* **2020**, *15*, e0228857. [CrossRef]
7. Yue, Y.; Chen, X.; Zhang, S.; Xu, Z.; Wang, Q. Commutability of Reference Materials for α-Fetoprotein in Human Serum. *Arch. Pathol. Lab. Med.* **2017**, *141*, 1421–1427. [CrossRef]
8. Lee, Y.J.; Lee, S.-W. Regression of Hepatocarcinoma Cells using RNA Aptamer Specific to alpha-Fetoprotein. *Biochem. Biophys. Res. Commun.* **2012**, *417*, 521–527. [CrossRef]
9. Huang, C.-J.; Lin, H.-I.; Shiesh, S.-C.; Lee, G.-B. An Integrated Microfluidic System for Rapid Screening of alpha-Fetoprotein-specific Aptamers. *Biosens. Bioelectron.* **2012**, *35*, 50–55. [CrossRef]
10. Dong, L.; Tan, Q.; Ye, W.; Liu, D.; Chen, H.; Hu, H.; Wen, D.; Liu, Y.; Cao, Y.; Kang, J.; et al. Screening and Identifying a Novel ssDNA Aptamer against Alpha-fetoprotein Using CE-SELEX. *Sci. Rep.* **2015**, *5*, 15552. [CrossRef]
11. Wu, X.; Fu, P.; Ma, W.; Xu, L.; Kuang, H.; Xu, C. SERS-active Silver Nanoparticle Trimers for sub-attomolar Detection of alpha-Fetoprotein. *RSC Adv.* **2015**, *5*, 73395–73398. [CrossRef]
12. Gui, R.; He, W.; Jin, H.; Sun, J.; Wang, Y. DNA Assembly of Carbon Dots and 5-Fluorouracil used for Room-Remperature Phosphorescence turn-on Sensing of AFP and AFP-triggered Simultaneous Release of Dual-drug. *Sens. Actuators B Chem.* **2018**, *255*, 1623–1630. [CrossRef]
13. Heiat, M.; Negahdary, M. Sensitive Diagnosis of alpha-Fetoprotein by a Label-free Nanoaptasensor designed by modified Au Electrode with Spindle-shaped Gold Nanostructure. *Microchem. J.* **2019**, *148*, 456–466. [CrossRef]

14. Li, G.; Li, S.; Wang, Z.; Xue, Y.; Dong, C.; Zeng, J.; Huang, Y.; Liang, J.; Zhou, Z. Label-free Electrochemical Aptasensor for Detection of alpha-Fetoprotein based on AFP-aptamer and Thionin/Reduced Graphene Oxide/Gold Nanoparticles. *Anal. Biochem.* **2018**, *547*, 37–44. [CrossRef]
15. De Ávila, B.E.F.; Watkins, H.M.; Pingarrón, J.M.; Plaxco, K.W.; Palleschi, G.; Ricci, F. Determinants of the Detection Limit and Specificity of Surface-Based Biosensors. *Anal. Chem.* **2013**, *85*, 6593–6597. [CrossRef]
16. Wu, Y.; Tilley, R.D.; Gooding, J.J. Challenges and Solutions in Developing Ultrasensitive Biosensors. *J. Am. Chem. Soc.* **2018**, *141*, 1162–1170. [CrossRef]
17. Lorenzo-Gómez, R.; Miranda-Castro, R.; de-los-Santos-Álvarez, N.; Lobo-Castañón, M.J. Electrochemical Aptamer-based Assays Coupled to Isothermal Nucleic Acid Amplification Techniques: New Tools for Cancer Diagnosis. *Curr. Opin. Electrochem.* **2019**, *14*, 32–43. [CrossRef]
18. Li, F.; Zhang, H.; Wang, Z.; Newbigging, A.M.; Reid, M.S.; Li, X.-F.; Le, X.C. Aptamers Facilitating Amplified Detection of Biomolecules. *Anal. Chem.* **2014**, *87*, 274–292. [CrossRef]
19. Barreda-García, S.; Miranda-Castro, R.; de-los-Santos-Álvarez, N.; Miranda-Ordieres, A.J.; Lobo-Castañón, M.J. Solid-Phase Helicase Dependent Amplification and Electrochemical Detection of Salmonella on highly Stable Oligonucleotide-modified ITO Electrodes. *Chem. Commun.* **2017**, *53*, 9721–9724. [CrossRef]
20. Barreda-Garcia, S.; Miranda-Castro, R.; de-los-Santos-Álvarez, N.; Miranda-Ordieres, A.J.; Lobo-Castañón, M.J. Helicase-Dependent Isothermal Amplification: A Novel Tool in the Development of Molecular-based Analytical Systems for Rapid Pathogen Detection. *Anal. Bioanal. Chem.* **2017**, *410*, 679–693. [CrossRef]
21. Sánchez-Salcedo, R.; Miranda-Castro, R.; de-los-Santos-Álvarez, N.; Lobo-Castañón, M.J. On-Gold Recombinase Polymerase Primer Elongation for Electrochemical Detection of Bacterial Genome: Mechanism Insights and Influencing Factors. *ChemElectroChem* **2018**, *6*, 793–800. [CrossRef]
22. Deng, H.; Gao, Z. Bioanalytical Applications of Isothermal Nucleic Acid Amplification Techniques. *Anal. Chim. Acta* **2015**, *853*, 30–45. [CrossRef]
23. Anne, A.; Bonnaudat, C.; Demaille, C.; Wang, K. Enzymatic Redox 3′-End-Labeling of DNA Oligonucleotide Monolayers on Gold Surfaces Using Terminal Deoxynucleotidyl Transferase (TdT)-Mediated Single Base Extension. *J. Am. Chem. Soc.* **2007**, *129*, 2734–2735. [CrossRef]
24. Motea, E.A.; Berdis, A.J. Terminal Deoxynucleotidyl Transferase: The Story of a Misguided DNA Polymerase. *Biochim. Biophys. Acta (BBA) Proteins Proteom.* **2009**, *1804*, 1151–1166. [CrossRef]
25. Schweitzer, B.; Wiltshire, S.; Lambert, J.; O'Malley, S.; Kukanskis, K.; Zhu, Z.; Kingsmore, S.F.; Lizardi, P.M.; Ward, D.C. Immunoassays with Rolling Circle DNA Amplification: A versatile Platform for Ultrasensitive Antigen Detection. *Proc. Natl. Acad. Sci. USA* **2000**, *97*, 10113–10119. [CrossRef]
26. Zhao, H.; Bian, S.; Yang, Y.; Wu, X. Chiroplasmonic Assemblies of Gold Nanoparticles as a Novel Method for Sensitive Detection of alpha-Fetoprotein. *Microchim. Acta* **2017**, *10*, 871–1862. [CrossRef]
27. Gao, T.; Zhi, J.; Mu, C.; Gu, S.; Xiao, J.; Yang, J.; Wang, Z.; Xiang, Y. One-Step Detection for Two Serological Biomarker Species to improve the Diagnostic Accuracy of Hepatocellular Carcinoma. *Talanta* **2018**, *178*, 89–93. [CrossRef]
28. Fowler, J.D.; Suo, Z. Biochemical, Structural, and Physiological Characterization of Terminal Deoxynucleotidyl Transferase. *Chem. Rev.* **2006**, *106*, 2092–2110. [CrossRef]
29. Jensen, M.; Davis, R.W. Template-Independent Enzymatic Oligonucleotide Synthesis (TiEOS): Its History, Prospects, and Challenges. *Biochemistry* **2018**, *57*, 1821–1832. [CrossRef]
30. Hermanová, M.; Havranová-Vidláková, P.; Ondráčková, A.; Kumar, S.S.; Bowater, R.; Fojta, M. Label-free Voltammetric Detection of Products of Terminal Deoxynucleotidyl Transferase Tailing Reaction. *Electroanalysis* **2018**, *31*, 246–255. [CrossRef]
31. Lorenzo-Gómez, R.; Fernández-Alonso, N.; Miranda-Castro, R.; de-los-Santos-Álvarez, N.; Lobo-Castañón, M.J. Unravelling the Lipocalin 2 Interaction with Aptamers: May Rolling Circle Amplification improve Their Functional Affinity? *Talanta* **2019**, *197*, 406–412. [CrossRef] [PubMed]
32. González-Fernández, E.; de-los-Santos-Álvarez, N.; Miranda-Ordieres, A.J.; Lobo-Castañón, M.J. SPR Evaluation of Binding Kinetics and Affinity study of Modified RNA Aptamers Towards Small Molecules. *Talanta* **2012**, *99*, 767–773. [CrossRef] [PubMed]

biosensors

MDPI

Article

Screen-Printed Sensor for Low-Cost Chloride Analysis in Sweat for Rapid Diagnosis and Monitoring of Cystic Fibrosis

Alicia Hauke [1,*], Susanne Oertel [1], Leona Knoke [2], Vanessa Fein [2], Christoph Maier [2], Folke Brinkmann [2,†] and Michael P. M. Jank [1,†]

1 Fraunhofer Institute for Integrated Systems and Device Technology, 91058 Erlangen, Germany; susanne.oertel@iisb.fraunhofer.de (S.O.); michael.jank@iisb.fraunhofer.de (M.P.M.J.)

2 Department of Paediatric Pneumology, University Children's Hospital, Ruhr-University Bochum, 44791 Bochum, Germany; leona.knoke@rub.de (L.K.); vanessa.fein@rub.de (V.F.); christoph.maier@rub.de (C.M.); folke.brinkmann@rub.de (F.B.)

* Correspondence: alicia.hauke@iisb.fraunhofer.de

† These authors contributed equally to this publication.

Received: 31 July 2020; Accepted: 7 September 2020; Published: 11 September 2020

check for updates

Abstract: Analysis of sweat chloride levels in cystic fibrosis (CF) patients is essential not only for diagnosis but also for the monitoring of therapeutic responses to new drugs, such as cystic fibrosis transmembrane conductance regulator (CFTR) modulators and potentiators. Using iontophoresis as the gold standard can cause complications like burns, is uncomfortable, and requires repetitive hospital visits, which can be particularly problematic during a pandemic, where distancing and hygiene requirements are increased; therefore, it is necessary to develop fast and simple measures for the diagnosis and monitoring of CF. A screen-printed, low-cost chloride sensor was developed to remotely monitor CF patients. Using potentiometric measurements, the performance of the sensor was tested. It showed good sensitivity and a detection limit of 2.7×10^{-5} mol/L, which covered more than the complete concentration range of interest for CF diagnosis. Due to its fast response of 30 s, it competes well with standard sensor systems. It also offers significantly reduced costs and can be used as a portable device. The analysis of real sweat samples from healthy subjects, as well as CF patients, demonstrates a proper distinction using the screen-printed sensor. This approach presents an attractive remote measurement alternative for fast, simple, and low-cost CF diagnosis and monitoring

Keywords: screen-printed sensor; sweat analysis; ion-selective electrode; chloride; cystic fibrosis

1. Introduction

Sweat analysis is an emerging field that provides insights into human health, and over the last few years has become an attractive, non-invasive alternative to blood analysis [1–5]. Sweat analysis is more comfortable for patients, has good accessibility, and is able to continuously monitor body parameters, which makes sweat analysis useful for monitoring health and fitness [4].

The development of so-called wearable devices increases rapidly due to new techniques of miniaturization, flexibility, and low-cost production. The number of publications in the field of sweat analysis rises steadily every year. Most of them use electrochemical sensors, especially ion-selective electrodes (ISEs), to analyze specific ions in sweat. Due to new developments in printing paste technologies, it is possible to produce screen-printed, ion-selective electrodes on flexible polymeric substrates, which are also inexpensive, due to mass fabrication [6–8]. Thus, screen-printing technology is preferred for the production of disposable, low-cost sweat sensors with small dimensions, in contrast to commonly used glass or tube ion-selective electrodes, which need high maintenance, high sample volume, and are rigid and expensive.

Sweat sensors integrating ion-selective electrodes can be used to collect vital data parameters like fatigue or fluid balance during sports, and can also support or replace diagnostic or monitoring measures. In addition to water, sweat contains characteristic plasma-related electrolytes, namely sodium (Na^+), potassium (K^+), and chloride (Cl^-) [1,9]. For example, the sodium concentration is linked to dehydration and electrolyte imbalance [10–12], whereas the potassium concentration in sweat is related to hypo-/hyperkalemia [13,14].

The detection of cystic fibrosis (CF) is one of the most used sweat diagnostics [15,16]. CF is a genetic disease, which causes a defective ion transfer through the epithelial cellular membranes (cystic fibrosis transmembrane conductance regulator (CFTR) chloride channels). As a result, chloride reabsorption from sweat is reduced, leading to a sticky mucus formation, which, if untreated, can be fatal. Therefore, an early diagnosis is crucial for symptomatic treatment to improve the quality of life and life expectancy of patients. As the gold standard for CF diagnosis, chloride concentration is determined using sweat tests [16,17]. In addition to this, a new generation of CF drugs, called CFTR modulators and potentiators, changes the chloride current across the epithelial barrier, and therefore, the chloride concentration in the sweat, which can be used as therapeutic monitoring.

During these tests, sweating is induced by so-called pilocarpine iontophoresis. In this procedure, a low current is applied between two electrodes attached to the forearm of the patient. For sweat inducement, a solution of pilocarpine is applied under the anode. Through the low-voltage induced electric current, pilocarpine is carried into the skin, where it stimulates the sweat glands. The sweat is then collected and analyzed regarding its chloride concentration.

Especially in terms of the current COVID-19 pandemic, the risk of infection for risk groups has increased, including CF-patients. Therefore, there is a high demand for an easy outpatient approach for sweat chloride measurement. The chloride concentration can be acquired by using a screen-printed chloride sensor, which can be used as a portable device and is user-friendly.

Beyond this, the use of portable sensors can help increase our understanding of CF. Due to the time-intensive setting of pilocarpine iontophoresis, which can only be carried out under medical supervision in clinics, the procedure is usually conducted on the patients in intervals of several months. As a result, there is almost no understanding of the development of the chloride concentration in the everyday life of the patient, and how it is affected by factors like CFTR modulators and other drugs, physical activity, rest, or sleep. A new, portable sweat sensor would allow patients to measure the chloride concentration in their sweat more regularly and, in doing so, it would provide data that would be useful for understanding the CF disease.

Within our study, a screen-printed chloride sensor was developed, and its performance was tested to show its usability as a CF sensor. The sensor was tested for its sensitivity, detection limit, and response time. As the sensor achieved the necessary performances under laboratory conditions, real sweat samples were measured for the determination of chloride concentrations, to validate the correct detection of CF. Sweat samples of CF patients were collected under stable clinical conditions, in the absence of acute infections.

2. Materials and Methods

2.1. Sensor Fabrication

For the analysis of sweat samples, a screen-printed potentiometric chloride sensor was developed. Sensor electrodes were fabricated using a semiautomatic screen and pattern printer from Ekra (series XH STS). Each electrode, consisting of different layers (see Figure 1a), was printed using a polyester screen. Between each printing step, the layer was annealed at 120 °C to 130 °C, for 3 min and up to 10 min, depending on the used printing paste. PET foils were used as substrates to yield a flexible sensor that could later be attached to the skin (see Figure 1b). The working electrode of the potentiometric sensor was printed using a silver paste as a conducting layer, covered by a carbon layer as solid contact. Additionally, a reference electrode was printed using a silver/silver chloride paste. For encapsulation of the electrodes, an insulating layer was printed on top of both electrodes, leaving the active parts of the working and the reference electrodes open. All printing pastes were purchased from Du Pont (UK) Limited.

(**a**)

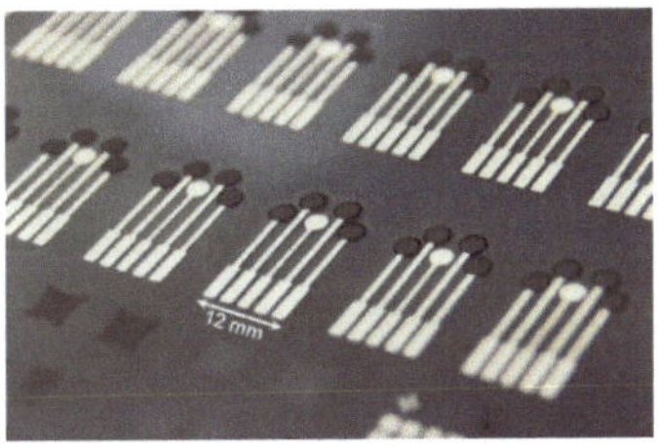

(**b**)

Figure 1. (**a**) Explosion view of screen-printed chloride sensor depicting the layers. (i) Silver (dark grey), (ii) silver/silver chloride (light grey), (iii) carbon (black), and (iv) insulation/encapsulation (green). (**b**) Manufactured sample sheet of sensors before deposition of the selective and reference membranes.

To functionalize the electrodes, an ion-selective membrane was drop-casted on top of the working electrode. It consisted of a chloride ionophore I - cocktail A from Sigma Aldrich, mixed with polyvinyl chloride as polymer and tetrahydrofuran as solvent. The resulting composition of the membrane is listed in Table 1. To achieve a constant potential of the reference electrode, a membrane consisting of polyvinyl butyral, methanol, and sodium chloride was used. The procedure was already described in previous studies [18,19].

Table 1. Membrane compounds of the chloride-selective membrane using the chloride ionophore I –cocktail A with its corresponding portion in parentheses.

Chloride Ionophore I—Cocktail A			Polymer	Solvent
Ionophore	**Plasticizer**	**Anionic Site**		
Chloride ionophore I (3.5%)	2-nitrophenyl octyl ether + 1-Decanol (65.8%)	Tridodecylmethyl-ammonium chloride (0.7%)	PVC (30.0%)	1 mL THF on 100 mg compounds

2.2. Potentiometric Measurements

The performance of the fabricated sensors was tested using potentiometry; a high resolution, high-input impedance, multi-channel potentiometer (EMF 16, Lawson Labs, Malvern, PA, USA) was utilized. The functionality of the sensors was validated in solutions of NaCl and NH_4Cl with increasing concentrations from 10^{-6} mol/L to 10^{-1} mol/L measuring the potential of the ion-selective electrode versus the reference electrode. For the measurement of sweat samples, the calibration range was extended until 1 mol/L. The potential response of the ISEs was recorded for a representative response of Cl^-. The resulting calibration plot was used to determine the limit of detection and the slope of the presented electrodes, as well as the response time.

2.3. Sweat Inducement and Collection

For the withdrawal of sweat and analysis of the chloride concentrations, we applied the most commonly used method of pilocarpine iontophoresis. Sweat excretion was induced by a small current between two electrodes attached to the forearm of the test subject, which drove pilocarpine into the epidermis (Figure 2a). After five minutes, the electrodes were removed (Figure 2b) and a test tube was pressed onto the forearm, in the former position of the pilocarpine electrode, to collect sweat (Figure 2c,d). The collection time took 20 min; thereafter, the chloride concentration of the sweat samples was measured using a Kreienbaum FKGO Chloridmeter, which is the established method in CF diagnosis at the University Children's Hospital, Ruhr-University Bochum. The remaining sweat was frozen at −20 °C and stored until measurements were taken using the printed chloride sensor. To collect a sufficient amount of sweat for both methods, iontophoresis was conducted on both forearms, after which the samples were combined.

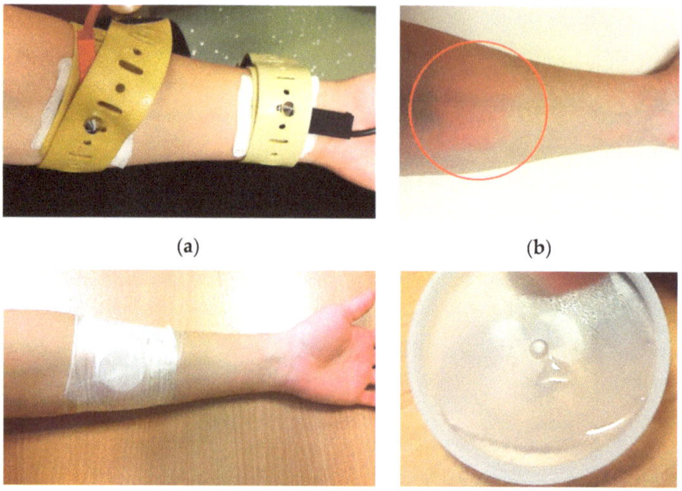

(a) (b) (c) (d)

Figure 2. (**a**) Sweat inducement by pilocarpine iontophoresis using two electrodes attached to the forearm; (**b**) irritation of skin after iontophoresis due to the applied electric current; (**c**) test tube pressed onto the forearm in the former position of the pilocarpine electrode to collect sweat; and (**d**) small amount of collected sweat with test tube after pilocarpine iontophoresis.

We included patients with genetically confirmed CF, alongside healthy patients, as test subjects for this pilot study. All tests were carried out under professional medical monitoring. All human trial regulations were met.

3. Results and Discussion

3.1. Sensor Performance

Screen-printing technology facilitates the production of large-scale reproducible sensor electrodes. However, the functionalization of the electrodes through drop-casting of membranes creates slight deviations between the ion-selective electrodes. Therefore, it is crucial to calibrate each sensor and characterize its performance regarding its sensitivity, the limit of detection, the linear range, and its response time.

To apply the chloride sensor for the initial diagnosis of CF, it is mandatory that the sensor covers the possible concentration range of chloride in sweat (typically 1 mM to 200 mM) and has a reasonable response time. Figure 3a shows a typical potentiometric response curve for a chloride sensor in NH_4Cl standard solution with varying concentrations. As depicted, the sensor reaches a stable potential over time, for concentrations above 10^{-5} mol/L. The evaluated response time in this region was as low as 30 s. Below 10^{-5} mol/L, the potential decreased continuously after reaching an initial peak indicative of the actual concentration. However, this was not crucial for the application in CF diagnostics, as the lowest possible concentration of chloride in sweat only reached about 10^{-3} mol/L [1].

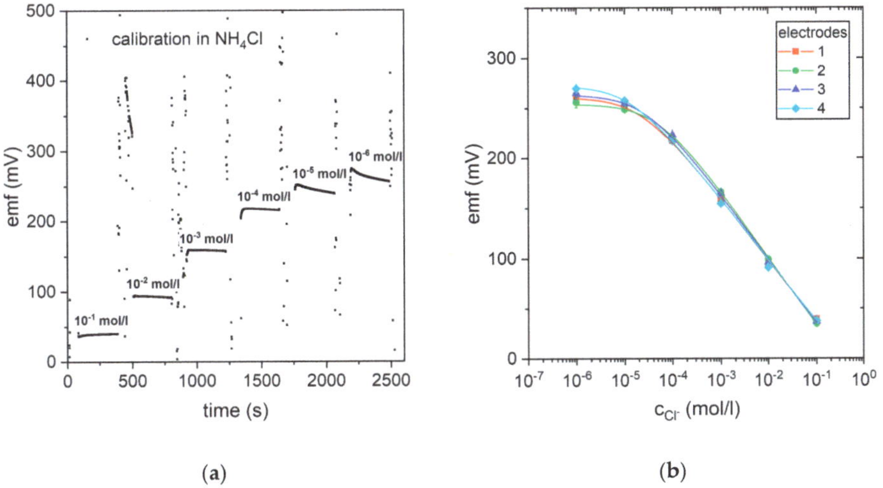

Figure 3. (**a**) Potentiometric response curve of the screen-printed chloride sensor measured in NH_4Cl solutions with different concentrations. (**b**) Resulting calibration plot measured in NH_4Cl solutions of four different chloride-selective electrodes sharing one reference electrode.

Calculating the mean value of each potentiometric step after reaching a constant potential and using the Nernstian equation [20], the calibration plot could be fitted. Figure 3b shows the calibration plot of four different chloride ISEs sharing one reference electrode in the NH_4Cl solution, utilizing the sensor layout shown in Figure 1. Figure 3b shows that all four electrodes achieved a reproducible dependence on the chloride concentration with a sensitivity of 62.9 mV/dec. This follows the ideal Nernstian slope of 59.6 mV. Using the fitted calibration plot, a detection limit of 2.7×10^{-5} mol/L and a linear behavior between 10^{-4} mol/L and 10^{-1} mol/L could be determined.

These results showed that the sensor had excellent performance, covering the complete concentration range of chloride in sweat, showing fast response time and good potential stability, meaning that this sensor was highly applicable. As the sensor could be produced at a low cost, and was small and flexible, it could be used as a portable device. This sensor had significant advantages over comparable methods. A comparison to the standard systems used in clinical diagnostics is listed in Table 2.

Table 2. Comparison of screen-printed chloride sensor to the standard sensor system.

Parameters	Screen-Printed Chloride Sensor	Coulometric Titration (FKGO Chloridmeter) [1]
Detection limit	0.03 mmol/L	10 mmol/L
Response time	<1 min	<1 min
System preparation	5 min	5 min
Sample preparation	-	+
Cost of equipment	<50 €	8200 €
Cost of assay	<3 €	ca. 5 €
Portability	+	-
Usability	+	-

[1] by Kreienbaum.

3.2. Measurement of Sweat Samples

Sweat samples were collected from healthy subjects as well as from patients with CF, according to the pilocarpine iontophoresis procedure described above. The samples were split to validate the suitability of the evaluation by the printed chloride sensors, against the gold standard. Before measurement, the sensor was calibrated in NaCl solutions with concentrations from 10^{-6} mol/L up to 1 mol/L. An NaCl solution was chosen along with water. Sweat consists mostly of NaCl, which means that this calibration appropriately mimics the real conditions. After defrosting the samples, each sample was dropped onto the sensor so that both electrodes, the ISE and reference electrode, were completely covered with a continuous sweat film. Using the recorded calibration plot, the concentration of each sweat sample was calculated and is listed in Table 3. According to Gibson et al. [21] and Collie et al. [21,22], a concentration level of chloride above 60 mmol/L indicates a CF disease. As pointed out in Table 3, all samples could be correctly determined and categorized in healthy subjects and CF patients. Figure 4 shows two contrasting samples, healthy (sample 1, 2, and 3) and CF (sample 9), overlaid with the corresponding calibration plot. The dashed line indicates the threshold of 60 mmol/L of chloride.

Table 3. The determined chloride concentration of sweat samples and categorization in health status.

Sample	$c(Cl^-)$ in mmol/L	Status [1] (Healthy or CF)
1	0.9	healthy
2	1.0	healthy
3	1.9	healthy
4	3.0	healthy
5	5.7	healthy
6	0.4	healthy
7	0.6	healthy
8	1.7	healthy
9	129.2	CF
10	100.0	CF
11	63.0	CF

[1] Derived from different accepted test protocols.

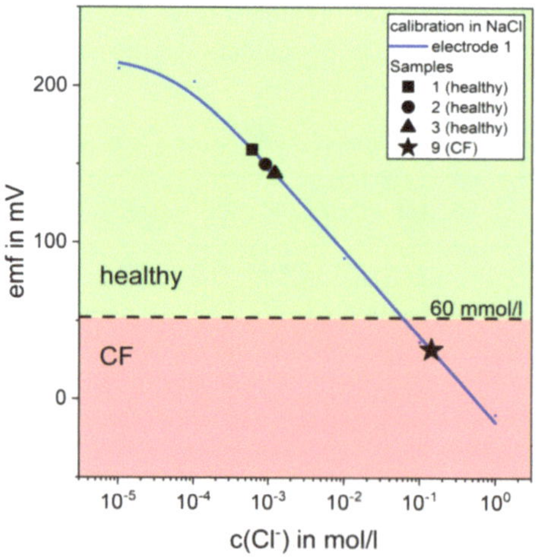

Figure 4. Measurement of sweat samples (samples 1, 2, 3, and 9) using the screen-printed chloride sensor for the detection of cystic fibrosis (CF). The dashed line displays the threshold of chloride concentration. Concentrations >60 mmol/L indicates CF disease.

4. Conclusions

In this paper, a screen-printed, low-cost chloride-selective sensor was presented. The sensor showed excellent sensitivity and covered the complete possible concentration range of chloride in sweat. More importantly, the concentration range, which is of high interest for CF diagnosis and the therapeutic monitoring of CF, lies within a linear range, allowing for a good determination of the chloride concentration. As the response time of the screen-printed sensor was only 30 s, fast and simple CF detection and monitoring could be achieved. To further characterize the performance of the sensor, possible disturbance due to interfering ions (e.g., HCO_3^-) will be investigated in future work.

To prove proper functionality and the applicability of the chloride sensor for CF diagnosis, measurements of real sweat samples were demonstrated. Comparison against the standard procedure successfully validated a correct distinction between healthy test subjects and patients suffering from CF. With this, the screen-printed sensor could be utilized for the remote monitoring of CF patients, to avoid contact with clinics and therefore to reduce the risk of infections, which is especially useful for maintaining hygiene and distancing regulations in times of a pandemic. Further, as the chloride sensor could be inexpensively manufactured, thanks to screen-printing technology, it can be used as a disposable sensor. In combination with the already published evaluation board for data acquisition and transfer [18], it can be used as a small handheld device. Additionally, due to its flexible design and small dimensions, it is also possible to attach the sensor directly onto the skin. With this, only a low amount of sweat is necessary. The integration of sweat inducement into the sensor system will be investigated in future work. Achieving this would mean that the therapeutic monitoring of CF could be made easy, enabling quick diagnosis and monitoring, with increased comfort for the patients. With this device, the close monitoring of intraindividual changes in sweat chloride levels is possible.

Author Contributions: Conceptualization, A.H., S.O., and M.P.M.J.; Formal analysis, A.H.; Investigation, A.H., S.O., L.K., V.F., and C.M.; Methodology, A.H., S.O., F.B., and M.P.M.J.; Supervision, C.M., F.B., and M.P.M.J.; Validation, A.H., S.O., F.B., and M.P.M.J.; Visualization, A.H. and L.K.; Writing—original draft, A.H., V.F. and F.B.; Writing—review & editing, S.O. and M.P.M.J. All authors have read and agreed to the published version of the manuscript.

Funding: No external funding was received.

Acknowledgments: Thank you to the volunteers and CF patients participating in the study.

Conflicts of Interest: The authors declare no conflict of interest.

References

1. Baker, L.B. Physiology of sweat gland function: The roles of sweating and sweat composition in human health. *Temperature* **2019**, *6*, 211–259. [CrossRef] [PubMed]
2. Chung, M.; Fortunato, G.; Radacsi, N. Wearable flexible sweat sensors for healthcare monitoring: A review. *J. R. Soc. Interface* **2019**, *16*, 20190217. [CrossRef] [PubMed]
3. Harshman, S.W.; Pitsch, R.L.; Smith, Z.K.; O'Connor, M.L.; Geier, B.A.; Qualley, A.V.; Schaeublin, N.M.; Fischer, M.V.; Eckerle, J.J.; Strang, A.J.; et al. The proteomic and metabolomic characterization of exercise-induced sweat for human performance monitoring: A pilot investigation. *PLoS ONE* **2018**, *13*, e0203133. [CrossRef] [PubMed]
4. Mena-Bravo, A.; Luque de Castro, M.D. Sweat: A sample with limited present applications and promising future in metabolomics. *J. Pharm. Biomed. Anal.* **2014**, *90*, 139–147. [CrossRef] [PubMed]
5. Jadoon, S.; Karim, S.; Akram, M.R.; Kalsoom Khan, A.; Zia, M.A.; Siddiqi, A.R.; Murtaza, G. Recent developments in sweat analysis and its applications. *Int. J. Anal. Chem.* **2015**, *2015*, 164974. [CrossRef]
6. Bandodkar, A.J.; Nunez-Flores, R.; Jia, W.; Wang, J. All-printed stretchable electrochemical devices. *Adv. Mater.* **2015**, *27*, 3060–3065. [CrossRef]
7. Bandodkar, A.J.; Jeerapan, I.; You, J.-M.; Nuñez-Flores, R.; Wang, J. Highly Stretchable Fully-Printed CNT-Based Electrochemical Sensors and Biofuel Cells: Combining Intrinsic and Design-Induced Stretchability. *Nano Lett.* **2016**, *16*, 721–727. [CrossRef]

8. Xu, H.; Yang, X.; Wang, Y.; Zheng, J.; Luo, Z.; Li, G. Disposable blood potassium sensors based on screen-printed thick film electrodes. *Meas. Sci. Technol.* **2010**, *21*, 55802. [CrossRef]
9. Sato, K.; Kang, W.H.; Saga, K.; Sato, K.T. Biology of sweat glands and their disorders. I. Normal sweat gland function. *J. Am. Acad. Dermatol.* **1989**, *20*, 537–563. [CrossRef]
10. Baker, L.B. Sweating Rate and Sweat Sodium Concentration in Athletes: A Review of Methodology and Intra/Interindividual Variability. *Sports Med.* **2017**, *47*, 111–128. [CrossRef]
11. Bandodkar, A.J.; Molinnus, D.; Mirza, O.; Guinovart, T.; Windmiller, J.R.; Valdes-Ramirez, G.; Andrade, F.J.; Schoning, M.J.; Wang, J. Epidermal tattoo potentiometric sodium sensors with wireless signal transduction for continuous non-invasive sweat monitoring. *Biosens. Bioelectron.* **2014**, *54*, 603–609. [CrossRef] [PubMed]
12. Coyle, S.; Lau, K.-T.; Moyna, N.; O'Gorman, D.; Diamond, D.; Di Francesco, F.; Costanzo, D.; Salvo, P.; Trivella, M.G.; De Rossi, D.E.; et al. BIOTEX—Biosensing textiles for personalised healthcare management. *IEEE Trans. Inf. Technol. Biomed.* **2010**, *14*, 364–370. [CrossRef]
13. Alizadeh, A.; Burns, A.; Lenigk, R.; Gettings, R.; Ashe, J.; Porter, A.; McCaul, M.; Barrett, R.; Diamond, D.; White, P.; et al. A wearable patch for continuous monitoring of sweat electrolytes during exertion. *Lab Chip* **2018**, *18*, 2632–2641. [CrossRef] [PubMed]
14. Worth, H.G.J. Measurement of sodium and potassium in clinical chemistry: A review. *Analyst* **1988**, *113*, 373. [CrossRef] [PubMed]
15. Choi, D.-H.; Kitchen, G.B.; Jennings, M.T.; Cutting, G.R.; Searson, P.C. Out-of-clinic measurement of sweat chloride using a wearable sensor during low-intensity exercise. *NPJ Digit. Med.* **2020**, *3*, 49. [CrossRef] [PubMed]
16. Esteves, C.Z.; de Aguiar Dias, L.; de Oliveira Lima, E.; de Oliveira, D.N.; Rodrigues Melo, C.F.O.; Delafiori, J.; Souza Gomez, C.C.; Ribeiro, J.D.; Ribeiro, A.F.; Levy, C.E.; et al. Skin Biomarkers for Cystic Fibrosis: A Potential Non-Invasive Approach for Patient Screening. *Front. Pediatrics* **2017**, *5*, 290. [CrossRef]
17. Ďurč, P.; Foret, F.; Pokojová, E.; Homola, L.; Skřičková, J.; Herout, V.; Dastych, M.; Vinohradská, H.; Kubáň, P. New approach for cystic fibrosis diagnosis based on chloride/potassium ratio analyzed in non-invasively obtained skin-wipe sweat samples by capillary electrophoresis with contactless conductometric detection. *Anal. Bioanal. Chem.* **2017**, *409*, 3507–3514. [CrossRef]
18. Zoerner, A.; Oertel, S.; Jank, M.P.M.; Frey, L.; Langenstein, B.; Bertsch, T. Human Sweat Analysis Using a Portable Device Based on a Screen-printed Electrolyte Sensor. *Electroanalysis* **2018**, *30*, 665–671. [CrossRef]
19. Zörner, A.; Oertel, S.; Schmitz, B.; Lang, N.; Jank, M.P.M.; Frey, L. Determination of the Selectivity of Printed Wearable Sweat Sensors. In Proceedings of the 10th International Joint Conference on Biomedical Engineering Systems and Technologies, (BIOSTEC 2017), Porto, Portugal, 21–23 February 2017; SCITEPRESS—Science and Technology Publications: Setúbal, Portugal, 2017; Volume 1, pp. 81–87, ISBN 978-989-758-216-5.
20. Buck, R.P.; Lindner, E. Recommendations for nomenclature of ionselective electrodes (IUPAC Recommendations 1994). *Pure Appl. Chem.* **1994**, *66*, 2527–2536. [CrossRef]
21. Gibson, L.E.; Cooke, R.E. A test for concentration of electrolytes in sweat in cystic fibrosis of the pancreas utilizing pilocarpine by iontophoresis. *Pediatrics* **1959**, *23*, 545–549.
22. Collie, J.T.B.; Massie, R.J.; Jones, O.A.H.; LeGrys, V.A.; Greaves, R.F. Sixty-five years since the New York heat wave: Advances in sweat testing for cystic fibrosis. *Pediatric Pulmonol.* **2014**, *49*, 106–117. [CrossRef] [PubMed]

biosensors

MDPI

Review

Electrochemiluminescence Biosensors Using Screen-Printed Electrodes

Emiliano Martínez-Periñán [1], Cristina Gutiérrez-Sánchez [1], Tania García-Mendiola [1,2,3] and Encarnación Lorenzo [1,2,3,*]

1 Departamento de Química Analítica y Análisis Instrumental Universidad Autónoma de Madrid, 28049 Madrid, Spain; emiliano.martinez@uam.es (E.M.-P.); cristina.gutierrezs@uam.es (C.G.-S.); tania.garcia@uam.es (T.G.-M.)

2 Institute for Advanced Research in Chemical Sciences (IAdChem) Universidad Autónoma de Madrid, 28049 Madrid, Spain

3 IMDEA-Nanociencia, Ciudad Universitaria de Cantoblanco, 28049 Madrid, Spain

* Correspondence: encarnacion.lorenzo@uam.es; Tel.: +34-91-497-4488

Received: 31 July 2020; Accepted: 7 September 2020; Published: 9 September 2020

check for updates

Abstract: Electrogenerated chemiluminescence (also called electrochemiluminescence (ECL)) has become a great focus of attention in different fields of analysis, mainly as a consequence of the potential remarkably high sensitivity and wide dynamic range. In the particular case of sensing applications, ECL biosensor unites the benefits of the high selectivity of biological recognition elements and the high sensitivity of ECL analysis methods. Hence, it is a powerful analytical device for sensitive detection of different analytes of interest in medical prognosis and diagnosis, food control and environment. These wide range of applications are increased by the introduction of screen-printed electrodes (SPEs). Disposable SPE-based biosensors cover the need to perform in-situ measurements with portable devices quickly and accurately. In this review, we sum up the latest biosensing applications and current progress on ECL bioanalysis combined with disposable SPEs in the field of bio affinity ECL sensors including immunosensors, DNA analysis and catalytic ECL sensors. Furthermore, the integration of nanomaterials with particular physical and chemical properties in the ECL biosensing systems has improved tremendously their sensitivity and overall performance, being one of the most appropriates research fields for the development of highly sensitive ECL biosensor devices.

Keywords: screen-printed electrodes; nanomaterials; enzymatic biosensor; immunosensor; DNA sensor

1. Introduction

Electrochemiluminescence (ECL) is a chemiluminescence phenomenon resulting from the electrochemical excitation of a luminescence system (luminophore) that emits light when it returns to its fundamental state [1]. The mechanisms associated with these phenomena are well known and described in the literature, being classified into two main types [2], as can be seen in Scheme 1.

The annihilation pathway: A reduced specie and an oxidized specie (charged radical ions) are simultaneously generated at the electrode surface by applying alternating pulse potentials. These two species react between them generating an excited form, which in the relaxation process to the ground state emits a photon [3].

Co-reactant pathway: A co-reactant is a chemical specie that is reduced or oxidized at the electrode surface, generating a very reactive intermediates that react with the reduced or oxidized luminophore (specie capable of emit light) present in the solution to produce the excited state. Finally, the excited state returns to the ground state to cause chemiluminescence. Employing a co-reactant is especially useful when either radical charged ions are not stable enough for the ECL annihilation reaction, or radical ions cannot both be formed because of the solvent has a narrow potential window. With a co-reactant ECL can be generated by applying a potential in one direction. There are two reaction paths to produce the excited state of the ECL emitter, reductive-oxidation or oxidative-reduction ECL. For instance, oxalate ion ($C_2O_4{}^{2-}$) [4,5] and several amines [6–9] can be used for oxidative-reduction ECL where an oxidative step produces a strong reductant, whereas peroxidisulfate ion ($S_2O_8{}^{2-}$) [10–12] is frequently used for reductive-oxidation ECL.

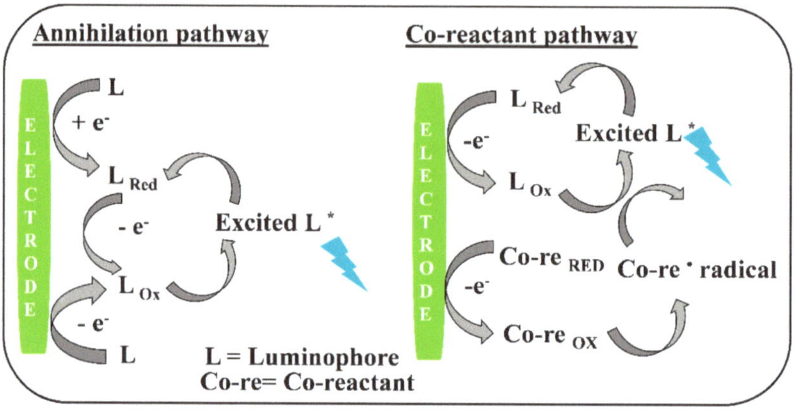

Scheme 1. ECL mechanisms: Annihilation and co-reactant pathway.

As described above, ECL reactions require a luminophore. Even though many compounds have been demonstrated to participate in ECL reactions, most of them require aprotic and deoxygenated solution conditions. Therefore, only few compounds and their derivatives are primarily utilized for aqueous-based ECL bioanalytical detection methods. These are luminol (5-amino-2,3-dihydrophthalazine-1,4-dione) and ruthenium (II) chelates $[RuL_3]^{2+}$. Nowadays, new luminophores such as semiconductor nanomaterials are being widely used with great results. This fact is one of the main reasons ECL sensor and biosensor are having a great and successful advance.

Bioanalytical methods based on $[RuL_3]^{2+}$ ECL were not developed until the co-reactant pathway was reported between Tris(bipyridine) ruthenium (II) $[Ru(bpy)_3]^{2+}$ and oxalate [5] in aqueous media and unaffected by the presence of oxygen was reported. $[Ru(bpy)_3]^{2+}$ is a model luminophore that is largely used nowadays; the discovery of its ECL emission in aqueous media with efficient co-reactants such as tri-n-propylamine (TPrA) [8,13] has led to successfully bioassays for clinical diagnosis. Probably one of the most relevant co-reactant pathways is the "oxidative–reductive" system between aliphatic amines and $[RuL_3]^{2+}$ as $[Ru(bpy)_3]^{2+}$. In this mechanism, both $[RuL_3]^{2+}$ and the co-reactant are oxidized (Equations (1) and (2)). The TPrA radical cation is unstable on the time frame of the experiment and quickly deprotonates, forming a free radical, TPrA* (Equation (3)). This free radical behaves as a reducing agent and transfers an electron to $[RuL_3]^{3+}$, producing an excited state (Equation (4)).

$$[RuL_3]^{2+} + 1e^- \rightarrow [RuL_3]^{3+} \quad (1)$$

$$TPrA \rightarrow TPrA^{+*} + 1e^- \quad (2)$$

$$TPrA^{+*} \rightarrow TPrA^* + H^+ \quad (3)$$

$$TPrA^* + [RuL_3]^{3+} \rightarrow [RuL_3]^{2+*} + products \quad (4)$$

As $[Ru(bpy)_3]^{2+}$, luminol is one of the most used reagents in ECL applications [1,14] in aqueous alkaline solutions. Luminol is oxidized at the electrode surface and forms a diazaquinone intermediate, which reacts quantitatively with hydrogen peroxide to produce the 3-aminophtalate in an excited state due to O–O bond cleavage in the endoperoxide form. 3-Aminophthalate then emits a characteristic blue light at 425 nm (see Figure 1) [14,15].

Different mechanistic pathways have been suggested depending on the applied electrode potentials. In addition, the ECL intensity correlates directly with the amount of hydrogen peroxide. The luminol ECL method can be used to determine either luminol or species labeled with luminol or peroxides. Since hydrogen peroxide is an analyte of interest in various biological applications, this luminophore can measure reactive oxygen species. In addition, as we describe below, the enzymatic activity of oxidase-type enzymes, which generate hydrogen peroxide in the presence of their substrates, can be monitored using luminol.

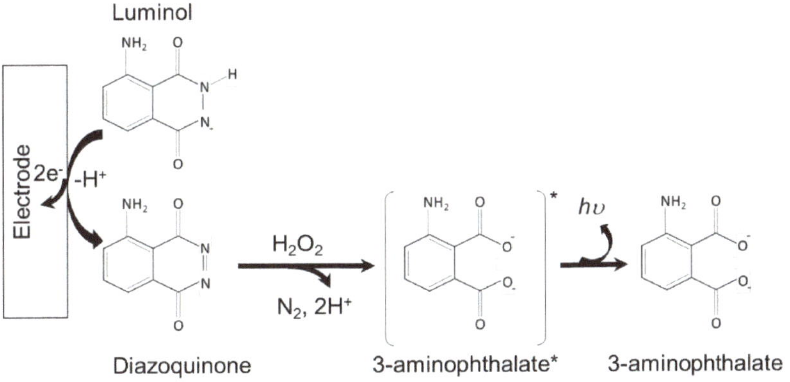

Figure 1. Scheme showing the ECL mechanism of luminol with hydrogen peroxide. Reproduced from reference [14]. Reproduced and adapted with permission of MDPI.

In electrochemiluminescence assays, the energy of the excitation source (electricity) and the detection signal (light) are totally different, which is why ECL presents a lower background signal, resulting in higher sensitivity than electrochemical detection. Hence, it has been widely employed as analytical technique, in particular as a detection system in sensors and biosensors. Besides its high sensitivity, other advantages are the broad dynamic range; the simplicity of the analysis methodology (similar to conventional ELISA), the great flexibility, due to the high stability of luminophores; and the facility of combining with biological systems. Due to its combination of electrochemical and optical advantages, ECL has been widely used in the detection of biomolecules in a variety of samples, including plasma, whole blood, serum and cell supernatant. ECL has been also successfully used as detector [16] of high-performance liquid chromatography (HPLC) [17], flow injection analysis (FIA) [18], micro total analysis (μTAS) [19] and capillary electrophoresis [20]. The great success of ECL immunosensor due to the minimal background signals, better specificity and higher sensitivity is increasing the use of meso scale discovery (MSD) ECL assays [21]. MSD assays work on the same principle as ELISA but use ECL as the detection method. Moreover, it shows the ability to measure multiple analytes in a single well, having also the possibilities of analysis high amounts of samples in short times thanks to the well-plate configuration.

Developing ECL biosensors is now a hot topic that is increasing the interest of the analytical chemistry community. In addition to the advantages already mentioned, the scalable instrumentation required and the possibility of combining them successfully with screen-printed electrodes (SPEs) technologies are promoting the development of new devices, most of them adaptable to the new trend of point of care (POC) systems.

SPEs' manufacturing process is characterized by the fast and easy mass production of reproducible electrochemical platforms, which offer true potential for application in the field of sensor [22]. SPEs are disposable electrochemical devices that avoid tedious pre-treatment steps. Hence, they are suitable for on-site and real-time sensing [23]. Moreover, the use of screen-printing technology also offers other interesting advantages, such as electrodes with different spatial distributions, appropriated for miniaturized devices [24] and the use of small sample volumes. Thus, SPEs have been widely employed in enzymatic biosensor [25–27], immunosensors [28–32] and DNA sensor fabrication [26,30,33–38].

The development of new miniaturized instruments integrating SPE and ECL has expanded the use of the technique to routine analysis. Among these, the *SpectroECL* and *μStat ECL*, commercial instruments from Metrohm-Dropsens, stand out for their ease of use. These instruments are small, portable and the SPEs are perfectly integrated into the device, being a convenient and a great affordable alternative.

In the last two decades, nanomaterials have burst into the field of ECL biosensors. Combining the bioselectivity and specificity of the biorecognition element with the numerous advantageous chemical and physical properties of nanomaterials has allowed the development of a whole new subset of sensitive biosensor devices. Therefore, great effort has been made in the development of new synthetic strategies to prepare a great variety of nanomaterials with highly controllable size, shape, surface charge and physicochemical properties [39]. In addition, functionalized nanomaterials show excellent properties for interfacing biological recognition events with electronic signal transduction in the design of a new generation of devices that exhibit new functionalities [40]. The appropriate use of nanomaterials clearly enhances the analytical properties of biosensors, increasing the sensitivity and lowering the detection limits in several orders of magnitudes.

A well-known advantage of nanomaterials is they provide a large effective surface area, which already enables the immobilization of higher amounts of bioreceptor units on the electrode surface [41]. They can also be used as a support of the biorecognition layer. Nanomaterials and biomolecules are in the same size range, facilitating the formation of hybrids with synergistic properties. They can act as labels of the recognition event or as catalyst for a reaction, as luminophore or as the energy acceptor to lead to an effective ECL resonance energy transfer (ECL-RET) [42]. Hence, the efficiency of ECL biosensors will be increased due to the use of different nanomaterials, such as graphene, carbon dots, quantum dots, metal nanoclusters, etc. combined with the recognition element and electrode modification strategy more appropriate [43].

ECL sensor and biosensor miniaturization is another area with great growth expectation in combination with microfluidic platforms or microarray technology and nanomaterials. All these areas together will improve the development of ultra-sensitive lab-on-a-chip systems for a variety of biosensing devices applied in clinical, food and environmental industries.

This review focuses on the use of screen-printed electrodes combined with nanomaterials in ECL biosensors, specifically how the biorecognition elements are immobilized on SPE and how nanomaterials are incorporated into the biosensor devices. ECL immunosensors, ECL enzyme-based biosensors and ECL DNA-based biosensors are discussed.

We believe this review will provide a base of knowledge in the development of ECL biosensor devices based on disposable SPEs, which promise selective and sensitive determinations of analytes in a wide range of samples, thanks to high affinity/biocatalytic interactions of the analytes with bioreceptors and the great advantages of ECL analysis methods. Thus, these devices will be marketed and widely used as rapid sensing and POC systems.

Screen Printed Electrodes (SPEs)

During the last decades, screen-printing technology, highly developed in microelectronics applications, has offered extraordinarily high-volume production of extremely cheap and highly reproducible and reliable single-use electrodes [44]. SPEs have been traditionally produced by printing different inks on plastic or ceramic substrates. The composition of the various inks used for SPEs manufacturing are of great importance for the electrochemical properties of the electrodes.

Carbon materials and graphite derivate in concrete are preferred in the ink composition due to their simple technological processing and low-cost. However, the use of metal particles in different ink compositions is also widespread in SPE fabrication. Among the most used metals, we can highlight gold, silver and platinum. Furthermore, materials such as Indium Tin Oxide (ITO) and poly(3,4-ethylenedioxythiophene) (PEDOT), which have the great advantages of being optically transparent for visible range, have also been employed to manufacture SPEs.

The great versatility of SPEs is due to the wide range of strategies in which the working electrode surface can be modified. Apart from the different ink composition, including the possibility of using a great variety of compounds to modify it with the aim of controlling its electrochemical behavior, the uncountable different strategies to modify SPE surface have allowed the SPEs employment in a multitude of sensors and biosensors. The great biocompatibility of some of the ink materials together with the successive modification by different strategies in a layer-by-layer system have been decisive for the appearance of a huge number of electrochemical and ECL biosensors. Another key factor of SPE modification was the appearance of a great variety of nanomaterials. This fact can be considered a great revolution in the field of electrodes modification due to the numerous and advantageous chemical and physical properties of nanomaterials. Nanomaterials have also demonstrated extraordinary properties and new interaction ways with biological elements and have been widely employed in biosensor development.

The disposable nature of SPEs is related with fundamental properties such as portability, low-cost, ease of use and mass production [45]. During the last decades, it has been a strong trend for disposables to be built with low-polluting materials, to minimize the amount of waste during the manufacture procedure, and the entire device must be recyclable for new uses [45]. In this way, paper is the most used material for the development of disposable sensors. It is a suitable material with high compatibility with inkjet [46] and screen printing techniques [44]. Moreover, it can be successfully combined with different surface modifiers such as polymers [47], nanomaterials [48], redox mediators [49,50] and biological element [51]. Different techniques such as photolithography, wax printing and chemical vapor-phase deposition have been successfully applied to transform paper platform into SPE. Apart from these sophisticated techniques for paper electrode modification, the creation of pencil drawn electrodes (PDEs) for electroanalytical applications has also been demonstrated, as it has previously been demonstrated in platforms such as paper [52], polyester [53] and PVC [54]. The rich surface chemistry of cellulose-based materials is a great advantage during the immobilization of bio-receptors by physisorption [55], bio-affinity interaction, by covalent bond due to a variety of chemical reactions (active esterification, maleimide cycloaddition, click chemistry, diazonium chemistry, etc.) [56]. All these excellent properties of the paper-based electrodes are making them a really useful option for competitive future electrochemical and ECL applications.

The great ambition to be able to monitor and measure in real time the presence of different analytes in different fields such as the environment, crops, industrial production chains, food production lines as well as in the human body itself have led to a great development of SPEs. The great diversity of different supports materials for SPEs has allowed its application in all the mentioned fields, each with particular requirements that can be supplied by the combination of different electrode supports and different inks for working electrode impression. A great advantage of some used materials as SPE supports is its flexibility and adaptability, which has been a great advance in the development of wearable devices [57]. The point-of-care diagnosis (POC) has been revolutionized by the development of these SPEs as a consequence of their disposability. All these great advantages make SPEs real competitors compared with conventional electrodes; consequently, they have been extensively employed in electrochemical and ECL biosensor development.

2. ECL Immunosensors

ECL has also become an important and powerful analytical technique in the field of immunosensors. Significant features such as high sensitivity, high reproducibility, versatility, simple optical technology and low background signal make ECL a good option for immunosensor development. In particular, SPEs are employed as electrochemical platforms because of their advantages in terms of low cost and rapid mass production, portability, high sensitivity, low sample volume requirement and easy handling [29]. The main drawbacks of the use of SPEs are the difficult of electrode surface regeneration and their reusability. These disadvantages are usually assumable if the immunosensor platform production is affordable.

During the last years, huge effort has been done to obtain higher sensitivity and develop new applications of ECL immunosensors based on different strategies. The complexity of the luminescence mechanism, which involves mass transport and electron transfer dynamics of abundant radical intermediates electrogenerated on the surface carry to low ECL efficiency [13,58]. Therefore, numerous ECL systems have been designed to enhance the ECL emission of luminophore through co-reactant pathways.

Different immunosensor strategies have been followed using ECL developments. Among the most common strategies sandwich-type immunosensor are widely used. Moreover, in the literature, there are many examples including label-free (direct or indirect immunosensor) and competitive immunosensors. The immunosensor classification is based on the most common immunosensors types reported on the literature for ECL immunosensors:

- The label-free configuration is usually the simplest. They are based on the immobilization of a capture antibody over the electrode surface followed by the specific analyte recognition by the capture specific antibody, resulting in the retention of the analyte over the electrode surface. This simplest change over the electrode surface affects the ECL signal, generating an increase of the ECL emission if the analyte acts as co-reactant of the ECL used system or in the case that the diffusion of luminophore or co-reactant will be favored as a consequence of less steric hindrance.
- Competitive configuration is based on the competition of labeled and unlabeled analytes for a limited number of antibody binding sites [59]. Only one antibody is used in a competitive configuration, which is usually attached to the electrode surface. Labeled analytes are usually modified with luminophore species, obtaining ECL signals which decrease with the higher concentration of the analyte of interest, as a consequence of fewer labeled analytes attached over the electrode surface.
- Sandwich type configuration is one of the most widely used on ECL immunosensor. It is based on the immobilization of a capture-specific antibody on the electrode surface (or over magnetic beads in some cases). After that, the specific antigen (analyte) is bonded over the capture antibody. Then, a secondary labeled antibody (detection antibody) reacts with the previous immobilized analyte. Considering the different labels attached to secondary antibodies, different configurations of sandwich-type ECL immunosensors have been developed:
    - The traditional sandwich type immunosensor is based on the linkage to the secondary antibody of a luminophore molecule or a nanomaterial, which acts as luminophore itself or is used as support to attach a great number of luminophore molecules.
    - The quench-type electrochemiluminescence immunosensor uses a secondary antibody linked to an element capable of generating resonance energy transfer (RET) phenomena of the emitting specie [60,61]. The element capable of quenching the ECL emission is usually a nanomaterial or a composite nanomaterial, decreasing the ECL emission when higher amounts of analytes are present in the sample.
    - A kind of sandwich-type ECL immunosensor that has recently appeared is named Faraday-cage-type electrochemiluminescence immunosensor [62,63]. The more important difference from the traditional sandwich-type immunosensor is that faraday-cage-type immunosensor uses a conductive two-dimensional nanomaterial (e.g., graphene, among others) simultaneously coated with a luminophore and a recognition component such as detection antibody, which could directly overlap on the electrode surface. In that configuration, electrons could flow freely from the working electrode surface to the detection

element, extending the outer Helmholtz plane (OHP) of the electrode. This strategy allows the two-dimensional nanomaterial coated by thousands of luminophore molecules, being all of them electrochemically "effective" and really close to the working electrode surface.

2.1. $[Ru(bpy)_3]^{2+}$ ECL Systems

Many ECL immunosensors with SPEs as electrochemical platform use ECL complex Tris(bipyridine) ruthenium (II) $[Ru(bpy)_3]^{2+}$, together with different co-reactants. The ECL signals are controlled by the greater or lesser steric hindrance of the ECL probe to reach the SPE surface. Usually, the antigen–antibody union generates a stiffer structure that causes a slight steric hindrance allowing the complex $[Ru(bpy)_3]^{2+}$ and the co-reactant to reach the SPE surface. As result, a higher ECL signal is obtained. In this sense, Rizwan et al. [64] have reported a label-free immunosensor for beta 2-microglobulin based on a composite material made of CdSe quantum dots and gold nanoparticles (AuNPs) decorated carbon nano-onions. The nanocomposite is employed together with the widely used co-reactant tri-n-propylamine (TPrA). CdSe quantum dots acts as an additional co-reactant, while gold nanoparticles decorated carbon nano-onions amplified the ECL signals due to their high conductivity. This immunosensor shows a really wide linear range from 1 fg/mL to 100 ng/mL.

In a completely opposite approach, ECL immunosensor response is based on the higher steric hindrance caused by addition of a secondary specific antibody in a sandwich configuration. This high protein electrode coating prevents the ECL probe or the co-reactant to reach the SPE, giving rise to the ECL signal decrease as the antigen concentration increases. Breast cancer antigen 15-3 (CA15-3) is determined from the ECL signal decrease on increasing the biomarker concentration [65].

The complex $[Ru(bpy)_3]^{2+}$ can be either in solution or immobilized on the electrode surface. Pu et al. [66] reported a composite nanomaterial including $[Ru(bpy)_3]^{2+}$ onto Pt nanoparticles, which is used to modify by dropcasting the working electrode surface. The developed immunosensor is specific to *Clostridium Perfringens*, an anaerobic bacillus that often causes gas gangrene. Anti-*C. Perfringens* antibody is immobilized on the working electrode surface, acting as capture antibody. As detection label, a composite nanomaterial based on type A antibody anti-*C. Perfringens* and Glucose dehydrogenase is used, being both of them adsorbed over ferrite nanoparticles covered by a gold layer. The ECL signal measurement is carried out in a solution containing gluconolactone and nicotinamide adenine dinucleotide (NAD), which is reduced to nicotinamide adenine dinucleotide hydride (NADH) and acts as co-reactant of $[Ru(bpy)_3]^{2+}$, generating ECL signals directly proportional to *C. perfringens* concentration. A great advantage of this ECL immunosensor is the analysis only requires 1 h, while other methods used for *C. perfringens* determination such as anaerobic culture (three days), real time PCR (2 h) or gel electrophoresis with fluorescent labels for DNA hybridization detection (9 h) are time consuming.

In other strategies, $[Ru(bpy)_3]^{2+}$ is covalently bounded to the detection antibody. Li, C. et al. [67], synthesized Ruthenium bis(2,2-bipyridine) (2,2-bipyridine-4,4-dicarboxylic acid) N-hydroxysuccinimide ester ($[Ru(bpy)_3]^{2+}$-NHS ester) that is bounded to different detection antibodies, through the reaction of N-hydroxysuccinimide with the carboxylic groups of the antibody structure, to develop a multiplexed immunoassay. Goat, rabbit and human immunoglobulins are attached to three of four electrodes on a screen-printed electrochemical array, maintaining one unmodified electrode as a control. In a competitive immunoassay, a solution containing the sample (consisting on different amounts of goat, rabbit and human immunoglobulins) and the different anti-immunoglobulins antibodies modified with $[Ru(bpy)_3]^{2+}$ is dropped over the four working electrodes, letting the specific recognition reaction takes place. Figure 2 summarizes the developed system. The ECL intensity decreases on increasing the concentration of immunoglobulins in the serum samples.

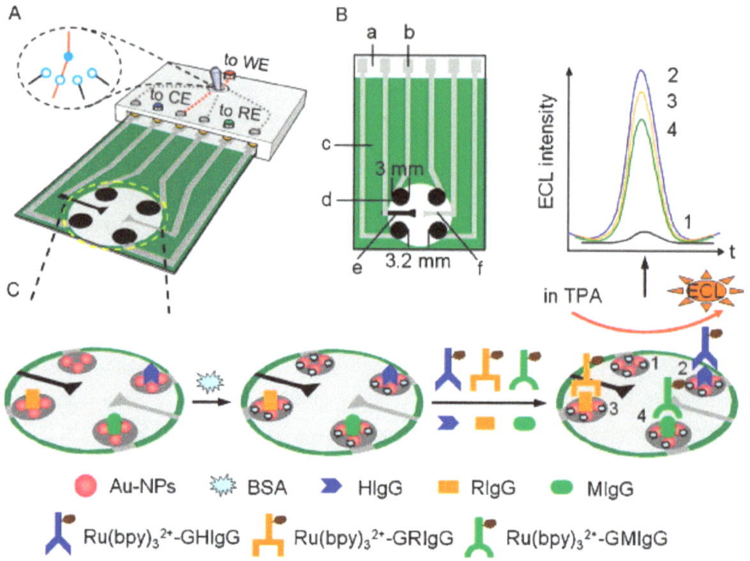

Figure 2. Schematic drawings of: (**A**) the immunosensor array for goat, rabbit and human immunoglobulins detection reported by Li et al. [67], held and controlled by a single-pore-four-throw switch; (**B**) SPE array containing four working electrodes, one reference electrode and one counter; and (**C**) immunosensor array preparation and immunoassay procedure. (**a**) Polyethylene terephthalate; (**b**) Ag conductive channels; (**c**) insulating layer; (**d**) carbon working electrode (WE); (**e**) carbon counter electrode (CE); and (**f**) Ag/AgCl reference electrode (RE). Reproduced and adapted with permission of Elsevier.

In a similar strategy, $[Ru(bpy)_3]^{2+}$ is linked to a streptavidin derivative to design an immunosensor for celiac disease diagnosis [68]. The immunosensor is specific to the celiac disease biomarker antitransglutaminase type-2 antibodies (anti-tTG). It consists of an ECL platform based on membrane-templated gold nanoelectrode ensembles (NEEs). In this work, an innovative sensing strategy is followed, based on the spatially separation of the initial electrochemical reaction and the immobilized biomolecules' location, where ECL light is generated. As a recognition element, the authors employed tissue transglutaminase (tTG) immobilized on the polycarbonate (PC) surface of the track-etched templating membrane. Transglutaminase antibody (anti-tTG) is retained on the surface, thanks to its reaction with tTG. This mechanism permits the immobilization of streptavidin-modified ruthenium complex luminophore via reaction with an appropriate biotinylated detection antibody. An oxidizing potential is applied when the electrodes are immersed in a tri-n-propylamine (TPrA) solution. The electrochemical oxidation of TPrA generates $TPrA^{*+}$ and $TPrA^{*}$ radicals that react with the $[Ru(bpy)_3]^{2+}$ ECL label, resulting in a great ECL signal, which is proportional to anti-tTG concentration with a linearity range between 1.5 ng/mL and 10 μg/mL and a detection limit of 0.5 ng/mL. A particular advantage of this strategy is that the ECL emission is obtained by applying a potential of 0.88 V versus Ag/AgCl, which is about 0.3 V lower than in the case when ECL is initiated by the electrochemical oxidation of $[Ru(bpy)_3]^{2+}$.

The use of nanomaterials to support $[Ru(bpy)_3]^{2+}$ is also widespread. Yang, H. et al. [52] developed an immunosensor to determine carbohydrate antigen 199 (CA199), a well-known tumor marker for early diagnosis of colon and pancreas cancer. AuNPs covered by $[Ru(bpy)_3]^{2+}$, just by adsorption, are linked to the detection antibody and used as label for ECL detection. The ECL signal is proportional to the logarithm of the CA199 concentration. An interesting feature of this immunosensor is the use of hand-drawn written pen-on-paper electrochemical platforms, which make it an affordable and easy way to prepare the device.

Among nanomaterials employed to support $[Ru(bpy)_3]^{2+}$ as ECL label, silica nanoparticles are very common. The immunosensor for whole cells of *Francisella tularensis* detection (responsible for tularemia infectious disease) is an illustrative example [69]. In this work, silica-encapsulated $[Ru(bpy)_3]^{2+}$ nanoparticles are synthesized and coated with polydiallyldimethylammonium. The detection antibody is covalently bounded through the carboxylic groups provided by addition of polyacrylic acid. ECL signals are directly proportional to *Francisella tularensis* concentration. This immunosensor shows a highly competitive result, with a limit of detection of 70 CFU/mL, whereas the reported limit of detection of the ELISA kit to determine this microbiological agent is 10^3 CFU/mL. Silica nanoparticles coated with $[Ru(bpy)_3]^{2+}$ bounded to the detection antibody have also been used in an immunoarray for different prostate cancer biomarkers (prostate specific antigen (PSA), prostate specific membrane antigen (PSMA) and platelet factor-4 (PF-4)) [70].

Nanocomposite materials using a combination of silica and gold nanoparticles bounded to $[Ru(bpy)_3]^{2+}$ have also been employed in different immunosensors, e.g., one developed for cell cancer detection [71] and the alpha-fetoprotein immunosensor based on a magnetic nanocomposite material developed by Gan, N. et al. [72] The advantages of using SiO_2 is they can load a large amount of $[Ru(bpy)_3]^{2+}$ while gold particles could provide large active surface to incorporate large amount of biological recognition element.

The use of a composite nanomaterial based on $[Ru(bpy)_3]^{2+}$@silica@AuNPs conjugated with the detection antibody has also been developed in combination with magnetic molecularly imprinted polymers (MMIP) as capture probes. Zhou, J. et al. [73] followed this strategy to construct a "single antibody sandwich" type immunosensor capable of ultratrace detection of hemoglobin. A similar strategy using molecularly imprinted polymers (MIP) instead of an antibody as capture probe has been used for carcinoembryonic antigen (CEA) and carbohydrate antigen-199 (CA199) immunosensors (Figure 3). As signal tag, Feng, X. et al. [74] employed a nanocomposite material based on $[Ru(bpy)_3]^{2+}$-silica@poly-L-lysine-AuNPs linked to the detection antibody.

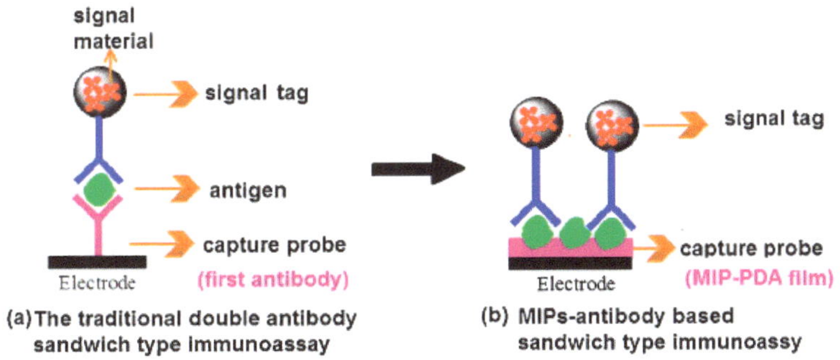

Figure 3. (**a**) The traditional double antibody sandwich type immunoassay; and (**b**) MIPs-antibody based sandwich type immunoassay developed by Feng, X. et al. [74]. Reproduced and adapted with permission of Elsevier.

Carbon nanomaterials are very common in the composite material used to prepare ECL immunosensor labels. Hou, J. et al. [75] employed a composite nanomaterial, based on multiwall carbon nanotubes, to adsorb the ECL probe $[Ru(bpy)_3]^{2+}$. The composite nanomaterial is them cover with AuNPs, which are used to link the detection antibody by adsorption. This new composite nanomaterial is used as ECL tag, amplifying the ECL signal and acting as immobilization substrates for detection antibody. Another example of the use of carbon nanomaterials is the mouse IgG (MIgG) immunosensor developed by Zhou, H. et al. [76]. The capture antibody (anti-MIgG) modified with the

ECL probe $[Ru(bpy)_3]^{2+}$ is covalently linked to the working electrode. The ECL signal is not affected by the reaction between the target analyte (MIgG) and the capture antibody; thus, they developed a competitive immunosensor. The authors prepared amorphous carbon nanoparticles (ACNPs), which have a quenching effect on the ECL signal of $[Ru(bpy)_3]^{2+}$. These ACNPs are modified with the MIgG (ACNPs-MIgG), so the unoccupied capture antibodies immobilized on the working electrode after incubation with the sample, react with ACNPs-MIgG, generating a decrease of the ECL signal when higher quantity of ACNPs-MIgG are retained. Therefore, the ECL signal is directly proportional to the concentration of MIgG at the sample. The method shows a linear range of 0.50–400 ng/mL and a detection limit of 0.35 ng/mL.

$[Ru(bpy)_3]^{2+}$ is widely used for anodic ECL, but some examples have demonstrated that, in combination with $S_2O_8{}^{2-}$, it can also be used for cathodic ECL. Fang, D. et al. [77] developed an immunosensor for thyroglobulin, an important biomarker for postoperative tumor recurrence or persistence in patients with thyroid cancer. The immunosensor is based on a TiO_2 nanodots modified electrode, where the capture antibody is immobilized. A bioconjugate formed by MXenes layers covered by $[Ru(bpy)_3]^{2+}$ and ZnO quantum dots, modified with the detection antibody are used as recognition element. The ECL signal is proportional to the logarithm of biomarker concentration. A specific characteristic of this immunosensor is the photothermal ECL amplification, which provides a low limit of detection and wide linear range. The amplification is due to ZnO quantum dots, which raises the electrode surface temperature by converting laser energy into heat.

2.2. Luminol ECL Systems

Luminol (2,3-aminophthalhydrazide) is a widely used ECL luminophore on immunosensor development. It has been used for anodic ECL together with oxidant species as co-reactant such as H_2O_2, O_2, $C_2O_4{}^{2-}$, etc. Luminol and its co-reactant can be used in solution, as in the case of the human immunodeficiency virus (HIV) type 1 antibody immunosensor developed by Zhou, J. et al. [78]. The immunosensor is based on the use of MMIP as capture probes. Human immunoglobulin G (HIgG) is used as the template for the MIP, as it exhibits the same antibody Fc region but a different Fab region as the anti-HIV-1. MMIP recognize the anti-HIV-1 present in the sample. The conjugated horseradish peroxidase-antigen (HRP-HIV-1) is used as label. HRP catalyze the oxidation of luminol in the presence of H_2O_2, generating an increase of ECL signal when higher amounts of anti-HIV-1 are retained on the MMIP particles. To generate the electrochemical required process, carbon SPEs are used together with a magnet under the working electrode, in order to transfer the modified MMIP to the electrode surface. A linear range of the anti-HIV-1 dilution ratio (standard positive serum) was obtained from 1:20,000 to 1:50, with a detection limit of 1:60,000. The developed method provides a low-cost, simple and sensitive way for the early diagnosis of HIV infection.

Luminol can also be covalently attached to detection antibodies. In the alpha-fetoprotein (AFP) immunosensor developed by Su, M. et al. [79], macroporous Au-paper working electrodes (Au-PWE) are modified with the capture antibody (anti-AFP), which traps the breast cancer cells (MCF-7) through the AFP molecules of their membrane. The labeled detection antibody (anti-AFP-Luminol) interacts with the trapped MCF-7 cells. The anodic ECL signal obtained, using H_2O_2 as co-reactant, is directly proportional to the logarithm concentration of AFP present over MCF-7 surface in the range of 0.01 until 200 ng/mL. In another study, carcinoma antigen 125 (CA125) is detected in clinical serum samples using carbon ink over previously wax-patterned cellulose paper to fabricate disposable SPEs [80]. As illustrated in Figure 4, after electrode modification with adsorbed gold nanoparticles (AuNPs), L-cysteine is chemisorbed through the thiol group. 1-ethyl-3-(3-dimethylaminopropyl) carbodiimide/N-hydroxysuccinimide (EDC/NHS) activated carboxylic groups of chemisorbed L-cysteine allow covalent link of anti-CA125 capture antibody through the reaction of amino groups from Fc antibody region with the activated carboxylic acid of L-cysteine. As ECL label, anti-CA125 secondary antibodies are modified with covalently linked luminol-AuNPs [81]. The labeled antibody is retained by the CA125 antigen previously attached over the working electrode surface. The ECL measurement is carried out using a cathodic scan and the signal increases proportionally to the logarithm of CA125 concentration.

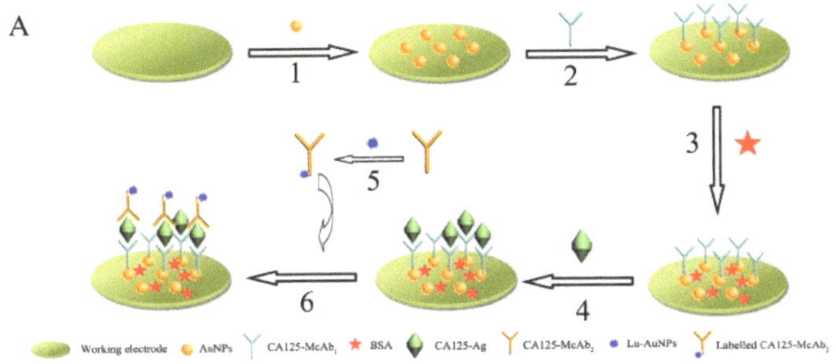

Figure 4. Schematic representation of the fabrication of the ECL CA125 immunosensor and assay procedure described by Wang, S. et al. [80]. Reproduced and adapted with permission of Elsevier.

A label-free ECL immunosensor for transferrin (TRF) has been developed using luminol-reduced gold nanoparticles (Lu-Re-AuNPs) modified carbon SPE [82]. Lu-Re-AuNPs composite is prepared in a chitosan polymeric film. AuNPs are used to adsorb Anti-TRF antibodies. Finally, the immunosensor is incubated with the sample containing TRF. The ECL signal is measured in anodic scan, using a carbonate solution buffer containing as co-reactant H_2O_2. The ECL signal decreases linearly with TRF concentration, as nonconductive proteins (as TRF) may interrupt the interfacial electron transfer and hinder the diffusion of the electrochemically active molecules. This method shows better sensitivity than other label-free immunoassays, such as immune precipitation and immune turbidity. It is also easily manipulated, affordable and the recognition activity of the immobilized antibody is highly stable.

2.3. $S_2O_8{}^{2-}$ ECL Systems

$S_2O_8{}^{2-}$ is the most used co-reactant employed on immunosensor development together with semiconductor nanomaterials responsible of ECL emission. In this case, the ECL signal is obtained at the cathodic scan. The $S_2O_8{}^{2-}$ is reduced to $SO_4{}^{*-}$ and $SO_4{}^{2-}$; $SO_4{}^{*-}$ reacts with the excited form of the semiconductor nanomaterial, which consequently falls down to the ground state generating the ECL signal. $S_2O_8{}^{2-}$ is usually in the solution, while the semiconductor material can be attached to the electrochemical platform or be part of the ECL label used as detection element.

Accordingly, Wu et al. developed a label-free immunosensor for determination of the tumor marker carbohydrate antigen 125 (CA125) [83]. It is based on $S_2O_8{}^{2-}$ as co-reactant and the semiconductor graphite-like carbon nitride (g-C_3N_4) modified with carboxylic groups and immobilized on the electrochemical platform. Aminated Fe_3O_4 nanoparticles and anti-CA125 (capture antibody) are linked to g-C_3N_4 through the reaction between activated carboxylic groups of g-C_3N_4 and amine groups of nanoparticles and antibody. As authors explained that carboxylated g-C_3N_4 transfers electrons from its conduction band to Fe_3O_4 nanoparticles, preventing carboxylated g-C_3N_4 from electrochemical degradation and simultaneously catalyzing the reduction of $S_2O_8{}^{2-}$ into $SO_4{}^{*-}$, which leads to the enhancement of ECL emission of g-C_3N_4.

Other strategies using $S_2O_8{}^{2-}$ as co-reactant are based on the use of semiconductor nanomaterials as label of the recognition event between the antigen and a detection antibody. Zhang, M. et al. [84] used semiconductor carbon nanocrystals (CNCs) as ECL transmitter for a PSA immunosensor (Figure 5). CNCs are also modified with PtAg alloy, generating a composite nanomaterial (PtAg@CNCs), which is linked with the detection antibody (anti-PSA). The resulting ECL label reacts with the PSA antigen, previously attached by the capture antibody to the working electrode. In a first step, carbon SPE is modified with carbon nanotubes covered with chitosan polymer and AuNPs (CNT-CHIT/AuNPs composite). The capture antibody is immobilized onto the CNT-CHIT/AuNPs. In a sandwich immunosensor configuration, $S_2O_8^{2-}$ is electroreduced due to the electrocatalytic activity of PtAg alloy, generating SO_4^{*-} radicals that react with CNCs electrochemically excited by electroreduction, transmitting an ECL signal directly proportional to the PSA concentration logarithm.

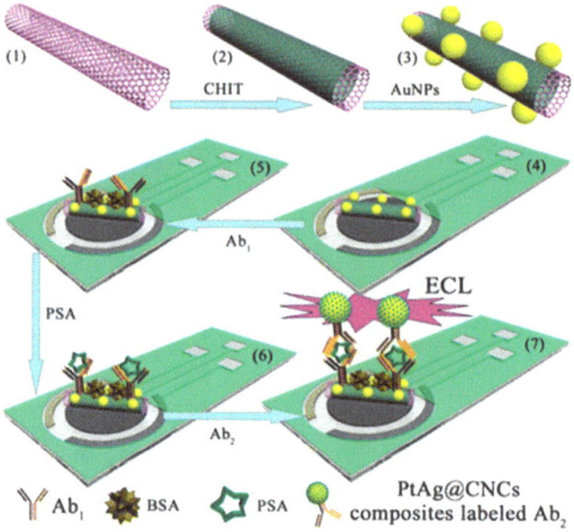

Figure 5. Schematic representation of the fabrication of the ECL immunosensor for PSA, developed by Zhang, M. et al. [84]: (**1**) CNT; (**2**) CNT-CHIT; (**3**) CNT-CHIT/AuNPs composite; (**4**) CNT-CHIT/AuNPs composite modified SPCEs; (**5**) after immobilization of Ab_1; (**6**) capture with PSA; and (**7**) immobilization with the PtAg@CNCs composites labeled Ab_2. Reproduced and adapted with permission of Royal Society of Chemistry.

Quantum dots have also been employed as ECL labels as they can behave as ECL transmitter, due to their semiconductor properties. In this way, CdTe quantum dots supported on porous silver nanoparticles (QDs/AgNPs) are used as ECL transmitter together with the co-reactant $S_2O_8^{2-}$ [85]. Silver nanoparticles act as $S_2O_8^{2-}$ reduction electrocatalyst, amplifying the ECL signal of the immunosensor designed for the determination of a tumor marker, carcinoembryonic antigen (CEA). The detection antibody (anti carcinoembryonic antigen) is covalently linked to QDs/AgNPs, acting as an ECL label of the sandwich-type immunosensor. It is interesting that the SPE electrochemical platform is based on a microfluidic origami device fabricated directly on wax-patterned cellulose paper, which make it a really affordable system. It is additionally modified with graphene nanosheets to covalent bond the capture antibody. The ECL signal is proportional to the CEA concentration logarithm.

Another similar development described by Zhang, Y. et al. is based on the use of carbon SPE modified with graphene nanosheets previously modified with silver and gold nanoparticles [86]. This composite nanomaterial acts as support of the capture antibody specific to CA 125 and facilitates the electron transfer. CdTe quantum dots coated carbon microspheres (QD@CM) are used as ECL label. QD@CM composite are modified with the detection antibody using EDC/NHS. The bioconjugate QD@CM/Ab_2 acts as ECL label of the sandwich-type immunosensor generating an ECL signal directly proportional to the CA 125 concentration logarithm. This tumor biomarker immunosensor has been applied to serum samples, obtaining results in agreement with those obtained with the reference method (commercially available Electrochemiluminescent Analyser ROCHE E601, Switzerland) and relative standard deviations about 2.84–7.31%.

The ECL behavior of semiconductor SnO_2, with lower biotoxicity and better stability, has seldom been investigated. Ma, C. et al. [87] reported a carcinoembryonic antigen (CEA) immunosensor, where SnO_2 nanocrystal (NC) is applied as a novel ECL signal reporter. Carbon SPEs are modified by dropcasting with nanoporous silver (obtained by dealloying method from $Ag_{23}Al_{77}$ alloy foils). The capture antibody is adsorbed on the resulting modified electrode to link CEA molecules present in the sample. Herein, SnO_2 nanocrystals are combined with PtRu alloy, which enhances the ECL signal. For this porpoise, SnO_2 is first functionalized with amino groups by reaction with (3-Aminopropyl) triethoxysilane and then bonded to PtRu alloy nanoparticles, being the amino groups coordinated with Pt atoms. Finally, the nanocomposite materials are functionalized with the detection antibody, by coordination of amino group of the antibody with the Pt atom. This nanocomposite material is used as ECL probe through the reaction of SO_4^{*-} radicals generated with reduced SnO_2 nanocrystals. Herein, the reduction of $S_2O_8^{2-}$ to SO_4^{*-} radicals is electrocatalyzed by PtRu alloy, increasing the ECL signal linearly with CEA concentration. The immunosensor has a limit of detection as low as 0.72 pg/mL, which is lower than of the other methods.

In a similar approach, ZnO quantum dots are employed for PSA determination [88]. Pt/AuNPs are previously electrodeposited on the carbon SPE. In a sandwich-type configuration, the capture antibody is attached to the Pt/AuNPs. As recognition label, the authors used a composite nanomaterial consisting on ZnO quantum dots synthesized on carbon nanotubes (ZnO@CNT) that is linked to the specific detection antibody. The ECL signal is obtained in presence of $S_2O_8^{2-}$, which generates SO_4^{*-} that reacts with reduced ZnO^{*-} emitting an ECL signal directly proportional to the concentration of PSA.

2.4. Other ECL System

Over recent years, besides the traditional ECL systems described above, new ECL systems based on nanomaterials have been used to develop immunosensors based on SPEs. Most of these alternative systems are based on the use of quantum dots as ECL emitters. Wang, S. et al. [89] developed a CEA immunosensor based on indium tin oxide (ITO) SPE. As shown in Figure 6, gold-coated magnetic iron nanoparticles (Fe@Au MNPs) are employed as solid support to CEA primary monoclonal antibody (CEA-$McAb_1$). CdTe quantum dots (QDs) are used to label CEA-secondary antibodies (CEA-$McAb_2$). By the help of a magnet, the sandwich-type immunoassay that takes place over the magnetic particles is transferred to the ITO SPE surface. The ECL signal is obtained during the anodic scan, when CdTe quantum dots are oxidized (CdTe($hole^+$)), being the electron rapidly transferred into an O_2 molecule dissolved in the buffer. The oxygen radical generated (O_2^{*-}) react with CdTe quantum, generating CdTe(e^-), which reacts with CdTe($hole^+$) emitting ECL. This ITO ECL device is an interesting alternative to ceramic- and polyethylene terephthalate (PET)-based carbon SPEs. ITO as transparent material improves the transmittance of ECL emission, enhancing the sensitivity.

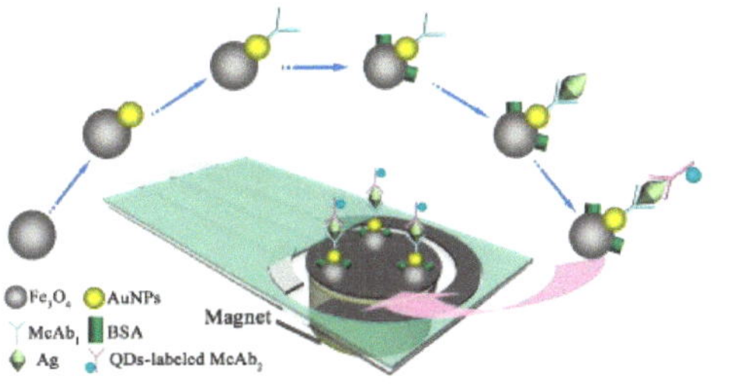

Figure 6. Schematic representation of the CEA ECL immunosensor developed by Wang, S. et al. [89]. Reproduced and adapted with permission of Elsevier.

A CEA immunosensor was also developed by Li, L. et al. [90] using in this case Au@Pt core-shell nanoparticles modified with graphene quantum dots (GQDs) as supporting for the detection antibody labeled with Glucose oxidase (GOx). The immunosensor is developed using nanoporous gold/chitosan modified paper electrode (NGC-PWE) as sensor platform. The ECL signal is based on the detection of H_2O_2 generated in the enzymatic reaction, which is instable and is transformed into (O_2^{*-}). This oxygen specie reacts with electrochemically oxidized graphene quantum dots ($GQDs^{*+}$), generating excited

GQDs* that emits an ECL signal when it returns to its fundamental state. The ECL intensity is proportional to the logarithm of CEA concentration. NGC-PWE improves electron transfer and allows the immobilization of a high number of antibodies. The loading of GQDs on Au@Pt gives rise to an amplified ECL signal and to the consequent improving of the sensitivity.

Carbon dots are new nanomaterials that are being used in ECL immunosensor. Liu, W. et al. [91] developed a CA125 immunosensor employing porous Ag-paper working electrode (Ag-PWE) to immobilize the capture antibody. In a sandwich-type configuration, the detection antibody is bounded to carbon nanodots modified nanoporous silica nanoparticles (MSNs). The ECL signal is obtained in the presence of triethylamine (TEA). At anodic scan, carbon nanodots and TEA are oxidized, generating radical species that react between them, generating excited carbon dots that emit ECL signal when they return to the ground state. The immunosensor present a wide linear range from 0.01 to 50 U/mL with a detection limit of 4.3 mU/mL.

Among others ECL luminophores, phenyleneethynylene derivatives because of its chemical stability and high efficiency of luminescence, have drawn intense attention in immunosensor field. A carcinoembryonic antigen (CEA) immunosensor has been developed using phenyleneethynylene derivatives ((4,4′-(2,5-dimethoxy-1,4-phenylene) bis(ethyne-2,1-diyl) dibenzoic acid (P-acid)) modified with mesoporous Pt–Ag alloy nanoparticles (P-acid/Pt–AgANPs) [92]. The working electrode is previously modified with graphene and a dense gold layer is formed through the growth of AuNPs on the graphene, which in turn will be beneficial to the sensitivity, stability and effective surface area of the SPE. Anti-CEA capture antibodies are immobilized onto the Au-Graphene layer. Finally, the P-acid/Pt–AgANPs composite is applied to the immobilization of a secondary antibody. According to sandwiched immunoreactions, once the sandwich-type immunosensor is incubated in the sample, CEA reacts with capture antibodies and the secondary antibody labeled with the composite nanomaterial, obtaining the ECL signal as the potential is scanned toward anodic values in a solution containing TEA. P-acid and TEA are oxidized during anodic scan, generating radical species that react between them. The ECL signal generated is directly proportional to the CEA concentration logarithm. This ECL immunosensor has two advantages: (1) the gold/graphene layer provides an effective antibody immobilization matrix with high stability and bioactivity; and (2) the P-acid/Pt–AgANPs system improves the ECL signal, obtaining better sensitivity.

ECL Immunosensors based on the use of SPE technology described above and their analytical properties are summarized in Table 1.

Table 1. Analytical parameters of SPE-based ECL Immunosensors.

Luminophore	Analyte	Sensing Type	Detection Limit	Reference
$[Ru(bpy)_3]^{2+}$	beta 2-microglobulin	Direct immunosensor-Label-free immunosensor CdSe QDs-Co-reactant AuNPs decorated carbon nano-onions- amplified the ECL signals TPrA-Co-reactant	1 fg/mL	[64]
$[Ru(bpy)_3]^{2+}$	*Clostridium Perfringens*	Sandwich type immunosensor Pt Nps as luminophore support Ferrite particles support detection antibody and Glucose Dehydrogenase	10^2 CFU/mL	[66]
$[Ru(bpy)_3]^{2+}$	human, rabbit and Goat immunoglobulins	Competitive immunoassay Detection antibody covalently bonded to $[Ru(bpy)_3]^{2+}$	2.9, 6.1 and 6.5 ng/mL	[67]
$[Ru(bpy)_3]^{2+}$	antitransglutaminase type-2 antibodies	Indirect immunosensor Biotinylated Detection antibody linked to streptavidin-$[Ru(bpy)_3]^{2+}$	0.47 ng/mL	[68]
$[Ru(bpy)_3]^{2+}$	carbohydrate antigen 199	Sandwich type immunosensor Detection antibody bonded to AuNPs covered by $[Ru(bpy)_3]^{2+}$	0.0055 U/mL	[52]
$[Ru(bpy)_3]^{2+}$	*Francisella tularensis*	Sandwich type immunosensor silica-encapsulated $[Ru(bpy)_3]^{2+}$ bonded to detection antibody	70 CFU/mL	[69]
$[Ru(bpy)_3]^{2+}$	prostate specific antigen (PSA), prostate specific membrane antigen (PSMA) and platelet factor-4 (PF-4)	Sandwich type immunosensor Silica nanoparticles coated with $[Ru(bpy)_3]^{2+}$ bounded to the detection antibody	300–500 fg/mL	[70]
$[Ru(bpy)_3]^{2+}$	MCF-7 cell	Direct immunosensor Silica and gold nanoparticles bounded to $[Ru(bpy)_3]^{2+}$	30 Cell/mL	[71]
$[Ru(bpy)_3]^{2+}$	alpha-fetoprotein	Sandwich type immunosensor Silica and gold nanoparticles bounded to $[Ru(bpy)_3]^{2+}$ Biotinylated capture antibody linked to streptavidin-coated magnetic particles	0.02 ng/mL	[72]

Table 1. *Cont.*

Luminophore	Analyte	Sensing Type	Detection Limit	Reference
$[Ru(bpy)_3]^{2+}$	hemoglobin	Molecularly imprinted polymer- antibody Sandwich type immunosensor $[Ru(bpy)_3]^{2+}$@silica@AuNPs conjugated with the detection antibody magnetic molecularly imprinted polymers	0.023 pg/mL	[73]
$[Ru(bpy)_3]^{2+}$	carbohydrate antigen-199 and carcinoembryonic antigen	Molecularly imprinted polymer- antibody Sandwich type immunosensor $[Ru(bpy)_3]^{2+}$@silica@AuNPs conjugated with the detection antibody	0.01 U/L and 0.02 pg/mL	[74]
$[Ru(bpy)_3]^{2+}$	alpha-fetoprotein	Sandwich type immunosensor $[Ru(bpy)_3]^{2+}$-MWCNT-AuNPs-Detection antibody Fe3O4@Au nanoparticles linked to capture antibody	3 pg/mL	[75]
$[Ru(bpy)_3]^{2+}$	mouse IgG	Competitive immunosensor amorphous carbon nanoparticles with quenching effect covalently linked to mouse IgG	0.35 ng/mL	[76]
$[Ru(bpy)_3]^{2+}$	thyroglobulin	Sandwich type immunosensor TiO_2 nanodots modified electrode, where the capture antibody is immobilized. Bioconjugate formed by MXenes layers covered by $[Ru(bpy)_3]^{2+}$ and ZnO quantum dots, linked to the detection antibody	1 fg/mL	[77]
Luminol	human immunodeficiency virus type 1 antibody (HIV-1)	Competitive immunosensor Molecularly imprinted polymer as capture probe Horse radish peroxidase-HIV-1	1:60,000 dilution ratio of standard positive serum	[78]
Luminol	alpha-fetoprotein	Sandwich type immunosensor Luminol groups adsorber over AuPd NPs linked to detection antibody	0.005 ng/mL	[79]
Luminol	carcinoma antigen 125	Sandwich type immunosensor Luminol labeled detection antibody	0.0074 U/mL	[80]

Table 1. *Cont.*

Luminophore	Analyte	Sensing Type	Detection Limit	Reference
Luminol	transferrin (TRF)	Label-free immunosensor Enhacement of luninol ECL when TRF is attached to the capture antibody	0.033 ng/mL	[82]
Semiconductor graphite-like carbon nitride/$S_2O_8^{2-}$ co-reactant	carbohydrate antigen 125	Label-free immunosensor Carboxilate g-C_3N_4/aminated Fe_3O_4/capture antibody	0.4 mU/mL	[83]
Semiconductor carbon nanocrystals (CNCs)/$S_2O_8^{2-}$ co-reactant	prostate specific antigen (PSA)	Sandwich type immunosensor PtAg@CNCs linked to detection antibody	0.6 pg/mL	[84]
CdTe quantum dots/$S_2O_8^{2-}$ co-reactant	carcinoembryonic antigen	Sandwich type immunosensor Detection antibody linked to CdTe quantum dots/AgNPs	2.5 mU/mL	[86]
SnO_2 nanocrystal/$S_2O_8^{2-}$ co-reactant	carcinoembryonic antigen	Sandwich type immunosensor SnO_2@PtRu linked to detection antibody	0.72 pg/mL	[87]
ZnO quantum dots/$S_2O_8^{2-}$ co-reactant	prostate specific antigen (PSA)	Sandwich type immunosensor ZnO@CNT linked to detection antibody	0.61 pg/mL	[88]
CdTe quantum dots (QDs)	carcinoma embryonic antigen	Sandwich type immunosensor CdTe QDs linked to detection antibody Capture antibody linked to AuNPs@Fe_3O_4 particles.	0.38 pg/mL	[89]
graphene quantum dots (GQDs)/H_2O_2 as co-reactant	carcinoma embryonic antigen	Sandwich type immunosensor Au@Pt core–shell nanoparticles modified with graphene quantum dots (GQDs)	0.6 pg/mL	[90]
Carbon dots/TEA as co-reactant	cancer antigen 125	Sandwich type immunosensor Detection antibody linked to carbon nanodots modified nanoporous silica nanoparticles	4.3 mU/mL	[91]
phenyleneethynylene derivatives (P-acid)/TEA as co-reactant	carcinoembryonic antigen	Sandwich type immunosensor P-acid/Pt–AgANPs linked to detection antibody	0.3 pg/mL	[92]

3. ECL Enzymatic Biosensors

Enzymes are proteins that catalyze chemical reactions in living organisms, behaving as highly selective catalysts that allow recognition of a wide variety of substrates of analytical interest. Therefore, enzymes immobilized on the electrode surfaces have been widely employed for the development of biosensors.

In general, enzyme immobilization methods are often classified in two broad categories: chemical bonding and physical retention. The covalent binding of enzymes to supports and cross-linking are the more prominent methods by chemical bonding [93]. The adsorption and entrapment of enzymes in porous substrates or the confinement of enzymes in semipermeable membranes are the main methods of immobilization by physical retention [94]. The most popular methods to build biosensors are cross-linking [95] and entrapment [96].

Among other factors, the choice of immobilization method must take into account the type of substrate that has to be processed and the conditions of the reaction. Another factor to consider in the case of ECL biosensors is the luminophore, which can be trapped together with the enzymes on the electrode surface or found in solution. Three main luminophores are used in enzyme biosensor such as $[Ru(bpy)_3]^{2+}$, luminol [97] and quantum dots (QDs) [98], being luminol the most used in oxidase enzyme biosensors. The light emission of luminol at the electrodes was first used in 1929 [99]. Luminol ECL is usually produced in alkaline solution with H_2O_2, when the anodic electrochemical oxidation of the electrogenerated excited monoanionic form of 3-aminophthalic acid occurs at a carbon or Pt electrode. The conditions under which ECL takes place are 0.6 V vs. saturated calomel electrode (SCE) and λ_{ECL} = 420 nm [100]. Because many oxide reductase enzymes can produce H_2O_2 during their catalytic activity, ECL enzyme biosensors can be carried out by coupling the light-emitting reaction produced by luminol with enzyme-catalyzed reactions that generate H_2O_2. Hence, luminol and its derivatives have been coupled with oxidase enzymes for the generation of ECL [97].

Clark and Lyons developed the first enzymatic biosensor in the 1960s, by immobilizing Glucose oxidase (GOx) on a platinum electrode to determine glucose. It revealed the possibility of enzyme immobilization on an electrode, resulting in enzyme-based amperometric biosensors that allow to detect electroactives species involved in the enzymatic reaction. Years later, Wilkinsont, J. S. developed an enzymatic ECL biosensor [101]. The glass waveguide is coated with indium tin oxide and modified with covalently attached GOx. The range of detection for glucose obtained was 0–10 mM with a detection limit of 0.3 mM. Since GOx is well-known and commercially available enzyme and diabetes mellitus is a non-infectious disease that has proliferated throughout the world, this enzyme is one of the most employed in biosensor development. Although it is not an acute disease, it can lead to complications such as kidney disease, blindness and foot disease or nerve damage. Tracking the diabetes related indices such as blood glucose concentration are very important to control and disease treatment. Therefore, enzymatic glucose biosensors have been a great deal of interest and new designs focus on the improvement of the sensitivity and selectivity have been described. ECL biosensors are one of the recent designs and GOx are the standard enzymes used for the majority of them. For this purpose, researchers have developed different strategies for the immobilization of GOx in the different supports. A sol–gel derived ceramic carbon composite electrode was used to develop a new type of biosensor based on luminol ECL for glucose [102]. The detection limit obtained was 8.16 mM. Guonan Chena et al. encapsulated GOx [103] in the film composed of Nafion and carbon nanotubes (CNTs). They reached an excellent electrocatalytic activity toward luminol, obtaining a detection limit of 2.0 μM glucose. Liu, L., et al. designed an ECL biosensor based on CdTe QDs and AuNPs [104]. The limit of detection for glucose was 5.28 μM. Other extremely sensitive glucose ECL biosensors were developed, where the authors utilized a membrane of poly(diallyldimethylammoniumchloride) doped with chitosan to immobilize GOx [105], obtaining a detection limit of 0.1 nM. Another combination of nanomaterials employed to fabricated a 3D platform was developed by electropolymerization of nickel (II) tetrasulfophthalocyanine (NiTSPc) on MWCNTs-modified glassy carbon electrode [106]. The detection limit obtained was 8.0×10^{-8} M. A high-performance three-dimensional 3D bio-platform of GOx adsorbed on AuNPs assembled polyaniline nanowires network was constructed. The ECL biosensor enabled the sensitive detection of glucose by an extremely low detection limit of 0.05 μM [107].

ECL Enzymatic Biosensors Based on SPEs

As can be inferred from the approaches described above, several glucose ECL biosensors have been developed so far using conventional electrodes. SPEs are recently employed as they can provide additional advantages, in particular to fabricate disposable biosensors, making them important tools for the development of POC biomedical applications. In this sense, Cheng, L. et al. [108] developed a ECL biosensor immobilizing GOx and surface-unpassivated CdTe QDs on SPE (Figure 7). The glucose biosensor showed rapid response. The biosensor linear range extends from 0.8 to 100 mM, and the detection limit of it is around 0.3 mM. The detection of glucose in real human serum samples demonstrated acceptable sensitivity and selectivity. Yu, L. et al. [109] developed another disposable glucose ECL biosensor using Au and TiO_2 nano-composite. GOx was cross-linked with BSA thanks to the use of glutaraldehyde, while Nafion is used for enzyme immobilization on the ITO surface. The biosensor has a linear range for the detection of glucose from 7.0 to 100 mM with a detection limit of 0.22 mM.

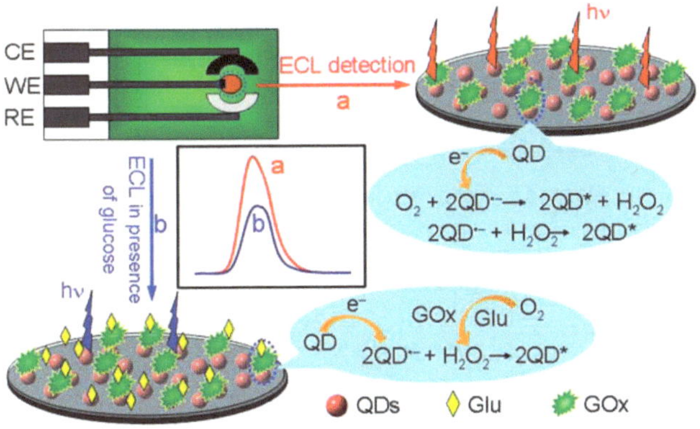

Figure 7. Schematic display of the ECL glucose biosensor developed by Cheng, L. et al. [108] employing bidentate-chelated CdTe QDs as emitters of ECL for glucose detection. Reproduced and adapted with permission of Royal Society of Chemistry.

Uricase is an enzyme that catalyzes the oxidation of uric acid to 5-hydroxyisourate, which subsequently forms (S)-allantoin. It is involved in purine metabolism, urate degradation and (S)-allantoin formation. Its main use is the determination of uric acid in biological fluids. Hence, it is important for the development of disposable biosensors. Uricase has been immobilized for the fabrication of a uric acid ECL biosensor using two different strategies: trapped in a pyrrole matrix [110] or in a double-layer design of luminol as a copolymer with 3,3,5,5-tetramethylbenzidine (TBM) and chitosan on gold SPE [111]. The fabricated biosensor showed a sensitive response to uric acid with a low detection limit of 4.4×10^{-7} M and wide linear response (from 1.5×10^{-6} to 1.0×10^{-4} M). The biosensor turned out to be suitable for uric acid determination in 24-h urine samples compared to a reference procedure.

Another enzyme of interest is Lactate oxidase (LOx). Lactate is a chemical compound that plays important roles in various biochemical processes and is found in various biological fluids. Measurements of blood lactate concentrations are common in exercise and in clinical settings, as they can reveal information about the participant's fitness. Blood lactate measurement is used as a marker of exercise intensity and training status. Therefore, methods with high sensitivity and good selectivity for the fast and dynamic lactate concentration response of lactate in human serum or other body fluids are urgently demanded. Lox-based biosensors for the determination of lactate in serum have been reported [112]. An ECL biosensor based on LOx and luminol with a higher sensitivity has been developed for the determination of this analyte in sweat [113]. Recently, Ballesta-Claver, J. et al. [114] developed a disposable lactate ECL biosensor based on LOx and luminol immobilized on a graphite SPE with Methocel membrane. The biosensor was applied to the analysis of lactate in saliva as an alternative procedure for obtaining the lactate level in a non-invasive way.

Other ECL enzyme biosensors have been developed so far, immobilizing different oxide reductase enzymes, such as cholesterol oxidase [115], glutamate oxidase [116], glycosylated hemoglobin [117], choline oxidase [118] and alcohol dehydrogenase [119], on conventional electrodes such as indium tin oxide glass, carbon and glassy carbon electrodes. Therefore, the possibility of developing SPE based disposable biosensors, benefiting from the advantages provided by this type of electrodes is open, as discussed in the Introduction.

Besides, screen printing technology allows the fabrication of platforms with several working electrodes that share auxiliary and reference electrodes. Different enzymes can be immobilized on these devices to the development of selective and disposable biosensors for multiple analysis. Blum, L.J. et al. [120] were the first to describe an SPE multi-parametric ECL biosensor, based on ECL detection principles of enzymatically produced H_2O_2. Years later, GOx and LOx were trapped in PVA–SbQ (poly(vinyl alcohol), bearing styrylpyridinium groups photopolymer deposited on the surface of SPE array (Figure 8) [121]. The achieved multiple biosensor allowed the simultaneous detection and quantification of lactate and glucose with detection limits of 3 and 10 µM, respectively. These results demonstrate the analytical possibilities of SPE arrays to produce multi-parameter biosensors, based on ECL detection.

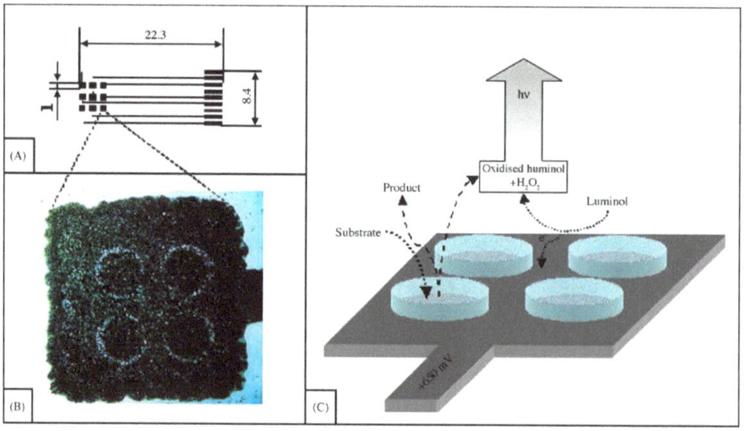

Figure 8. SPE multi-parametric glucose and lactate ECL Biosensor developed by Corgier, B.P. et al. [121]. (**A**) Schematic representation of the nine screen-printed array for ECL measurements. Size in mm. (**B**) Optical micrograph of the electrode surface (1 mm × 1 mm) supporting the four spots sensing layer. (**C**) Schematic representation of the sensing layer organization which is formed by a hybrid layer of photopolymer-Sepharose beads PVA-SbQ and reaction at a carbon electrode surface (1 mm × 1 mm). Reproduced and adapted with permission of Elsevier.

Enzymatic biosensors have many advantages compared to conventional analysis techniques, such as high specificity provided by the biological element, high sensitivity, short analysis times and the possibility of reusing the biological component, which, together with the advantages that ECL presents, makes enzyme ECL biosensors excellent candidates to continue investigating in this area.

In recent years, a huge boost has been observed in areas such as nanotechnology and nanoscience, being important the advances obtained in biomedicine [122], energy [123], catalysis [124] and electronic [125]. To a large extent, it has been due to the development and incorporation of new nanomaterials. This advance has also benefited Analytical Chemistry. In particular, in the field of ECL enzymatic biosensors, nanomaterials have provided significant improvements in the performance of these devices. By coupling nanomaterials and ECL, the signal-to-noise ratio increases compared with photoluminescence, since light scattering side effects are minimized. The development of new systems and strategies based on nanomaterials have allowed the determination of new analytes, even in very complex matrices [126]. Therefore, researchers have investigated the improvements that nanomaterials produce in ECL (bio)sensors.

In general, ECL enzymatic biosensors based on nanomaterials show a higher selectivity, sensitivity and stability. Furthermore, nanomaterials have provided improvements in the catalytic properties, by increasing the electronic transfer between the co-reactant and luminophores, as well as the conductivity, which lead to a high ECL response. In addition to the advantages explained above, nanomaterial have a great biocompatibility with enzymes, since they are in the same size range. Thus, they can form bioconjugates with synergistic properties. Finally, nanomaterials provide a large surface area, allowing the immobilization of a large quantity of enzyme molecules, as we shown in more detail below.

There are many nanomaterials used now for the development of enzymatic biosensors and their nature is broad, such as metallic nanoparticles of silver (AgNPs) [127] and gold (AuNPs) [128]; carbon nanomaterials as carbon nanodots (CNDs) [112], graphene [129] and carbon nanotubes (CNTs) [130]; and biochar [131]. Semiconductors such as quantum dots (QDs) [132] and silicates [133] are also investigated. All these nanomaterials can be deposited on the SPE and conventional electrodes surface for using as sensor platforms. There are different techniques for incorporating nanomaterials such as electrodeposition, as is the case of metallic nanoparticles [134] and CNDs [135], where more controlled coverage and reproducibility are obtained, through electrostatic interactions or formation of covalent bonds [93], click chemistry [136] with modified surfaces, electrospray deposition [137] and dropcasting [138], being the last one the most used, simplest and quickest way to prepare a sensing surface.

Some of the most used QDs in ECL biosensors are CdTe, CdSe, CdS and graphene quantum dots. These nanomaterials can act by themselves as luminophores generating an increase in the ECL signal, due to its good electronic, optical and photophysical properties. CdTe QDs have been used for the development of a disposable carbon SPE glucose biosensor [108]. In addition, these nanomaterials are widely applied in the design of other biosensors such as ECL immunosensors [132]. The QDs can also be managed together other semiconductors to gain synergetic system at the electrode where they can work as ECL enhancer since they act as efficient emitters of ECL [139]. ZnO is another kind of semiconducting nanomaterial that improves the ECL sensor response, which has been deposited on carbon and gold SPE for the determination of glyphosate [140] and taurine [141], respectively.

CNDs are water-soluble carbon nanomaterials with low toxicity and facile functionalization. Due to core-related quantum confinement effects of CNDs structure and edge effects near surfaces, they show electro-optical properties. These advantages make the CNDs suitable nanomaterial for the development of ECL biosensors, producing a response increase. CNDs have disclosed interesting features as nanomaterials for photoinduced electron transfer and ECL. In the last one, CNDs have demonstrated their ability to act as co-reactants [142], which is why this nanomaterial is widely used in the development of ECL biosensors [143].

AuNPs are characterized by, in addition to their stability and large surface area, excellent conductive properties, which allow improving the ECL responses by increasing their intensity. AuNPs can be trapped with hydrolyzed (3-aminopropyl) trimethoxysilane as linker to the electrode ITO surface [144] or deposited directly on the electrode surface [145]. AuNPs have not only provided a surface area increase but also formed a bioconjugate that improves the analytical performance of the ECL biosensor. Metal nanoparticles with the central core of Au and Ag have attracted great interest, due to their catalytic and plasmonic applications. Li, J. et al. [146] used them on glassy carbon electrodes to develop a ECL sensor for the detection of highly upregulated in liver cancer (HULC), obtaining extremely low detection limits.

The current trend in ECL biosensors is the combination of several nanomaterials, where the advantages of both nanomaterials are blended, leading to the formation of new materials with synergistic properties [147]. The use of these nanomaterials produces enhanced ECL signal intensity. Zhang, C. et al. [148] employed mixed nanomaterials as Au nanoclusters and CeO_2 nanowires for the development of a biosensor. This ECL biosensor was based on the enzyme acetylcholinesterase. The use of CeO_2 nanowires on glassy carbon electrodes improve the stability of Au nanoclusters. In this case, the authors proposed another mixture of nanomaterials, AgNPs were combined with carbon QDs obtaining a nanocomposite, that showed superb ECL activity in the development of an enzymatic glucose biosensor [149]. Du, X. et al. [150] proposed the use of CeO_2 nanocrystallines assembled on the surface of graphene sheets doped with nitrogen (CeO_2-NG) for the development of an ECL enzymatic biosensor based on cholesterol oxidase. The results display that the use of CeO_2-NG nanocomposites facilitates the electrochemical redox process of co-reactant, improving the intensity of ECL. Numerous combinations of nanomaterials have been used until now and researchers continue to investigate new nanomaterials and their applications. The use of deposited nanomaterials, either on conventional electrodes or SPEs, produces considerable improvements in ECL biosensors. SPEs offer an additional advantage since they are disposable electrodes.

Enzymatic biosensors based on the use of SPE technology described above and their analytical properties are summarized in Table 2.

Table 2. Analytical parameters of SPE-based ECL enzymatic biosensors.

Luminophore	Enzyme	Analyte	Sensing Type	Detection Limit	Reference
Luminol	Glucose oxidase	Glucose	Surface-unpassivated CdTe QDs	0.3 mM	[108]
Luminol	Glucose oxidase	Glucose	Au and TiO_2 nano-composite	0.22 mM	[109]
Luminol	Uricase	Uric acid	3,3,5,5-tetramethylbenzidine and chitosan	0.44 µM	[111]
Luminol	Lactate oxidase	Lactate	Methocel membrane	5 µM	[114]
Luminol	Choline oxidase	Choline	Poly(vinyl alcohol) bearing styrylpyridinium groups (PVA–SbQ) photocrosslinked polymer	0.2 µM	[120]
Luminol	Glucose oxidase and Lactate oxidase	Glucose Lactate	PVA–SbQ (poly(vinyl alcohol)	3 µM 10 µM	[121]

4. ECL DNA Biosensors

DNA assays have many applications in different areas such as medical diagnostics, gene expression analysis, environmental investigations, biological warfare agent detection and pharmaceutical studies. DNA biosensors have been widely developed in different areas such as medicine and have brought enormous challenges to the early diagnosis of diseases [35,151–153]. In general, this kind of devices must have simplicity, fast response, good stability and high sensitivity to meet the requirements of clinical application and commercialization. In this sense, among the different transducing signal techniques, ECL is considered a promising one to improve the sensitivity and selectivity by lowering the background signal for DNA probe assays. ECL DNA biosensors use specific biological interactions to recognize and detect nucleic acids, producing a luminescent signal. An ECL DNA biosensor is a kind of biosensor in which a single-stranded DNA probe is immobilized on the electrode surface, which acts as a signal transducer of the hybridization event in an ECL signal by using different luminophores (Figure 9). Bard et al. [154,155] studied for the first time the ECL behavior of ruthenium and osmium pyridine complexes and DNA, paving the way for the ECL application in biological DNA analysis.

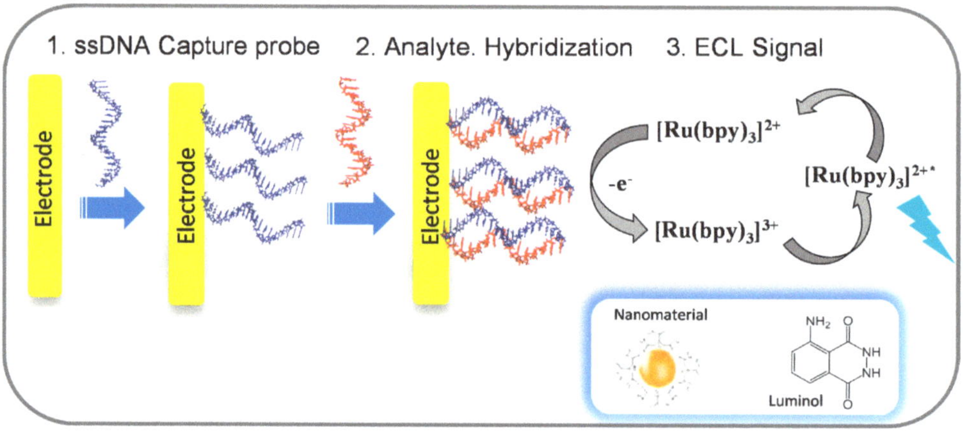

Figure 9. General scheme of ECL DNA Biosensor based on $[Ru(bpy)_3]^{2+}$.

4.1. *$[Ru(bpy)_3]^{2+}$, Luminol and Nanomaterials ECL Systems*

As in the case of ECL immuno- or enzymatic biosensors described above, the generating ECL signals in ECL DNA biosensors are principally based on the use of luminophores as ruthenium complexes, luminol or nanomaterials. In the case of using ruthenium complexes as luminophore, a selective and highly sensitive detection of miRNA-21, based on the toehold-mediated strand displacement (TMSD) amplification with $[Ru(phen)_3]^{2+}$ loaded DNA nanoclews (NCs–$[Ru(phen)_3]^{2+}$) as signal tags, has recently been reported [156]. In this work, the stable DNA nanoclews, synthesized by a simple rolling circle amplification reaction, were employed with $[Ru(phen)_3]^{2+}$ efficiently as ECL signal tags to amplify the signals. The ECL intensity of the system act as a function of the logarithm values of microRNA-21 with a limit of detection of 0.65 fM. The strategy was further applied to detect miRNA-21 in complex samples (different cells containing the HEK-293, HeLa cells and MCF-7) and the result was consistent with the real-time quantitative reverse transcription PCR (qRT-PCR).

Concerning ECL DNA biosensors using luminol as luminophore, Gao, W. et al. [157] developed a new ECL DNA biosensor with high sensitivity and throughput to ECL imaging of nucleolin in a single HeLa cell. In brief, mesoporous silica nanoparticles (MSN) loaded with doxorubicin (DOX) and phorbol 12-myristate 13-acetate (PMA) were employed as drug carriers and can be specifically opened by nucleolin in a HeLa cell. Then, PMA induced the HeLa cell to generate reactive oxygen species (ROS) and realized ECL imaging of nucleolin. Afterwards, ROS damages DNA and proteins of the tumor cell and doxorubicin induces the apoptosis of HeLa cells by inhibiting the synthesis of genetic material. Finally, DOX and ROS in a synergetic pathway kill HeLa cells efficiently. Kiani et al. [158] reported a bipolar electrode array based on luminol-platinum nanoparticles. These nanoparticles were modified with various monobases and served as ECL probes for detection of mismatches or single nucleotide polymorphisms (SNPs) associated with human diseases. The DNA probe was attached to the anodic poles of the array. When the electrodes were exposed to the monobase-luminol-platinum nanoparticle, hybridization occur. The ECL signal was measured with a digital camera. This biosensor can detect SNPs in the range of 2–600 pM. This simple instrument was applied for the discrimination between Polymerase Chain reaction (PCR) DNA samples of normal, heterozygous and homozygous beta thalassemia genetic disorders. It showed promise as an inexpensive and efficient method for detecting genetic disorders.

As described above, a strategy recently used to improve the ECL response is the introduction of nanomaterials is to increase the sensitivity and therefore extend the applications of ECL sensing.

Nanomaterials present different sizes, morphologies and chemical composition with excellent chemical and optical properties, converting them to good candidates for their use in ECL biosensors. They can be incorporated into the device to improve efficiency, but they also represent a new class of ECL emitters. A novel highly efficient ECL sensor based on reductive Cu(I) particles catalyzed Zn-doped MoS_2 QDs for human papilloma virus (HPV16 DNA) or cervical cancer detection has been recently developed [159]. In this work, the sulfur vacancies controlled with zinc doping led to the adsorption and coordination with transition metals. Cu(I) particles were prepared to catalyze H_2O_2 (co-reactant) in the ECL system. A 4.5-fold enhancement of the signal of Zn-doped MoS_2 QDs was obtained with the assistance of reductive Cu(I) particles. The biosensor achieved HPV 16 DNA sensitive determination with a limit of detection of 0.03 nmol/L. A smartphone can capture and process the ECL signal by self-developed software into high-resolution imaging, which provides the possibility of developing a point-of-care HPV 16 DNA determination in the future.

Yuan, R. et al. also reported a signal "off–on" microRNA-155 ECL platform based on carboxyl functionalized-poly (9,9-di-n-octylfluorenyl-2,7-diyl) polymer dots (PFO Pdots) [160]. Initially, PFO Pdots were immobilized onto the electrode trough their carboxylic groups to capture DNA duplex track-locker. In the presence of H_2O_2, the ECL signal of PFO Pdots was quenched to obtain a signal-off state. Then, the DNA walker, obtained through the target miRNA-155-triggered catalytic hairpin assembly (CHA), walked along the DNA duplex track-locker to output amounts of G-rich short chain, forming a hemin/G-quadruplex. The ECL signal is restored to a signal-on state with the consumption of H_2O_2 by hemin/G-quadruplex, achieving an ultrasensitive miRNA-155 detection. The integration of excellent ECL performance of PFO Pdots have provided an ECL emission and an attractive ECL platform for clinical diagnosis with highly efficient quenching effect of H_2O_2 without O_2 as co-reactant or exogenous species.

4.2. ECL DNA Biosensors Based on SPEs

As described above, ECL DNA biosensors have a wide range of applications in different areas as medical diagnosis, food analysis and biowarfare agent detection (virus and bacteria), among others. Screen-printing technology for ECL DNA biosensors is one of the most promising approaches towards simple, rapid and inexpensive production of disposable sensing devices intended for use at point of care. In recent years, SPEs with low cost and mass production using thick film technology have been extensively employed for developing novel sensing platforms with improved performance and application in different areas.

Currently, medical diagnosis is a research area of great interest. Reviewing the ECL DNA biosensors trends in this area, it is evident that there is a continued increase of reports in the last years. In this sense, cancer biomarker as miRNA detection is one of the hot topics.

MicroRNAs (miRNAs) are a class of endogenous non code small molecules (18–22 nt), which play important roles as significant regulators of fundamental cellular procedures. Recently, abnormal expression of miRNAs has been found in a variety of tumors, which is closely related to the progress, clinical treatment and prognosis of tumors [161,162]. Therefore, miRNAs are becoming as new targets for early cancer diagnosis and treatment [163–165]. Currently, simple and sensitive methods for miRNAs detection are still highly demanded. Figure 10 illustrates a novel $[Ru(bpy)_3]^{2+}$-based ECL immunosensor for rapid and sensitive detection of miRNA-155 utilizing a microfluidic system based on a paper-based closed Au-bipolar electrode (BPE) [166]. Wax-printing technology, screen printing method and in-situ AuNPs growth were used for the microfluidic platform fabrication. DNA (S1)–AuPd nanoparticles was used to modify the cathode of bipolar electrode with a by hybridization chain reaction, in which the target initiated multiple cycles reaction to load high amounts of AuPd nanoparticles that in turn catalyzed H_2O_2 reduction. Additionally, the anode of the bipolar electrode is coupled to the $[Ru(bpy)_3]^{2+}$/TPrA ECL system. Due to the charge balance between the anode and the cathode of BPE, the ECL signal was enhanced. The ECL signal correlated quantitatively with the concentration of miRNA-155 up to 10 µM with a detection limit of 0.67 pM providing a novel way for highly sensitive miRNA-155 detection in clinical application.

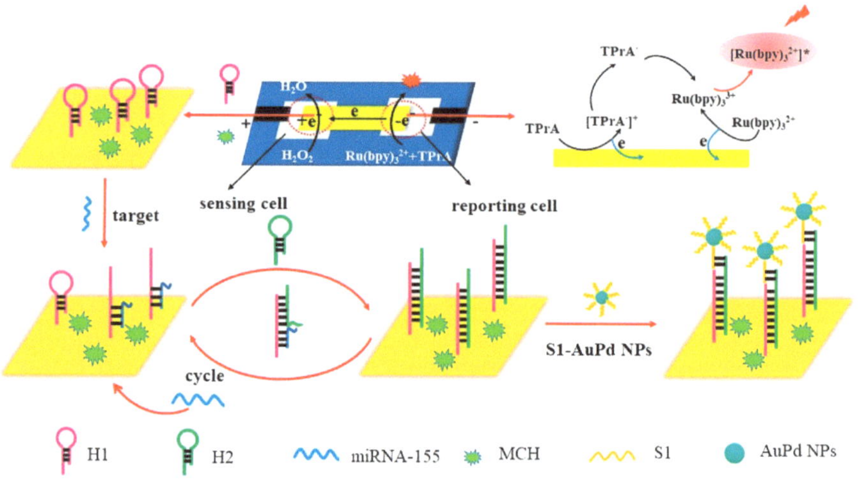

Figure 10. Scheme of ECL biosensor development for the detection of miRNA-155 using a microfluidic system based on a paper-based closed Au-bipolar electrode proposed by Yu et al. [166]. Reproduced and adapted with permission of Elsevier.

Another important analyte for medical porpoises is the p16INK4a gene, which is a classical tumor suppressor gene that inhibits the cyclin D-dependent protein kinases. Xu et al. [167] reported on an ECL biosensing system using functional paste-like nanofibers composites-modified carbon SPE for sensitive detection of p16INK4a gene. Figure 11 shows a schematic representation of the ECL DNA biosensor development. Figure 11A shows the dsDNA formation. Firstly, the $[Ru(bpy)_3]^{2+}$/silver nanoparticles (AgNPs) doped gold core–shell luminescent composite nanoparticles were labeled with ssDNA2 (RuAg@AuNPs-ssDNA2). Afterwards, hybridization reaction of ssDNA1 with the corresponding sequence of p16INK4a gene takes place forming RuAg@AuNPs-dsDNA particles. Figure 11B shows the paste-like nanofibers composites formation using electrospun nanofibers, graphene (GR) and chitosan (CS), which serves as the nanosized backbones for pyrrole (Py) electropolymerization on the modified carbon SPE surface. The composites were used as a substrate for dsDNA immobilization. The ECL intensity is linear to the p16INK4a gene concentration in the range from 0.1 pM to 1 nM, with a detection limit of 0.05 pM.

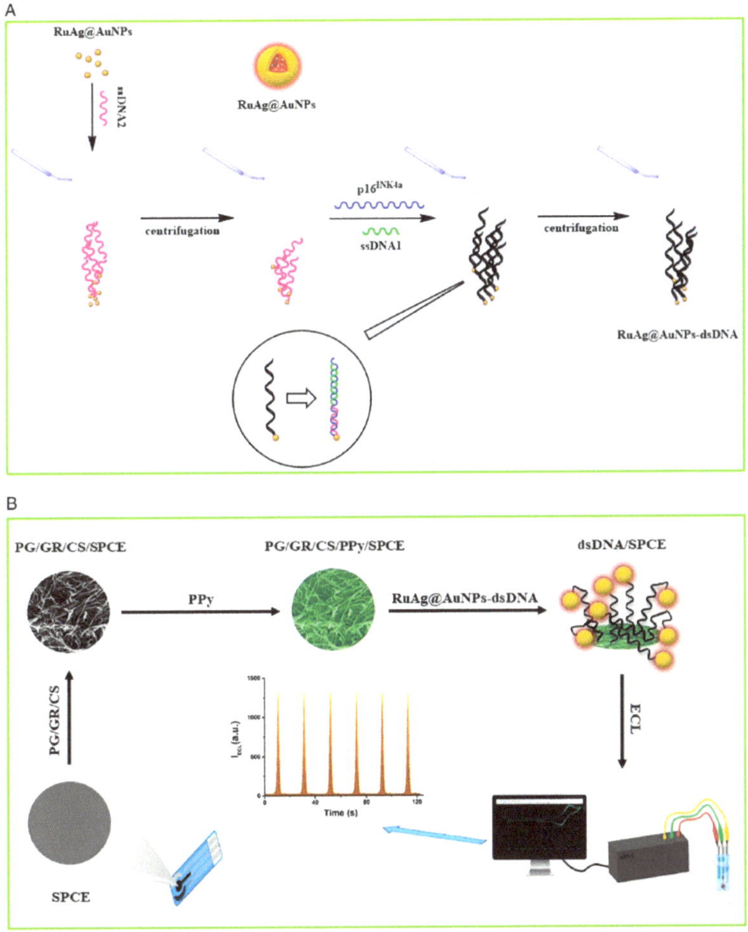

Figure 11. Schematic representation of the DNA-based ECL biosensor for detection of p16INK4a gene using functional paste-like nanofibers composites-modified carbon SPE, developed by Xu et al. [167]. Reproduced and adapted with permission of Elsevier. (**A**) RuAg@AuNPs-dsDNA preparation; (**B**) Paste-like nanofibers composites formation, electropolymerization on the modified carbon SPE surface and dsDNA immobilization.

Nowadays, there is a growing concern of consumers on the authenticity of food ingredients including adulteration. For this reason, the development of low-cost bioanalytical devices that enable high throughput and performance assays in small point-of-analysis will be a priority to improve the quality of food and health. Therefore, researchers have been focusing on developing new detection procedures and tools [168]. In this sense, the combination of an oligonucleotide as recognition element for a specific analyte and of ECL as a readout method has proven to be a valuable strategy for sensitive and specific analytical detection. Figure 12 illustrates the development of a rapid, simple and sensitive luminol-based ECL biosensors for Susscrofa (Porcine) DNA detection [169]. Firstly, porcine DNA was amplified using the loop-mediated isothermal amplification (LAMP) method and then added to luminol solution for the ECL analysis quantification. The DNA-luminol complexes formed slow down the diffusion of luminol towards the electrode surface resulting in low luminol intensity. The LAMP-ECL biosensor shows a sensitive and highly specific detection of Susscrofa (Porcine) DNA detection with a rapid response (around 5 min) and a detection limit of 0.1 pg/μL This approach used in combination with carbon SPE has a great potential to compact biosensors development for food authenticity control.

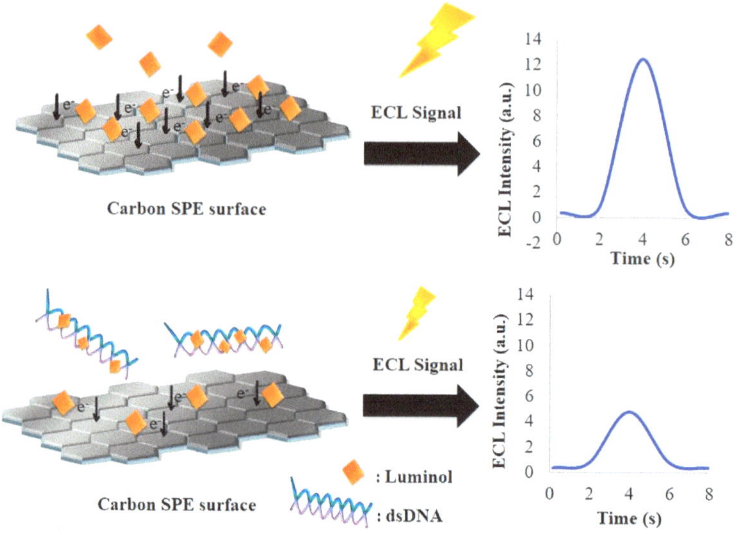

Figure 12. Schematic diagram of DNA ECL immunosensor development for detection of Susscrofa (Porcine) reported by Azam, N.F.N. et al. [169]. Reproduced and adapted with permission of Elsevier.

A new reported application of ECL DNA biosensors based on SPEs is biowarfare agent detection. These agents, which include bacteria, viruses and ortoxins derived from living organisms, are used for bioterrorism purposes and pose an ever-increasing security risk to military and civilian populations [170–172]. For this reason, the development of multiplex ECL assays for biowarfare agent is of great interest. O´Sullivan et al. [173] reported the develop of a multiplex detection ECL DNA sensor for the simultaneous detection of six pathogens: *Brucella, melitensis, Bacillus anthracis, Burkholderia mallei, Francisella tularensis, Bacillus thuringiensis var. kurstaki* and *Coxiella burnetiid*. A carbon SPE array was used to develop the assay. This array comprised 42 individual working electrodes with shared counter and reference electrodes. Activated-carbon electrodes were modified covalently with DNA capture probes and then hybridized to target strands. Afterwards, pathogens detection was achieved via sandwich-type format, using $[Ru(bpy)_3]^{2+}$ labeled reporter probes that were hybridized to the probe–target complexes. An automated microsystem made in a custom designed ECL detection box with integrated electronics, movable photomultiplier detector and fluidics was performed. The detection limits were found to be 0.6–1.2 nmol/L for six targets (from 50 to 122 base pairs) and linear ranges up to 15 nmol/L. The developed system allows the detection in single chip of six targets at sub-nanomolar concentrations.

In conclusion, ECL DNA biosensors based on screen printed technology are successful analytical platforms with high sensitivity and excellent selectivity that are starting to be used for the detection of important analytes, such as viruses, toxins and bacteria (see Table 3). However, the development of new improved devices can be considered as a new research line that is at its beginnings, since most of the reported works have been recently published. There are still several challenges that must be addressed in order to develop devices that have the analytical characteristics required to solve the problems that society demands today.

Table 3. Analytical parameters of SPE-based ECL DNA biosensors.

Luminophore	Analyte	Sensing Type	Detection Limit	Reference
$[Ru(phen)_3]^{2+}$	mRNA-21	Toehold-mediated strand displacement (TMSD)	0.65 fM	[156]
Luminol	nucleolin in a single HeLa cell	Mesoporous silica nanoparticles (MSN) loaded with doxorubicin (DOX) and phorbol 12-myristate 13-acetate (PMA)		[157]
Luminol	SNP detection	Luminol-platinum nanoparticles	2–600 pM	[158]
Nanomaterials: MoS2 QDs	HPV16 DNA	Cu(I) reductive particles catalyzed Zn-doped MoS2 QDs	0.03 nmol/L	[159]
Nanomaterials: PFO Pdots	microRNA-155	Quenching effect of H_2O_2	12.2 aM	[160]
$[Ru(phen)_3]^{2+}$/TPrA	mRNA-21	Wax-printing technology, screen printing method and in-situ AuNPs growth	0.67 pM	[166]
$[Ru(bpy)_3]^{2+}$/silver nanoparticles (AgNPs)	p16INK4a	Paste-like nanofibers composites-modified carbon SPE	0.05 pM	[167]
Luminol	Susscrofa (Porcine) DNA	Loop-mediated isothermal amplification	0.1 pg/μL	[169]
$[Ru(bpy)_3]^{2+}$	pathogens	Sandwich-type assay	0.6–1.2 nmol/L	[173]

5. Conclusions

ECL biosensors based on screen-printed electrodes are currently commercially successful analytical platforms with high sensitivity and excellent selectivity that are used for the detection of a wide range of analytes, such as biomarkers, toxins, viruses, bacteria and metal ions. The unique and attractive properties of the ECL technique and the popularization of the commercial instrumentation, together with the recent introduction of screen-printed electrodes as electrochemical platforms have paved the way for the development of new designs and applications in clinical prognostics and diagnostics, food and environmental control, biodefense, etc. We overview different applications of ECL biosensing systems in which the electrode employed is a disposable screen-printed electrode. We discuss ECL immunosensors, enzymatic and DNA biosensor. The development of new ECL light-emitting molecules and co-reactants, new ECL mechanisms, disposable electrodes with diverse configurations, miniaturization of instruments and ECL imaging techniques demonstrates that in the future the development of ECL technology will benefit from the rapid and continuous current progress. This in part is due to the advantages that the disposable screen-printed electrodes provide to the system. In particular, recent developments of ECL include fabrication of portable devices for medical point-of-care and field instruments for use in environmental research. Furthermore, the recent incorporation of nanotechnology will improve the efficiency of ECL biosensors based on screen-printed electrodes, since they can act as electrode modifiers giving nanostructured surfaces or as co-reactants in the ECL process.

Author Contributions: E.M.-P., Conceptualization, Resources and Writing Introduction and ECL Immunosensors original draft preparation; C.G.-S., Conceptualization, Resources and Writing ECL Enzymatic Biosensors original draft preparation.; T.G.-M., Conceptualization, Resources and Writing ECL DNA Biosensors original draft preparation; and E.L., Conceptualization, Supervision, Writing Abstract and Conclusions and Review and Editing. All authors have read and agreed to the published version of the manuscript.

Funding: This research was founded by the Ministerio de Ciencia, Innovación, Universidades of Spain (CTQ2017-84309-C2-1-R and RED2018-102412-T), Comunidad Autónoma de Madrid (TRANSNANOAVANSENS Program). C.G.-S. also acknowledges the financial support from the Comunidad Autónoma de Madrid, Atracción de Talento Program (2017-T1/BIO-5435).

Conflicts of Interest: The authors declare no conflict of interest.

References

1. Miao, W. Electrogenerated Chemiluminescence and Its Biorelated Applications. *Chem. Rev.* **2008**, *108*, 2506–2553. [CrossRef]
2. Sojic, N.; Arbault, S.; Bouffier, L.; Kuhn, A. Applications of electrogenerated chemiluminescence in analytical chemistry. In *Luminescence in Electrochemistry: Applications in Analytical Chemistry, Physics and Biology*; Miomandre, F., Audebert, P., Eds.; Springer International Publishing: Cham, Switzerland, 2017; pp. 257–291. [CrossRef]
3. Choi, J.-P.; Bard, A.J. Electrogenerated chemiluminescence (ECL) 79.: Reductive-oxidation ECL of tris(2,2′ -bipyridine)ruthenium(II) using hydrogen peroxide as a coreactant in pH 7.5 phosphate buffer solution. *Anal. Chim. Acta* **2005**, *541*, 141–148. [CrossRef]
4. Chang, M.-M.; Saji, T.; Bard, A.J. Electrogenerated chemiluminescence. 30. Electrochemical oxidation of oxalate ion in the presence of luminescers in acetonitrile solutions. *J. Am. Chem. Soc.* **1977**, *99*, 5399–5403. [CrossRef]
5. Rubinstein, I.; Bard, A.J. Electrogenerated chemiluminescence. 37. Aqueous ecl systems based on tris(2,2′-bipyridine)ruthenium(2+) and oxalate or organic acids. *J. Am. Chem. Soc.* **1981**, *103*, 512–516. [CrossRef]
6. Noffsinger, J.B.; Danielson, N.D. Generation of chemiluminescence upon reaction of aliphatic amines with tris(2,2′-bipyridine)ruthenium(III). *Anal. Chem.* **1987**, *59*, 865–868. [CrossRef]
7. Leland, J.K.; Powell, M.J. Electrogenerated Chemiluminescence: An Oxidative-Reduction Type ECL Reaction Sequence Using Tripropyl Amine. *J. Electrochem. Soc.* **2019**, *137*, 3127–3131. [CrossRef]

8. Zu, Y.; Bard, A.J. Electrogenerated Chemiluminescence. 66. The Role of Direct Coreactant Oxidation in the Ruthenium Tris(2,2′)bipyridyl/Tripropylamine System and the Effect of Halide Ions on the Emission Intensity. *Anal. Chem.* **2000**, *72*, 3223–3232. [CrossRef]
9. Kanoufi, F.; Zu, Y.; Bard, A.J. Homogeneous Oxidation of Trialkylamines by Metal Complexes and Its Impact on Electrogenerated Chemiluminescence in the Trialkylamine/Ru(bpy)32+ System. *J. Phys. Chem. B* **2001**, *105*, 210–216. [CrossRef]
10. White, H.S.; Bard, A.J. Electrogenerated chemiluminescence. 41. Electrogenerated chemiluminescence and chemiluminescence of the Ru(2,21-bpy)32+-S2O82- system in acetonitrile-water solutions. *J. Am. Chem. Soc.* **1982**, *104*, 6891–6895. [CrossRef]
11. Becker, W.G.; Seung, H.S.; Bard, A.J. Electrogenerated chemiluminescence: Part XLIII. Aromatic hydrocarbon /peroxydisulfate systems in acetonitrile-benzene solutions. *J. Electroanal. Chem. Interfacial Electrochem.* **1984**, *167*, 127–140. [CrossRef]
12. Fabrizio, E.F.; Prieto, I.; Bard, A.J. Hydrocarbon Cation Radical Formation by Reduction of Peroxydisulfate. *J. Am. Chem. Soc.* **2000**, *122*, 4996–4997. [CrossRef]
13. Miao, W.; Choi, J.-P.; Bard, A.J. Electrogenerated Chemiluminescence 69: The Tris(2,2′-bipyridine) ruthenium(II), (Ru(bpy)32+)/Tri-n-propylamine (TPrA) System RevisitedA New Route Involving TPrA•+ Cation Radicals. *J. Am. Chem. Soc.* **2002**, *124*, 14478–14485. [CrossRef] [PubMed]
14. Hiramoto, K.; Villani, E.; Iwama, T.; Komatsu, K.; Inagi, S.; Inoue, K.Y.; Nashimoto, Y.; Ino, K.; Shiku, H. Recent Advances in Electrochemiluminescence-Based Systems for Mammalian Cell Analysis. *Micromachines* **2020**, *11*, 530. [CrossRef]
15. Ye, R.; Huang, L.; Qiu, B.; Song, Z.; Lin, Z.; Chen, G. Cathodic electrochemiluminescent behavior of luminol at nafion–nano-TiO2 modified glassy carbon electrode. *Luminescence* **2011**, *26*, 531–535. [CrossRef]
16. Yin, X.-B.; Dong, S.; Wang, E. Analytical applications of the electrochemiluminescence of tris (2,2′-bipyridyl) ruthenium and its derivatives. *TrAC Trends Anal. Chem.* **2004**, *23*, 432–441. [CrossRef]
17. Morita, H.; Konishi, M. Electrogenerated Chemiluminescence Derivatization Reagents for Carboxylic Acids and Amines in High-Performance Liquid Chromatography Using Tris(2,2′-bipyridine)ruthenium(II). *Anal. Chem.* **2002**, *74*, 1584–1589. [CrossRef]
18. Li, F.; Cui, H.; Lin, X.-Q. Determination of adrenaline by using inhibited Ru(bpy)32+ electrochemiluminescence. *Anal. Chim. Acta* **2002**, *471*, 187–194. [CrossRef]
19. Forster, R.J.; Hogan, C.F. Electrochemiluminescent Metallopolymer Coatings: Combined Light and Current Detection in Flow Injection Analysis. *Anal. Chem.* **2000**, *72*, 5576–5582. [CrossRef]
20. Liu, J.; Cao, W.; Yang, X.; Wang, E. Determination of diphenhydramine by capillary electrophoresis with tris(2,2′-bipyridyl)ruthenium(II) electrochemiluminescence detection. *Talanta* **2003**, *59*, 453–459. [CrossRef]
21. Marchese, R.D.; Puchalski, D.; Miller, P.; Antonello, J.; Hammond, O.; Green, T.; Rubinstein, L.J.; Caulfield, M.J.; Sikkema, D. Optimization and Validation of a Multiplex, Electrochemiluminescence-Based Detection Assay for the Quantitation of Immunoglobulin G Serotype-Specific Antipneumococcal Antibodies in Human Serum. *Clin. Vaccine Immunol.* **2009**, *16*, 387–396. [CrossRef]
22. Metters, J.P.; Kadara, R.O.; Banks, C.E. New directions in screen printed electroanalytical sensors: An overview of recent developments. *Analyst* **2011**, *136*, 1067–1076. [CrossRef] [PubMed]
23. Li, M.; Li, D.-W.; Xiu, G.; Long, Y.-T. Applications of screen-printed electrodes in current environmental analysis. *Curr. Opin. Electrochem.* **2017**, *3*, 137–143. [CrossRef]
24. Yáñez-Sedeño, P.; Campuzano, S.; Pingarrón, J.M. Integrated Affinity Biosensing Platforms on Screen-Printed Electrodes Electrografted with Diazonium Salts. *Sensors* **2018**, *18*, 675. [CrossRef] [PubMed]
25. Arduini, F.; Micheli, L.; Moscone, D.; Palleschi, G.; Piermarini, S.; Ricci, F.; Volpe, G. Electrochemical biosensors based on nanomodified screen-printed electrodes: Recent applications in clinical analysis. *TrAC Trends Anal. Chem.* **2016**, *79*, 114–126. [CrossRef]
26. Yamanaka, K.; Vestergaard, M.C.; Tamiya, E. Printable Electrochemical Biosensors: A Focus on Screen-Printed Electrodes and Their Application. *Sensors* **2016**, *16*, 1761. [CrossRef]
27. Gómez-Anquela, C.; García-Mendiola, T.; Abad, J.M.; Pita, M.; Pariente, F.; Lorenzo, E. Scaffold electrodes based on thioctic acid-capped gold nanoparticles coordinated Alcohol Dehydrogenase and Azure A films for high performance biosensor. *Bioelectrochemistry* **2015**, *106*, 335–342. [CrossRef]
28. Rama, E.C.; Costa-García, A. Screen-printed Electrochemical Immunosensors for the Detection of Cancer and Cardiovascular Biomarkers. *Electroanalysis* **2016**, *28*, 1700–1715. [CrossRef]

29. Mistry, K.K.; Layek, K.; Mahapatra, A.; RoyChaudhuri, C.; Saha, H. A review on amperometric-type immunosensors based on screen-printed electrodes. *Analyst* **2014**, *139*, 2289–2311. [CrossRef]
30. Taleat, Z.; Khoshroo, A.; Mazloum-Ardakani, M. Screen-printed electrodes for biosensing: A review (2008–2013). *Microchim. Acta* **2014**, *181*, 865–891. [CrossRef]
31. Serafín, V.; Valverde, A.; Martínez-García, G.; Martínez-Periñán, E.; Comba, F.; Garranzo-Asensio, M.; Barderas, R.; Yáñez-Sedeño, P.; Campuzano, S.; Pingarrón, J.M. Graphene quantum dots-functionalized multi-walled carbon nanotubes as nanocarriers in electrochemical immunosensing. Determination of IL-13 receptor $\alpha 2$ in colorectal cells and tumor tissues with different metastatic potential. *Sens. Actuators B Chem.* **2019**, *284*, 711–722. [CrossRef]
32. Martínez-Periñán, E.; Sánchez-Tirado, E.; González-Cortés, A.; Barderas, R.; Sánchez-Puelles, J.M.; Martínez-Santamaría, L.; Campuzano, S.; Yáñez-Sedeño, P.; Pingarrón, J.M. Amperometric determination of endoglin in human serum using disposable immunosensors constructed with poly(pyrrolepropionic) acid-modified electrodes. *Electrochim. Acta* **2018**, *292*, 887–894. [CrossRef]
33. García, T.; Fernández-Barrena, M.G.; Revenga-Parra, M.; Núñez, A.; Casero, E.; Pariente, F.; Prieto, J.; Lorenzo, E. Disposable sensors for rapid screening of mutated genes. *Anal. Bioanal. Chem.* **2010**, *398*, 1385–1393. [CrossRef] [PubMed]
34. García-Mendiola, T.; Cerro, M.R.; López-Moreno, J.M.; Pariente, F.; Lorenzo, E. Dyes as bifunctional markers of DNA hybridization on surfaces and mutation detection. *Bioelectrochemistry* **2016**, *111*, 115–122. [CrossRef]
35. García-Mendiola, T.; Bayon-Pizarro, V.; Zaulet, A.; Fuentes, I.; Pariente, F.; Teixidor, F.; Viñas, C.; Lorenzo, E. Metallacarboranes as tunable redox potential electrochemical indicators for screening of gene mutation. *Chem. Sci.* **2016**, *7*, 5786–5797. [CrossRef]
36. García-Mendiola, T.; Bravo, I.; López-Moreno, J.M.; Pariente, F.; Wannemacher, R.; Weber, K.; Popp, J.; Lorenzo, E. Carbon nanodots based biosensors for gene mutation detection. *Sens. Actuators B Chem.* **2018**, *256*, 226–233. [CrossRef]
37. García-Mendiola, T.; Requena-Sanz, S.; Martínez-Periñán, E.; Bravo, I.; Pariente, F.; Lorenzo, E. Influence of carbon nanodots on DNA-Thionine interaction. Application to breast cancer diagnosis. *Electrochim. Acta* **2020**, *353*, 136522. [CrossRef]
38. García-Mendiola, T.; Gutiérrez-Sánchez, C.; Gibaja, C.; Torres, I.; Busó-Rogero, C.; Pariente, F.; Solera, J.; Razavifar, Z.; Palacios, J.J.; Zamora, F.; et al. Functionalization of a Few-Layer Antimonene with Oligonucleotides for DNA Sensing. *ACS Appl. Nano Mater.* **2020**, *3*, 3625–3633. [CrossRef]
39. Chen, A.; Chatterjee, S. Nanomaterials based electrochemical sensors for biomedical applications. *Chem. Soc. Rev.* **2013**, *42*, 5425–5438. [CrossRef]
40. Song, S.; Qin, Y.; He, Y.; Huang, Q.; Fan, C.; Chen, H.-Y. Functional nanoprobes for ultrasensitive detection of biomolecules. *Chem. Soc. Rev.* **2010**, *39*, 4234–4243. [CrossRef]
41. Holzinger, M.; Le Goff, A.; Cosnier, S. Nanomaterials for biosensing applications: A review. *Front. Chem.* **2014**, *2*, 63. [CrossRef] [PubMed]
42. Zhai, Q.; Li, J.; Wang, E. Recent Advances Based on Nanomaterials as Electrochemiluminescence Probes for the Fabrication of Sensors. *ChemElectroChem* **2017**, *4*, 1639–1650. [CrossRef]
43. Shen, J.; Zhou, T.; Huang, R. Recent Advances in Electrochemiluminescence Sensors for Pathogenic Bacteria Detection. *Micromachines* **2019**, *10*, 532. [CrossRef] [PubMed]
44. Renedo, O.D.; Alonso-Lomillo, M.A.; Martínez, M.J.A. Recent developments in the field of screen-printed electrodes and their related applications. *Talanta* **2007**, *73*, 202–219. [CrossRef] [PubMed]
45. Moro, G.; Bottari, F.; Van Loon, J.; Du Bois, E.; De Wael, K.; Moretto, L.M. Disposable electrodes from waste materials and renewable sources for (bio)electroanalytical applications. *Biosens. Bioelectron.* **2019**, *146*, 111758. [CrossRef] [PubMed]
46. Moya, A.; Gabriel, G.; Villa, R.; Javier del Campo, F. Inkjet-printed electrochemical sensors. *Curr. Opin. Electrochem.* **2017**, *3*, 29–39. [CrossRef]
47. Silva, N.F.D.; Almeida, C.M.R.; Magalhães, J.M.C.S.; Gonçalves, M.P.; Freire, C.; Delerue-Matos, C. Development of a disposable paper-based potentiometric immunosensor for real-time detection of a foodborne pathogen. *Biosens. Bioelectron.* **2019**, *141*, 111317. [CrossRef]

48. Rubio-Govea, R.; Hickey, D.P.; García-Morales, R.; Rodriguez-Delgado, M.; Domínguez-Rovira, M.A.; Minteer, S.D.; Ornelas-Soto, N.; García-García, A. MoS2 nanostructured materials for electrode modification in the development of a laccase based amperometric biosensor for non-invasive dopamine detection. *Microchem. J.* **2020**, *155*, 104792. [CrossRef]
49. Rungsawang, T.; Punrat, E.; Adkins, J.; Henry, C.; Chailapakul, O. Development of Electrochemical Paper-based Glucose Sensor Using Cellulose-4-aminophenylboronic Acid-modified Screen-printed Carbon Electrode. *Electroanalysis* **2016**, *28*, 462–468. [CrossRef]
50. Sánchez-Calvo, A.; Costa-García, A.; Blanco-López, M.C. Paper-based electrodes modified with cobalt phthalocyanine colloid for the determination of hydrogen peroxide and glucose. *Analyst* **2020**, *145*, 2716–2724. [CrossRef]
51. Tao, C.; Yen, C.-S.; Liu, J.-T.; Chen, C.-J. Analytical performance of paper electro-biosensor detection platform for point-of-care diagnosis. *Cellulose* **2016**, *23*, 3799–3808. [CrossRef]
52. Yang, H.; Kong, Q.; Wang, S.; Xu, J.; Bian, Z.; Zheng, X.; Ma, C.; Ge, S.; Yu, J. Hand-drawn&written pen-on-paper electrochemiluminescence immunodevice powered by rechargeable battery for low-cost point-of-care testing. *Biosens. Bioelectron.* **2014**, *61*, 21–27. [CrossRef] [PubMed]
53. Bernalte, E.; Foster, C.W.; Brownson, D.A.; Mosna, M.; Smith, G.C.; Banks, C.E. Pencil It in: Exploring the Feasibility of Hand-Drawn Pencil Electrochemical Sensors and Their Direct Comparison to Screen-Printed Electrodes. *Biosensors* **2016**, *6*, 45. [CrossRef] [PubMed]
54. Foster, C.W.; Brownson, D.A.C.; Ruas de Souza, A.P.; Bernalte, E.; Iniesta, J.; Bertotti, M.; Banks, C.E. Pencil it in: Pencil drawn electrochemical sensing platforms. *Analyst* **2016**, *141*, 4055–4064. [CrossRef] [PubMed]
55. Kannan, B.; Jahanshahi-Anbuhi, S.; Pelton, R.H.; Li, Y.; Filipe, C.D.M.; Brennan, J.D. Printed Paper Sensors for Serum Lactate Dehydrogenase using Pullulan-Based Inks to Immobilize Reagents. *Anal. Chem.* **2015**, *87*, 9288–9293. [CrossRef]
56. Credou, J.; Berthelot, T. Cellulose: From biocompatible to bioactive material. *J. Mater. Chem. B* **2014**, *2*, 4767–4788. [CrossRef] [PubMed]
57. Yáñez-Sedeño, P.; Campuzano, S.A.-O.; Pingarrón, J.A.-O. Screen-Printed Electrodes: Promising Paper and Wearable Transducers for (Bio)Sensing. *Biosensors* **2020**, *10*, 76. [CrossRef]
58. Wang, T.; Wang, D.; Padelford, J.W.; Jiang, J.; Wang, G. Near-Infrared Electrogenerated Chemiluminescence from Aqueous Soluble Lipoic Acid Au Nanoclusters. *J. Am. Chem. Soc.* **2016**, *138*, 6380–6383. [CrossRef]
59. Cox, K.L.; Devanarayan, V.; Kriauciunas, A.; Manetta, J.; Montrose, C.; Sittampalam, S. Immunoassay Methods. Available online: https://www.ncbi.nlm.nih.gov/books/NBK92434/#:~{}:text=Sandwich%20Immunoassay%20(ELISA),attached%20to%20a%20solid%20surface (accessed on 18 August 2020).
60. Song, C.; Li, X.; Hu, L.; Shi, T.; Wu, D.; Ma, H.; Zhang, Y.; Fan, D.; Wei, Q.; Ju, H. Quench-Type Electrochemiluminescence Immunosensor Based on Resonance Energy Transfer from Carbon Nanotubes and Au-Nanoparticles-Enhanced g-C3N4 to CuO@Polydopamine for Procalcitonin Detection. *ACS Appl. Mater. Interfaces* **2020**, *12*, 8006–8015. [CrossRef]
61. Xue, J.; Yang, L.; Wang, H.; Yan, T.; Fan, D.; Feng, R.; Du, B.; Wei, Q.; Ju, H. Quench-type electrochemiluminescence immunosensor for detection of amyloid β-protein based on resonance energy transfer from luminol@SnS2-Pd to Cu doped WO3 nanoparticles. *Biosens. Bioelectron.* **2019**, *133*, 192–198. [CrossRef]
62. Kannan, P.; Chen, J.; Su, F.; Guo, Z.; Huang, Y. Faraday-Cage-Type Electrochemiluminescence Immunoassay: A Rise of Advanced Biosensing Strategy. *Anal. Chem.* **2019**, *91*, 14792–14802. [CrossRef]
63. Guo, Z.; Sha, Y.; Hu, Y.; Yu, Z.; Tao, Y.; Wu, Y.; Zeng, M.; Wang, S.; Li, X.; Zhou, J.; et al. Faraday cage-type electrochemiluminescence immunosensor for ultrasensitive detection of Vibrio vulnificus based on multi-functionalized graphene oxide. *Anal. Bioanal. Chem.* **2016**, *408*, 7203–7211. [CrossRef] [PubMed]
64. Rizwan, M.; Mohd-Naim, N.F.; Keasberry, N.A.; Ahmed, M.U. A highly sensitive and label-free electrochemiluminescence immunosensor for beta 2-microglobulin. *Anal. Methods* **2017**, *9*, 2570–2577. [CrossRef]
65. Li, X.; Li, Y.; Feng, R.; Wu, D.; Zhang, Y.; Li, H.; Du, B.; Wei, Q. Ultrasensitive electrochemiluminescence immunosensor based on Ru(bpy)32+ and Ag nanoparticles doped SBA-15 for detection of cancer antigen 15-3. *Sens. Actuators B Chem.* **2013**, *188*, 462–468. [CrossRef]

66. Jiang, D.; Zhang, L.; Liu, F.; Liu, C.; Liu, L.; Pu, X. An electrochemiluminescence sensor with dual signal amplification of Ru(bpy)32+ based on PtNPs and glucose dehydrogenase for diagnosis of gas gangrene. *RSC Adv.* **2016**, *6*, 19676–19685. [CrossRef]
67. Li, C.; Fu, Z.; Li, Z.; Wang, Z.; Wei, W. Cross-talk-free multiplexed immunoassay using a disposable electrochemiluminescent immunosensor array coupled with a non-array detector. *Biosens. Bioelectron.* **2011**, *27*, 141–147. [CrossRef] [PubMed]
68. Habtamu, H.B.; Sentic, M.; Silvestrini, M.; De Leo, L.; Not, T.; Arbault, S.; Manojlovic, D.; Sojic, N.; Ugo, P. A Sensitive Electrochemiluminescence Immunosensor for Celiac Disease Diagnosis Based on Nanoelectrode Ensembles. *Anal. Chem.* **2015**, *87*, 12080–12087. [CrossRef] [PubMed]
69. Spehar-Délèze, A.-M.; Julich, S.; Gransee, R.; Tomaso, H.; Dulay, S.B.; O'Sullivan, C.K. Electrochemiluminescence (ECL) immunosensor for detection of Francisella tularensis on screen-printed gold electrode array. *Anal. Bioanal. Chem.* **2016**, *408*, 7147–7153. [CrossRef]
70. Kadimisetty, K.; Mosa, I.M.; Malla, S.; Satterwhite-Warden, J.E.; Kuhns, T.M.; Faria, R.C.; Lee, N.H.; Rusling, J.F. 3D-printed supercapacitor-powered electrochemiluminescent protein immunoarray. *Biosens. Bioelectron.* **2016**, *77*, 188–193. [CrossRef] [PubMed]
71. Jian, Y.; Wang, H.; Sun, X.; Zhang, L.; Cui, K.; Ge, S.; Yu, J. Electrochemiluminescence cytosensing platform based on Ru(bpy)32+@silica-Au nanocomposite as luminophore and AuPd nanoparticles as coreaction accelerator for in situ evaluation of intracellular H2O2. *Talanta* **2019**, *199*, 485–490. [CrossRef] [PubMed]
72. Gan, N.; Hou, J.; Hu, F.; Cao, Y.; Li, T.; Guo, Z.; Wang, J. A Renewable and Ultrasensitive Electrochemiluminescence Immunosenor Based on Magnetic RuL@SiO2-Au~RuL-Ab2 Sandwich-Type Nano-Immunocomplexes. *Sensors* **2011**, *11*, 7749–7762. [CrossRef] [PubMed]
73. Zhou, J.; Gan, N.; Hu, F.; Li, T.; Zhou, H.; Li, X.; Zheng, L. A single antibody sandwich electrochemiluminescence immunosensor based on protein magnetic molecularly imprinted polymers mimicking capture probes. *Sens. Actuators B Chem.* **2013**, *186*, 300–307. [CrossRef]
74. Feng, X.; Gan, N.; Zhou, J.; Li, T.; Cao, Y.; Hu, F.; Yu, H.; Jiang, Q. A novel dual-template molecularly imprinted electrochemiluminescence immunosensor array using Ru(bpy)32+-Silica@Poly-L-lysine-Au composite nanoparticles as labels for near-simultaneous detection of tumor markers. *Electrochim. Acta* **2014**, *139*, 127–136. [CrossRef]
75. Hou, J.; Gan, N.; Hu, F.; Zheng, L.; Cao, Y.; Li, T. One Renewable and Magnetic Electrochemiluminescence Immunosenor Based on Tris(2,2′-bipyridine) ruthenium(II) Modified Magnetic Composite Nanoparticles Labeled Anti-AFP. *Int. J. Electrochem. Sci.* **2011**, *6*, 2845–2858.
76. Zhou, H.; Yue, H.; Zhou, Y.; Wang, L.; Fu, Z. A novel disposable immunosensor based on quenching of electrochemiluminescence emission of Ru(bpy)32+ by amorphous carbon nanoparticles. *Sens. Actuators B Chem.* **2015**, *209*, 744–750. [CrossRef]
77. Fang, D.; Ren, H.; Huang, Y.; Dai, H.; Huang, D.; Lin, Y. Photothermal amplified cathodic ZnO quantum dots/Ru(bpy)32+/S2O82- ternary system for ultrasensitive electrochemiluminescence detection of thyroglobulin. *Sens. Actuators B Chem.* **2020**, *312*, 127950. [CrossRef]
78. Zhou, J.; Gan, N.; Li, T.; Hu, F.; Li, X.; Wang, L.; Zheng, L. A cost-effective sandwich electrochemiluminescence immunosensor for ultrasensitive detection of HIV-1 antibody using magnetic molecularly imprinted polymers as capture probes. *Biosens. Bioelectron.* **2014**, *54*, 199–206. [CrossRef] [PubMed]
79. Su, M.; Liu, H.; Ge, S.; Ren, N.; Ding, L.; Yu, J.; Song, X. An electrochemiluminescence lab-on-paper device for sensitive detection of two antigens at the MCF-7 cell surface based on porous bimetallic AuPd nanoparticles. *RSC Adv.* **2016**, *6*, 16500–16506. [CrossRef]
80. Wang, S.; Ge, L.; Yan, M.; Yu, J.; Song, X.; Ge, S.; Huang, J. 3D microfluidic origami electrochemiluminescence immunodevice for sensitive point-of-care testing of carcinoma antigen 125. *Sens. Actuators B Chem.* **2013**, *176*, 1–8. [CrossRef]
81. Cui, H.; Wang, W.; Duan, C.-F.; Dong, Y.-P.; Guo, J.-Z. Synthesis, Characterization, and Electrochemiluminescence of Luminol-Reduced Gold Nanoparticles and Their Application in a Hydrogen Peroxide Sensor. *Chem. Eur. J.* **2007**, *13*, 6975–6984. [CrossRef]
82. Kong, W.; Zhou, H.; Ouyang, H.; Li, Z.; Fu, Z. A disposable label-free electrochemiluminescent immunosensor for transferrin detection based on a luminol-reduced gold nanoparticle-modified screen-printed carbon electrode. *Anal. Methods* **2014**, *6*, 2959–2964. [CrossRef]

83. Wu, L.; Sha, Y.; Li, W.; Wang, S.; Guo, Z.; Zhou, J.; Su, X.; Jiang, X. One-step preparation of disposable multi-functionalized g-C3N4 based electrochemiluminescence immunosensor for the detection of CA125. *Sens. Actuators B Chem.* **2016**, *226*, 62–68. [CrossRef]
84. Zhang, M.; Dai, W.; Yan, M.; Ge, S.; Yu, J.; Song, X.; Xu, W. Ultrasensitive electrochemiluminescence immunosensor using PtAg@carbon nanocrystals composites as labels and carbon nanotubes-chitosan/gold nanoparticles as enhancer. *Analyst* **2012**, *137*, 2112–2118. [CrossRef] [PubMed]
85. Li, W.; Li, L.; Li, S.; Wang, X.; Li, M.; Wang, S.; Yu, J. 3D origami electrochemiluminescence immunodevice based on porous silver-paper electrode and nanoporous silver double-assisted signal amplification. *Sens. Actuators B Chem.* **2013**, *188*, 417–424. [CrossRef]
86. Zhang, Y.; Li, L.; Yang, H.; Ding, Y.-N.; Su, M.; Zhu, J.; Yan, M.; Yu, J.; Song, X. Gold–silver nanocomposite-functionalized graphene sensing platform for an electrochemiluminescent immunoassay of a tumor marker. *RSC Adv.* **2013**, *3*, 14701–14709. [CrossRef]
87. Ma, C.; Liu, F.; Yang, H.; Ge, S.; Yu, J.; Yan, M.; Song, X. Application of SnO2 nanocrystal as novel electrochemiluminescence signal reporter for sensitive immunoassay with nanoporous PtRu alloy enhancement. *Sens. Actuators B Chem.* **2014**, *195*, 423–430. [CrossRef]
88. Liu, F.; Deng, W.; Zhang, Y.; Ge, S.; Yu, J.; Song, X. Application of ZnO quantum dots dotted carbon nanotube for sensitive electrochemiluminescence immunoassay based on simply electrochemical reduced Pt/Au alloy and a disposable device. *Anal. Chim. Acta* **2014**, *818*, 46–53. [CrossRef]
89. Wang, S.; Zhang, Y.; Yu, J.; Song, X.; Ge, S.; Yan, M. Application of indium tin oxide device in gold-coated magnetic iron solid support enhanced electrochemiluminescent immunosensor for determination of carcinoma embryonic antigen. *Sens. Actuators B Chem.* **2012**, *171–172*, 891–898. [CrossRef]
90. Li, L.; Li, W.; Ma, C.; Yang, H.; Ge, S.; Yu, J. Paper-based electrochemiluminescence immunodevice for carcinoembryonic antigen using nanoporous gold-chitosan hybrids and graphene quantum dots functionalized Au@Pt. *Sens. Actuators B Chem.* **2014**, *202*, 314–322. [CrossRef]
91. Liu, W.; Ma, C.; Yang, H.; Zhang, Y.; Yan, M.; Ge, S.; Yu, J.; Song, X. Electrochemiluminescence immunoassay using a paper electrode incorporating porous silver and modified with mesoporous silica nanoparticles functionalized with blue-luminescent carbon dots. *Microchim. Acta* **2014**, *181*, 1415–1422. [CrossRef]
92. Yan, J.; Yan, M.; Ge, L.; Ge, S.; Yu, J. An origami electrochemiluminescence immunosensor based on gold/graphene for specific, sensitive point-of-care testing of carcinoembryonic antigen. *Sens. Actuators B Chem.* **2014**, *193*, 247–254. [CrossRef]
93. Gutiérrez-Sánchez, C.; Pita, M.; Vaz-Domínguez, C.; Shleev, S.; De Lacey, A.L. Gold Nanoparticles as Electronic Bridges for Laccase-Based Biocathodes. *J. Am. Chem. Soc.* **2012**, *134*, 17212–17220. [CrossRef] [PubMed]
94. Dhanjai; Lu, X.; Wu, L.; Chen, J.; Lu, Y. Robust Single-Molecule Enzyme Nanocapsules for Biosensing with Significantly Improved Biosensor Stability. *Anal. Chem.* **2020**, *92*, 5830–5837. [CrossRef]
95. Miao, Y.; Tan, S.N. Amperometric hydrogen peroxide biosensor based on immobilization of peroxidase in chitosan matrix crosslinked with glutaraldehyde. *Analyst* **2000**, *125*, 1591–1594. [CrossRef]
96. Cen, Y.-K.; Liu, Y.-X.; Xue, Y.-P.; Zheng, Y.-G. Immobilization of Enzymes in/on Membranes and their Applications. *Adv. Synth. Catal.* **2019**, *361*, 5500–5515. [CrossRef]
97. Fang, C.; Li, H.; Yan, J.; Guo, H.; Yifeng, T. Progress of the Electrochemiluminescence Biosensing Strategy for Clinical Diagnosis with Luminol as the Sensing Probe. *ChemElectroChem* **2017**, *4*, 1587–1593. [CrossRef]
98. Da Silva, B.F.; Pérez, S.; Gardinalli, P.; Singhal, R.K.; Mozeto, A.A.; Barceló, D. Analytical chemistry of metallic nanoparticles in natural environments. *TrAC Trends Anal. Chem.* **2011**, *30*, 528–540. [CrossRef]
99. Harvey, N. Luminescence during Electrolysis. *J. Phys. Chem.* **1929**, *33*, 1456–1459. [CrossRef]
100. Richter, M.M. Electrochemiluminescence (ECL). *Chem. Rev.* **2004**, *104*, 3003–3036. [CrossRef]
101. Kremeskotter, J.; Wilson, R.; Schiffrin, D.J.; Luff, B.J.; Wilkinson, J.S. Detection of glucose via electrochemiluminescence in a thin-layer cell with a planar optical waveguide. *Meas. Sci. Technol.* **1995**, *6*, 1325–1328. [CrossRef]
102. Zhu, L.; Li, Y.; Tian, F.; Xu, B.; Zhu, G. Electrochemiluminescent determination of glucose with a sol–gel derived ceramic–carbon composite electrode as a renewable optical fiber biosensor. *Sens. Actuators B Chem.* **2002**, *84*, 265–270. [CrossRef]
103. Lin, Z.; Chen, J.; Chen, G. An ECL biosensor for glucose based on carbon-nanotube/Nafion film modified glass carbon electrode. *Electrochim. Acta* **2008**, *53*, 2396–2401. [CrossRef]

104. Liu, L.; Ma, Q.; Li, Y.; Liu, Z.; Su, X. A novel signal-off electrochemiluminescence biosensor for the determination of glucose based on double nanoparticles. *Biosens. Bioelectron.* **2015**, *63*, 519–524. [CrossRef] [PubMed]
105. Dai, H.; Wu, X.; Xu, H.; Wang, Y.; Chi, Y.; Chen, G. A highly performing electrochemiluminescent biosensor for glucose based on a polyelectrolyte-chitosan modified electrode. *Electrochim. Acta* **2009**, *54*, 4582–4586. [CrossRef]
106. Qiu, B.; Lin, Z.; Wang, J.; Chen, Z.; Chen, J.; Chen, G. An electrochemiluminescent biosensor for glucose based on the electrochemiluminescence of luminol on the nafion/glucose oxidase/poly(nickel(II) tetrasulfophthalocyanine)/multi-walled carbon nanotubes modified electrode. *Talanta* **2009**, *78*, 76–80. [CrossRef]
107. Lou, F.; Lu, Z.; Hu, F.; Li, C.M. A 3D bio-platform constructed by glucose oxidase adsorbed on Au nanoparticles assembled polyaniline nanowires to sensitively detect glucose by electrochemiluminescence. *J. Electroanal. Chem.* **2017**, *787*, 125–131. [CrossRef]
108. Cheng, L.; Deng, S.; Lei, J.; Ju, H. Disposable electrochemiluminescent biosensor using bidentate-chelated CdTe quantum dots as emitters for sensitive detection of glucose. *Analyst* **2012**, *137*, 140–144. [CrossRef]
109. Yu, L.; Wei, X.; Fang, C.; Tu, Y. A disposable biosensor for noninvasive diabetic diagnosis rest on the Au/TiO2 nano-composite intensified electrochemiluminescence. *Electrochim. Acta* **2016**, *211*, 27–35. [CrossRef]
110. Chu, H.; Wei, X.; Wu, M.; Yan, J.; Tu, Y. An electrochemiluminescent biosensor based on polypyrrole immobilized uricase for ultrasensitive uric acid detection. *Sens. Actuators B Chem.* **2012**, *163*, 247–252. [CrossRef]
111. Ballesta-Claver, J.; Díaz Ortega, I.F.; Valencia-Mirón, M.C.; Capitán-Vallvey, L.F. Disposable luminol copolymer-based biosensor for uric acid in urine. *Anal. Chim. Acta* **2011**, *702*, 254–261. [CrossRef]
112. Bravo, I.; Gutiérrez-Sánchez, C.; García-Mendiola, T.; Revenga-Parra, M.; Pariente, F.; Lorenzo, E. Enhanced Performance of Reagent-Less Carbon Nanodots Based Enzyme Electrochemical Biosensors. *Sensors* **2019**, *19*, 5576. [CrossRef]
113. Cai, X.; Yan, J.; Chu, H.; Wu, M.; Tu, Y. An exercise degree monitoring biosensor based on electrochemiluminescent detection of lactate in sweat. *Sens. Actuators B Chem.* **2010**, *143*, 655–659. [CrossRef]
114. Claver, J.B.; Mirón, M.C.V.; Capitán-Vallvey, L.F. Disposable electrochemiluminescent biosensor for lactate determination in saliva. *Analyst* **2009**, *134*, 1423–1432. [CrossRef] [PubMed]
115. Marquette, C.A.; Ravaud, S.; Blum, L.J. Luminol Electrochemiluminescence-Based Biosensor for Total Cholesterol Determination in Natural Samples. *Anal. Lett.* **2000**, *33*, 1779–1796. [CrossRef]
116. Zhu, S.; Lin, X.; Ran, P.; Mo, F.; Xia, Q.; Fu, Y. A glassy carbon electrode modified with C-dots and silver nanoparticles for enzymatic electrochemiluminescent detection of glutamate enantiomers. *Microchim. Acta* **2017**, *184*, 4679–4684. [CrossRef]
117. Zhao, Q.; Tang, S.; Fang, C.; Tu, Y.-F. Titania nanotubes decorated with gold nanoparticles for electrochemiluminescent biosensing of glycosylated hemoglobin. *Anal. Chim. Acta* **2016**, *936*, 83–90. [CrossRef]
118. Dai, H.; Chi, Y.; Wu, X.; Wang, Y.; Wei, M.; Chen, G. Biocompatible electrochemiluminescent biosensor for choline based on enzyme/titanate nanotubes/chitosan composite modified electrode. *Biosens. Bioelectron.* **2010**, *25*, 1414–1419. [CrossRef]
119. Gao, W.; Chen, Y.; Xi, J.; Lin, S.; Chen, Y.; Lin, Y.; Chen, Z. A novel electrochemiluminescence ethanol biosensor based on tris(2,2′-bipyridine) ruthenium (II) and alcohol dehydrogenase immobilized in graphene/bovine serum albumin composite film. *Biosens. Bioelectron.* **2013**, *41*, 776–782. [CrossRef]
120. Leca, B.; Blum, L.J. Luminol electrochemiluminescence with screen-printed electrodes for low-cost disposable oxidase-based optical sensors. *Analyst* **2000**, *125*, 789–791. [CrossRef]
121. Corgier, B.P.; Marquette, C.A.; Blum, L.J. Screen-printed electrode microarray for electrochemiluminescent measurements. *Anal. Chim. Acta* **2005**, *538*, 1–7. [CrossRef]
122. Wang, L.; Wang, Y.; Xu, T.; Liao, H.; Yao, C.; Liu, Y.; Li, Z.; Chen, Z.; Pan, D.; Sun, L.; et al. Gram-scale synthesis of single-crystalline graphene quantum dots with superior optical properties. *Nat. Commun.* **2014**, *5*, 5357. [CrossRef]
123. Mo, R.; Li, F.; Tan, X.; Xu, P.; Tao, R.; Shen, G.; Lu, X.; Liu, F.; Shen, L.; Xu, B.; et al. High-quality mesoporous graphene particles as high-energy and fast-charging anodes for lithium-ion batteries. *Nat. Commun.* **2019**, *10*, 1474. [CrossRef]

124. Chen, H.; Dong, F.; Minteer, S.D. The progress and outlook of bioelectrocatalysis for the production of chemicals, fuels and materials. *Nat. Catal.* **2020**, *3*, 225–244. [CrossRef]
125. Deshmukh, M.A.; Park, S.-J.; Kang, B.-C.; Ha, T.-J. Carbon Nanohybrids for Advanced Electronic Applications. *Phys. Status Solidi (a)* **2020**. [CrossRef]
126. Valenti, G.; Rampazzo, E.; Kesarkar, S.; Genovese, D.; Fiorani, A.; Zanut, A.; Palomba, F.; Marcaccio, M.; Paolucci, F.; Prodi, L. Electrogenerated chemiluminescence from metal complexes-based nanoparticles for highly sensitive sensors applications. *Coord. Chem. Rev.* **2018**, *367*, 65–81. [CrossRef]
127. Molinero-Abad, B.; Alonso-Lomillo, M.A.; Domínguez-Renedo, O.; Arcos-Martínez, M.J. Amperometric determination of sulfite using screen-printed electrodes modified with metallic nanoparticles. *Microchim. Acta* **2013**, *180*, 1351–1355. [CrossRef]
128. Pita, M.; Gutierrez-Sanchez, C.; Toscano, M.D.; Shleev, S.; De Lacey, A.L. Oxygen biosensor based on bilirubin oxidase immobilized on a nanostructured gold electrode. *Bioelectrochemistry* **2013**, *94*, 69–74. [CrossRef]
129. Fritea, L.; Tertis, M.; Sandulescu, R.; Cristea, C. Chapter Eleven—Enzyme–Graphene platforms for electrochemical biosensor design with biomedical applications. In *Methods in Enzymology*; Kumar, C.V., Ed.; Academic Press: Cambridge, MA, USA, 2018; Volume 609, pp. 293–333.
130. Martinez-Perinan, E.; Revenga-Parra, M.; Gennari, M.; Pariente, F.; Mas-Balleste, R.; Zamora, F.; Lorenzo, E. Insulin sensor based on nanoparticle-decorated multiwalled carbon nanotubes modified electrodes. *Sens. Actuators B-Chem.* **2016**, *222*, 331–338. [CrossRef]
131. Cancelliere, R.; Carbone, K.; Pagano, M.; Cacciotti, I.; Micheli, L. Biochar from Brewers' Spent Grain: A Green and Low-Cost Smart Material to Modify Screen-Printed Electrodes. *Biosensors* **2019**, *9*, 139. [CrossRef]
132. Bertoncello, P.; Ugo, P. Recent Advances in Electrochemiluminescence with Quantum Dots and Arrays of Nanoelectrodes. *ChemElectroChem* **2017**, *4*, 1663–1676. [CrossRef]
133. Soldatkina, O.V.; Soldatkin, O.O.; Kasap, B.O.; Kucherenko, D.Y.; Kucherenko, I.S.; Kurc, B.A.; Dzyadevych, S.V. A Novel Amperometric Glutamate Biosensor Based on Glutamate Oxidase Adsorbed on Silicalite. *Nanoscale Res. Lett.* **2017**, *12*, 260. [CrossRef]
134. Hezard, T.; Fajerwerg, K.; Evrard, D.; Collière, V.; Behra, P.; Gros, P. Gold nanoparticles electrodeposited on glassy carbon using cyclic voltammetry: Application to Hg(II) trace analysis. *J. Electroanal. Chem.* **2012**, *664*, 46–52. [CrossRef]
135. Gutiérrez-Sánchez, C.; Mediavilla, M.; Guerrero-Esteban, T.; Revenga-Parra, M.; Pariente, F.; Lorenzo, E. Direct covalent immobilization of new nitrogen-doped carbon nanodots by electrografting for sensing applications. *Carbon* **2020**, *159*, 303–310. [CrossRef]
136. Bravo, I.; García-Mendiola, T.; Revenga-Parra, M.; Pariente, F.; Lorenzo, E. Diazonium salt click chemistry based multiwall carbon nanotube electrocatalytic platforms. *Sens. Actuators B Chem.* **2015**, *211*, 559–568. [CrossRef]
137. Castrovilli, M.C.; Bolognesi, P.; Chiarinelli, J.; Avaldi, L.; Cartoni, A.; Calandra, P.; Tempesta, E.; Giardi, M.T.; Antonacci, A.; Arduini, F.; et al. Electrospray deposition as a smart technique for laccase immobilisation on carbon black-nanomodified screen-printed electrodes. *Biosens. Bioelectron.* **2020**, *163*, 112299. [CrossRef] [PubMed]
138. Antuña-Jiménez, D.; González-García, M.B.; Hernández-Santos, D.; Fanjul-Bolado, P. Screen-Printed Electrodes Modified with Metal Nanoparticles for Small Molecule Sensing. *Biosensors* **2020**, *10*, 9. [CrossRef]
139. Tian, C.; Wang, L.; Luan, F.; Zhuang, X. An electrochemiluminescence sensor for the detection of prostate protein antigen based on the graphene quantum dots infilled TiO2 nanotube arrays. *Talanta* **2019**, *191*, 103–108. [CrossRef]
140. Habekost, A. Rapid and sensitive spectroelectrochemical and electrochemical detection of glyphosate and AMPA with screen-printed electrodes. *Talanta* **2017**, *162*, 583–588. [CrossRef]
141. Guerrero-Esteban, T.; Gutiérrez-Sánchez, C.; Revenga-Parra, M.; Pau, J.L.; Pariente, F.; Lorenzo, E. Enhanced electrochemiluminescence by ZnO nanowires for taurine determination. *Talanta* **2019**, *204*, 63–69. [CrossRef]
142. Long, Y.-M.; Bao, L.; Peng, Y.; Zhang, Z.-L.; Pang, D.-W. Self-co-reactant and ion-annihilation electrogenerated chemiluminescence of carbon nanodots. *Carbon* **2018**, *129*, 168–174. [CrossRef]
143. Chen, Y.; Cao, Y.; Ma, C.; Zhu, J.-J. Carbon-based dots for electrochemiluminescence sensing. *Mater. Chem. Front.* **2020**, *4*, 369–385. [CrossRef]

144. Liu, X.; Fang, C.; Yan, J.; Li, H.; Tu, Y. A sensitive electrochemiluminescent biosensor based on AuNP-functionalized ITO for a label-free immunoassay of C-peptide. *Bioelectrochemistry* **2018**, *123*, 211–218. [CrossRef] [PubMed]
145. Zhang, M.; Yuan, R.; Chai, Y.; Chen, S.; Zhong, H.; Wang, C.; Cheng, Y. A biosensor for cholesterol based on gold nanoparticles-catalyzed luminol electrogenerated chemiluminescence. *Biosens. Bioelectron.* **2012**, *32*, 288–292. [CrossRef]
146. Li, J.-J.; Shang, L.; Jia, L.-P.; Ma, R.-N.; Zhang, W.; Jia, W.-L.; Wang, H.-S.; Xu, K.-H. An ultrasensitive electrochemiluminescence sensor for the detection of HULC based on Au@Ag/GQDs as a signal indicator. *J. Electroanal. Chem.* **2018**, *824*, 114–120. [CrossRef]
147. Ma, F.; Zhang, Q.; Zhang, C.-Y. Nanomaterial-based biosensors for DNA methyltransferase assay. *J. Mater. Chem. B* **2020**, *8*, 3488–3501. [CrossRef] [PubMed]
148. Zhang, C.; Fan, Y.; Zhang, H.; Chen, S.; Yuan, R. An ultrasensitive signal-on electrochemiluminescence biosensor based on Au nanoclusters for detecting acetylthiocholine. *Anal. Bioanal. Chem.* **2019**, *411*, 905–913. [CrossRef] [PubMed]
149. Salehnia, F.; Hosseini, M.; Ganjali, M.R. Enhanced electrochemiluminescence of luminol by an in situ silver nanoparticle-decorated graphene dot for glucose analysis. *Anal. Methods* **2018**, *10*, 508–514. [CrossRef]
150. Du, X.; Jiang, D.; Chen, S.; Dai, L.; Zhou, L.; Hao, N.; You, T.; Mao, H.; Wang, K. CeO2 nanocrystallines ensemble-on-nitrogen-doped graphene nanocomposites: One-pot, rapid synthesis and excellent electrocatalytic activity for enzymatic biosensing. *Biosens. Bioelectron.* **2017**, *89*, 681–688. [CrossRef]
151. García, T.; Revenga-Parra, M.; Sobrino, B.; Carracedo, A.; Alonso, C.; Lorenzo, E.; Pariente, F. Electrochemical DNA base pairs quantification and endonuclease cleavage detection. *Biosens. Bioelectron.* **2011**, *27*, 40–45. [CrossRef]
152. García-Mendiola, T.; Barreiro Martínez, T.; Pariente, F.; Molano, J.; Lorenzo, E. Screening of Specific Gene Mutations Associated with Cystic Fibrosis. *Electroanalysis* **2014**, *26*, 1362–1372. [CrossRef]
153. Hasanzadeh, M.; Shadjou, N. Electrochemical nanobiosensing in whole blood: Recent advances. *TrAC Trends Anal. Chem.* **2016**, *80*, 167–176. [CrossRef]
154. Carter, M.T.; Bard, A.J. Electrochemical investigations of the interaction of metal chelates with DNA. 3. Electrogenerated chemiluminescent investigation of the interaction of tris(1,10-phenanthroline)ruthenium(II) with DNA. *Bioconjugate Chem.* **1990**, *1*, 257–263. [CrossRef] [PubMed]
155. Rodriguez, M.; Bard, A.J. Electrochemical studies of the interaction of metal chelates with DNA. 4. Voltammetric and electrogenerated chemiluminescent studies of the interaction of tris(2,2′-bipyridine) osmium(II) with DNA. *Anal. Chem.* **1990**, *62*, 2658–2662. [CrossRef] [PubMed]
156. Zhang, Y.; Xu, G.; Lian, G.; Luo, F.; Xie, Q.; Lin, Z.; Chen, G. Electrochemiluminescence biosensor for miRNA-21 based on toehold-mediated strand displacement amplification with Ru(phen)32+ loaded DNA nanoclews as signal tags. *Biosens. Bioelectron.* **2020**, *147*, 111789. [CrossRef] [PubMed]
157. Gao, W.; Liu, Y.; Zhang, H.; Wang, Z. Electrochemiluminescence Biosensor for Nucleolin Imaging in a Single Tumor Cell Combined with Synergetic Therapy of Tumor. *ACS Sens.* **2020**, *5*, 1216–1222. [CrossRef] [PubMed]
158. Khoshfetrat, S.M.; Ranjbari, M.; Shayan, M.; Mehrgardi, M.A.; Kiani, A. Wireless Electrochemiluminescence Bipolar Electrode Array for Visualized Genotyping of Single Nucleotide Polymorphism. *Anal. Chem.* **2015**, *87*, 8123–8131. [CrossRef]
159. Nie, Y.; Zhang, X.; Zhang, Q.; Liang, Z.; Ma, Q.; Su, X. A novel high efficient electrochemiluminescence sensor based on reductive Cu(I) particles catalyzed Zn-doped MoS2 QDs for HPV 16 DNA determination. *Biosens. Bioelectron.* **2020**, *160*, 112217. [CrossRef]
160. Liu, D.; Zhang, X.; Zhao, J.; Chen, S.; Yuan, R. An ultrasensitive sensing platform for microRNA-155 based on H2O2 quenched hydroxide-dependent ECL emission of PFO Pdots. *Biosens. Bioelectron.* **2020**, *150*, 111872. [CrossRef]
161. Li, J.; Tan, S.; Kooger, R.; Zhang, C.; Zhang, Y. MicroRNAs as novel biological targets for detection and regulation. *Chem. Soc. Rev.* **2014**, *43*, 506–517. [CrossRef]
162. Xia, N.; Zhang, Y.; Wei, X.; Huang, Y.; Liu, L. An electrochemical microRNAs biosensor with the signal amplification of alkaline phosphatase and electrochemical–chemical–chemical redox cycling. *Anal. Chim. Acta* **2015**, *878*, 95–101. [CrossRef]

163. Liu, L.; Gao, Y.; Liu, H.; Xia, N. An ultrasensitive electrochemical miRNAs sensor based on miRNAs-initiated cleavage of DNA by duplex-specific nuclease and signal amplification of enzyme plus redox cycling reaction. *Sens. Actuators B Chem.* **2015**, *208*, 137–142. [CrossRef]
164. Xia, N.; Liu, K.; Zhou, Y.; Li, Y.; Yi, X. Sensitive detection of microRNAs based on the conversion of colorimetric assay into electrochemical analysis with duplex-specific nuclease-assisted signal amplification. *Int. J. Nanomed.* **2017**, *12*, 5013–5022. [CrossRef] [PubMed]
165. Liu, L.; Xia, N.; Liu, H.; Kang, X.; Liu, X.; Xue, C.; He, X. Highly sensitive and label-free electrochemical detection of microRNAs based on triple signal amplification of multifunctional gold nanoparticles, enzymes and redox-cycling reaction. *Biosens. Bioelectron.* **2014**, *53*, 399–405. [CrossRef] [PubMed]
166. Wang, F.; Fu, C.; Huang, C.; Li, N.; Wang, Y.; Ge, S.; Yu, J. Paper-based closed Au-Bipolar electrode electrochemiluminescence sensing platform for the detection of miRNA-155. *Biosens. Bioelectron.* **2020**, *150*, 111917. [CrossRef] [PubMed]
167. Wang, X.; Wang, Y.; Shan, Y.; Jiang, M.; Jin, X.; Gong, M.; Xu, J. A novel and sensitive electrogenerated chemiluminescence biosensor for detection of p16INK4a gene based on the functional paste-like nanofibers composites-modified screen-printed carbon electrode. *J. Electroanal. Chem.* **2018**, *823*, 368–377. [CrossRef]
168. Neethirajan, S.; Kobayashi, I.; Nakajima, M.; Wu, D.; Nandagopal, S.; Lin, F. Microfluidics for food, agriculture and biosystems industries. *Lab Chip* **2011**, *11*, 1574–1586. [CrossRef]
169. Azam, N.F.N.; Roy, S.; Lim, S.A.; Uddin Ahmed, M. Meat species identification using DNA-luminol interaction and their slow diffusion onto the biochip surface. *Food Chem.* **2018**, *248*, 29–36. [CrossRef]
170. Anderson, P.D.; Bokor, G. Bioterrorism: Pathogens as Weapons. *J. Pharm. Pract.* **2012**, *25*, 521–529. [CrossRef]
171. Klietmann, W.F.; Ruoff, K.L. Bioterrorism: Implications for the clinical microbiologist. *Clin. Microbiol. Rev.* **2001**, *14*, 364–381. [CrossRef]
172. Shah, J.; Wilkins, E. Electrochemical Biosensors for Detection of Biological Warfare Agents. *Electroanalysis* **2003**, *15*, 157–167. [CrossRef]
173. Spehar-Délèze, A.-M.; Gransee, R.; Martinez-Montequin, S.; Bejarano-Nosas, D.; Dulay, S.; Julich, S.; Tomaso, H.; O'Sullivan, C.K. Electrochemiluminescence DNA sensor array for multiplex detection of biowarfare agents. *Anal. Bioanal. Chem.* **2015**, *407*, 6657–6667. [CrossRef]

biosensors

MDPI

Article

Impedimetric Sensing of Factor V Leiden Mutation by Zip Nucleic Acid Probe and Electrochemical Array

Arzum Erdem * and **Ece Eksin**

Analytical Chemistry Department, Faculty of Pharmacy, Ege University, Bornova, Izmir 35100, Turkey; eceksin@hotmail.com

* Correspondence: arzum.erdem@ege.edu.tr or arzume@hotmail.com

Received: 29 July 2020; Accepted: 3 September 2020; Published: 7 September 2020

check for updates

Abstract: A carbon nanofiber enriched 8-channel screen-printed electrochemical array was used for the impedimetric detection of SNP related to Factor V Leiden (FV Leiden) mutation, which is the most common inherited form of thrombophilia. FV Leiden mutation sensing was carried out in three steps: solution-phase nucleic acid hybridization between zip nucleic acid probe (Z-probe) and mutant type DNA target, followed by the immobilization of the hybrid on the working electrode area of array, and measurement by electrochemical impedance spectroscopy (EIS). The selectivity of the assay was tested against mutation-free DNA sequences and synthetic polymerase chain reaction (PCR) samples. The developed biosensor was a trustful assay for FV Leiden mutation diagnosis, which can effectively discriminate wild type and mutant type even in PCR samples.

Keywords: 8-channel screen-printed electrochemical arrays; zip nucleic acids; SNP; electrochemical impedance spectroscopy

1. Introduction

Laboratory analyzers are suited for a hospital setting and requires trained personnel to operate the analyzer and interpret the results accurately. However, biosensors have many advantages since they are user-friendly and provide results rapidly with high sensitivity. A great demand for accurate monitoring of biomarkers related to important diseases is the driving force toward the development of novel analytical tools for diagnostics [1,2].

Many researchers are interested in the development of fast screening tools for clinical use. These tools are mostly based on optical or spectroscopic techniques. However, electrochemical methods have many strong points with regard to other techniques such as simple instrumentation, easy to use, low cost, and low sample necessity [3]. Electrochemical biosensors provide direct analysis of various analytes within minutes. Therefore, electrochemical biosensors can be considered as the most appropriate tool for clinical diagnosis by means of their superior features as above-mentioned. Electrochemical DNA biosensor arrays, as a member of the electrochemical biosensor family, are widely used as powerful tools for the diagnosis of genetic and infectious diseases [4,5]. Many recent works have demonstrated the usage of screen-printed electrode arrays in order to carry out multiple measurements at the same time [5–9]. Recently, our group developed sensitive biosensors for the detection of protein, microRNA, and SNPs [10–15].

FV Leiden, with the most common inherited prothrombotic conditions, occurs due to a single point mutation of the coagulation factor V gene in the chromosome [16]. Between 3% and 8% of Europeans carry the one copy of the Factor V Leiden mutation, and about 1 in 5000 people have two copies of the FV Leiden mutation. Inheriting one copy slightly increases the risk of developing blood clots. Furthermore, inheriting two copies—one from each parent—significantly increases the risk of developing blood clots. These abnormal clots can lead to long-term health problems or become life-threatening [17]. Due to the importance of FV Leiden mutation sensing, there is an urgent need to develop sensitive, reliable, and fast detection protocols for the FV Leiden mutation. Under this scope, the quantitative analysis of Factor V Leiden was carried out by different methodologies such as the immunosorbent assay [18], fluorescent assay [19], and sandwich-optical sensing method [20]. However, these methods require expensive facilities and complex procedures in combination with the use of radioactive/fluorescent tags. Therefore, these methods are not suitable for the development of simple and low-cost point-of-care (PoC) devices.

Zip nucleic acids contain a cationic compound, spermine, that has an impact on the affinity between the oligonucleotide and its target nucleic acid. Therefore, single base-mismatched sequences or single nucleotide polymorphisms (SNPs) can be successfully discriminated by using zip nucleic acid probes (Z-probes) [21]. Recent studies have confirmed the reliability of Z-probes such as primers, real-time PCR probes, and splice switching oligonucleotides (SSOs) [22–25]. Z-probe applications also include miRNA detection, detection of AT-rich sequences, in situ hybridization, etc. [26,27]. Furthermore, Z-probes were used as an efficient probe for the development of the impedimetric detection protocol of single nucleotide mutation related to FV Leiden in our previous study [28]. Before and after solution-phase hybridization occurred between Z-probes and its mutant type DNA target, the impedimetric measurement was performed by carbon nanofiber enriched screen-printed electrodes. The impedimetric detection of different single point mutations such as G to A, G to C, and G to T in short DNA oligonucleotides was successfully carried out. The discrimination between mutant type DNA (G to A) and wild type DNA was explored successfully even though the target sequence with a mutation (G to A) was at the 3′-end position of both PCR products at the length of 143 nt or 220 nt. In addition, the detection of any other SNPs (G to C, or G to T) was performed in solution phase hybridization more selectively by using Z-probes in contrast to the DNA probe. Please note that this is a follow-up study of our previous study.

Herein, EIS based sensing protocol for of the FV Leiden mutation was performed by 8-channel screen-printed electrochemical arrays. In order to improve the efficiency of solution-phase nucleic acid hybridization between Z-probes and target sequence, the experimental parameters were optimized. The selectivity of the assay was tested against wild type DNA sequence and synthetic PCR samples.

2. Materials and Methods

2.1. Instruments and Chemicals

AUTOLAB-302 PGSTAT with the GPES 4.9.007 software package (Eco Chemie, Utrecht, The Netherlands) was used for electrochemical impedance (EIS) measurements.

The detailed information about oligonucleotides, PCR products, and carbon nanofibers enriched 8-channel screen-printed electrochemical arrays can be found in the Supplementary Materials.

2.2. Methodology

The following steps were carried out for EIS-based sensing of the FV Leiden mutation.

(i). Hybridization of the Z-probe with mutant type DNA, or wild type DNA, or C-mutant type DNA, T-mutant type DNA, ODN-1, ODN-2, mutant type, and wild type PCR in the solution phase;
(ii). Immobilization of the hybridization products on the working electrode area of the array electrode;
(iii). Impedimetric measurements.

The desired concentrations of the Z-probe (or DNA probe) and mutant type DNA (or any other of oligonucleotides (ODNs)) were prepared in phosphate buffer saline (PBS, pH, 7.4) and mixed in the ratio of 1:1 (v:v). This mixture was allowed to undergo solution-phase hybridization for 10 min with gentle mixing at 400 rpm under room temperature using a Thermo-Shaker (Biosan, Latvia).

The solution containing hybrids of the Z-probe-DNA target or DNA probe-DNA target was dropped onto the surface of the working electrode of the array system and incubated for 15 min. Immobilization of the hybrids was performed according to the drop-casting method. After that, the electrodes were washed with PBS before measurement. The representative scheme is given in Scheme 1.

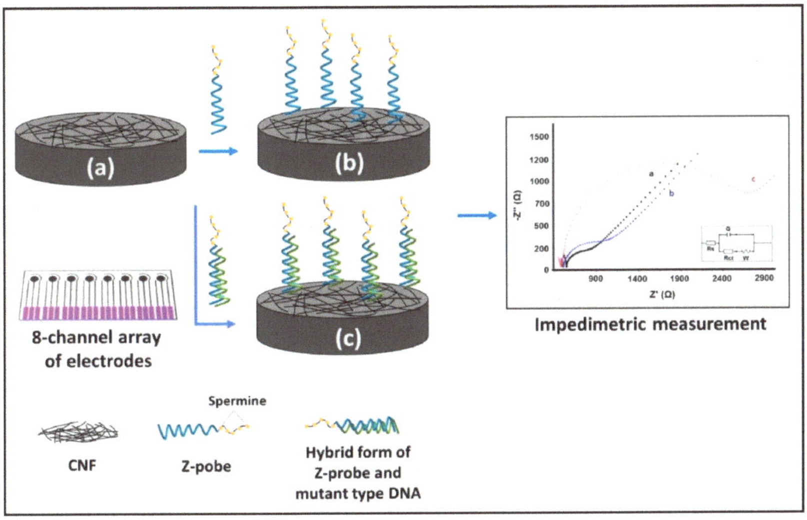

Scheme 1. The representative experimental scheme related to electrochemical impedance spectroscopy based sensing of the FV Leiden mutation. (**a**) Electrode control, (**b**) Z-probe in the absence of target, (**c**) the hybrid form of Z-probe and mutant type DNA target.

2.3. Electrochemical Impedance Spectroscopy Measurements

EIS measurements were performed as reported in our previous work [28]. The Randles circuit was used as the equivalent circuit model used for fitting EIS data, which is shown as the inset in all Nyquist diagrams.

3. Results

We aimed to apply our previous assay [28] to the 8-channel array of electrodes in order to perform multiple simultaneous analysis. This array of electrodes has been successfully employed in numerous applications based on electrochemical biosensors [10–15,29,30] while presenting their great advantages such as low sample requirement and easy implementation to point of care (PoC) system. The characterization of the carbon nanofiber enriched electrodes has been given in earlier reports [28,31–33].

The impedimetric sensing performance and operational characteristics of the array biosensor were evaluated after optimizing the experimental conditions (Figures S1–S7). The optimized variables are given in Table 1 and the Nyquist diagrams related to the hybridization of the Z-probe and mutant type DNA target under optimized conditions are given in Figure 1.

The effect of mutant type DNA target concentration on the hybridization process was investigated. Figure S8 shows the line graph for the tested mutant type DNA target concentrations. There was an increase at R_{ct} up to 12.0 μg mL^{-1}, then a decrease at R_{ct} was recorded up to 16.0 μg mL^{-1}. As shown in Figure S8, the highest R_{ct} value was measured at 12.0 μg mL^{-1} in mutant type DNA target of 3226.0 ± 456.6 Ω with the RSD%, 14.2% (n = 3); the upper limit of the linear range was chosen as 12.0 μg mL^{-1}. A linearity in the response based on the R_{ct} value was obtained in the mutant type DNA concentration range varying from 2.0 to 10.0 μg mL^{-1}. The representative Nyquist diagrams are shown in Figure 2. The limit of detection was estimated by using the Miller and Miller technique [34] and was found to be 1.9 μg mL^{-1} (equal to 266.0 nM, 2.6 pmol in the 10.0 μL sample) according to the calibration plot shown in the Figure S8 inset with the equation of $y = 208.96x + 235.66$. In addition, the sensitivity was calculated and found to be 2149.8 Ω·mL/μg·cm^2.

Table 1. Experimental variables optimized for the impedimetric sensing performance of the array biosensor.

Variable	Evaluated Conditions	Optimum Condition
Hybridization temperature, °C	25, 50, 75	25
Hybridization buffer, pH	ABS (pH, 4.8) PBS (pH, 7.4) CBS (pH, 9.5)	PBS (pH 7.4)
[Mg^{2+}] in hybridization buffer, mM	NA *, 0.5, 1	NA *
Hybridization time, minute	5, 10, 15	10

* NA: There is no Mg^{2+} available in hybridization buffer.

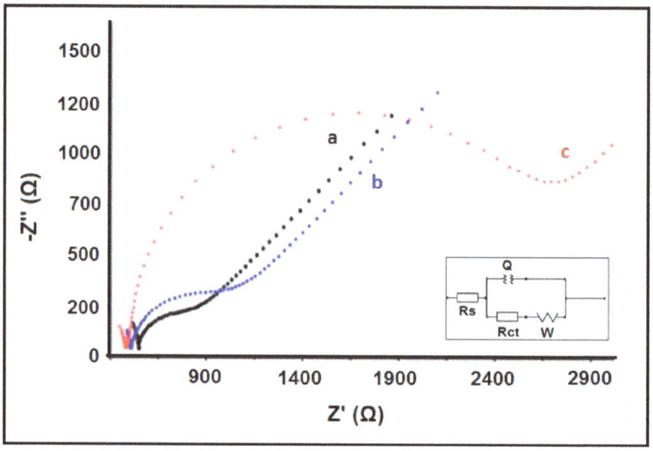

Figure 1. The Nyquist diagrams obtained after the hybridization of the Z-probe and mutant type DNA target under optimized conditions. (**a**) Electrode control, (**b**) Z-probe in the absence of target, (**c**) the hybrid form of Z-probe and mutant type DNA target.

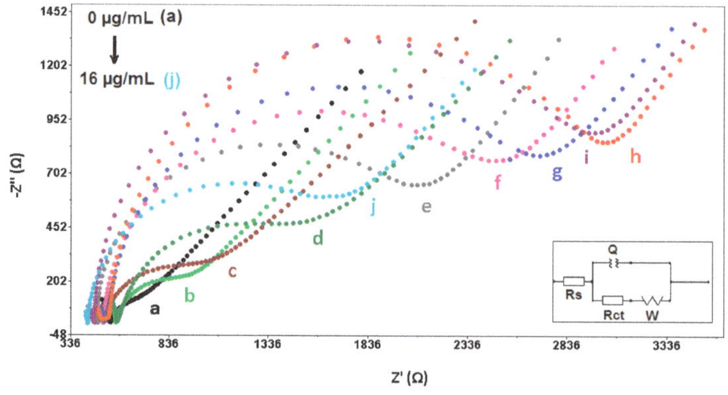

Figure 2. Nyquist diagrams related to (**a**) electrode control, (**b**) 1.0 µg mL^{-1} Z-probe in the absence of target, the hybrid form of Z-probe and (**c**) 2.0, (**d**) 4.0, (**e**) 6.0, (**f**) 8.0, (**g**) 10.0, (**h**) 12.0, (**i**) 14.0, (**j**) 16.0 µg mL^{-1} mutant type DNA target.

The selectivity of the Z-probe to mutant type DNA target was then tested against the wild type DNA target. The same experiment was performed with the DNA probe instead of the Z-probe in order to compare the performance of the Z-probe.

The average R_{ct} value was obtained of 3219.0 ± 373.0 Ω (RSD%, 11.6%, n = 3) after hybridization of the Z-probe and mutant type DNA target (Figure 3A). However, it was 1083.0 ± 65.0 Ω (RSD%, 6.0%, n = 3) after the hybridization of the Z-probe with the wild type DNA target. On the other hand, the average R_{ct} value was obtained of 2504.0 ± 629.3 Ω (RSD%, 25.1%, n = 3) in the presence of hybrid of the DNA probe and mutant type DNA target (Figure 3B), and 2327.0 ± 89.1 Ω (RSD%, 3.8%, n = 3) with the hybrid of the DNA probe and wild type DNA target. According to the efficiency of hybridization% (H_{Eff}%), it can be said that the Z-probe exhibited a selective behavior to its wild type DNA target (shown in Table 2). Nevertheless, the DNA probe was not selective enough to the wild type target.

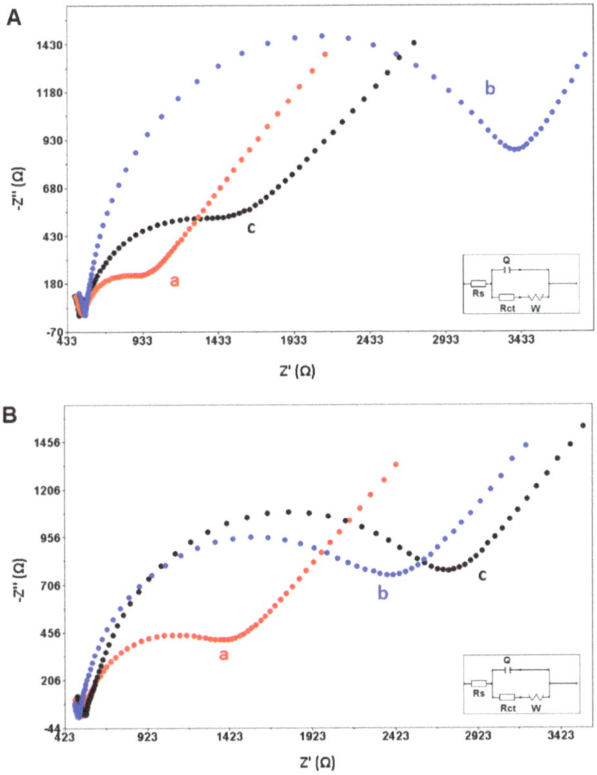

Figure 3. The hybridization between 1.0 μg mL^{-1} (**A**) Z-probe, or (**B**) DNA probe and 12.0 μg mL^{-1}. mutant type DNA or wild type DNA target. Nyquist diagrams related to (a) Z-probe or DNA probe in the absence of target, the hybrid form of Z-probe, or DNA probe with (b) mutant type DNA target, (c) wild type DNA target.

Table 2. The average R_{ct} values measured with the hybrid form of the Z-probe or DNA probe with mutant type DNA target/wild type DNA target and H_{Eff}% values.

	R_{ct} (Ω)	H_{Eff}%
Z-probe	369.4 ± 59.9	-
Z-probe and mutant type DNA target	3219.0 ± 373.0	89.0
Z-probe and wild type DNA target	1083.0 ± 65.0	65.0
DNA probe	848.0 ± 209.0	-
DNA probe and mutant type DNA Target	2504.0 ± 629.3	66.0
DNA probe and wild type DNA target	2327.0 ± 89.1	63.0

The selectivity of the impedimetric assay based on the Z-probe and nanomaterials enriched array biosensor was tested over different oligonucleotides with different single-base mutations located at mutant type DNA target sequence, or noncomplementary sequences, C-mutant type DNA or T-mutant type DNA or ODN-1 or ODN-2 (Figures S9 and S10). The average R_{ct} with the H_{Eff}% is given in Table S1. The highest H_{Eff}% was found to be 89% with the hybrid form of the Z-probe and mutant type DNA. A selective behavior of the Z-probe was monitored even in the presence of DNA oligonucleotides with different single-base mutation or noncomplementary ODNs.

The impedimetric sensing of FV Leiden mutation in the PCR products with the length of 143 nt was analyzed using the Z-probe accordingly, and presented comparatively with the DNA probe. The hybridization of the 1.0 μg mL^{-1} Z-probe and 12.0 μg mL^{-1} (equals to 0.3 μM) mutant type PCR, or wild type PCR was performed under optimum experimental conditions and the average R_{ct} with H_{Eff}% values is shown in Table 3.

Table 3. The average R_{ct} values measured with the hybrid form of the Z-probe or DNA probe with mutant type PCR or wild type PCR with the values of H_{Eff}%.

	R_{ct} (Ω)	H_{Eff}%
Z-probe	369.4 ± 59.9	-
Z-probe and mutant type PCR	2608.0 ± 361.3	86.0
Z-probe and wild type PCR	1923.3 ± 516.4	81.0
DNA probe	848.0 ± 209.0	-
DNA probe and mutant type PCR	1795.0 ± 481.2	53.0
DNA probe and wild type PCR	1848.5 ± 379.7	54.0

The H_{Eff}% value obtained by the Z-probe (86.0%) was found to be higher than the one obtained by the DNA probe (53.0%). Hence, the discrimination of a single-base mutation was selectively and sensitively explored in the presence of the Z-probe, even when the target sequence was part of the PCR product with the length of 143 nt.

The earlier studies related to the detection of DNA using the array of electrodes are listed in Table 4 and compared to the present study.

4. Conclusions

The impedimetric analysis of the FV Leiden mutation was carried out by 8-channel screen-printed electrochemical arrays in a relatively shorter time (i.e., 30 min) compared to previous works performed by different arrays of electrodes [13–15,35–47] (see Table 4). Using screen-printed electrochemical arrays, a single-base mutation successfully discriminated complementary target DNAs by means of the Z-probe. The LODs were calculated and found to be 1.9 μg mL^{-1} (266.0 nM). Based on our experiences, the Z-probe based impedimetric biosensors will be at the front of furthering the expansion of a new generation of nucleic acids on the development of PoC devices.

Table 4. The earlier studies developed for the detection of DNA by the array of electrodes in contrast to the present study.

Electrode	Modification	Analite	Technique	Assay Time	LOD	Reference
8-channel array of electrodes	Carbon nanofiber	DNA (FV Leiden mutation)	DPV	135 min	1.6 µM	[13]
8-channel array of electrodes	Carbon nanofiber	DNA (FV Leiden mutation)	DPV	35 min	0.4 µM	[14]
8-channel array of electrodes	Carbon nanofiber	DNA (FV Leiden mutation)	EIS	50 min	133.0 nM	[15]
Gold film electrode array chip	-	DNA	SWV	NA	NA	[35]
Three-dimensional interdigitated electrode array	Silane	DNA	EIS	24 h	NA	[36]
ITO electrode array	Graphene-mesoporous silica hybrid nanosheets	DNA	DPV	70 min	10.0 fM	[37]
nanodisk-array electrodes	polystyrene-block-poly(methylmethacrylate)-derived thin films	DNA	CV	2 h	0.4 nM–4.2 nM	[38]
32 microelectrode Chips (gold)	-	DNA	CV	30 min		[39]
ITO electrode array on glass wafer	-	H1N1 influenza virus DNA	Capacitance	3.5 h	3.9 nM	[40]
16 thru-hole array on printed circuit board	-	Hepatitis A-B-C virus DNA	ECL	3.5 h	NA	[41]
Gold electrode microarray	-	DNA	EIS	22 h	1.0 pM	[42]
16-gold electrode sensor arrays	-	DNA	CV, Amperometry	3 h	NA	[43]
electrode array housed within the microfluidic cell	-	*Karlodinium armiger* DNA	SWV, Chronoamperometry	17 h	277.0 aM	[44]
120-channel gold microelectrode array chip	-	miRNA	CV	1 h	140.0 zmol	[45]
multi-electrode array (6-gold electrode)	-	HIV-1, HIV-2 DNA	SWV	7.5 h	0.1 nM	[46]
16 microwells- column electrode	-	DNA	Amperometry	2 h	30.0 nM	[47]
8-channel array of electrodes	Carbon nanofiber	DNA (FV Leiden mutation)	EIS	25 min	266.0 nM	Present work

Supplementary Materials: The following are available online at http://www.mdpi.com/2079-6374/10/9/116/s1. Figure S1. The Nyquist diagrams obtained after the hybridization of 2.0 μg mL^{-1} (A) DNA probe, (B) 3′Z-probe, (C) 5′Z-probe and 10.0 μg mL^{-1} mutant type DNA target at 25 °C. (a) electrode itself, (b) the pseudo-hybridization of DNA probe, 3′Z-probe or 5′Z-probe, (c) the hybridization of DNA probe, 3′Z-probe or 5′Z-probe and mutant type DNA target. (D) Histograms representing the average R_{ct} values obtained by (a) electrode itself, the pseudo-hybridization of 2.0 μg mL^{-1} (b) DNA probe, (d) 3′Z-probe or (f) 5′Z-probe, the hybridization between 2.0 μg mL^{-1} (c) DNA probe, (e) 3′Z-probe or (g) 5′Z-probe and 10.0 μg mL^{-1} mutant type DNA target. Figure S2. Nyquist diagrams obtained by (a) electrode itself, the immobilization of 2.0 μg mL^{-1} of (b) spermine or (c) Z-probe, and the interaction/hybridization of 2.0 μg mL^{-1} of (d) spermine or (e) Z-probe with 10.0 μg mL^{-1} mutant type DNA target. Figure S3. The Nyquist diagrams obtained after the hybridization of 2.0 μg mL^{-1} Z-probe and 10.0 μg mL^{-1} mutant type DNA target at (A) 25 °C, (B) 50 °C and (C) 75 °C. (a) electrode itself, (b) the pseudo-hybridization of Z-probe at 25 °C, 50 °C and 75 °C, (c) the hybridization of Z-probe and mutant type DNA target at 25 °C, 50 °C and 75 °C. Figure S4. The Nyquist diagrams obtained after the hybridization of 2 μg mL^{-1} Z-probe and 10.0 μg mL^{-1} mutant type DNA target in (A) ABS (pH 4.8), (B) PBS (pH 7.4) and (C) CBS (pH 9.5). (a) electrode itself, (b) the pseudo-hybridization of Z-probe in ABS (pH 4.8), PBS (pH 7.4) or CBS (pH 9.5). (c) the hybridization of Z-probe and mutant type DNA target in ABS (pH 4.8), PBS (pH 7.4) or CBS (pH 9.5). Inset was the equivalent circuit model used for fitting of the impedance datas. Figure S5. The Nyquist diagrams obtained after the hybridization of 2.0 μg mL^{-1} Z-probe and 10.0 μg mL^{-1} mutant type DNA target in (A) PBS (pH 7.4) or (B) 0.5 mM and (C) 1.0 mM Mg^{2+} contained PBS (pH 7.4). (a) electrode itself, (b) the pseudo-hybridization of Z-probe in PBS (pH 7.4), or PBS (pH 7.4) containing 0.5 mM or 1.0 mM Mg^{2+}. (c) the hybridization of Z-probe and mutant type DNA target in PBS (pH 7.4), or PBS (pH 7.4) containing 0.5 mM or 1.0 mM Mg^{2+}. Figure S6. The Nyquist diagrams obtained after the hybridization of 2.0 μg mL^{-1} Z-probe and 10.0 μg mL^{-1} mutant type DNA target during 5 min, 10 min and 15 min. (a) electrode itself, the pseudo-hybridization of Z-probe during (b) 5 min, (c) 10 min, (d) 15min, the hybridization of Z-probe and mutant type DNA target during (b′) 5 min, (c′) 10 min and (d′) 15 min. Figure S7. Nyquist diagrams of (a) electrode itself, before (A) and after (B) the hybridization of (b) 0.25, (c) 0.5, (d) 1.0, (e) 2.0 and (f) 4.0 μg mL^{-1} Z-probe and 10.0 μg mL^{-1} mutant type DNA target. Figure S8. Line graph representing the R_{ct} values recorded by the hybridization of 1.0 μg mL^{-1} Z-probe and mutant type DNA target at the concentration level from 2.0 to 16.0 μg/mL. Inset: Calibration graph based on the average R_{ct} values ($n = 3$) obtained after the hybridization of Z-probe with mutant type DNA target in the concentration range from 2.0 to 10.0 μg mL^{-1}. Figure S9. The hybridization of 1.0 μg mL^{-1} Z-probe and 12.0 μg mL^{-1} mutant type DNA target or C-mutant type DNA or T-mutant type DNA. (A) Nyquist diagrams, (B) histograms representing the R_{ct} values obtained by (a) electrode itself, (b) the pseudo-hybridization of Z-probe, after the hybridization of Z-probe and (c) mutant type DNA target, (d) C-mutant type DNA, (e) T-mutant type DNA. Figure S10. The hybridization of Z-probe and mutant type DNA target or ODN-1 or ODN-2. (A) Nyquist diagrams, (B) histograms representing the R_{ct} values obtained by (a) electrode itself, (b) presudo-hybridization of Z-probe, after the hybridization of Z-probe and (c) mutant type DNA target, (d) ODN-1, (e) ODN-2 ($n = 3$). Table S1. H_{Eff}% calculated based on the average R_{ct} value obtained after the hybridization of Z-probe with mutant type DNA target/C-mutant type DNA/T-mutant type DNA/ODN-1/ODN-2 in contrast to the average R_{ct} value obtained in the presence of pseudo hybridization.

Author Contributions: A.E. designed, assisted, analyzed the experiments, and wrote and revised the manuscript. E.E. performed the experiments and wrote and revised the draft manuscript. All authors have read and agreed to the published version of the manuscript.

Funding: The financial support was obtained from the Turkish Scientific and Technological Research Council (TUBITAK; Project no. 114Z400).

Acknowledgments: Arzum Erdem acknowledges the financial support from the Turkish Scientific and Technological Research Council (TUBITAK; Project no. 114Z400) as a project investigator, and she would also like to express her gratitude to the Turkish Academy of Sciences (TUBA) as a Principal Member for its partial support. Ece Eksin acknowledges a project PhD scholarship through project (TUBITAK Project no. 114Z400).

Conflicts of Interest: The authors declare no conflict of interest.

References

1. Song, S.; Xu, H.; Fan, C. Potential diagnostic applications of biosensors: Current and future directions. *Int. J. Nanomed.* **2006**, *1*, 433–440. [CrossRef]
2. Sang, S.; Li, Y.; Guo, X.; Zhang, B.; Xue, X.; Zhuo, K.; Zhao, C.; Zhang, W.; Yuan, Z. A Portable Device for Rapid Detection of Human Serum Albumin using an immunoglobulin-coating-based Magnetoelastic Biosensors. *Biosens. Bioelectron.* **2019**, *141*, 111399. [CrossRef]
3. Paleček, E.; Bartošík, M. Electrochemistry of Nucleic Acids. *Chem. Rev.* **2012**, *112*, 3427–3481. [CrossRef]
4. Beaudet, A.L.; Belmont, J.W. Array-based DNA diagnostics: Let the revolution begin. *Annu. Rev. Med.* **2008**, *59*, 113–129. [CrossRef]

5. Wakai, J.; Takagi, A.; Nakayama, M.; Miya, T.; Miyahara, T.; Iwanaga, T.; Takenaka, S.; Ikeda, Y.; Amano, M. A novel method of identifying genetic mutations using an electrochemical DNA array. *Nucleic Acids Res.* **2004**, *32*, e141. [CrossRef]
6. Arduini, F.; Micheli, L.; Moscone, D.; Palleschi, G.; Piermarini, S.; Ricci, F.; Volpe, G. Electrochemical biosensors based on nanomodified screen-printed electrodes: Recent applications in clinical analysis. *Trends Anal. Chem.* **2016**, *79*, 114–126. [CrossRef]
7. Mu, S.; Wang, X.; Li, Y.T.; Wang, Y.; Li, D.W.; Long, Y.T. A novel screen-printed electrode array for rapid high-throughput detection. *Analyst* **2012**, *137*, 3220–3223. [CrossRef]
8. Apostolou, T.; Pascual, N.; Marco, M.P.; Moschos, A.; Petropoulos, A.; Kaltsas, G.; Kintzios, S. Extraction-less, rapid assay for the direct detection of 2,4,6-trichloroanisole (TCA) in cork samples. *Talanta* **2014**, *125*, 336–340. [CrossRef]
9. Biscay, J.; Begoña, M.; García, G.; García, A.C. Electrochemical Biotin determination based on a screen printed carbon electrode array and magnetic beads. *Sens. Actuators B* **2014**, *205*, 426–432. [CrossRef]
10. Erdem, A.; Congur, G.; Eksin, E. Multi channel screen printed array of electrodes for enzyme-linked voltammetric detection of MicroRNAs. *Sens. Actuators B* **2013**, *188*, 1089–1095. [CrossRef]
11. Erdem, A.; Congur, G. Dendrimer modified 8-channel screen-printed electrochemical array system for impedimetric detection of activated protein C. *Sens. Actuators B* **2014**, *196*, 168–174. [CrossRef]
12. Erdem, A.; Congur, G. Label-free voltammetric detection of MicroRNAs at multi-channel screen printed array of electrodes comparison to graphite sensors. *Talanta* **2014**, *118*, 7–13. [CrossRef] [PubMed]
13. Erdem, A.; Eksin, E. Magnetic beads assay based on Zip nucleic acid for electrochemical detection of Factor V Leiden mutation. *Int. J. Biol. Macromol.* **2019**, *125*, 839–846. [CrossRef] [PubMed]
14. Erdem, A.; Eksin, E. Zip nucleic acid based single-use biosensor for electrochemical detection of Factor V Leiden mutation. *Sens. Actuators B* **2019**, *288*, 634–640. [CrossRef]
15. Erdem, A.; Eksin, E. ZNA probe immobilized single-use electrodes for impedimetric detection of nucleic acid hybridization related to single nucleotide mutation. *Anal. Chim. Acta* **2019**, *1071*, 78–85. [CrossRef] [PubMed]
16. Bertina, R.M.; Koeleman, B.P.; Koster, T.; Rosendaal, F.R.; Dirven, R.J.; de Ronde, H.; van der Velden, P.A.; Reitsma, P.H. Mutation in blood coagulation factor V associated with resistance to activated protein C. *Nature* **1994**, *369*, 64–67. [CrossRef]
17. Kujovich, J.L. Factor V Leiden thrombophilia. *Genet. Med.* **2011**, *13*, 1–16. [CrossRef]
18. Ren, Y.; Rezania, S.; Kang, K.A. Biosensor for Diagnosing Factor V Leiden, A Single Amino Acid Mutated Abnormality of Factor V. In *Advances in Experimental Medicine and Biology*; Kyung, A.K., David, K.H., Duane, F.B., Eds.; Springer: Boston, MA, USA, 2008; pp. 245–252.
19. Vlachou, M.A.; Glynou, K.M.; Ioannou, P.C.; Christopoulos, T.K.; Vartholomatos, G. Development of a three-biosensor panel for the visual detection of thrombophilia associated mutations. *Biosens. Bioelectron.* **2010**, *26*, 228–234. [CrossRef]
20. Kang, K.A.; Ren, Y.; Sharma, V.R.; Peiper, S.C. Near real-time immuno-optical sensor for diagnosing single point mutation: A model system: Sensor for factor V Leiden diagnosis. *Biosens. Bioelectron.* **2009**, *24*, 2785–2790. [CrossRef]
21. Noir, R.; Kotera, M.; Pons, B.; Remy, J.S.; Behr, J.P. Oligonucleotide- Oligospermine Conjugates (Zip Nucleic Acids): A Convenient Means of Finely Tuning Hybridization Temperatures. *J. Am. Chem. Soc.* **2008**, *9*, 13500–13505. [CrossRef]
22. Moreau, V.; Voirin, E.; Paris, C.; Kotera, M.; Nothisen, M.; Rémy, J.S.; Behr, J.P.; Erbacher, P.; Lenne-Samuel, N. Zip nucleic acids (ZNAs): New high affinity oligonucleotides as potent primers for PCR and reverse transcription. *Nucleic Acids Res.* **2009**, *37*, e130. [CrossRef] [PubMed]
23. Paris, C.; Moreau, V.; Deglane, G.; Voirin, E.; Erbacher, P.; Lenne-Samuel, N. Zip nucleic acids are potent hydrolysis probes for quantitative PCR. *Nucleic Acids Res.* **2010**, *38*, e95. [CrossRef] [PubMed]
24. Alvandi, E.; Koohdani, F. Zip nucleic acid: A new reliable method to increase the melting temperature of real-time PCR probes. *J. Diabetes Metab. Disord.* **2014**, *13*, 26–29. [CrossRef] [PubMed]
25. Nothisen, M.; Perche-Letuvee, P.; Behr, J.-P.; Remy, J.-S.; Kotera, M. Cationic Oligospermine-Oligonucleotide Conjugates Provide Carrier-free Splice Switching in Monolayer Cells and Spheroids. *Cell* **2018**, *13*, 483–492. [CrossRef]

26. Kandemir, H.; Erdal, M.E.; Selek, S.; Ay, Ö.İ.; Karababa, İ.F.; Kandemir, S.B.; Ay, M.E.; Yılmaz, Ş.G.; Bayazıt, H.; Taşdelen, B. Evaluation of several micro RNA (miRNA) levels in children and adolescents with attention deficit hyperactivity disorder. *Neurosci. Lett.* **2014**, *580*, 158–162. [CrossRef]
27. Begheldo, M.; Ditengou, F.A.; Cimoli, G.; Trevisan, S.; Quaggiotti, S.; Nois, A.; Palme, K.; Ruperti, B. Whole-mount in situ detection of microRNAs on Arabidopsis tissues using Zip Nucleic Acid probes. *Anal. Biochem.* **2013**, *434*, 60–66. [CrossRef]
28. Erdem, A.; Eksin, E. Electrochemical Detection of Solution Phase Hybridization Related to Single Nucleotide Mutation by Carbon Nanofibers Enriched Electrodes. *Materials* **2019**, *12*, 3377. [CrossRef]
29. Martín-Yerga, D.; Costa-García, A. Towards a blocking-free electrochemical immunosensing strategy for anti-transglutaminase antibodies using screen-printed electrodes. *Bioelectrochemistry* **2015**, *105*, 88–94. [CrossRef]
30. Bartosik, M.; Durikova, H.; Vojtesek, B.; Anton, M.; Jandakova, E.; Hrstka, R. Electrochemical chip-based genomagnetic assay for detection of high-risk human papillomavirus DNA. *Biosens. Bioelectron.* **2016**, *83*, 300–305. [CrossRef]
31. Erdem, A.; Eksin, E.; Congur, G. Indicator-free electrochemical biosensor for microRNA detection based on carbon nanofibers modified screen printed electrodes. *J. Electroanal. Chem.* **2015**, *755*, 167–173. [CrossRef]
32. Perez-Rafols, C.; Serrano, N.; Díaz-Cruz, J.M.; Arino, C.; Esteban, M. New approaches to antimony film screen-printed electrodes using carbon-based nanomaterials substrates. *Anal. Chim. Acta* **2016**, *916*, 17–23. [CrossRef] [PubMed]
33. Eissa, S.; Alshehri, N.; Rahman, A.M.A.; Dasouki, M.; Abu-Salah, K.M.; Zourob, M. Electrochemical immunosensors for the detection of survival motor neuron (SMN) protein using different carbon nanomaterials-modified electrodes. *Biosens. Bioelectron.* **2018**, *101*, 282–289. [CrossRef] [PubMed]
34. Miller, J.N.; Miller, J.C. *In Statistics and Chemometrics for Analytical Chemistry*, 5th ed.; Pearson Education: Essex, UK, 2005; p. 121.
35. Aoki, H.; Sukegawa, T.; Torimura, M.; Nakazato, T. Nonlabeling and Nonexternal Indicator DNA Sensing Based on Ferrocene-terminated Probes Immobilized on Gold Film Electrode Arrays with Plasma and Acid Treatments. *Sens. Mater.* **2020**, *32*, 1079–1090. [CrossRef]
36. Bratov, A.; Rmon_Axcon, J.; Abramova, N.; Merlos, A.; Adrian, J.; Sancez-Baeza, F.; Marco, M.P.; Dominguez, C. Three-dimensional interdigitated electrode array as a transducer for label-free biosensors. *Biosens. Bioelectron.* **2008**, *24*, 729–735. [CrossRef]
37. Du, Y.; Guo, S.; Dong, S.; Wang, E. An integrated sensing system for detection of DNA using new parallel-motif DNA triplex system and graphene mesoporous silica gold nanoparticle hybrids. *Biomaterials* **2011**, *32*, 8584–8592. [CrossRef] [PubMed]
38. Harandizadeh, Z.; Ito, T. Block Copolymer-Derived Recessed Nanodisk-Array Electrodes as Platforms for Folding-Based Electrochemical DNA Sensors. *ChemElectroChem* **2019**, *6*, 5627–5632. [CrossRef]
39. Jambrec, D.; Kayran, Y.U.; Schuhmann, W. Controlling DNA/Surface Interactions for Potential Pulse-Assisted Preparation of Multi-Probe DNA Microarrays. *Electroanalysis* **2019**, *31*, 1943–1951. [CrossRef]
40. Lee, J.Y.; Won, B.Y.; Park, H.G. Label-Free Multiplex DNA Detection Utilizing Projected Capacitive Touchscreen. *Biotechnol. J.* **2018**, *13*, 1700362. [CrossRef]
41. Lee, J.G.; Yun, K.; Lim, G.S.; Lee, S.E.; Kim, S.; Park, J.K. DNA biosensor based on the electrochemiluminescence of Ru (bpy) 3 2+ with DNA-binding intercalators. *Bioelectrochemistry* **2007**, *70*, 228–234. [CrossRef]
42. Li, X.; Lee, J.S.; Kraatz, H.B. Electrochemical Detection of Single-Nucleotide Mismatches Using an Electrode Microarray. *Anal. Chem.* **2006**, *78*, 6096–6101. [CrossRef]
43. Liu, G.; Lao, R.; Xu, Q.; Li, L.; Zhang, M.; Song, S.; Fan, C. Single-nucleotide polymorphism genotyping using a novel multiplexed electrochemical biosensor with nonfouling surface. *Biosens. Bioelectron.* **2013**, *42*, 516–521. [CrossRef] [PubMed]
44. Magrina, I.; Toldra, A.; Campas, M.; Ortiz, M.; Simonova, A.; Katakis, I.; Hocek, M.; O'Sullivan, C.K. Electrochemical genosensor for the direct detection of tailed PCR amplicons incorporating ferrocene labelled dATP. *Biosens. Bioelectron.* **2019**, *134*, 76–82. [CrossRef] [PubMed]
45. Takase, S.; Miyagawa, K.; Ikeda, H. Label-Free Detection of Zeptomol miRNA via Peptide Nucleic Acid Hybridization Using Novel Cyclic Voltammetry Method. *Sensors* **2020**, *20*, 836. [CrossRef] [PubMed]

46. Zhang, D.; Peng, Y.; Qi, H.; Gao, Q.; Zhang, C. Label-free electrochemical DNA biosensor array for simultaneous detection of the HIV-1 and HIV-2 oligonucleotides incorporating different hairpin-DNA probes and redox indicator. *Biosens. Bioelectron.* **2010**, *25*, 1088–1094. [CrossRef] [PubMed]
47. Zhu, X.; Ino, K.; Lin, Z.; Shiku, H.; Chen, G.; Matsue, T. Amperometric detection of DNA hybridization using a multi-point, addressable electrochemical device. *Sens. Actuators B* **2011**, *160*, 923–928. [CrossRef]

biosensors

MDPI

Article

Magnetic Bead-Based Electrochemical Immunoassays On-Drop and On-Chip for Procalcitonin Determination: Disposable Tools for Clinical Sepsis Diagnosis

Águeda Molinero-Fernández [1,†], María Moreno-Guzmán [2,†], Miguel Ángel López [1,3,*] and Alberto Escarpa [1,3,*]

1 Department of Analytical Chemistry, Physical Chemistry and Chemical Engineering, Universidad de Alcalá, Ctra. Madrid-Barcelona, Km. 33.600, Alcalá de Henares, 28871 Madrid, Spain; agueda.molinero@edu.uah.es
2 Department of Chemistry in Pharmaceutical Sciences, Analytical Chemistry, Faculty of Pharmacy, Universidad Complutense de Madrid, Avenida Complutense, s/n, 28040 Madrid, Spain; marimore@ucm.es
3 Chemical Research Institute "Andres M. Del Rio", Universidad de Alcalá, Alcalá de Henares, 28871 Madrid, Spain
* Correspondence: miguelan.lopez@uah.es (M.Á.L.); alberto.escarpa@uah.es (A.E.)
† Both authors contributed equally to this work.

Received: 27 April 2020; Accepted: 11 June 2020; Published: 17 June 2020

check for updates

Abstract: Procalcitonin (PCT) is a known protein biomarker clinically used for the early stages of sepsis diagnosis and therapy guidance. For its reliable determination, sandwich format magnetic bead-based immunoassays with two different electrochemical detection approaches are described: (i) disposable screen-printed carbon electrodes (SPE-C, on-drop detection); (ii) electro-kinetically driven microfluidic chips with integrated Au electrodes (EMC-Au, on-chip detection). Both approaches exhibited enough sensitivity (limit of detection (LOD) of 0.1 and 0.04 ng mL^{-1} for SPE-C and EMC-Au, respectively; cutoff 0.5 ng mL^{-1}), an adequate working range for the clinically relevant concentrations (0.5–1000 and 0.1–20 ng mL^{-1} for SPE-C and EMC-Au, respectively), and good precision (RSD < 9%), using low sample volumes (25 µL) with total assay times less than 20 min. The suitability of both approaches was successfully demonstrated by the analysis of human serum and plasma samples, for which good recoveries were obtained (89–120%). Furthermore, the EMC-Au approach enabled the easy automation of the process, constituting a reliable alternative diagnostic tool for on-site/bed-site clinical analysis.

Keywords: electrochemical immunoassays; microfluidic chips; screen-printed; sepsis

1. Introduction

Nowadays, the burden of sepsis on health care is highly significant. It is estimated that approximately 13 million people worldwide become septic every year and four million people die of sepsis. Mortality rates for severe sepsis range between 30 to 50%, and higher than 50% for septic shock [1]. Furthermore, despite advances in health care, the incidence of sepsis is increasing every year and a continuous increment is expected as the population ages [2]. With this in mind, new diagnostic tests that help clinicians to diagnose and manage this disease can significantly yield improvements in their patients' outcomes.

Procalcitonin (PCT) is a protein precursor of the calcitonin hormone, composed of 116 amino acids, with a molecular weight of 13 kDa. It is considered a very specific biomarker in early clinical diagnosis for severe infection diseases, including sepsis [3]. Under normal physiological conditions,

PCT levels in human blood is lower than 0.25 ng mL^{-1}, but it can rapidly increase in response to pro-inflammatory stimulation, with a half-life of 24 h [4,5]. Measurements of PCT levels could be used for diagnosis, as well as for evaluations of the treatment effectiveness, determining the appropriate dosage and duration of antibiotic therapy [6–8].

Several methods have been reported for PCT determination, mainly based in the selective recognition provided by immunological interactions. These methodologies include immunoturbidimetric assays [9], chemiluminescent immunoassays [10–16], immunochromatographic assays [17–22], surface plasmon resonance biosensors [23–25], fluorescence immunosensors [26–32], ellipsometry immunosensors [33], colorimetric immunoassays [34–36], and electrochemical immunoassays [37–56]. Even different immunoanalytical methodologies are commercially available by Brahams GmbH (Henningsdorf, Germany). Although some of them reach impressive sensitivity, the vast majority present several limitations, such as their high complexity, sophisticated instrumentation requirements and/or unproven applicability in the real clinical scenario.

Furthermore, among all the detection systems, electrochemistry stands out due to its inherent miniaturization, portability, low cost and ability to tailor electrode materials [57] The coupling of this detection principle has been widely explored in immunoassays for clinical diagnoses. In particular, electrochemical magneto-immunosensors have proven to be sensitive, accurate, fast and inexpensive, capable of achieving adequate limits of detection and very suitable as point-of-care tools for decentralized analysis. The use of magnetic beads has been largely exploited in recent years due to their well-known properties such as improved assay kinetics and easy manipulation, while they are perfectly coupled to electrochemical detection [58]. However, magneto-immunosensors [16,17,35,36] in general, and particularly those using electrochemical detection [40], have not been widely explored for PCT determination.

Keeping in mind the aforementioned characteristics of electrochemistry, it can be considered as a very suitable detection technique to be interfaced with microfluidics as well. In addition, microfluidics enable lower sample consumption, as well as finely controlled and automated (electro)-chemical reactions in pocket-sized devices containing microchannels to provide point-of-care applications [59,60]. In this sense, they constitute an ideal platform to perform integrated microscale immunoassays [59–64]. Despite these inherent advantages, few microfluidic immunosensors for PCT determination have been previously published, and most of them are based on optical techniques [26,28–30,39].

In this work, we demonstrate the use of magnetic bead-based electrochemical immunoassays for PCT determination in human serum samples using two approaches: (i) disposable screen-printed carbon electrodes (SPE-C, on-drop detection) and (ii) electro-kinetically driven microfluidic chips with integrated Au electrodes (EMC-Au, on-chip detection).

2. Materials and Methods

2.1. Reagent and Solutions

The analyte, procalcitonin (PCT) (8PC5), and biotinylated and horseradish peroxidase (HRP)-conjugated monoclonal anti-PCT antibodies(18B7,44D9) were purchased from HyTest (Turku, Finland).

The lyophilized PCT was dissolved in deionized water (Millipore Milli-Q purification system). Further PCT dilutions were carried out with 0.1 M phosphate buffered saline and 0.01% Tween-20 (PBST) buffer, pH 7.5. Antibody solutions were prepared in 0.1 M phosphate buffered saline and 0.01% Tween-20 (PBST) buffer, pH 7.5.

Streptavidin-coated superparamagnetic beads (Dynabeads® M-280 Streptavidin) (10 mg mL^{-1}) were obtained from Invitrogen (Carlsbad, CA, USA), and bovine serum albumin (BSA), hydroquinone (HQ) and 30% H_2O_2 PERDROGENTM (w/w) were purchased from Sigma-Aldrich (Madrid, Spain). Hydroquinone and hydrogen peroxide solutions for the electrochemical detection were prepared in 0.1 M of phosphate-buffered (PB) solution, pH 7.0.

2.2. Apparatus and Electrodes

Magnetic racks for magnetic bead immobilization onto the working electrode surface were purchased from Metrohm DropSens (Oviedo, Spain). A magnetic block, DynaMag™-2, and a Sample Rack for DynaMag™-2 for magnetic particle handling were purchased from ThermoFisher, (Carlsbad, CA, USA). For the incubation steps, a Vortex Mixer-ZX3 from Velp Scientifica and a Thermoshaker TS-100C from Biosan, (Riga, Latvia) were used.

Multi potentiostat/galvanostat µSTAT 8000 and "DropView 8400" software for measurement setup and data acquisition, handling, processing, and exporting was used for the on-drop amperometric measurements. Screen-printed carbon electrodes (SPCE) DRP-110, with a carbon working electrode (ø = 4 mm), carbon counter electrode and silver reference electrode, from Metrohm DropSens (Oviedo, Spain), were used (Figure S1A). A holder (MCE-HOLDER-DC02) and microchips (MCE-SU8-Au002T) (38 × 13 × 0.75 mm) from MicruX Technologies S.L. (Oviedo, Spain) were used for on-chip measurements. The microchips integrate three Au electrodes of 100 µm (working electrode (WE), auxiliary electrode (AE) and counter electrode (CE)) with a separation channel length of 30 mm and an injection channel length of 5 mm. The width of the microchannel is 50 µm and their depth is 20 µm. The microchip was placed into a holder, where all electric contacts and reservoirs were pre-defined (Figure S1B).

For microchip pre-treatment, 0.1 M NaOH solution was flushed through the channels for 20 min, followed by rinsing with deionized water for 10 min and PBST 0.1 mM (running buffer) for 10 min.

The Bi-potentiostat HVSTAT2010, for applying the high-voltages and recording the amperometric measurements, was obtained from MicruX Technologies S.L. (Oviedo, Spain).

2.3. Samples

Human samples from healthy volunteers, with undetectable PCT levels, were obtained after their written informed consent and authorization.

This study was conducted in accordance with the Declaration of Helsinki Ethical Principles, and was approved by the Ethics Committee of the Hospital Clínico San Carlos (Spain) (reference code: C.P.—C.I. 16/161-E. Date of approval: 23 May 2016).

2.4. Immunoassay Procedures

Based on a typical sandwich ELISA protocol, 2 µL of the commercial streptavidin-coated magnetic bead (MBs) suspension was placed into a microcentrifuge tube, and subjected to a washing step according to the manufacturer's protocol. These beads were incubated in 50 µL (5 µg mL^{-1}) of biotinylated anti-PCT solution in PBST buffer at room temperature and stirred for 5 min. After that, the microcentrifuge tube was placed on the magnetic block and the supernatant was removed, followed by two washing steps with 100 µL of PBST buffer. Then, the MBs functionalized with anti-PCT antibody were re-suspended in 25 µL of sample or PCT standard solutions, plus 25 µL of HRP-conjugated anti-PCT antibody solution (0.36 µg mL^{-1}, final concentration) in PBST with 0.1% BSA. After the suspension incubation at room temperature for 15 min, the supernatant was removed and three washing steps were carried out.

Once the immunoreaction was carried out, electrochemical detection was performed using both approaches: (i) on-drop onto screen-printed carbon electrodes (SPE-C); (ii) on-chip into electro-kinetically driven microfluidic chips with integrated Au electrodes (EMC-Au) (Figure 1).

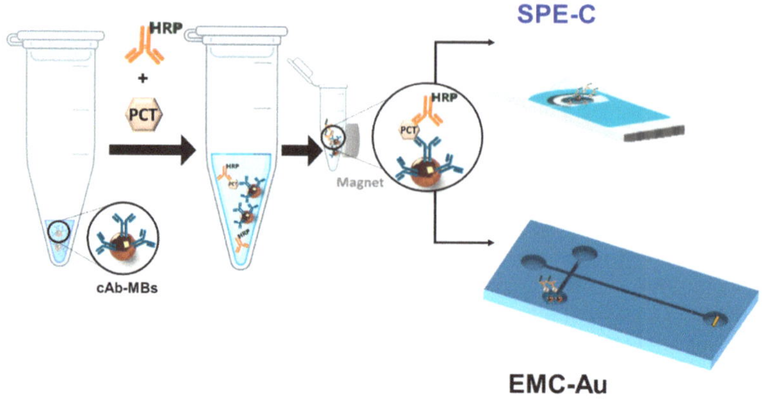

Figure 1. Magnetic bead-based electrochemical immunoassays—on-drop screen-printed carbon electrodes (SPE-C) and on-chip electro-kinetically driven microfluidic chips with integrated Au electrodes (EMC-Au).

2.5. *Electrochemical Detection On-Drop onto SPE-C*

After the immunocomplex formation, the MBs were re-suspended in 1-mM (45-µL) hydroquinone solution and transferred to the SPCE, where they were placed onto the working electrode surface using a magnet. Finally, amperometric measurements were performed at an applied potential of −0.20 V. After current stabilization, 5 µL of hydrogen peroxide solution (final concentration = 5 mM) was added, and the current was recorded.

The amperometric signals were calculated as the difference between the steady-state and the background currents at 200 s (Figure S2A). The signals were then fit to a four-parameter logistic regression using SigmaPlot 10.0 (Equation (1)).

$$\text{ip} = \left(\frac{i_{max} - i_{min}}{1 + \left(\frac{EC_{50}}{x}\right)^h} + i_{min} \right) \quad (1)$$

where i_{max} and i_{min} are the maximum and minimum current values of the calibration graph; EC_{50} value is the analyte concentration corresponding to 50% of the maximum signal; h is the hill slope.

2.6. *Electrochemical Detection into EMC-Au*

In this case, the MB immunocomplexes were re-suspended in 10 µL of PBST buffer for their subsequent electro-kinetical introduction into the microfluidic chip. Therefore, this suspension was deposited into the sample reservoir (SR) of the microfluidic chip (Figure 2). In addition, microchannels, the running buffer and detection reservoirs (RB and DR) were filled with PBST, while the enzymatic substrate reservoir (ER) was filled with a mixture of 45 µL of 1 mM HQ plus 5 µL of 50 mM H_2O_2.

An electrokinetic injection protocol was optimized for the EMC-Au electrochemical detection (Figure 2). MBs were dragged to the longitudinal channel, applying a voltage of +1500 V between reservoirs SR and DR for three pulses of 25 s, while other reservoirs were left floating. They were retained within the microchannel by the aid of a magnet situated on the top. After a washing step with PBST (10 s applying +1500 V from RB to DR reservoirs) the enzymatic substrates were injected and pumped to cross through the particle bed (200 s applying +1500 V from ES to DR reservoirs). In-channel amperometric measurements were taken at an applied potential of −0.20 V on the Au working electrode. The amperometric signals were calculated as the difference between the steady-state and the background currents at 200 s (Figure S2B) and fit to a four-parameter logistic regression (Equation 1) using SigmaPlot 10.0. After the measurement, MBs were removed from the main channel by taking off the magnet and washing the channel by injection of buffer for 200 s (+1500 V) from RB to DR.

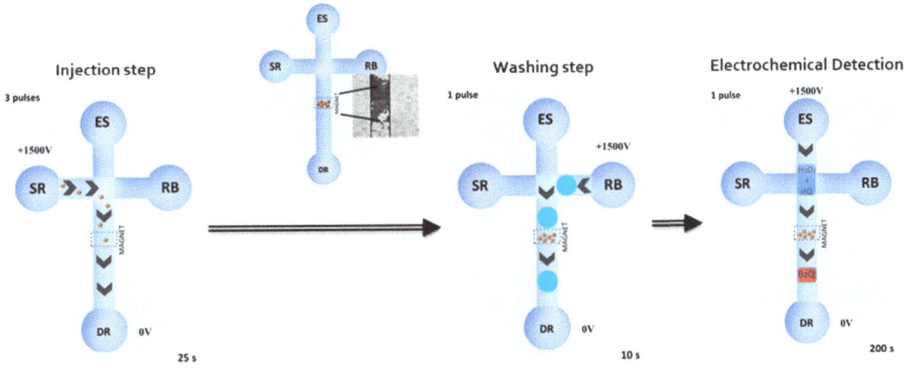

Figure 2. Electrokinetic protocol and electrochemical detection on EMC-Au. Sample reservoir (SR), running buffer (RB), enzymatic substrate reservoir (ER) and detection reservoirs (DR).

Taking into account that only a small fraction of the MBs deposited into the sample reservoir are introduced into the main channel, the analysis can be automatically repeated several times without the need for manual intervention or conditioning of the microchip.

3. Results and Discussion

3.1. Optimization of the Immunoassay

The functionalization of the MBs with the biotinylated captured antibody (cAb) was evaluated in a concentration of antibodies between zero and 7.5 μg mL^{-1}. The amount of cAb depends on the number of MBs used and the number of streptavidin molecules immobilized onto them. The maximum current intensity was obtained using a concentration of 5.0 μg mL^{-1}, followed by a plateau that denotes the saturation of the binding sites (Figure S3A). A similar selection protocol was followed for the determination of the optimal concentration of the detection antibody. Titration was performed for concentrations ranging from 0.04 to 0.7 μg mL^{-1}, where the maximum intensity current was reached for 0.36 μg mL^{-1} of anti-PCT-HRP producing the saturation of the antigen/capture-antibody binding sites (sandwich format) (Figure S3B). Incubation times were also studied for different stages. Times of 5 min for the immobilization of captured antibodies to modified magnetic beads produced 85% of the maximum intensity current (Figure S4A). Moreover, the simultaneous or sequential incubation of the analyte and detection antibody was also considered. In total, 97% of the maximum current was obtained when simultaneous incubation of both species was performed for 15 min (Figure S4B). Non-specific adsorption was almost negligible (<1%) when adding 0.1% BSA to the dilution buffer during the incubation stages.

Once the immunorecognition was performed, the electrochemical detection was carefully studied using two different approaches: (i) SPE-C, on-drop detection; (ii) EMC-Au, on-chip detection.

MB immunocomplexes were deposited onto the surface of the SPE-C and retained by a magnet, while the enzymatic substrate and electrochemical mediator (H_2O_2 and HQ) were added to perform the amperometric detection at −0.20 V. The detection potential was evaluated between zero and −0.3 V. The signal increased up to −0.2 V, keeping constant for larger negative potentials, as this one was the one we selected for the amperometric measurements.

For the EMC-Au approach, electrokinetic protocol for the injection of modified MB immunocomplexes and enzymatic substrate/electrochemical mediators as well as electrochemical detection were carefully studied as well (see Table 1).

Table 1. Electrokinetics and electrochemical detection, EMC-Au optimizations.

Step	Parameter	Studied Range	Selected Value
Immunocomplex-MB Injection	Immunocomplex-MB dilution (*v*/*v*)	1:10–1:200	1:10
	Applied Voltage (V)	+1000–2000	+1500
	Number of pulses	1–5	3
	Pulse time (s)	10–50	25
Washing	Applied Voltage (V)	+1000–2000	1500
	Number of pulses	1–5	1
	Pulse time (s)	10–50	25
Enzyme substrates pumping/driven	Applied Voltage (V)	+1000–2000	+1500
	Time (s)	—	200 s
Detection	E (V)	−0.10–(−0.30)	−0.20

The applied voltage, number of pulses and pulse time for the MB immunocomplex injection from the SR were assayed to place the optimum amount of MBs in the microchannel, in order to obtain the highest signal without clogging the channel. Electrokinetic conditions for the washing step were also studied to eliminate non-magnetically retained MBs, improving the assay precision. Then, the enzymatic substrates were continuously driven at +1500 V for 200 s. Under these optimal electrokinetic conditions, the detection potential was assayed between −0.10 V and −0.30 V. The highest signal/noise features were obtained at −0.20 V, which was chosen as the optimum detection potential.

3.2. *Analytical Characteristics*

Analytical performance was carefully evaluated in both SPE-C and EMC-Au approaches. The calibration curves are depicted in Figure 3, while their corresponding analytical characteristics are summarized in Table 2.

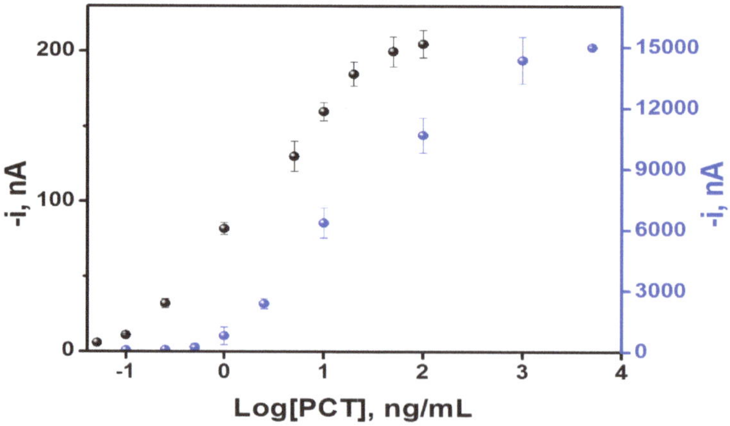

Figure 3. Calibration curves for PCT determination using both approaches: on-drop SPE-C and on-chip EMC-Au. The results are expressed as Mean values ± standard deviation (n = 3). The error bar corresponds with the standard deviation.

Table 2. Analytical characteristics for procalcitonin (PCT) determination using both approaches: SPE-C and EMC-Au.

Analytical Characteristic	SPE-C	EMC-Au
EC_{50}, ng mL^{-1}	20.2	2.2
Working range, ng mL^{-1}	0.5–1000	0.1–20
r	0.990	0.990
LOD, ng mL^{-1}	0.1	0.04
Intra-assay, CV%	<7.5%	5%
Inter-assay, CV%	8%	9%

From the obtained results, it is important to remark that both immunoassay detection approaches (SPE-C and EMC-Au) enabled PCT determination at the clinical significance levels needed for sepsis diagnosis and monitoring (LOD < cutoff). Interestingly, EMC-Au detection provides a lower LOD (calculated with a 3 S/m criteria where S is the standard deviation of the lowest assayed concentration (n = 10) and m is the slope of the calibration plot), while the SPE-C detection offers a wider working range (see Figure 3 and Table 2). This aspect can be attributed to the difference in the number of MBs trapped onto the SPE-C and the dimensions and material of the working electrode (C electrode, Ø = 4 mm) compared to that of the microfluidic chip (Au electrode, w = 100 μm). Moreover, a relevant comparative aspect deals with the shorter analysis time in the microfluidic chip, together with the possibility to automate the process, which enhances its potential as a point-of-care (POC) device. Indeed, taking into account that only a small fraction of the MBs from the sample reservoir are introduced into the main channel for detection each time, the analysis can be automatically repeated several times by programming the corresponding electro-kinetically driven protocol; (washing out the MBs from the previous run and placing a new batch into the central channel without any external intervention for a next run). The same MB immunocomplex batch (using 25 μL of sample) can be consecutively measured five times with good intra-assay precision (CV < 5%) in just 40 min. However, in the case of the SPE-C approach, to obtain five replicates, 125 μL of the sample and around 100 min would be needed.

PCT concentrations at two levels were used to evaluate the intra-assay (0.5 and 0.1 ng mL^{-1}) and inter-assay precision (1000 and 20 ng mL^{-1}) for on-drop SPE-C and on-chip EMC-Au, respectively (Table 2). In the case of SPE-C detection, the intra-assay and inter-assay precisions (n = 5) gave CV values below 8% in both cases. For EMC-Au, the intra-assay precision gave CV values of 5% (n = 5, same batch of MB immunocomplexes). The inter-assay precision for different MBs batches gave CV values of 9%.

The selectivity of the immunosensing configuration for PCT analysis was checked in the presence of a large excess of C-reactive protein (CRP) (16 μg mL^{-1}, another biomarker usually determined for sepsis diagnosis), heparin (1 mg mL^{-1}), ethylenedinitrilotetraacetic acid (EDTA) (1 mg mL^{-1}) and citrate (0.15 M), as other relevant molecules that can coexist in blood samples. Without exception, cross-reactivity percentages lower than 1% were obtained. These results demonstrate the excellent selectivity of the PCT immunoassay.

Due to its potential use as a POC, in order to simplify the entire procedure and, in turn, to reduce the final analysis times, the stability of the MB–captured antibody complexes was studied to be used as stock "reagents". Their stability was studied at 4 °C during a period of 1 month using the on-drop SPE-C immunosensor approach. The control chart of the stability assay is shown in Figure 4, where each point corresponds to the mean value for three successive measurements performed in the same day (intra-day immunoassays). As can be seen, the immunosensor response remained inside the control limits placed at ±three times the standard deviation value calculated for the whole set of experiments, during the entire period of time checked (inter-day immunoassays, n = 14). These results demonstrate the excellent stability of the MB–cAb complexes.

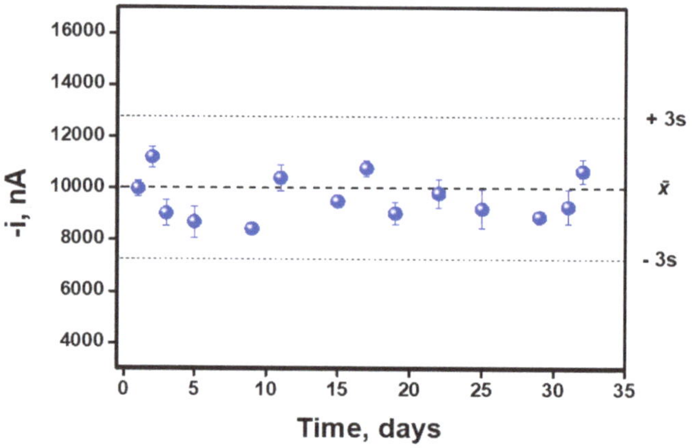

Figure 4. Stability of the MB–cAb complexes. Central and limit lines correspond to average ± three times the standard deviation (n = 14) obtained for inter-day immunoassays. Each individual point corresponds to the average ± standard deviation (n = 3) obtained for the intra-day immunoassays.

3.3. Analysis of Human Serum and Plasma Samples

Analytical capabilities for PCT determination in clinical samples were also evaluated. Calibration curves were carried out in human serum and plasma from healthy individuals, using the on-drop SPE-C immunosensor. No matrix effect was observed after a comparison of the slope calibration plots obtained in PBS buffer with those obtained in both kinds of matrices. Indeed, identical slope calibration values (sensitivities) of 4300 ± 200, 4400 ± 400 and 4500 ± 200 nA ng^{-1} mL were obtained in buffer, serum and plasma, respectively.

Then, accuracy was carefully studied using both detection approaches by recovery experiments conducted on both kinds of matrices spiked with relevant clinical levels of PCT. Table 3 demonstrates the suitability of the developed on-drop SPE-C and on-chip EMC-Au for PCT determination in human blood matrices at clinically relevant levels. It is important to remark that no single sample pretreatment was needed, due to the absence of matrix effects and the adequate working range of the immunoassay in both detection schemes. This aspect, which enhances the ease of use, together with the automation of the detection step and the portable characteristics of the devices, make the developed approaches suitable for potential POC tools for PCT determination and its use for sepsis diagnosis.

Table 3. PCT determination in human plasma and serum samples.

	SPE-C				EMC-Au	
	Serum		Plasma		Serum	
PCT_{added} (ng/mL)	PCT_{found} (ng/mL)	Recovery (%)	PCT_{found} (ng/mL)	Recovery (%)	PCT_{found} (ng/mL)	Recovery (%)
Before spiked	<0.1	—	<0.1	—	<0.04	—
1.0	0.9	90 ± 3	1.2	120 ± 6	1.1	110 ± 5
10.0	8.9	89 ± 14	10.4	104 ± 7	9.6	96 ± 2
100.0	99	99 ± 7	101	101 ± 2	— *	—

* Sample was not directly measured, since its concentration is beyond the working range.

As previously mentioned in the introduction section, in recent years, significant effort has been focused on the development of new approaches for PCT determination. However, the potential of immunosensors based on magnetic beads has not been widely explored for PCT determination, which is apparent in the low number of publications (Table 4). In comparison with those works, our SPE-C approach offers enough sensitivity (similar to our previous work [40]) to perform reliable PCT detection, but with a significant reduction in the analysis time and sample volumes [16,17,35,36]. Moreover, the easy automation of the electrochemical transduction and the improved sensitivity [40] bring our new approach (on-chip EMC-Au) closer to the POC concept.

On the other hand, our on-chip EMC-Au approach has demonstrated to be a promising analytical strategy for PCT determination. It couples a magneto immunoassay and an electrochemical microfluidic chip under controlled electrokinetics. This approach presents advantages such as the easy miniaturization and integration of all system elements, fulfilling the POC requirements. As can be observed in Table 5, our approach is highly competitive in terms of sensitivity, analysis time and sample volume with previous on-chip non electrochemical-based approaches reported in the literature.

Table 4. Overview of magneto-immunosensors for PCT determination.

Technique	cAb Immobilization	Assay Format	dAb Label	WR	LOD	Analysis Time *	Sample Volume	Sample	Ref.
Chemiluminescence	MB–(anti-FITC–Ab)/FITC–cAb	Double Sandwich	dAb–ABEI	0.09–600 ng mL^{-1}	30 pg mL^{-1}	25 min	40 μL	Serum	16
Chemiluminescence	MB–COOH/cAb	Sandwich	dAb–(PS-ALP)	1–10^4 pg mL^{-1}	0.045 pg mL^{-1}	1 h	800 μL	Serum	17
UV-vis Spectroscopy	MB–COOH/cAb	Sandwich	dAb–HRP	0.1–10 ng mL^{-1}	40 pg mL^{-1}	1.5 h	100 μL	Serum	35
UV-vis Spectroscopy	MB–COOH/cAb	Sandwich	dAb–(AuNPs-HRP)	0.02–20 ng mL^{-1}	20 pg mL^{-1}	1.5 h	50 μL	Serum	36
Amperometry	MB–Streptavidin/Biotin–cAb	Sandwich	dAb–HRP	0.25–100 ng mL^{-1}	50 pg mL^{-1}	20 min	25 μL	Neonates plasma	40
Amperometry	MB–Streptavidin/Biotin–cAb	Sandwich	dAb–HRP	0.5–1000 ng mL^{-1}	100 pg mL^{-1}	20 min	25 μL	Serum Plasma	Our work (SPE-C)

* Analysis time is measured after captured Ab immobilization stage. Abbreviations used: antibody (Ab); captured antibody (cAb); detection antibody (dAb); working range (WR); limit of detection (LOD); magnetic beads (MBs); horseradish peroxidase (HRP); fluorescein isothiocyanate (FITC); (aminobutyl)-N-(ethylisoluminol) (ABEI-N); alkaline phosphatase (ALP); polystyrene microsphere (PS); gold nanoparticles (AuNPs).

Table 5. Overview of microfluidics immunoassays for PCT determination.

Technique	cAb Immobilization	Assay Format	dAb Label	WR	LOD	Analysis Time *	Sample Volume	Sample	Ref.
Fluorescence	Covalent cAb	Sandwich	dAb–DY647	0.7–25 ng mL^{-1}	0.2 ng mL^{-1}	23 min	100 μL	Serum	26
Reflection Fluorescence	Covalent cAb	Sandwich	dAb–DY647	5–500 ng mL^{-1}	1 ng mL^{-1}	11 min	10–75 μL	Serum Plasma	28
Fluorescence	Adsorption cAb	Sandwich	dAb–DY647	640–3400 ng mL^{-1}	50 ng mL^{-1}	22 min	280 μL	Serum Diluted 1:10	29
Reflection Fluorescence	Adsorption cAb	Sandwich	dAb–cyanine	0.06–7.18 ng mL^{-1}	0.02 ng mL^{-1}	<9 min	50 μL	Serum Plasma Whole blood	30
Nanoplasmonic	Covalent cAb	Sandwich	dAb–AuNPs	1 pg mL^{-1}–100 ng mL^{-1}	95 fg mL^{-1}	<15 min	—	Serum	39
Amperometry	MB–Streptavidin/Biotin–cAb	Sandwich	dAb–HRP	0.1–20 ng mL^{-1}	40 pg mL^{-1}	20 min	25 μL	Serum Plasma	Our work (EMC-Au)

* Analysis time is measured after captured Ab immobilization stage. Abbreviations used: antibody (Ab); captured antibody (cAb); detection antibody (dAb); working range (WR); limit of detection (LOD); magnetic beads (MBs); horseradish peroxidase (HRP); fluorescein isothiocyanate (FITC); (aminobutyl)-N-(ethylisoluminol) (ABEI-N); alkaline phosphatase (ALP); polystyrene microsphere (PS); gold nanoparticles (AuNPs).

4. Conclusions

A magnetic bead-based immunoassay using both on-drop SPE-C and on-chip EMC-Au electrochemical detection approaches have exhibited an excellent analytical performance for PCT determination, allowing for its determination in the clinically relevant concentration range, using very short analysis times and a low volume of serum and plasma samples. Both detection technologies are complementary. While SPE-C was simpler, the EMC-Au approach permits greater control and easier automation of the process, constituting an even more reliable alternative diagnostic tool for on-site/bed-site clinical analysis.

Both investigated approaches have demonstrated excellent biosensing capabilities for the simple and accurate determination of PCT in human samples when only small sample volumes are accessible. Therefore, these results reveal the analytical potential of highly miniaturized electrochemical devices in the field of PCT biosensing, one of the most important sepsis protein biomarkers.

Supplementary Materials: The following are available online at http://www.mdpi.com/2079-6374/10/6/66/s1, Figure S1: Electrochemical devices for immunoassay electrochemical detection: disposable SPE-C for on-drop approach and EMC-Au for on-chip approach; Figure S2: Amperometric detection of the mediated reduction of H_2O_2 with HQ using on-drop (A) and on-chip (B) approaches in the absence and presence of PCT; [PCT] = 100 ng mL^{-1} (A), 0.5 ng mL^{-1} (B); Figure S3: Current signals obtained for different concentrations of capture antibody (cAb) (A) and detection antibody (dAb) (B) using the on-drop SPE-C approach. Conditions: incubation time = 60 min each stage; PCT concentration 1000 ng mL^{-1}; Figure S4: Current signals obtained for different incubation times in the immobilization of capture antibody (cAb) (A) and binding of detection antibody (dAb) (B) using the on-drop SPE-C approach. Conditions: cAb = 5 μg mL^{-1}; dAb = 0.36 μg mL^{-1}; PCT concentration 1000 ng mL^{-1}.

Author Contributions: Conceptualization, Á.M.-F., M.M.-G., M.Á.L. and A.E.; methodology, Á.M.-F., M.M.-G., M.Á.L. and A.E.; software, Á.M.-F. and M.M.-G.; validation, Á.M.-F., M.M.-G., M.Á.L. and A.E.; formal analysis, Á.M.-F. and M.M.-G..; investigation, Á.M.-F. and M.M.-G.; resources, M.Á.L. and A.E.; data curation, Á.M.-F. and M.M.-G.; writing—original draft preparation, Á.M.-F., M.M.-G., M.Á.L. and A.E.; writing—review and editing, Á.M.-F., M.M.-G., M.Á.L. and A.E.; visualization, Á.M.-F., M.M.-G., M.Á.L. and A.E.; supervision, M.Á.L. and A.E.; project administration, M.Á.L. and A.E.; funding acquisition, M.Á.L. and A.E. All authors have read and agreed to the published version of the manuscript.

Funding: This research was funded by the Spanish Ministry of Economy, Industry and Competitiveness (CTQ2017-86441-C2-1-R) and the TRANSNANOAVANSENS program (S2018/NMT-4349) from the Community of Madrid and LaCaixaImpulse program (CI017-00038) (A.E., M.Á.L., M.M.-G. and Á.M.-F.). Á.M.-F. acknowledges an FPU fellowship from the Spanish Ministry of Education, Culture and Sports.

Conflicts of Interest: The authors declare no conflict of interest.

References

1. Paoli, C.J.; Reynolds, M.A.; Sinha, M.; Gitlin, M.; Crouser, E. Epidemiology and costs of sepsis in the United States—An analysis based on timing of diagnosis and severity level. *Crit. Care Med.* **2018**, *46*, 1889–1897. [CrossRef] [PubMed]
2. Singer, M.; Deutschman, C.S.; Seymour, C.W.; Shankar-Hari, M.; Annane, D.; Bauer, M.; Bellomo, R.; Bernard, G.R.; Chiche, J.D.; Coopersmith, C.M.; et al. The third International consensus definitions for sepsis and septic shock (sepsis-3). *JAMA* **2016**, *315*, 801–810. [CrossRef] [PubMed]
3. Assicot, M.; Gendrel, D.; Carsin, H.; Raymond, J.; Guilbaud, J.; Bohuon, C. High serum procalcitonin concentrations in patients with sepsis and infections. *Lancet* **1993**, *341*, 515–518. [CrossRef]
4. Bouadma, L.; Luyt, C.E.; Tubach, F.; Cracco, C.; Alvarez, A.; Schwebel, C.; Schortgen, F.; Lasocki, S.; Veber, B.; Dehoux, M.; et al. Use of procalcitonin to reduce patients'exposure to antibiotics in intensive care units (PRORATA trial): A multicentre randomised controlled trial. *Lancet* **2010**, *375*, 463–474. [CrossRef]
5. Meisner, M.; Tschaikowsky, K.; Schnabel, S.; Schmidt, J.; Katalinic, A.; Schuttler, J. Procalcitonine influence of temperature, storage, anticoagulation and arterial or venous asservation of blood samples on procalcitonin concentrations. *Eur. Clin. Chem. Lab. Med.* **1997**, *35*, 597–601. [CrossRef]
6. Sager, R.; Kutz, A.; Mueller, B.; Schuetz, P. Procalcitoninguided diagnosis andantibiotic stewardship revisited. *BMC Med.* **2017**, *15*, 15. [CrossRef]

7. Schuetz, P.; Birkhahn, R.; Sherwin, R.; Jones, A.E.; Singer, A.; Kline, J.A.; Runyon, M.S.; Self, W.H.; Courtney, D.M.; Nowak, R.M.; et al. Serial procalcitonin predicts mortality in severe sepsis patients: Results from the multicenter procalcitonin monitoring—SEpsis (MOSES) study. *Crit. Care Med.* **2017**, *45*, 781–789. [CrossRef]
8. Schuetz, P.; Wirz, Y.; Sager, R.; Christ-Crain, M.; Stolz, D.; Tamm, M.; Bouadma, L.; Luyt, C.E.; Wolff, M.; Chastre, J.; et al. PCT approved blood infection marker for guiding antibiotic therapy. *Lancet Infect. Dis.* **2018**, *18*, 95–107. [CrossRef]
9. Dipalo, M.; Guido, L.; Micca, G.; Pittalis, S.; Locatelli, M.; Motta, A.; Bianchi, V.; Callegari, T.; Aloe, R.; Rin, G.; et al. Multicenter comparison of automated procalcitonin immunoassays. *Pract. Lab. Med.* **2015**, *2*, 22–28. [CrossRef]
10. Xue, J.; Yang, L.; Jia, Y.; Wang, H.; Zhang, N.; Ren, X.; Ma, H.; Wei, Q.; Ju, H. Electrochemiluminescence double quenching system based on novel emitter GdPO4: Eu with low-excited positive potential for ultrasensitive procalcitonin detection. *ACS Sens.* **2019**, *4*, 2825–2831. [CrossRef]
11. Chen, P.; Xia, F.; Tian, D.; Zhou, C. A dual-coreactants electrochemiluminescent immunosensor for procalcitonin detection based on CdS-MoS_2 nanocomposites. *J. Electrochem. Soc.* **2018**, *165*, B196–B201. [CrossRef]
12. Jia, Y.; Yang, L.; Xue, J.; Zhang, N.; Fan, D.; Ma, H.; Ren, X.; Hu, L.; Wei, Q. Bioactivity-protected electrochemiluminescence biosensor using gold nanoclusters as the low-potential luminophor and Cu_2S snowflake as co-reaction accelerator for procalcitonin analysis. *ACS Sens.* **2019**, *4*, 1909–1916. [CrossRef] [PubMed]
13. Chen, P.; Qiao, X.; Liu, J.; Xia, F.; Tian, D.; Zhou, C. A dual-signals response electrochemiluminescence immunosensor based on PTC-DEPA/KCC-1 NCs for detection of procalcitonin. *Sens. Actuators B Chem.* **2018**, *267*, 525–532. [CrossRef]
14. Xu, X.; Song, X.; Nie, R.; Yang, Y.; Chen, Y.; Yang, L. Ultra-sensitive capillary immunosensor combining porous-layer surface modification and biotin-streptavidin nano-complex signal amplification: Application for sensing of procalcitonin in serum. *Talanta* **2019**, *205*. [CrossRef] [PubMed]
15. Li, H.; Sun, Y.; Elseviers, J.; Muyldermans, S.; Liu, S.; Wan, Y. A nanobody-based electrochemiluminescent immunosensor for sensitive detection of human procalcitonin. *Analyst* **2014**, *139*, 3718–3721. [CrossRef] [PubMed]
16. Qi, S.; Li, Q.; Rao, W.; Liu, X.; Yin, L.; Zhang, H. Determining the concentration of procalcitonin using a magnetic particles-based chemiluminescence assay for the clinical diagnosis of sepsis. *Anal. Sci.* **2013**, *29*, 805–810. [CrossRef]
17. Chen, Y.; Xianyu, Y.; Wu, J.; Dong, M.; Zheng, W.; Sun, J.; Jiang, X. Double-Enzymes-Mediated Bioluminescent Sensor for Quantitative and Ultrasensitive Point-of-Care Testing. *Anal. Chem.* **2017**, *89*, 5422–5427. [CrossRef]
18. Taranova, N.A.; Urusov, A.E.; Sadykhov, E.G.; Zherdev, A.V.; Dzantiev, B.B. Bifunctional gold nanoparticles as an agglomeration-enhancing tool for highly sensitive lateral flow tests: A case study with procalcitonin. *Microchim. Acta* **2017**, *184*, 4189–4195. [CrossRef]
19. Wang, H.; Wang, H.; Chen, S.; Dzakah, E.E.; Kang, K.; Wang, J.; Wang, J. Development of a fluorescent immnunochromatographic assay for the procalcitonin detection of clinical patients in China. *Clin. Chim. Acta* **2015**, *444*, 37–42. [CrossRef]
20. Zhao, Y.; Zhou, C.; Wu, R.; Li, L.; Shen, H.; Li, L.S. Preparation of multi-shell structured fluorescent composite nanoparticles for ultrasensitive human procalcitonin detection. *RSC Adv.* **2015**, *5*, 5988–5995. [CrossRef]
21. Tang, J.; Lei, L.; Feng, H.; Zhang, H.; Han, Y. Preparation of K+-Doped core-shell $NaYF_4$: Yb, Er upconversion nanoparticles and its application for fluorescence immunochromatographic assay of human procalcitonin. *J. Fluoresc.* **2016**, *26*, 2237. [CrossRef] [PubMed]
22. Shao, X.Y.; Wang, C.R.; Xie, C.M.; Wang, X.G.; Liang, R.L.; Xu, W.W. Rapid and sensitive lateral flow immunoassay method for procalcitonin (PCT) based on time-resolved immunochromatography. *Sensors* **2017**, *17*, 480. [CrossRef] [PubMed]
23. Nie, R.; Xu, X.; Cui, X.; Chen, Y.; Yang, L. A highly sensitive capillary-based immunosensor by combining with peroxidase nanocomplex-mediated signal amplification for detection of procalcitonin in human serum. *ACS Omega* **2019**, *4*, 6210–6217. [CrossRef]
24. Vashist, S.K.; Schneider, E.M.; Barth, E.; Luong, J.H.T. Surface plasmon resonance-based immunoassay for procalcitonin. *Anal. Chim. Acta* **2016**, *938*, 129–136. [CrossRef] [PubMed]

25. Jing, W.; Wang, Y.; Yang, Y.; Wang, Y.; Ma, G.; Wang, S.; Tao, N. Time-resolved digital immunoassay for rapid and sensitive quantitation of procalcitonin with plasmonic imaging. *ACS Nano* **2019**, *13*, 8609–8617. [CrossRef] [PubMed]
26. Baldini, F.; Bolzoni, L.; Giannetti, A.; Kess, M.; Krämer, P.M.; Kremmer, E.; Senesi, F.; Trono, C. A new procalcitonin optical immunosensor for POCT applications. *Anal. Bioanal. Chem.* **2009**, *393*, 1183–1190. [CrossRef] [PubMed]
27. Buchegger, P.; Preininger, C. Four assay designs and on-chip calibration: Gadgets for a sepsis protein array. *Anal. Chem.* **2014**, *86*, 3174–3180. [CrossRef]
28. Kemmler, M.; Sauer, U.; Schleicher, E.; Preininger, C.; Brandenburg, A. Biochip point-of-care device for sepsis diagnostics. *Sens. Actuators B Chem.* **2014**, *192*, 205–215. [CrossRef]
29. Kramer, P.M.; Kess, M.; Kremmer, E.; Schulte-Hostede, S. Multi-parameter determination of TNFα, PCT and CRP for point-of-care testing. *Analyst* **2011**, *136*, 692–695. [CrossRef]
30. Rascher, D.; Geerlo, A.; Kremmer, E.; Krämer, P.; Schmid, M.; Hartmann, A.; Rieger, M. Total internal reflection (TIRF)-based quantification of procalcitonin for sepsis diagnosis—A point-of-care testing application. *Biosens. Bioelectron.* **2014**, *59*, 251–258. [CrossRef]
31. Sauer, U.; Domnanich, P.; Preininger, C. Protein chip for the parallel quantification of high and low abundant biomarkers for sepsis. *Anal. Biochem.* **2011**, *419*, 46–52. [CrossRef] [PubMed]
32. Molinero-Fernández, Á.; Moreno-Guzmán, M.; Arruza, L.; López, M.Á.; Escarpa, A. Polymer-based micromotors fluorescence immunoassay for on the move sensitive procalcitonin determination in very low birth weight infants' plasma. *ACS Sens.* **2015**. [CrossRef] [PubMed]
33. Li, Y.; Liu, W.; Jin, G.; Niu, Y.; Chen, Y.; Xie, M. Label-free sandwich imaging ellipsometry immunosensor for serological detection of procalcitonin. *Anal. Chem.* **2018**, *90*, 8002–8010. [CrossRef] [PubMed]
34. Kremmer, E.; Meyer, K.; Grässer, F.A.; Flatley, A.; Kösters, M.; Luppa, P.B.; Krämer, P.M. A new strategy for the development of monoclonal antibodies for the determination of human procalcitonin in serum samples. *Anal. Bioanal. Chem.* **2012**, *402*, 989–995. [CrossRef]
35. Zhang, Y.; Si, X.; Zhang, M.; Yang, X.; Yuan, H.; Wang, X.; Zhang, Y.; Wang, H. Rapid colorimetric determination of procalcitonin using magnetic separation and enzymatic catalysis. *Anal. Lett.* **2019**, *52*, 602–612. [CrossRef]
36. Liao, T.; Yuan, F.; Yu, H.; Li, Z. An ultrasensitive ELISA method for the detection of procalcitonin based on magnetic beads and enzyme-antibody labeled gold nanoparticles. *Anal. Methods* **2016**, *8*, 1577–1585. [CrossRef]
37. Sener, G.; Ozgur, E.; Rad, A.Y.; Uzun, L.; Say, R.; Denizli, A. Rapid real-time detection of procalcitonin using a microcontact imprinted surface plasmon resonance biosensor. *Analyst* **2013**, *138*, 6422–6428. [CrossRef]
38. Yang, Z.-H.; Ren, S.; Zhou, Y.; Yuan, R.; Chai, Y.-Q. Cu/Mn double-doped CeO_2 nanocomposites as signal tags and signal amplifiers for sensitive electrochemical detection of procalcitonin. *Anal. Chem.* **2017**, *89*, 13349–13356. [CrossRef]
39. Chiang, C.Y.; Huang, T.T.; Wang, C.H.; Huang, C.J.; Tsai, T.H.; Yu, S.N.; Chen, Y.T.; Hong, S.W.; Hsu, C.W.; Chang, T.C.; et al. Fiber optic nanogold-linked immunosorbent assay for rapid detection of procalcitonin at femtomolar concentration level. *Biosens. Bioelectron.* **2019**, *151*, 111871. [CrossRef] [PubMed]
40. Molinero-Fernández, Á.; Moreno-Guzmán, M.; Arruza, L.; López, M.Á.; Escarpa, A. Toward early diagnosis of late-onset sepsis in preterm neonates: Dual magnetoimmunosensor for simultaneous procalcitonin and co-reactive protein determination in diagnosed clinical samples. *ACS Sens.* **2019**, *4*, 2117–2123. [CrossRef] [PubMed]
41. Lim, J.M.; Ryu, M.Y.; Kim, J.H.; Cho, C.H.; Park, T.J.; Park, J.P. An electrochemical biosensor for detection of the sepsis-related biomarker procalcitonin. *RSC Adv.* **2017**, *7*, 36562–36565. [CrossRef]
42. Fang, Y.; Hu, Q.; Yu, X.; Wang, L. Ultrasensitive electrochemical immunosensor for procalcitonin with signal enhancement based on zinc nanoparticles functionalized ordered mesoporous carbon-silica nanocomposites. *Sens. Actuators B Chem.* **2018**, *258*, 238–245. [CrossRef]
43. Seshadri, P.; Manoli, K.; Schneiderhan-Marra, N.; Anthes, U.; Wierzchowiec, P.; Bonrad, K.; Di Franco, C.; Torsi, L. Low-picomolar, label-free procalcitonin analytical detection with an electrolyte-gated organic field-effect transistor based electronic immunosensor. *Biosens. Bioelectron.* **2018**, *104*, 113–119. [CrossRef] [PubMed]

44. Liu, A.; Wang, X. Amperometric immunosensor of procalcitonin based on amplification strategy of ferrocene-modified gold nanoparticles. *Int. J. Electrochem. Sci.* **2015**, *10*, 9342–9350.
45. Sui, Y.; Xu, A.; Jin, X.; Zheng, J.; He, X.; Cheng, Y.; Xie, Q.; Liu, R. In situ enzymatic generation of gold for ultrasensitive amperometric sandwich immunoassay of procalcitonin. *Biosens. Bioelectron.* **2018**, *117*, 422–428. [CrossRef]
46. Zhang, T.; Ren, X.; Fan, D.; Kuang, X.; Wang, H.; Wu, D.; Wei, Q. Electrochemical procalcitonin immunoassay based on Au@Ag heterojunction nanorods as labels and CeO_2-CuO nanorods as enhancer. *Sens. Actuators B Chem.* **2019**, *297*, 126800. [CrossRef]
47. Gao, Z.; Li, Y.; Zhang, C.; Zhang, S.; Jia, Y.; Dong, Y. An enzyme-free immunosensor for sensitive determination of procalcitonin using NiFe PBA nanocubes@TB as the sensing matrix. *Anal. Chim. Acta* **2019**. [CrossRef]
48. Ghrera, A.S. Quantum dot modified interface for electrochemical immunosensing of procalcitonin for the detection of urinary tract infection. *Anal. Chim. Acta* **2019**, *1056*, 26–33. [CrossRef]
49. Qian, Y.; Feng, J.; Wang, H.; Fan, D.; Jiang, N.; Wei, Q.; Ju, H. Sandwich-type signal-off photoelectrochemical immunosensor based on dual suppression effect of PbS quantum dots/CO_3O_4 polyhedron as signal amplification for procalcitonin detection. *Sens. Actuators B Chem.* **2019**, *300*, 127001. [CrossRef]
50. Mahe, L.S.A.; Green, S.J.; Winlove, C.P.; Jenkins, A.T.A. Pyrene-wired antibodies on highly oriented pyrolytic graphite as a label-free impedance biosensor for the sepsis biomarker procalcitonin. *J. Solid State Electrochem.* **2014**, *18*, 3245–3249. [CrossRef]
51. Shen, W.J.; Zhuo, Y.; Chai, Y.Q.; Yang, Z.H.; Han, J.; Yuan, R. Enzyme-free electrochemical immunosensor based on host–guest Nanonets catalyzing amplification for Procalcitonin detection. *ACS Appl. Mater. Interfaces* **2015**, *7*, 4127–4134. [CrossRef] [PubMed]
52. Yang, Z.H.; Zhuo, Y.; Yuan, R.; Chai, Y.Q. Electrochemical activity and electrocatalytic property of cobalt phthalocyanine nanoparticles-based immunosensor for sensitive detection of procalcitonin. *Sens. Actuators B Chem.* **2016**, *227*, 212–219. [CrossRef]
53. Liu, F.; Xiang, G.; Yuan, R.; Chen, X.; Luo, F.; Jiang, D.; Huang, S.; Li, Y.; Pu, X. Procalcitonin sensitive detection based on graphene–gold nanocomposite film sensor platform and single-walled carbon nanohorns/hollow Pt chains complex as signal tags. *Biosens. Bioelectron.* **2014**, *60*, 210–217. [CrossRef] [PubMed]
54. Liu, F.; Xiang, G.; Chen, X.; Luo, F.; Jiang, D.; Huang, S.; Yi, L.; Pu, X. A novel strategy of procalcitonin detection based on multi-nanomaterials of single-walled carbon nanohorns–hollow Pt nanospheres/PAMAM as signal tags. *RSC Adv.* **2014**, *4*, 13934–13940. [CrossRef]
55. Fang, Y.S.; Wang, H.Y.; Wang, L.S.; Wang, J.F. Electrochemical immunoassay for procalcitonin antigen detection based on signal amplification strategy of multiple nanocomposites. *Biosens. Bioelectron.* **2014**, *51*, 310–316. [CrossRef]
56. Li, P.; Zhang, W.; Zhou, X.; Zhang, L. C60 carboxyfullerene-based functionalised nanohybrids as signal-amplifying tags for the ultrasensitive electrochemical detection of procalcitonin. *Clin. Biochem.* **2015**, *48*, 156–161. [CrossRef]
57. Tan, E.K.W.; Shrestha, P.K.; Pansare, A.V.; Chakrabarti, S.; Li, S.; Chu, D.; Lowe, C.R.; Nagarkar, A.A. Density modulation of embedded nanoparticles via spatial, temporal and chemical control elements. *Adv. Mater.* **2019**, *31*, 1901802. [CrossRef]
58. Yáñez-Sedeño, P.; Campuzano, S.; Pingarrón, J.M. Magnetic particles coupled to disposable screen printed transducers for electrochemical biosensing. *Sensors* **2016**, *16*, 1585. [CrossRef]
59. Fernández-la-Villa, A.; Pozo-Ayuso, D.F.; Castaño-Álvarez, M. Microfluidics and electrochemistry: An emerging tandem for next-generation analytical microsystems. *Curr. Opin. Electrochem.* **2019**, *15*, 175–185. [CrossRef]
60. Rackus, D.G.; Shamsi, M.H.; Wheeler, A.R. Electrochemistry, biosensors and microfluidics: A convergence of fields. *Chem. Soc. Rev.* **2015**, *4*, 5320–5340. [CrossRef]
61. Mou, L.; Jiang, X. Materials for microfluidic immunoassays: A review. *Adv. Healthc. Mater.* **2017**, *6*, 1601403. [CrossRef] [PubMed]
62. Wu, J.; Dong, M.; Rigatto, C.; Liu, Y.; Lin, F. Lab-on-chip technology for chronic disease diagnosis. *NPJ Digit. Med.* **2018**, *1*, 1–11. [CrossRef] [PubMed]

63. Sackmann, E.K.; Fulton, A.L.; Beebe, D.J. The present and future role of microfluidics in biomedical research. *Nature* **2014**, *507*, 181–189. [CrossRef] [PubMed]
64. Jung, W.; Han, J.; Choi, J.-W.; Ahn, C.H. Point-of-care testing (POCT) diagnostic systems using microfluidic lab-on-a-chip technologies. *Microelectron. Eng.* **2015**, *132*, 46–57. [CrossRef]

biosensors

MDPI

Article

Development of an Electrochemical Immunosensor for Specific Detection of Visceral Leishmaniasis Using Gold-Modified Screen-Printed Carbon Electrodes

Beatriz R. Martins [1], Yanne O. Barbosa [1], Cristhianne M. R. Andrade [2], Loren Q. Pereira [2], Guilherme F. Simão [3], Carlo J. de Oliveira [1,2], Dalmo Correia [2], Robson T. S. Oliveira, Jr. [1], Marcos V. da Silva [2], Anielle C. A. Silva [4], Noelio O. Dantas [4], Virmondes Rodrigues, Jr. [1,2], Rodrigo A. A. Muñoz [5,*] and Renata P. Alves-Balvedi [1,6,*]

1. Institute of Biological and Natural Sciences, Federal University of Triângulo Mineiro, Uberaba-MG 38025-180, Brazil; biaroma_95@hotmail.com (B.R.M.); yanne.way@hotmail.com (Y.O.B.); carlo.oliveira@uftm.edu.br (C.J.d.O.); robson.junior@uftm.edu.br (R.T.S.O.J.); virmondes.rodrigues@uftm.edu.br (V.R.J.)
2. Institute of Health Sciences, Federal University of Triângulo Mineiro, Uberaba-MG 38025-180, Brazil; cristhianne_m@hotmail.com (C.M.R.A.); lorenbiomedica@gmail.com (L.Q.P.); dalmo@mednet.com.br (D.C.); marcosuftm@gmail.com (M.V.d.S.)
3. Institute of Technological and Exact Sciences, Federal University of Triângulo Mineiro, Uberaba-MG 38025-180, Brazil; guilhermefelipesimao@gmail.com
4. Institute of Physics, Federal University of Alagoas, Maceio-AL 57072-970, Brazil; acalmeida@fis.ufal.br (A.C.A.S.); noelio@fis.ufal.br (N.O.D.)
5. Institute of Chemistry, Federal University of Uberlândia, Uberlândia-MG 38408-100, Brazil
6. Federal University of Triângulo Mineiro, Iturama-MG 38025-180, Brazil

* Correspondence: munoz@ufu.br (R.A.A.M.); renata.balvedi@uftm.edu.br (R.P.A.-B.)

Received: 16 June 2020; Accepted: 17 July 2020; Published: 23 July 2020

check for updates

Abstract: Visceral leishmaniasis is a reemerging neglected tropical disease with limitations for its diagnosis, including low concentration of antibodies in the serum of asymptomatic patients and cross-reactions. In this context, this work proposes an electrochemical immunosensor for the diagnosis of visceral leishmaniasis in a more sensitive way that is capable of avoiding cross-reaction with Chagas disease (CD). Crude *Leishmania infantum* antigens tested in the enzyme-linked immunosorbent assay (ELISA) were methodologically standardized to best engage to the sensor. The antibodies anti-*Trypanosoma cruzi* and anti-*Leishmania* sp. Present in serum from patients with diverse types of CD or leishmaniasis were chosen. A screen-printed carbon electrode modified with gold nanoparticles was the best platform to guarantee effective adsorption of all antigens so that the epitope of specific recognition for leishmaniasis occurred efficiently and without cross-reaction with the evaluated CD. The current peaks reduced linearly after the recognition, and still were able to notice the discrimination between different kinds of diseases (digestive, cardiac, undetermined Chagas/acute and visceral chronic leishmaniasis). Comparative analyses with ELISA were performed with the same groups, and a low specificity (44%) was verified due to cross-reactions (high number of false positives) on ELISA tests, while the proposed immunosensor presented high selectivity and specificity (100%) without any false positives or false negatives for the serum samples from isolated patients with different types of CD and visceral leishmaniasis. Furthermore, the biosensor was stable for 5 days and presented a detection limit of 200 ng mL^{-1}.

Keywords: electrochemical biosensor; visceral leishmaniasis; Chagas disease; gold nanoparticles; point-of-care; portable analysis

1. Introduction

Leishmaniasis is a complex of diseases caused by a protozoan of the genus *Leishmania* [1,2] that affects millions of people worldwide. Visceral leishmaniasis, caused by *Leishmania donovani and Leishmania infantum*, represents the most severe form and can lead to death if not treated [3,4]. The onset of the infection and clinical manifestations are dependent on many factors including environmental and host immunologic status, especially in the early stages of infection [5]. Visceral leishmaniasis represents a major health problem in some tropical areas of the world. The currently available serum diagnosis does not fit the proper criteria of sensitivity and specificity, especially for identification of asymptomatic and or low symptomatic patients due to the low concentration of antibodies in the serum, particularly in the case of asymptomatic patients, which results in high cross-reactions [6]. Furthermore, due to its epidemiological characteristics, a diagnostic test that is accessible in remote areas is a desired tool for precise diagnosis and early therapeutic intervention.

The diagnosis of visceral leishmaniasis is made by combining clinical signs with parasitological or serological tests; however, they depend on extremely equipped laboratories, qualified labor, and a long period of time to carry out the tests [7,8]. The parasitological diagnosis is the reference choice for exams for detecting the disease, which shows the parasite directly in tissues or in culture. Aspirates from the spleen, bone marrow, and lymph nodes are used, and liver biopsy can also be performed. These techniques have high specificity and variable sensitivity. Until 2014, the Ministry of Health used two tests: the fluorescence indirect antibody test (IFAT) for human leishmaniasis, developed by the Institute of Immunobiological Technology (Biomanguinhos), Fundação Oswaldo Cruz, Brazil, and the Kalazar Detect rapid test (InBios International, Seattle, WA, USA).

Several studies show a comparison between the techniques, such as rapid test rK39, direct agglutination test (DAT), and ELISA, with some even showing good sensitivities and specificities in these patients [9]. However, the identification of asymptomatic infection remains challenging, since it depends on the sensitivity and specificity of the employed technique [10]. Furthermore, there is no agreement among the available techniques, and thus it is necessary to search for the best method to measure visceral leishmaniasis positives not only in symptomatic patients [11]. Hence, the average linear range of asymptomatic patients is a concern.

In this context, the presentation of new methodologies that show good performance, easy handling, speed, and detection of asymptomatic patients is essential to the control and early treatment of leishmaniasis. In other words, the development of tools that contribute to the optimization of a portable platform of leishmaniasis is a priority. Thus, the development of electrochemical biosensors shows advantages in comparison to traditional techniques, such as fast execution, a small amount of sample utilization, portability, selectivity, and specificity to obtain diagnoses.

The physical principle of the biosensor is to turn the biological sign into an electrical sign, making it possible in this way to monitor and quantification of signals. The immobilization of antigens, which specifically recognize the antibodies, can be provided on the surface of the sensor. The biorecognition, in the case of the evaluated pathologies, involves the antibodies as the target of detection, and they can be also quantified [9,12].

Electrochemical biosensors were developed by some research groups seeking to solve the problem of diagnosis of visceral leishmaniasis, investigating novel platforms and modification procedures for the diagnostic needs of *Leishmania* sp. [13–23]. Amongst the platforms used in sensors, the carbon-based electrodes present conditions for immobilization (through adsorption) once they enable a random anchorage and orientation of the biomolecules on its surface. Meanwhile, gold electrodes enable oriented couplings that minimize the distance between the active biomolecule sites and the electrode surface, facilitating the electron transfer, and a greater number of antibodies can be immobilized on the electrode surface [24–26]. Considering the use of portable platforms for point-of-care diagnosis, screen-printed electrodes play a key role in the development of electrochemical biosensors for several applications, as reported in the literature [27–35].

In order to an immunosensor functions properly for the diagnosis of leishmaniasis, it should be highly sensitive, specific, fast, and simple, with potential application for the serological diagnosis of leishmaniasis, since the disease is directly related to cross-reactivity with Chagas disease and other diseases. There are some studies reporting sensors for the diagnostic of *Leishmania* sp. Infection. Mohan et. al. [36] developed a genosensor modified with NiO nanostructured on an indium-tin oxide conductive glass plate to distinguish DNA from parasites extracted from human DNA extracts; however, these tests have not yet been performed in clinical samples. Moradi et al. [13] also developed a genosensor based on gold nanoparticles immobilized on polycrystalline gold discs and found high sensitivity in cutaneous leishmaniasis. However, the DNA extraction process has a higher cost compared to the use of protein antigens. Facing the diagnostic difficulties of cross-reactions, in this study, we describe how the electrode allows for differentiated interactions of the same biomolecule using a gold nanoparticle-modified surface, starting from the principle of this being biologically compatible and non-toxic. Such properties have attracted attention in diagnostic application because the molecules should not have their biological properties altered [26,36,37].

In this context, this work shows the development of an immunosensor for the specific diagnosis of visceral leishmaniasis without the cross-reactivity with Chagas disease (CD). Linear response, sensitivity, selectivity, specificity, repeatability, reproducibility, and stability were researched. To our knowledge, the concentration of *Leishmania* antigens considered as potential risk to develop the disease is not accurately known, and for this reason the proposed biosensor provides a detection limit in very concentrations (ng mL^{-1}) to detect antigens before the appearance of disease symptoms.

2. Materials and Methods

2.1. Reagents and Biomolecules

All used reagents were of analytical grade and were used without further purification. Ultrapure water (MilliQ, Resistivity value greater than 18.2 MΩ, Millipore Corporation, Burlington, MA, USA) was used in the preparation of all solutions. The aqueous solution of the mixture of potassium ferricyanide/ferrocyanide ($[Fe(CN)_6]^{3-}/[Fe(CN)_6]^{4-}$) in KCl (5 mmol L^{-1}, 0.1 mol L^{-1}, pH 7.4, LabSynth, Brazil) used for the electrochemical characterization of the immunosensor was prepared immediately before the use. Gold (III) chloride was dissolved in sulfuric acid medium (1 g L^{-1} $AuCl_3$ in 10 mL of 0.5 mol L^{-1} sulfuric acid). All experiments were carried out at controlled room temperature (25 ± 1 °C).

Preparation of the leishmania infantum crude antigen: The PP75 strain of *Leishmania infantum* cultured in the Schneider medium, supplemented with 20% fetal bovine serum, in the exponential phase was centrifuged 2000× *g* at 25 °C for 20 min and then washed three times with phosphate buffer (PB) solution and discarded supernatant. The pellet was resuspended in PB containing 0.05% NP40 (Nonidet P-40 Substitute, Roche) with the COMPLETE protease inhibitor (ROCHE, SWI). The antigen was obtained by the method of freezing in liquid nitrogen and thawing in a 37 °C water bath and then centrifuging it at 10,000× *g* for 30 min, and then the supernatant containing the soluble crude antigen was stored at −80 °C until the moment of use. The protein concentration of the antigen was determined by the Lowry method [38]. Aliquots of the extract were stored with the total soluble antigens at −80 °C until further use. The preparation of crude antigens specific to *Leishmania infantum* was performed. This species is characteristic of visceral leishmaniasis [39]. The technique was performed and adapted on the basis of [40]. Moreover, in our experiments, the efficiency of soluble and membrane-free extract of the parasite (data not shown) was proven, since these *Leishmania infantum* antigens are able to specifically be recognized by the serum antibodies of patients with visceral leishmaniasis antibodies.

The real samples used in the experiments appeared favorable and were substantiated from CEP (Comitê de Ética em Pesquisa/Research Ethics Committee) by Plataforma Brasil. *Leishmania* sera have the CAAE 58301516.8.0000.5154 and were 1,846,584 in number. Chagas sera have the CAAE 64048117.3.0000.5154 and were 2,163,043 in number. The stock solutions of total antigens (0.01 µg mL^{-1})

and serum (visceral leishmaniasis = 1:100–0.202 mg mL^{-1}; CD = 1:100–0.146 mg mL^{-1}) were diluted in deionized water and frozen until the electrochemical experiment.

2.2. Devices

Screen-printed carbon electrodes (DPR-110) and screen-printed gold electrodes (DPR-220 BT) were purchased from DropSens (Oviedo, Asturias, Spain), which consist of a ceramic strip containing a three-electrode system (working, counter, and reference electrodes) for a single-drop analysis. The reference was made of a silver ink (known as silver pseudo-reference electrode) and the counter and working electrodes were made of carbon ink (in DPR-110) or of made of a gold ink (in DPR-220 BT). The working electrode of the screen-printed carbon electrode was modified with gold by electrodeposition (next described), which is the third electrochemical device evaluated in this work. Electrochemical analyses were performed by cyclic voltammetry using Em Stat 1 equipment (PalmSens BV, The Netherlands) connected to a notebook. The changes in the electrochemical signals of $[Fe(CN)_6]^{4-}/[Fe(CN)_6]^{3-}$ (5 mM) were evaluated (scan rate of 100 mV s^{-1}). For ELISA tests (EnSpire/PerkinElmer), optical density (OD) values were determined on a microtiter plate reader at 490 nm.

2.3. Indirect ELISA

The indirect ELISA for the detection of immunoglobulin G (IgG) antibodies against leishmania used high affinity plaques (Thermo Scientific Tm Nunc Tm, Waltham, MA, USA), which were sensitized with the antigens (1 μg mL^{-1}), diluted in 0.06 mol L^{-1} carbonate-bicarbonate buffer (pH 9.6), and incubated for 18 h at 4 °C. After this period, all plates were washed six times with PB containing 0.05% Tween 20 (PB-T) and blocked with PB containing 5% skimmed milk powder (Molico, Nestle, São Paulo, Brazil—PB-M5%) for 4 h at room temperature. After further washing, the serum samples were 1:40 diluted in 5% PB-M and incubated for 2 hours at room temperature. After six washes, the anti-human IgG antibody (1:2000) conjugated to peroxidase (IgG/horseradish peroxidase(HRP), Dako) was added and incubated for 2 h at room temperature. After further washing, the reaction was developed by addition of the enzymatic substrate 1,2-orthophenylenediamine (OPD, Dako) with 0.05% H_2O_2 and stopped with H_3PO_4. Positive and negative controls were included on the plate. The levels of antibodies were expressed in ELISA, according to the following formula: EI = Abs sample/cut-off, where cut-off is calculated as the mean of the Abs of negative control serum plus three standard deviations. EI values > 1.4 were considered positive.

2.4. Electrodeposition of Gold Nanoparticles on Carbon Electrodes

The electrodes were submitted to a 30-cycle cyclic voltammetry (CV) pre-treatment in 1 mol L^{-1} H_2SO_4 solution in the potential range between −0.3 and +1.2 V at 100 mV s^{-1} for surface cleaning and activation. After that, the electrodes were submitted to the deposition of gold nanoparticles by 15 voltammetric cycles in a gold chloride ($HauCl_4$, 1 g L^{-1}) solution prepared in 1 mol L^{-1} H_2SO_4 in the potential range between 0.3 and +1.0 V at scan rate of 0.1 V s^{-1} [41–44]. After the electrodeposition, it is possible to verify the color change of the working electrode evidencing the formation of gold nanoparticles.

The activation of the modified electrode was performed by 10 cycles in 1 mol L^{-1} H_2SO_4 solution to eliminate impurities that can hinder the adsorption of molecules, diminishing the reproducibility and stability of the modified surface [45–47].

2.5. Immunosensor

The first step involved the immobilization of the total soluble antigens on the surface of the working electrode (carbon, gold, and carbon modified with an electrodeposited gold nanoparticles) by drop-casting. The dissolution method of 4 μL lasted until the solution dried (15 min). In order to prevent nonspecific binding, we added 4 μL of 1% bovine serum albumin (BSA) as a blocking solution

(15 min) after the first step. At the end, the serum was made available until it dried. After each step, the electrodes were washed and dried in a desiccator. For the interaction investigation between total antigens immobilized on each working electrode and total soluble antigens recognition, we used the solution of $[Fe(CN)_6]^{4-}/[Fe(CN)_6]^{3-}$ (5 mM) as a redox probe indicator. Thus, 80 μL of this solution was dropped over the three electrodes, closing the working electrode circuit between the other two electrodes (counter electrode and reference). Reactions occurred at room temperature (25 ± 1 °C). Using the cyclic voltammetry (CV) technique, we evaluated the behavior of the electrochemical signal of the supporting electrolyte (indirect detection) on the sensor, as shown in Figure 1.

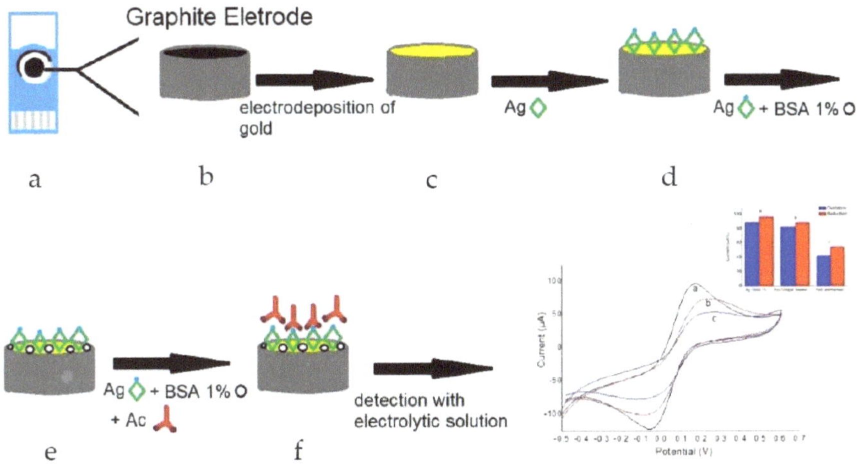

Figure 1. Scheme of the carbon immunosensor using a gold-modified electrode. The surface antigen probe was autonomized as presented in a voltammogram. The steps of preparation of the immunosensor are as follows: (**a**) the carbon electrode was selected as the base platform; (**b**) carbon was electrodeposited with gold nanoparticles; (**c**) the *Leishmania infantum* antigen was immobilized on the gold nanoparticle-modified surface; (**d**) the 1% bovine serum albumin (BSA) blocking solution was coupled to the platform as a blocking solution; (**e**) I after the preparations, the antibodies were coupled (real sample/serum); and (**f**) at the end, the electroanalytical solution was inserted and the process of transduction was initiated.

2.6. Specificity

An aliquot of 4 μL of positive serum for Chagas disease (1:100 diluted) was pipetted on the immunosensor and kept for 15 min at room temperature. Thereafter, a final wash occurred with MiliQ water (50 μL), and the electrode was dried. Using the other electrode, the same protocol with positive serum for leishmaniasis (1:100 diluted) was performed, and on a third electrode, it was performed with serum negative (diluted 1:100, protein concentration in 5 μg mL^{-1}). In all tests, including triplicates, the changes in the electrochemical signals of $[Fe(CN)_6]^{4-}/[Fe(CN)_6]^{3-}$ (5 mM) were evaluated (scan rate: 100 mV s^{-1}).

2.7. Sensor Stability

To evaluate the stability of the immunosensor, we stored modified electrodes containing the total soluble antigen at 4 °C for 5 days, protecting them from light and oxygen.

2.8. Calibration Curve

To validate the immunosensor sensitivity analyses, we added 4 μL of different serum dilutions (1:25, 1:50, 1:100, 1:250, 1:100,000) to the immunosensor. Incubation was for 15 min at 25 ± 1 °C.

2.9. Statistical Analysis

The analyses are descriptive and are based on the comparative study of the voltammograms and their reinterpretations in bar charts and linear graphs (calibration).

3. Results and Discussion

3.1. Screen-Printed Electrode

Considering the differential molecular interaction with electrodes, we proposed the evaluation of the antigens and their recognition by anti-*Leishmania* antibodies present in the serum using a screen-printed carbon electrode and a screen-printed gold electrode (results shown in Figure 2A,B). It is important to emphasize that the same antibody/antigen system was used in the proposed electrochemical biosensor. The analyses enabled the evaluation of which platform increased stability and maintenance of the biological activity of the antibody because the immobilization of the probe on the electrode surface is a crucial step in the development of the sensors. To homogenize the analyses, we assembled the column graphs from current peak data, extracted from CV measurements. The current percentages (oxidation and reduction currents of the redox probe) were calculated from the initial CV (without the biomolecule) counting 100%. As the immobilization of biomolecules occurs by physical adsorption, conducting sites of the working electrode are blocked and therefore a fall of the current occurs (fewer conducting sites are available for the redox probe undergoes the electron transfer). Thus, low percentage refers to high blocking surface due to proportional immobilization or molecular recognition (Figure 2). On carbon (Figure 2A), an affinity for adsorption of antigens was shown, but no recognition for antibodies present in serum from patients with CD and visceral leishmaniasis was found, which indicates the effect of cross-reactivity. On the other hand, when the gold electrode was used as a platform (Figure 2B), a greater affinity for CD occurred more effectively in comparison with visceral leishmaniasis. This result may be explained by the molecular organization of the antigens on the gold surface due to thiol group presented in the leishmaniasis antigens, as previously stated in the literature for the immobilization of visceral leishmaniasis antigens [48].

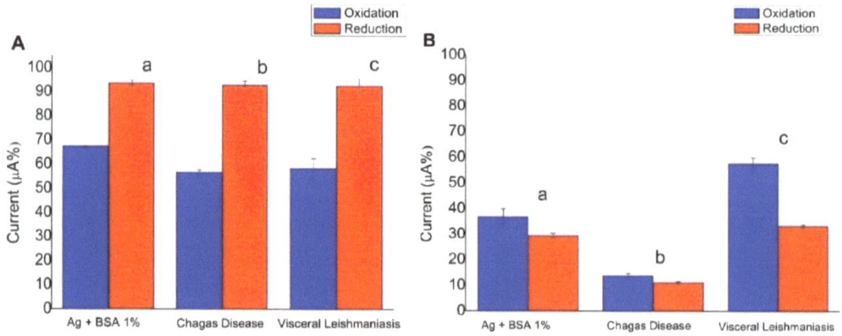

Figure 2. Column graphs extracted from cyclic voltammetry (CV) shows the variation of current peak percentages for (a) after immobilization of total soluble antigen followed by the addition of BSA, (b) after addition of Chagas disease (CD) antibodies, and (c) after addition of visceral leishmaniasis antibodies. The percentages were calculated from initial CV (without biomolecule) counting 100%. (**A**) Screen-printed carbon electrode used as a platform. (**B**) Screen-printed gold electrode used as a platform. The data are oxidation in blue and reduction in red. The electrochemical probe was 5 mmol L^{-1} $[Fe(CN)_6]^{4-}/[Fe(CN)_6]^{3-}$ and scan rate was 100 mV s^{-1}.

3.2. Gold-Modified Electrode Used as a Platform

Preliminary results using the screen-printed gold electrodes showed higher sensitivity than the screen-printed carbon electrodes; however, the results were not completely satisfactory when the immunosensor was evaluated in real samples (sensitivity was still moderate), and for this reason we investigated a novel platform—the gold-modified screen-printed carbon electrodes. This platform is well known for the formation of gold nanoparticles by electrodeposition [41], and thus the working electrode of the screen-printed carbon electrode strip was used as the electrode surface for modification. The modification with gold nanoparticles provided an increase in surface area and hence could potentially improve sensitivity. Figure 3A shows the CV recordings and the respective current percentages (bar plot beside the CVs) after the addition of (a) antigens, (b) CD, and (c) visceral leishmaniasis. This figure shows that the affinity for adsorption of antigens was effective and even more effective for the recognition of anti-leishmania antibodies than using the gold electrode because the current values of the redox probe decreased sequentially in (b) and (c). We can note the absence of cross-reactivity with Chagasic serum, proving the specificity of this sensor (Figure 3A). Moreover, on the basis of the higher efficiency of antigen immobilization without affecting its biorecognition site, we selected carbon electrodes modified with electrodeposited gold nanoparticles. Figure 3A also shows that the total soluble antigens could quantitatively discriminate (b) and (c) from each serum (CH 146, LSH 202). Figure 3B shows the response in the presence of groups of different clinical forms of Chagas diseases. Even under these conditions, the immunosensor responded only to anti-*Leishmania* antibodies, which indicated the absence of cross-reactivity. Figure 3C shows changes in current peaks from the different types of leishmaniasis, indicating the recognition of the immunosensor towards the different leishmaniasis antibodies.

3.3. Stability

The stability studies of the biosensor were evaluated under storage at 4 °C for 5 days, protected from light and oxygen. This experiment was performed with the same electrode previously optimized using a gold-modified screen-printed carbon electrode modified with leishmaniasis antigens. The biosensor was evaluated in the presence of anti-*Leishmania* antibodies and we observed a decrease of 46% in the detection capacity after 5 days. This signal decrease indicated lack of stability, which is a compromising feature of the proposed biosensor. Future experiments are required to investigate a condition to improve the stability of the immunosensor.

3.4. Calibration Curve

Figure 4 shows the preliminary analysis of the calibration curve using the immunosensor. Keeping in mind that the current of the redox probe is inversely proportional to the concentration of antibodies, we used diluted serum at 1:25, 1:50, 1:100, 1:250, and 1:100,000 ratios. Higher dilutions did not generate linearity on the results (tests performed in triplicates). This plot presents the correlation coefficient of 0.9746 (for the equation: i(%) = −727.5 × [serum dilution ratio] + 72.83), an estimated limit of detection of 202 ng mL^{-1}, and limit of quantification of 606 ng mL^{-1}. The inset shows the equation obtained from the linear regression of a current peak (%) vs. concentration of leishmaniasis.

In the screen-printed electrodes without alteration of their surfaces (carbon surface), it was possible to observe that there was reactivity and absence of specification. The choice of electrodeposition of gold nanoparticles provided to the carbon electrode new physical-chemical properties of the biomolecules that were immobilized on its surface. The results demonstrated that the electrodeposition of the gold nanoparticles was not only capable of promoting reactivity, but also the desired selectivity. In addition to improving the responses on biosensor platforms, it also allowed the interaction of the biological probes to the surface of the same ones [49]. Given the results, the electrodeposited gold on the screen-printed carbon electrode improved the sensitivity of the sensor by effectively increasing the surface area of the electrode, promoting a greater site of adsorption of total soluble antigens [50].

Moreover, the orientation of the total soluble antigen adsorption on gold nanoparticles may have contributed to the improved specificity not obtained when the total soluble antigens were immobilized on the unmodified carbon electrode. The interaction and orientation of total soluble antigens may occur through the S–H bonds of some amino acids of the total soluble antigens, knowing that these were strongly linked to gold (chemisorption/covalent) [51,52]. This combination generates a late rally organized system between the biomolecules in a spontaneous, stabilized, and oriented way, being well-known as a self-assembled monolayer [48]. This kind of modification has extra experimental advantages over the use of gold in biosensors. Furthermore, it is linked to ease of handling and preparation, low cost, accessibility, and stability without the need for an additional step involving the addition of a thiol monolayer on the electrode surface [53,54].

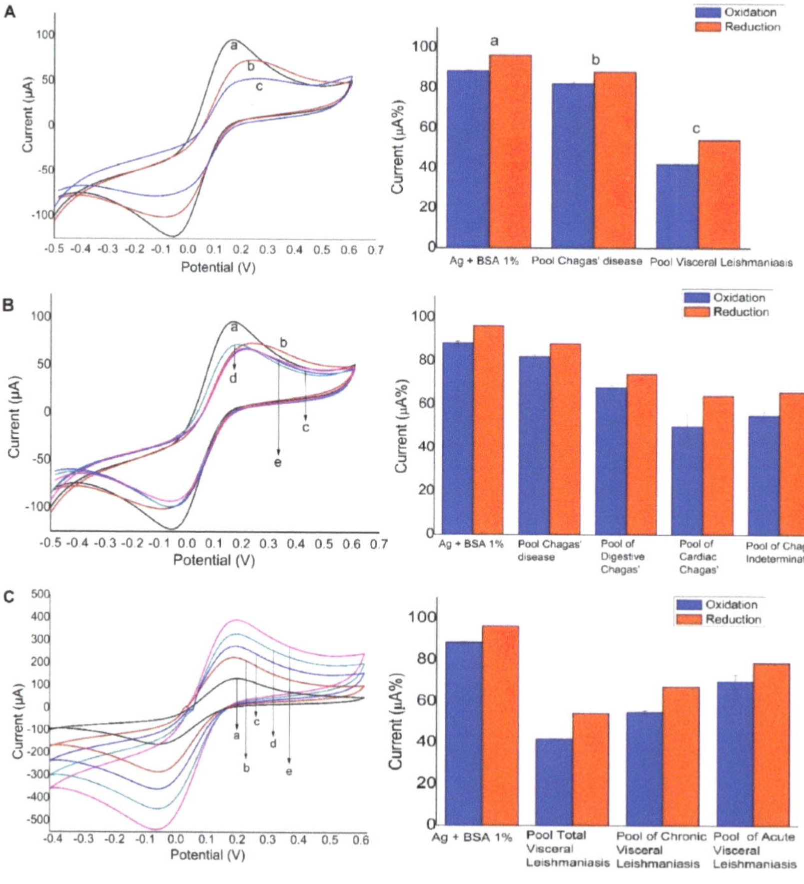

Figure 3. Cyclic voltammograms for the redox probe on the carbon electrode modified with gold nanoparticles that were electrodeposited and antigens; bar plots show the variation of peak current percentages. (**A**) Immunosensor (a), pool Chagas serum (b), pool visceral leishmaniarum (c). (**B**) Immunosensor (a) and different clinical forms by CD: Chagas serum pool (b), cardiac Chagas serum (c), digestive Chagas serum (d), Chagas undetermined serum (e). (**C**) Pool of total visceral leishmaniasis (a), pool of acute visceral leishmaniasis (b), pool of chronic visceral leishmaniasis (c), antigen *Leishmania infantum* + BSA1% (d), pool negative for leishmaniasis (e). The changes in the electrochemical signals of $[Fe(CN)_6]^{4-}/[Fe(CN)_6]^{3-}$ (5 mM) were evaluated (scan rate of 100 mV s^{-1}).

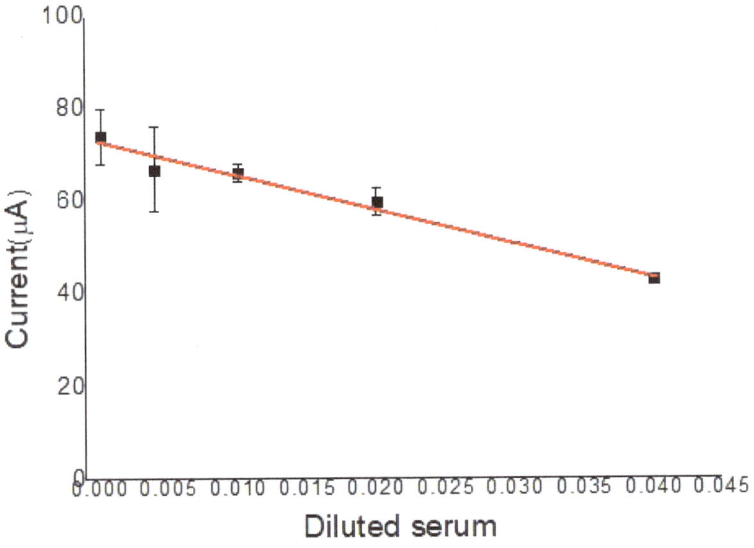

Figure 4. Calibration curve obtained from current peak percentages for triplicate measurements of the biosensors in the presence of diluted serum (1:25, 1:50, 1:100, 1:250, and 1:100,000) containing antibodies (stock serum solution of 20.2 mg mL^{-1}). The percentages were calculated from initial CV (without biomolecule) counting 100%. The changes in the electrochemical signals of $[Fe(CN)_6]^{4-}/[Fe(CN)_6]^{3-}$ (5 mM) were evaluated (scan rate of 100 mV s^{-1}).

3.5. Comparison with Indirect ELISA

In this study, a comparison of the results of the electrochemical immunosensor with indirect ELISA was carried out. The different diagnostic methods of visceral leishmaniasis present detection difficulties justified by the occurrence of cross-reactions with other trypanosomatids, explained by phylogenetic limitations existing among protozoa [1]. The results for indirect ELISA indicate low specificity (Table S1 shows the representation values from the ELISA plate tests shown in Figure S1). As expected, the cross-reaction occurred, with discrimination only in positive serum from patients with acute leishmaniasis (Figure S1). The lack of specificity that occurred in the ELISA test can be explained by the occurrence of affinities between the total antigen of visceral leishmaniasis and the antibody of Chagas disease. Figure 5A shows the distribution of ELISA index (EI) values obtained for the tests performed in different serum samples (data from Table S1), with the individual tests showing cross-reaction. From the distribution in Figure 5A, it is possible to observe 1 false negative and 15 false positive tests. Figure 5B shows the EI values and percentage current obtained by the proposed immunosensor obtained for the same serum samples (six different diseases and a pool of all of them). We found that the proposed biosensor detects Chagas disease as being a weak interaction, with this connection being due to the existence of some possible interactions between the total leishmaniasis antigen and the anti-*Trypanosoma cruzi* antibody that causes Chagas disease. In the detection of leishmaniasis, there is a strong interaction between the total visceral leishmaniasis antigen and the anti-*Leishmania infantum* antibody, a result that converges with what is expected in theory. While the electrochemical immunosensor was able to discriminate acute and chronic leishmaniasis from all the analyzed serum samples, the ELISA test showed values of EI higher than 1.4 for almost all cases presented in Figure 5B, which indicates false negative for cardiac, digestive, and indeterminate Chagas. Thus, the electrochemical biosensor can differentiate diseases, even with the occurrence of such affinities reported in the literature.

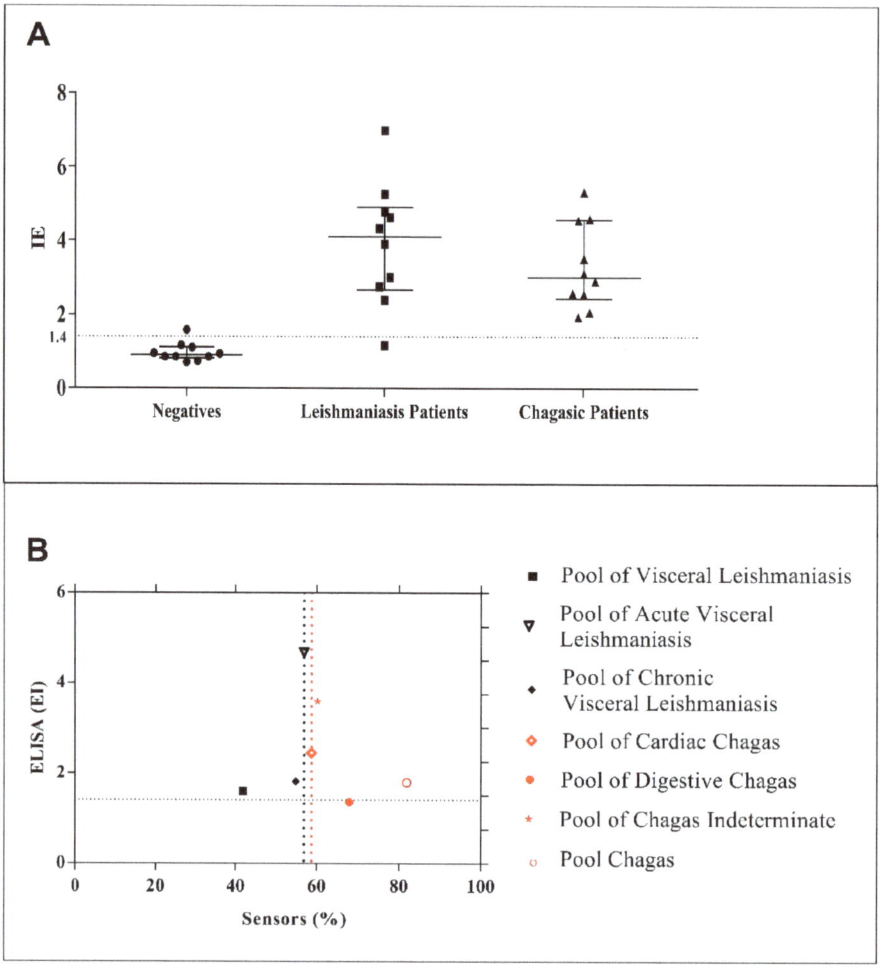

Figure 5. (**A**) Distribution of ELISA index (EI) values obtained by indirect ELISA (data from Table S1), with a line at EI = 1.4 to guide the reader (values above 1.4 are considered positive for the tests), for serum samples from patients with different types of diseases. (**B**) EI values (ELISA) and percentage current (proposed immunosensor) obtained for the same serum samples of different diseases (acute and chronic visceral leishmaniasis; cardiac, digestive, and indeterminate Chagas; and a pool of all). Current values below the black dotted line indicate positive tests for leishmaniasis, while current values above the red dotted line indicate negative tests.

On the basis of these results, we calculated the selectivity and specificity parameters for the electrochemical biosensor and ELISA using the following equations [55,56]:

$$(\text{Specificity}) = (\text{total of negative tests})/(\text{false positives} + \text{total negative tests}) \quad (1)$$

$$(\text{Sensitivity}) = (\text{total of positive tests})/(\text{false negatives} + \text{total positive tests}) \quad (2)$$

ELISA tests presented low specificity (44%) for a total of 38 serum samples analyzed due to the high number of false positives (15). Selectivity was calculated as 93.2% due to one false negative test.

On the other hand, the proposed electrochemical biosensor did not present any false positive or false negative for the total of 12 analysis of serum samples from patients of different types of CD and leishmaniasis, and consequently the calculated specificity and sensitivity values were both 100%. Therefore, the developed immunosensor presented impressive advantages over the current ELISA tests for the diagnosis of *Leishmania infantum*.

Finally, Table 1 compares the sensitivity and specificity of the proposed electrochemical biosensor and ELISA tests with data referring to the diagnostic tests of existing leishmaniasis. This table shows that the proposed biosensor presents the highest sensitivity and specificity values compared to traditional ELISA tests, as well as other diagnostic tests.

Table 1. Results of research conducted on patients with the aim of finding an efficient diagnosis of leishmaniasis.

Method	Material	Sensitivity/Specificity	Reference
IFAT	Serum	88–92%/83–88%	[57]
Kalazar Detect	Serum	84–88.1%/91%	[57]
IT LEISHBio-Rad	Blood/Serum	92–93%/92–98%	[57]
PCR	Blood	93%/96%	[8,57,58]
DAT-LPC	Blood	99%/98%	[57]
RIFI	Serum	0–100%/80%	[9,10]
Immunochromatographic tests	Serum	87%/94%	[11–13]
ELISA	Serum	80–99%/81–100%	[14–18]
Electrochemical	DNA extracted of blood	Not explained	[22]
ELISA	Serum	92.3%/44%	This work
Electrochemical	Serum	100%/100%	This work

4. Conclusions

The results showed that the electrodeposition of gold nanoparticles on the carbon electrode was efficient in immobilization of the antigens, providing better results than bare carbon electrode or gold electrode. This property allowed specific epitope recognition of antibodies present in serum from patients with visceral leishmaniasis. It should be noted that the use of serum from patients with CD solves a major problem in detection tests present in the market. We aim to improve and develop these sensors for application in public health, that is, in a large number of tests with high performance in sensitivity and specificity. Therefore, future studies will be extended in the selection of antigens purified for the diagnosis of asymptomatic visceral leishmaniasis patients and tests with extended periods in stability. In this way, such a tool may be feasible for practical applications and commercial purposes, with a reduced cost compared with Western blot, and improved performance in comparison with ELISA. Hence, electrochemical immunosensors offer great promises for specific and selective diagnosis of visceral leishmaniasis.

Supplementary Materials: The following are available online at http://www.mdpi.com/2079-6374/10/8/81/s1, Figure S1: ELISA plates for immunological tests. Table S1: Values of ELISA index (IE) obtained for all samples. Values above 1.4 are considered positive for the tests.

Author Contributions: Conceptualization, V.R.J., R.A.A.M., R.P.A.-B.; methodology/resources, C.J.d.O., D.C., R.T.S.O.J., M.V.d.S., A.C.A.S., N.O.D., V.R.J., R.A.A.M., R.P.A.-B.; validation/formal analysis/investigation, B.R.M., Y.O.B., C.M.R.A., L.Q.P., G.F.S.; data curation, C.J.d.O., D.C., R.T.S.O.J., M.V.d.S., A.C.A.S., N.O.D., V.R.J., R.A.A.M., R.P.A.-B.; writing—original draft preparation, R.P.A.-B.; writing—review and editing, R.A.A.M., R.P.A.-B.; visualization, R.A.A.M., R.P.A.-B.; supervision, R.A.A.M., R.P.A.-B.; project administration, R.P.A.-B. All authors have read and agreed to the published version of the manuscript.

Funding: This research was funded by the Research Foundation of the State of Minas Gerais (FAPEMIG, APQ–01083-16), the National Council for Scientific and Technological Development (CNPq, 465389/2014-7 INCTBio), the Coordination for the Improvement of Higher Education Personnel (CAPES, financial code 001).

Acknowledgments: We are grateful to Pro-Rectory for Research and Post-Graduation of UFTM and the Graduate Programs in Physiological Sciences (PPGCF) and in Tropical Medicine and Infectiology (PPGMEDTROP) of UFTM. We would also like to thank Luiz Fernando de Oliveira for the English grammar review of the manuscript.

Conflicts of Interest: The authors declare no conflict of interest.

References

1. Convit, J.; Ulrich, M.; Fernandez, C.T.; Tapia, F.J.; Caceres-Dittmar, G.; Castes, M.; Rondon, A.J. The clinical and immunological spectrum of American cutaneous leishmaniasis. *Trans. R. Soc. Trop. Med. Hyg.* **1993**, *87*, 444–448. [CrossRef]
2. Castellano, L.R.; Correia-Filho, D.; Agiro, L.; Dessein, H.; Prata, A.; Dessein, A.; Rodrigues, V. Th1/Th2 immune responses are associated with active cutaneous leishmaniasis and clinical cure is associated with strong interferon-γ production. *Hum. Immonol.* **2009**, *70*, 383–390. [CrossRef] [PubMed]
3. Pérez-Cabeza, B.; Cecílio, P.; Robalo, A.L.; Silvestre, R.; Carrilho, E.; Moreno, J.; San Martin, J.V.; Vasconcellos, R.; Cordeiro-da -Silva, A. Interleukin-27 Early Impacts Leishmania infantum Infection in Mice and Correlates with Active Visceral Disease in Humans. *Front Immunol.* **2016**, *7*, 478. [CrossRef] [PubMed]
4. Gutiérrez-Rebolledo, G.A.; Drier-Jonas, S.; Jiménez-Arellanes, M.A. Natural compounds and extracts from Mexican medicinal plants with anti-leishmaniasis activity: An update. *Asian Pac. J. Trop. Med.* **2017**, *10*, 1105–1110. [CrossRef]
5. Gollob, K.J.; Viana, A.G.; Dutra, W.O. Immunoregulation in human American leishmaniasis: Balancing pathology and protection. *Parasite Immunol.* **2014**, *36*, 367–376. [CrossRef]
6. Matos, H.J.d.; Pinto, A.Y.d.N.; Miranda, A.M.M.; Silva, F.L.C.; Ramos, F.L.P. Cross-reactivity in serological tests between Chagas disease and visceral leishmaniasis in endemic regions for both diseases. *Rev. Pan-Amaz. Saúde* **2015**, *6*, 65–68. [CrossRef]
7. Pan American Health Organization, Key facts on Neglected Infectious Diseases. Leishmaniasis. 2017. Available online: https://www.paho.org/hq/dmdocuments/2017/2017-cha-leishmaniasis-factsheet-work.pdf (accessed on 12 June 2020).
8. Freire, M.L.; Machado de Assis, T.; Oliveira, E.; Moreira de Avelar, D.; Siqueira, I.C.; Barral, A.; Rabello, A.; Cota, G. Performance of serological tests available in Brazil for the diagnosis of human visceral leishmaniasis. *PLoS Negl. Trop. Dis.* **2019**, *13*, e0007484. [CrossRef]
9. Lemos, A.J.G.; Balvedi, R.P.A.; Rodovalho, V.R.; Resende, L.O.; Castro, A.C.H.; Cuadros-Orellana, S.; Madurro, J.M.; Brito-Madurro, A.G. Immunosensor assembled on polymeric nanostructures for clinical diagnosis of C-reactive protein. *Microchem. J.* **2017**, *133*, 572–576. [CrossRef]
10. Pereira, S.V.; Bertolino, F.A.; Fernández-Baldo, M.A.; Messina, G.A.; Salinas, E.; Sanz, M.I.; Raba, J. A microfluidic device based on a screen-printed carbon electrode with electrodeposited gold nanoparticles for the detection of IgG anti-Trypanosoma cruzi antibodies. *Analyst* **2011**, *136*, 4745–4751. [CrossRef]
11. Moreno, E.C.; Gonçalves, A.V.; Chaves, A.V.; Melo, M.N.; Lambertucci, J.R.; Andrade, A.S.R.; Negrão-Corrêa, D.; Antunes, C.M.D.F.; Carneiro, M. Inaccuracy of enzyme-linked immunosorbent assay using soluble and recombinant antigens to detect asymptomatic infection by Leishmania infantum. *PLoS Negl. Trop. Dis.* **2009**, *3*, 1–7. [CrossRef]
12. Balvedi, R.P.A.; Castro, A.C.H.; Madurro, J.M.; Brito-Madurro, A.G. Detection of a specific biomarker for Epstein-Barr virus using a polymer-based genosensor. *Int. J. Mol. Sci.* **2014**, *15*, 9051–9066. [CrossRef] [PubMed]
13. Moradi, M.; Sattarahmady, N.; Rahi, A.; Hatam, G.R.; Sorkhabadi, S.M.R.; Heli, H. A label-free, PCR-free and signal-on electrochemical DNA biosensor for Leishmania major based on gold nanoleaves. *Talanta* **2016**, *161*, 48–53. [CrossRef] [PubMed]
14. Sakkas, H.; Gartzonika, C.; Levidiotou, S. Laboratory diagnosis of human visceral leishmaniasis. *J. Vector Borne Dis.* **2016**, *53*, 8–16. [PubMed]
15. Luciano, R.M.; Lucheis, S.B.; Troncarelli, M.Z.; Luciano, D.M.; Langoni, H. Leishmania sp. And Trypanossoma cruzi antigen cross reaction evaluation in dogs by indirect immunofluorescence diagnosis (RIIF). *Braz. J. Vet. Res. Anim. Sci.* **2009**, *46*, 181–187. [CrossRef]

16. Burns, J.M.; Shreffler, W.G.; Benson, D.R.; Ghalib, H.W.; Badaro, R.; Reed, S.G. Molecular characterization of a kinesin-related antigen of Leishmania Chagasic that detects specific antibody in African and American visceral leishmaniasis. *Proc. Natl. Acad. Sci. USA* **1993**, *90*, 775–779. [CrossRef] [PubMed]
17. Maia, Z.; Lírio, M.; Mistro, S.; Mendes, C.M.C.; Mehta, S.R.; Badaro, R. Comparative study of rK39 Leishmania antigen for serodiagnosis of visceral leishmaniasis: Systematic review with meta-analysis. *PLoS Negl. Trop. Dis.* **2012**, *6*, e1484. [CrossRef]
18. Saghrouni, F.; Gaïed-Meksi, S.; Fathallah, A.; Amri, F.; Ach, H.; Guizani, I.; Saïd, M. Ben Immunochromatographic rK39 strip test in the serodiagnosis of visceral leishmaniasis in Tunisia. *Trans. R. Soc. Trop. Med. Hyg.* **2009**, *103*, 1273–1278. [CrossRef]
19. Segatto, M.; Ribeiro, L.S.; Costa, D.L.; Costa, C.H.N.; de Oliveira, M.R.; Carvalho, S.F.G.; Macedo, A.M.; Valadares, H.M.S.; Dietze, R.; de Brito, C.F.A.; et al. Genetic diversity of Leishmania infantum field populations from Brazil. *Mem. Inst. Oswaldo Cruz* **2012**, *107*, 39–47. [CrossRef]
20. Arraes, S.M.A.A.; Marini, M.T.; Martello, D.; Silveira, T.G.V.; Lonardoni, M.V.C.; Nanni, M.R. Serological investigation of subclinical cutaneous leishmaniasis cases following an outbreak in an endemic area. *Rev. Soc. Bras. Med. Trop.* **2008**, *41*, 205–208. [CrossRef]
21. Sato, C.M.; Sanchez, M.C.A.; Celeste, B.J.; Duthie, M.S.; Guderian, J.; Reed, S.G.; De Brito, M.E.F.; Campos, M.B.; De Souza Encarnação, H.V.; Guerra, J.; et al. Use of recombinant antigens for sensitive serodiagnosis of American tegumentary leishmaniasis caused by different leishmania species. *J. Clin. Microbiol.* **2017**, *55*, 495–503. [CrossRef]
22. Ferreira, M.P.; Roselino, A.M.F.; Nascimento, M.M.P.; Machado Aires, J.; Figueiredo, J.F.C. Sensitivity of an immunoenzymatic test for the detection of anti-l. braziliensis antibodies compared to other tests used for the diagnosis of american cutaneous leishmaniasis. *Rev. Inst. Med. Trop. S. Paulo* **2006**, *48*, 1–3. [CrossRef] [PubMed]
23. Laurenti, M. Correlation between parasitological and serological diagnosis in canine american visceral leishmaniasis. *BEPA Bol. Epidemiológico Paul.* **2009**, *6*, 13–23.
24. Pereira, A.C.; Santos, A.D.S.; Kubota, L.T. Trends in amperometrics electrodes modification for electroanalytical applications. *Quim. Nova* **2002**, *25*, 1012–1021.
25. Freire, R.S.; Pessoa, C.A.; Kubota, L.T. Self-assembled monolayers applications for the development of electrochemical sensors. *Quim. Nova* **2003**, *26*, 381–389. [CrossRef]
26. Fonseca, R.A.S.; Ramos-Jesus, J.; Kubota, L.T.; Dutra, R.F. A nanostructured piezoelectric immunosensor for detection of human cardiac troponin T. *Sensors* **2011**, *11*, 10785–10797. [CrossRef]
27. Pérez-Fernández, B.; Costa-García, A.; Muñiz, A.D.L.E. Electrochemical (Bio)Sensors for Pesticides Detection Using Screen-Printed Electrodes. *Biosensors* **2020**, *10*, 32–58.
28. Shkodra, B.; Demelash Abera, B.; Cantarella, G.; Douaki, A.; Avancini, E.; Petti, L.; Lugli, P. Flexible and Printed Electrochemical Immunosensor Coated with Oxygen Plasma Treated SWCNTs for Histamine Detection. *Biosensors* **2020**, *10*, 35. [CrossRef]
29. Cancelliere, R.; Carbone, K.; Pagano, M.; Cacciotti, I.; Micheli, L. Biochar from brewers' spent grain: A green and low-cost smart material to modify screen-printed electrodes. *Biosensors* **2019**, *9*, 139. [CrossRef]
30. Chiticaru, E.A.; Pilan, L.; Damian, C.M.; Vasile, E.; Burns, J.S.; Ionita, M. Influence of graphene oxide concentration when fabricating an electrochemical biosensor for DNA detection. *Biosensors* **2019**, *9*, 113. [CrossRef]
31. Martínez-García, G.; Pérez-Julián, E.; Agüí, L.; Cabré, N.; Joven, J.; Yáñez-Sedeño, P.; Pingarrón, J.M. An electrochemical enzyme biosensor for 3-hydroxybutyrate detection using screen-printed electrodes modified by reduced graphene oxide and thionine. *Biosensors* **2017**, *7*, 50. [CrossRef]
32. Ahmad, N.M.; Abdullah, J.; Yusof, N.A.; Ab Rashid, A.H.; Rahman, S.A.; Hasan, M.R. Amperometric biosensor based on zirconium oxide/polyethylene glycol/tyrosinase composite film for the detection of phenolic compounds. *Biosensors* **2016**, *6*, 31. [CrossRef]
33. Badalyan, A.; Dierich, M.; Stiba, K.; Schwuchow, V.; Leimkühler, S.; Wollenberger, U. Electrical wiring of the aldehyde oxidoreductase PaoABC with a polymer containing osmium redox centers: Biosensors for benzaldehyde and GABA. *Biosensors* **2014**, *4*, 403–421. [CrossRef] [PubMed]
34. Setterington, E.B.; Alocilja, E.C. Electrochemical biosensor for rapid and sensitive detection of magnetically extracted bacterial pathogens. *Biosensors* **2012**, *2*, 15–31. [CrossRef] [PubMed]

35. Rama, E.C.; Costa-García, A. Screen-printed Electrochemical Immunosensors for the Detection of Cancer and Cardiovascular Biomarkers. *Electroanalysis* **2016**, *28*, 1700–1715. [CrossRef]
36. Mohan, S.; Srivastava, P.; Maheshwari, S.N.; Sundar, S.; Prakash, R. Nano-structured nickel oxide-based DNA biosensor for detection of visceral leishmaniasis (Kala-azar). *Analyst* **2011**, *136*, 2845–2851. [CrossRef] [PubMed]
37. Sharma, A.; Matharu, Z.; Sumana, G.; Solanki, P.R.; Kim, C.G.; Malhotra, B.D. Antibody immobilized cysteamine functionalized-gold nanoparticles for aflatoxin detection. *Thin Solid Film.* **2010**, *519*, 1213–1218. [CrossRef]
38. Lowry, O.H.; Rosebrough, N.J.; Farr, A.L.; Randall, R.J. Protein measurement with the Folin phenol reagent. *J. Biol. Chem.* **1951**, *193*, 265–275.
39. Gontijo, C.M.F.; Melo, M.N. Visceral Leishmaniasis in Brazil: Current status, challenges, and prospects. *Rev. Bras. Epidemiol.* **2004**, *7*, 338–349. [CrossRef]
40. Scott, P.; Pearce, E.; Natovitz, P.; Sher, A. Vaccination against cutaneous leishmaniasis in a murine model. I. Induction of protective immunity with a soluble extract of promastigotes. *J. Immunol.* **1987**, *139*, 221–227.
41. Dai, X.; Nekraseova, O.; Hyde, M.E.; Compton, R.G. Anodic stripping voltammetry of arsenic (III) using gold nanoparticle-modified electrodes. *Anal. Chem.* **2004**, *76*, 5924–5929. [CrossRef]
42. Chikae, M.; Fukuda, T.; Kerman, K.; Idegami, K.; Miura, Y.; Tamiya, E. Amyloid-β detection with saccharide immobilized gold nanoparticle on carbon electrode. *Bioelectrochemistry* **2008**, *74*, 118–123. [CrossRef]
43. Regiart, M.; Pereira, S.V.; Bertolino, F.A.; Garcia, C.D.; Raba, J.; Aranda, P.R. An electrochemical immunosensor for anti-T. cruzi IgM antibodies, a biomarker for congenital Chagas disease, using a screen-printed electrode modified with gold nanoparticles and functionalized with shed acute phase antigen. *Microchim. Acta* **2016**, *183*, 1203–1210. [CrossRef]
44. Bollella, P.; Gorton, L.; Ludwig, R.; Antiochia, R. A third-generation glucose biosensor based on cellobiose dehydrogenase immobilized on a glassy carbon electrode decorated with electrodeposited gold nanoparticles: Characterization and application in human saliva. *Sensors* **2017**, *17*, 1912. [CrossRef] [PubMed]
45. Almeida, E.S.; Richter, E.M.; Munoz, R.A.A. On-site fuel electroanalysis: Determination of lead, copper and mercury in fuel bioethanol by anodic stripping voltammetry using screen-printed gold electrodes. *Anal. Chim. Acta* **2014**, *837*, 38–43. [CrossRef] [PubMed]
46. Da Silva, S.M.; Squissato, A.L.; Rocha, D.P.; Vasconcellos, M.L.S.; Ferreira, R.D.; Richter, E.M.; Munoz, R.A.A. Improved anodic stripping voltammetric detection of zinc on a disposable screen-printed gold electrode. *Ionics* **2020**, *26*, 2611–2621. [CrossRef]
47. Vasconcellos, M.L.S.; Rocha, D.P.; Castro, S.V.F.; Silva, L.R.G.; Munoz, R.A.A.; Freitas, M.B.J.G.; Ferreira, R.Q. Electroanalytical Method for Determination of Trace Metals in Struvite Using Electrochemically Treated Screen-Printed Gold Electrodes. *J. Braz. Chem. Soc.* **2020**, *31*, 1873–1882. [CrossRef]
48. Cordeiro, T.A.R.; Gonçalves, M.V.C.; Franco, D.L.; Reis, A.B.; Martins, H.R.; Ferreira, L.F. Label-free electrochemical impedance immunosensor based on modified screen-printed gold electrodes for the diagnosis of canine visceral leishmaniasis. *Talanta* **2019**, *195*, 327–332. [CrossRef]
49. Janegitz, B.C.; Medeiros, R.A.; Rocha-Filho, R.C.; Fatibello-Filho, O. Direct electrochemistry of tyrosinase and biosensing for phenol based on gold nanoparticles electrodeposited on a boron-doped diamond electrode. *Diam. Relat. Mater.* **2012**, *25*, 128–133. [CrossRef]
50. Chu, X.; Zhao, Z.L.; Shen, G.L.; Yu, R.Q. Quartz crystal microbalance immunoassay with dendritic amplification using colloidal gold immunocomplex. *Sens. Actuators B Chem.* **2006**, *114*, 696–704. [CrossRef]
51. Godínez, L.A. Substratos modificados con monocapas autoensambladas: Dispositivos para fabricar sensores y estudiar procesos químicos y fisicoquímicos interfaciales. *J. Mex. Chem. Soc.* **1999**, *43*, 219–229.
52. Galal, A.; Atta, N.F.; El-Ads, E.H. Probing cysteine self-assembled monolayers over gold nanoparticles-Towards selective electrochemical sensors. *Talanta* **2012**, *93*, 264–273. [PubMed]
53. Chaki, N.K.; Vijayamohanan, K. Self-assembled monolayers as a tunable platform for biosensor applications. *Biosens. Bioelectron.* **2002**, *17*, 1–12. [CrossRef]
54. Gómez-Velasco, A.; León-Cortés, J.L.; Gordillo-Marroquín, C.; Sánchez-Pérez, H.J.; Alocilja, E.C.; Muñoz-Jiménez, S.G.; Enríquez-Ríos, N.; Bencomo-Alerm, A.; Jonapá-Gómez, L.; Gómez-Bustamante, A. Use of magnetic nanoparticles and a biosensor for the diagnosis and monitoring of emergent, re-emergent and neglected tropical infectious diseases. *Rev. Enf. Emerg.* **2019**, *18*, 23–31.

55. Cios, K.J.; William Moore, G.; Moore, G.W. Unique features of medical data mining and knowledge discovery; Ethical; Security and legal aspects of medical data mining. *Artif. Intell. Med.* **2002**, *26*, 1–24. [CrossRef]
56. Ortolani, C.; Ispano, M.; Pastorello, E.A.; Ansaloni, R.; Magri, G.C. Comparison of results of skin prick tests (with fresh foods and commercial food extracts) and RAST in 100 patients with oral allergy syndrome. *J. Allergy Clin. Immunol.* **1989**, *83*, 683–690. [CrossRef]
57. Assis, T.S.M.; Azevedo-da-Silva, A.L.F.; Oliveira, D.; Cota, G.; Werneck, G.L.; Rabello, A. Budgetary impact of diagnostic tests for visceral leishmaniasis in Brazil. *Cad. Saúde Publica* **2017**, *33*, 1–8. [CrossRef]
58. Faria, A.R.; de Andrade, H.M. Diagnosis of canine visceral leishmaniasis: Major technological advances and few practical applications. *Rev. Pan-Amaz Saúde* **2012**, *3*, 47–57. [CrossRef]

biosensors

MDPI

Review

Grafting of Diazonium Salts on Surfaces: Application to Biosensors

Dardan Hetemi [1], Vincent Noël [2] and Jean Pinson [2,*]

1 Pharmacy Department, Medical Faculty, University of Prishtina, "Hasan Prishtina", Rr. "Dëshmorët e Kombit" p.n., 10000 Prishtina, Kosovo; dardan.hetemi@uni-pr.edu
2 Université de Paris, ITODYS, CNRS, UMR 7086, 15 rue J-A de Baïf, F-75013 Paris, France; vincent.noel@univ-paris-diderot.fr
* Correspondence: jean.pinson@univ-paris-diderot.fr

Received: 23 December 2019; Accepted: 9 January 2020; Published: 15 January 2020

check for updates

Abstract: This review is divided into two parts; the first one summarizes the main features of surface modification by diazonium salts with a focus on most recent advances, while the second part deals with diazonium-based biosensors including small molecules of biological interest, proteins, and nucleic acids.

Keywords: diazonium; biosensor; surface modification

1. An Overview of the Reactions of Diazonium Salts with Surfaces

1.1. The Principle of the Reaction

The simplest way to represent the grafting of aryldiazonium salts on surfaces is shown in Scheme 1.

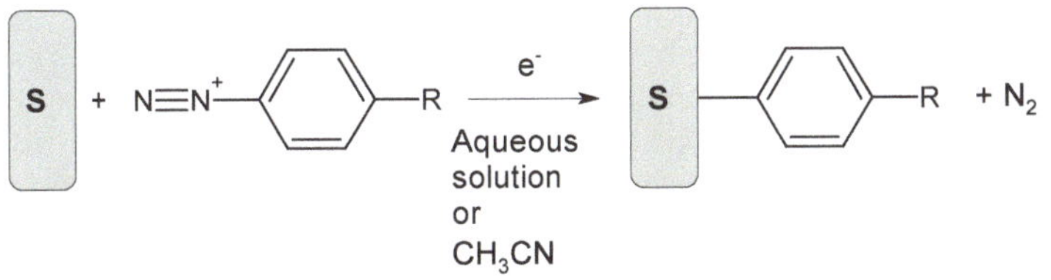

Scheme 1. Electrografting of diazonium salts.

A substituted aryldiazonium salt dissolved in an aqueous medium or in acetonitrile (ACN) is reduced by one electron; as a result, the electrode is modified by aryl groups [1]. This is a very simple reaction with many experimental alternatives concerning the process itself, the surfaces, and the choice of the substituent. The result is a modified surface with strongly bonded aryl groups [2–5].

The main characteristics of this reaction are as follows:

* Diazonium salts are easily synthetized (isolated or not) from aromatic amines, many of which are commercially available.
* All surfaces can be modified by this reaction, conductive or not.
* The reaction can be performed by electrochemistry, spontaneously, by photochemistry, and by other methods.
* The resulting modification is very stable due to the formation of a covalent bond between the surface and the aryl group.

* The key species of this reaction is an aryl radical, and this reaction presents the typical behavior of radical reactions.
* The reaction provides most often disordered oligomers ("multilayers").

Below, we review these different points with a focus on most recent examples.

1.2. Synthesis and Stability of Diazonium Salts

Surface modification of diazonium salts was achieved starting either from isolated salts or from solutions of in situ synthetized compounds [6]. The diazoniation was achieved with $NaNO_2$ in acidic aqueous solution, or with *t*-butylnitrite or $NOBF_4$ in ACN. As many aromatic amines are commercially available, the use of simple diazonium salts only necessitates a minimum effort of synthesis.

The stability of diazonium salts was measured; 4-nitrobenzenediazonium tosylate and tetrafluoroborate have half-life times close to 4.5 years, whereas 4-nitrobenzenediazonium triflate is much more stable with a half-life time of 46 years. These half-lives are more than sufficient to obtain reliable experiments [7]. However, chlorides and other salts are explosive.

When dissolved in ACN or aqueous acidic solution, the diazonium salts are present as $Ar\text{-}N{\equiv}N^+$; however, as the pH increases in aqueous solution, they exist as Ar-N=N-OH (diazohydroxide) and $Ar\text{-}N{=}N\text{-}O^-$ (diazoate). The first pKa of 4-nitrobenzenediazonium is 5.24. In ACN and aqueous acidic solution, diazonium salts are relatively stable ($k_{dec} = 8.26 \times 10^{-6}$ s^{-1} in aqueous acidic solution); however, as the pH increases, the rate of decomposition of 4-methylbenzenediazonium increases ($k_{dec} \approx 1.3 \times 10^{-2}$ and 3.5×10^{-2} s^{-1} at pH 4 and 8, respectively (60 °C, in EtOH/H_2O) [8,9]. The dediazoniation occurs either heterolytically or homolytically to give either a carbocation or a radical; as concerns the grafting reaction, there is only a very minor involvement of carbocations in the case of spontaneous reactions [10].

1.3. Different Grafting Methods of Diazonium Salts

As diazonium salts are easily reduced, a number of methods permit their homolytic dediazoniation; the most used ones are presented below.

By electrochemistry. As indicated in Scheme 1, triggering of the reaction is possible by electrochemistry. The voltammograms obtained for the reduction of diazonium salts are very characteristic, showing a broad irreversible wave at potentials close to 0 V/SCE (Saturated Calomel Electrode). The wave is irreversible due to the cleavage and loss of N_2; it is broad because the surface is modified during the voltammogram. Upon repetitive scanning, this wave decreases and finally disappears as an insulating organic film forms on the electrode (Figure 1) [11]. Repetitive cyclic voltammetry (5–10 cycles) and chronoamperometry at a potential equal or negative to the reduction peak are common methods for the modification of electrodes.

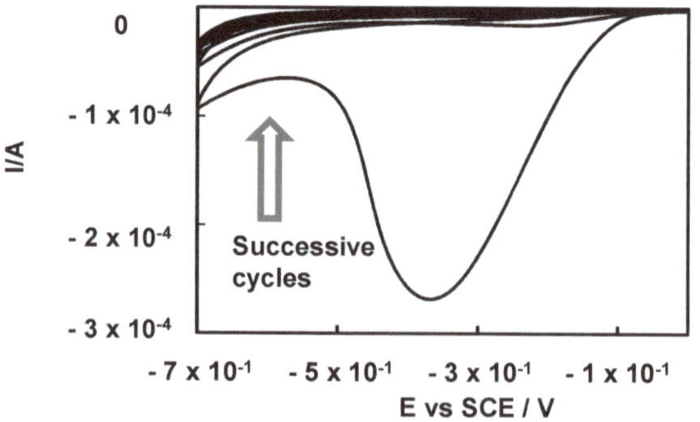

Figure 1. Cyclic voltammograms (scans 1–3) at fluorine tin oxide (FTO) electrode in HCl 0.5 M + 2×10^{-3} M 4-aminobenzoic acid + 2.2×10^{-3} M $NaNO_2$. Scan rate of 0.1 $V{\cdot}s^{-1}$. By courtesy of C. Cannizzo.

By spontaneous reaction. As the pH of an aqueous solution increases, the diazonium cation is transformed into the much more unstable diazohydroxide and diazoate. Consecutive homolytic cleavage of these species provides a radical that reacts with the surface. For example, a diazonium salt bearing an Iniferter initiator (a group acting as initiator, transfer, and terminator agent of controlled free radical polymerization) was grafted on isolating silica particles in basic medium to give a silica core@poly(acrylic acid) shell [12]. Trifluoromethylphenyl layers were grafted by immersing SnO_2 plates in an aqueous solution of trifluoromethylbenzenediazonium for 8 h, in the dark [13].

By reducing surfaces. As diazonium salts are very easily reduced (Figure 1), materials such as copper, iron, and even carbon are reducing enough to perform the spontaneous grafting reaction onto their surface. A nanostructured and oxidized copper surface was grafted spontaneously by a perfluorobenzenediazonium salt [14] or an Iniferter initiator [15] through a one-pot reaction; the diazonium was prepared from the corresponding aniline in ACN + *t*-butylnitrite, and grafting took place spontaneously in the same solution. In the first case, a super-hydrophobic surface was obtained with a water contact angle of 158°, and, in the second one, it was possible to grow polyacrylic brushes by Iniferter polymerization.

Silica nanoparticles were modified with a molecule terminated by an aniline group that could be diazotized and grafted spontaneously on an iron surface at room temperature [16]. Anthraquinonediazonium prepared in ACN reacts with carbon black just by overnight reaction at room temperature [17], and graphene was modified with 3,5-bis difluorobenzenediazonium in acidic medium (the same medium used for the preparation of the diazonium cation) [18].

By reducing reagents. Even very mild reducing agents such as hypophosphorous acid, ascorbic acid, and iron powder are able to reduce diazonium cations leading to aryl radical formation. Coal powder was modified by 4-nitrophenyl groups via reduction of the corresponding diazonium cation by hypophosphorous acid. This modified coal embedded in reverse osmosis membranes improved their performance [19]. TiN is a ceramic material used in microelectronic devices, solar cells, electrical, interconnections, orthopedic prostheses, and cardiac valves. Its surface was modified by reaction of the mono-diazonium salt of *p*-phenylenediamine (NH_2-C_6H_4-$N{\equiv}N^+$) in the presence of hypophosphorous acid or iron powder [20]. This reaction offers the first step of an alternative to the metallization of titanium nitride by direct electrochemical deposition.

By photochemistry. Diazonium salts were grafted on metals gold, copper, and iron under UV (UltraViolet) light by irradiation in the presence of a photosensitizer ($Ru(bipy)_3{}^{2+}$ or eosin Y). In this way, gold and polyvinylchloride surfaces were modified by 4-phenylacetic, 4-carboxy, 4-methoxy, and 3,5-bis trifluoromethyl phenyl groups [21]. The reaction also took place under visible light by irradiation of charge transfer complexes such as 1,4-dimethoxybenzene and pentafluoro benzenediazonium cation (λ_{max} ~400 nm) [22].

Surface modification can also be triggered by localized surface plasmon excitation [23–25]. Using this approach, 4-[1-(2-bisthienyl)], 4-carboxy 4-hydroxyethyl benzenediazonium cations were grafted on gold nanostructures. Upon irradiation, hot collective oscillation of the conductive electrons at the particle surface provides the localized surface plasmon resonances (LSPR); hot electrons are generated on nanostructures particularly at the extremities of nanorods, nanotriangles, etc. These hot electrons induce the reduction of diazonium salts, the homolytic dediazoniation, and grafting at the extremities of gold nanostructures. In this way, regioselectively modified gold nanodiscs were obtained by successive use of two different light polarizations in the presence of two diazonium salts (Figure 2). Recently, diazonium-modified graphene-protected metal thin films (Cu) SPR biochips were designed for the detection of toxins [26].

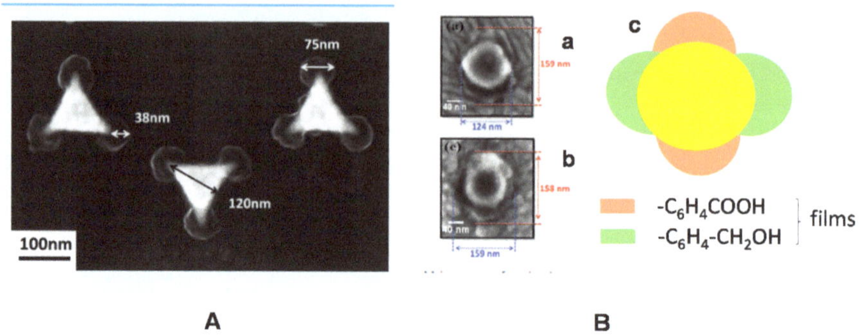

Figure 2. SEM image of (**A**) Au nanotriangles after irradiation with visible light in the presence of 4-[1-(2-bisthienyl)]benzenediazonium, (**B**) nanodiscs modified with visible light (Ba) with carboxyphenyl films grafted along the *Y*-direction, (Bb) additional hydroxyethyl phenyl films along the *X*-axis, (**B**c) Schematic presentation of the modified nanodisk. From References [23–25], with permission of the American Chemical Society (ACS) and the Royal Chemical Society, respectively.

1.4. The Different Surfaces That Can be Grafted

Many examples of grafting diazonium salts on glassy carbon, metals, and semiconductors are described [2]. Due to the current interest in nanoscience, more recent investigations examined one- and two-dimensional (1D and 2D) materials, as well as nanoparticles [27–29] (carbon nanotubes, graphene, graphene oxide, MoS_2) as substrate.

Carbon nanotubes (CNT). A detailed investigation of the grafting of 4-iodobenzenediazonium on SWCNTs (single-walled carbon nanotubes) showed that (i) the bonded aryl groups were very stable, as they cleaved above 200 °C, (ii) a logarithmic correlation between the degree of functionalization and diazonium concentration was observed, and (iii) the maximum surface concentration was measured as one aryl group per 100 carbon atoms [30].

In view of biological applications (antifouling properties), the surface of carbon nanotubes was modified in order to inhibit the growth of uropathogenic *Escherichia coli*. This was achieved by spontaneously grafting polyethyleneglycol (PEG) chains terminated by mannose at one and a benzenediazonium at the other end [31] (Figure 3).

Crosslinked assemblies of bonded CNTs were prepared using a molecule with three diazonium functional groups. The walls of the CNTs were firstly protected by wrapping DNA to limit their functionalization and favor the modification at nanotube end; then, the triple diazonium salt molecule was reacted spontaneously. The SWCNTs were bonded mainly through side-to-end junctions, and eventually through side-to-side interactions (Figure 4) [32].

Graphene; graphene oxide (GO), reduced graphene oxide (RGO), highly oriented pyrolytic graphite (HOPG), carbon dots. Graphene sheets were obtained by electrochemical exfoliation and functionalization of graphite using diazonium salts. Both functionalization and exfoliation occurred at the same time; in this way, mono- or few-layer graphene was functionalized and stabilized in situ before it aggregated; N_2 generated during in situ diazonium reduction favored the separation of functionalized graphene sheets [33]. In planes or edges, grafting was easily achieved with diazonium salts; GO and RGO were modified with sulfophenyl groups up to 12 wt% [34].

The planes and edges of graphene (respectively, sp^2 and sp^3 carbons) have very different structures, but both react with diazonium salts under electrochemical conditions.

The surface modification of HOPG was examined at the micrometer scale [35] using scanning electrochemical cell microscopy where a dual-barrel micro-pipet explored the localized electrochemistry of the surface. In the presence of diazonium salts, grafting occurred along with the re-hybridization of surface carbons from sp^2 to sp^3 as confirmed by Raman spectroscopy; after diazonium modification, the D-band developed at ~1350 cm^{-1}, diagnostic of the local sp^3 carbons.

An electrode was constructed with only graphene edges exposed by cutting a connected graphene monolayer embedded in a polymer with an atomically sharp microtome knife. This graphene-edge electrode was modified by electrochemical reduction of 4-nitrobenzenediazonium. The cyclic voltammetry of 4-nitrophenyl films was observed, and the signal of the redox probe $Fe(CN)_6^{3-/4-}$ was completely inhibited, indicating a blocking of the electrode by the grafted film; this was confirmed by the Raman spectrum that indicated an increase of the D-band [36].

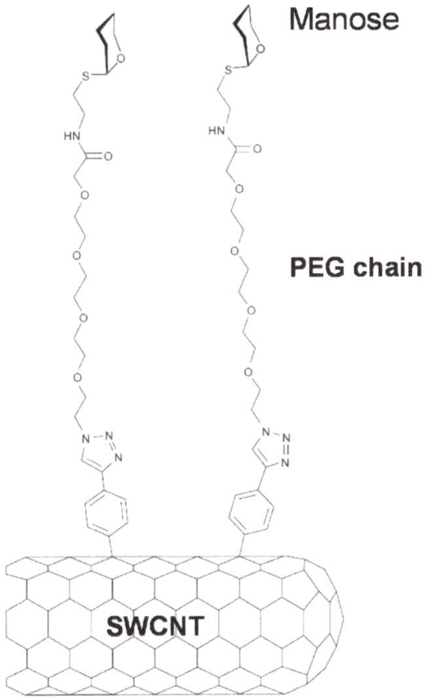

Figure 3. Single-walled carbon nanotube (SWCNT) modified with polyethyleneglycol (PEG) and mannose groups. From Reference [31] with permission of Elsevier.

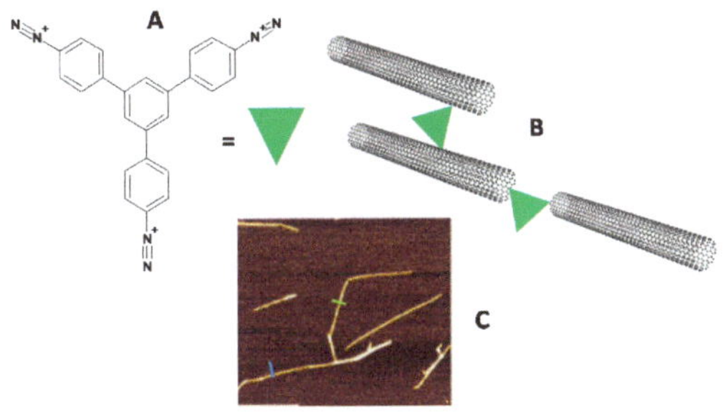

Figure 4. Arrays of SWCNTs bonded by reaction of diazonium salts.

The reactivity of graphene edges was harnessed to prepare covalently bonded graphene nanoflakes further assembled by noncovalent interactions to give nanopapers. Binding of nanoflakes was achieved by use of a bis-diazonium salt. This diazonium cation was prepared in situ and grafted by increasing the temperature (Figure 5). Modification led to a 20% enhancement of the thermal conductivity, while the cross-plane thermal conductivity was boosted by 190% [37].

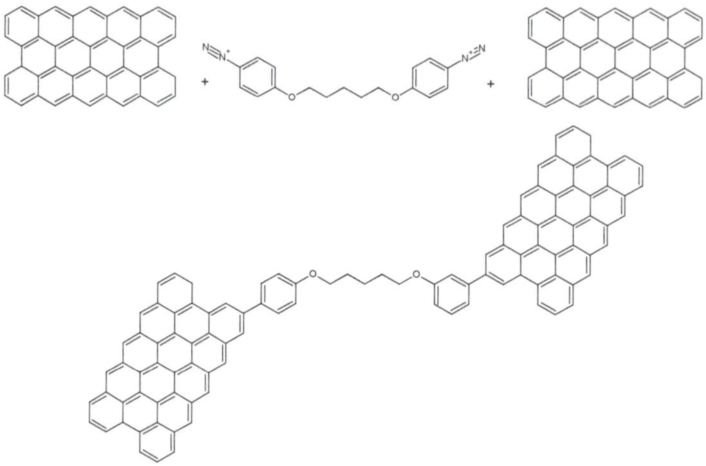

Figure 5. Binding graphene sheets using a bis-diazonium salt. From Reference [37] with permission of Wiley.

It is also possible to prepare graphene ribbons from diazonium salts. Molecular junctions using aryl oligomers were obtained from diazonium salts [38]. An organic oligomeric aryl layer was grown on a conducting substrate and capped with, for example, a gold layer; this provided a molecular junction between two conducting materials [39]. Such molecular junctions are used in molecular electronic devices. These aryl oligomers were replaced by five-carbon-wide graphene ribbons (GR) with lengths of 2–12 nm; their conductance was more than one hundred times that observed for other molecular junctions of similar thicknesses. These nanoribbons were obtained and grafted by electrochemical reduction of the 1,8-bis naphthalenediazonium salt, as presented in Figure 6 [40].

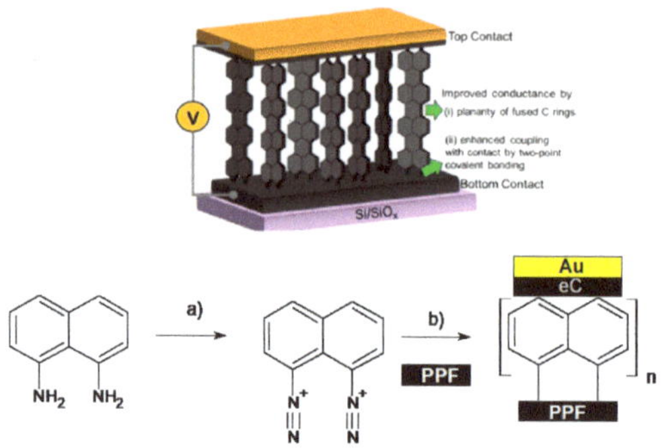

Figure 6. Molecular junction with carbon nanoribbons: (**a**) *t*-butylnitrite in ACN; (**b**) electrochemical reduction. PPF: pyrolyzed photoresist film, a very flat carbon surface, eC: electron-beam deposited carbon. From Reference [40] with permission of ACS.

Other 2D materials. Due to the interest in graphene, other 2D materials were modified with diazonium salts.

Few layers of black phosphorous were modified with a zinc phthalocyanine-based diazonium salt for applications in non-linear optics [41].

Metal dichalcogenides MX_2, where M is a transition metal (M = Mo, W, Nb, Ta, etc.) and X is a chalcogen (X = S, Se, or Te), were exfoliated into two-dimensional (2D) nanosheets. Among them, MoS_2, WS_2, $MoSe_2$, and WSe_2 were modified with diazonium salts (Figure 7). The pnictogen chalcogenides

Sb_2S_3 and Bi_2S_3 were exfoliated into one-dimensional (1D) nanoribbons and 2D nanosheets and derivatized [42].

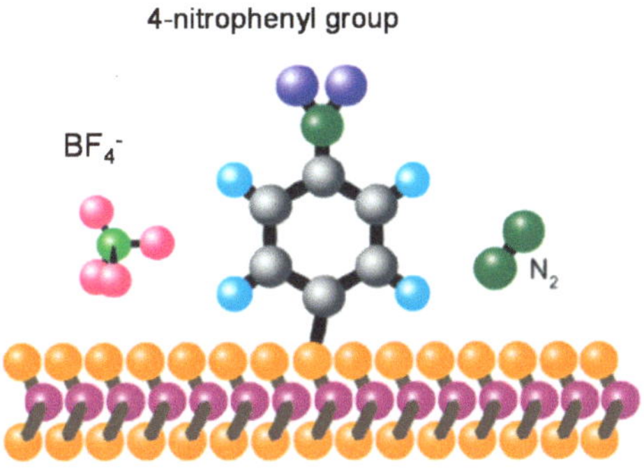

Figure 7. Modification of a MoS_2 sheet by 4-nitrobenzenediazonium (Mo violet, S orange). From Reference [42] with permission of the Royal Society of Chemistry.

MXenes are 2D nitrides and carbides; for example, Ti_3C_2 was intercalated by Na^+ ions and then grafted by reaction with 4-sulfonylbenzenediazonium to obtain enhanced super-capacitive performances [43].

Nanoobjects. Nanoparticles were capped with aryl groups through diazonium chemistry to imbue these objects with new properties such as catalysts, scavengers, and reagents [44]. It is also possible, with the same reaction, to decorate various surfaces with nanoparticles.

Nanoparticles can be stabilized. Iron oxide Fe_2O_3 nanoparticles could be capped with BF_4, N_2-C_6H_4-$(CH_2)_2$-OH by spontaneous reaction in basic aqueous medium. These nanoparticles retained their magnetic properties and were soluble in apolar organic solvents such as dichloromethane, tetrahydrofurane (THF), toluene, and chloroform, as well as in polar solvents such as methanol, ethanol, or water [45,46]. Cerium oxide (CeO_2) nanoparticles were grafted with 4-methyl-, 4-ethyl-, and 4-*n*-butyl benzenediazonium; the water contact angle increased from 34° to 63° and 125° as the chain length increased. With this modification, the cerium oxide NPs are more compatible with an electrolytic solution for the formation of coatings or a metal composite matrix [47].

Nanoparticles can be used as catalysts. TiO_2 nanoparticles were modified with 4-diphenylamine groups (DPA) by reaction of the corresponding diazonium salt; from this surface, polyaniline (PANI) was prepared by in situ polymerization of aniline. This bonding prevented polyaniline from leaching in polar solvents. This TiO_2–DPA–PANI assembly efficiently catalyzed the degradation of the dye methyl orange in aqueous media under UV light [48].

Nanoparticles can also be attached to surfaces for increasing analytical sensitivity. Gold nanoparticles were attached to the surface of screen-printed electrodes (SPE) derivatized with aminophenyl groups; the amino group was transformed into a diazonium that reacts with gold nanoparticles. In turn, these nanoparticles were modified by reaction of 4-carboxybenzenediazonium. With this system, it was possible to detect Pb(II) down to 2.5×10^{-9} M (Figure 8) [49].

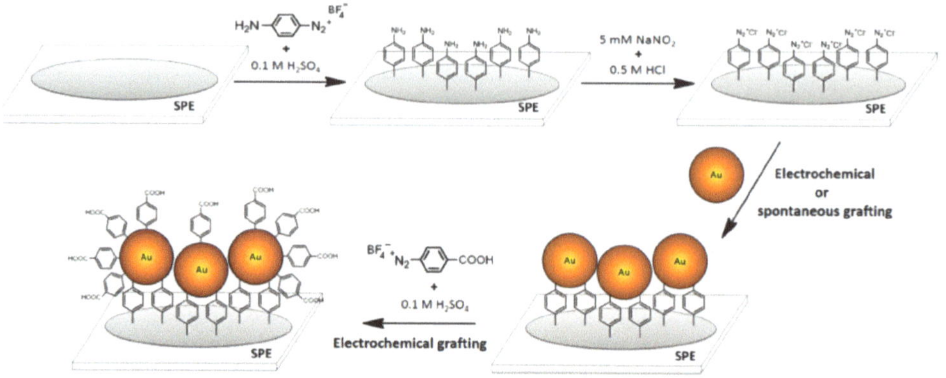

Figure 8. Gold nanoparticles covalently attached to a screen-printed electrode (SPE) and modified with caboxyphenyl groups for sensing metal ions. From Reference [49] with permission from Elsevier.

1.5. The Surface Aryl Bond

One of the most important features of diazonium grafting is the stability of the construct; this was demonstrated by several methods including (i) thermogravimetry, where exfoliated graphene lost only 7% of its coating at 400 °C [50], (ii) spectrometric methods, where a small Raman band at 412 cm^{-1} was assigned to the *Au(nP)*–C(aryl) bond of gold nanoparticles modified by 4-nitrobenzenediazonium [51]; grafting of an aryl group on graphene [52] and carbon nanotubes [53,54] transformed an sp^2 carbon into a sp^3, which translated into the growth of the D-band.

1.6. The Structure of the Grafted Film

The structure of the films obtained by dediazoniation of diazonium salts is quite complex and is not yet been completely elucidated. Upon dediazoniation, radicals are responsible for the grafting reaction and for the structure of the obtained nanometric films. The radicals react on the surface but also on the first grafted groups to produce "multilayered" films. This term is widely but somewhat abusively used in the literature as the structure of the film is not layered as, for example, layer-by-layer constructs. Figure 9 presents a very schematic presentation of a film obtained from diazonium salts.

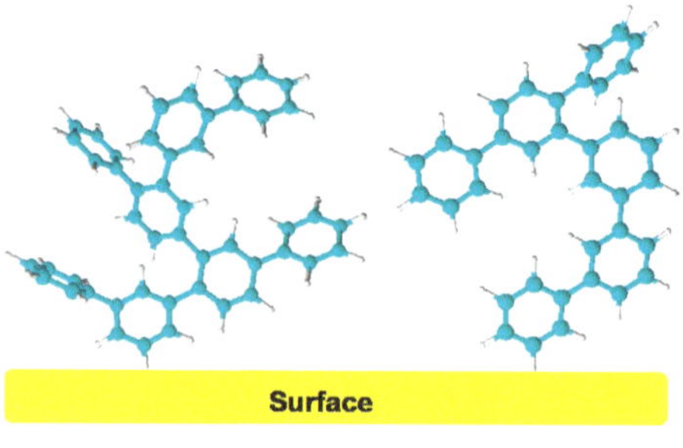

Figure 9. A schematic presentation of a film obtained from diazonium salts.

Aryl–aryl bonds permit a free rotation; optical absorption spectroscopy of thin (1–15 nm) oligomeric polyaromatic films attached to an atomically flat pyrolyzed photoresist film (PPF) permitted concluding that the molecular layers were composed of *n*-mers possessing very limited conjugation that extended only to one monomer [55]. However, by combining the electrografting of diazonium salts on Au and the oxidative electropolymerization of biphenyl in an ionic liquid, a regular poly(*para*-phenylene) film was obtained [56].

The thickness of diazonium derived films is measured by ellipsometry, AFM (atomic force microscopy), or STM (scanning tunneling microscopy). STM images (Figure 10A) showed a growing film of 4-nitrophenyl groups on the surface of HOPG; one can observe isolated oligomeric groups (up to 2 nm), indicating that, on this surface, the aryl radical reacts faster on the first grafted group than on the surface [57]. On the contrary, on PPF, a uniform monolayer was obtained [58], indicating that the reaction is faster on PPF than on the first grafted group (Figure 10C). This is related to the difference in reactivity of the two different carbons. This renders the control, a priori, of the thickness quite difficult.

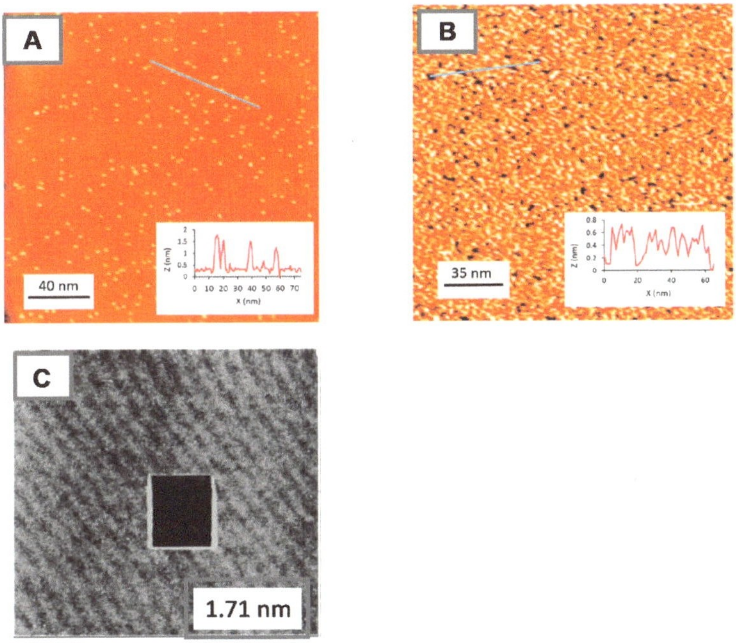

Figure 10. (**A**,**B**) STM (scanning tunneling microscopy) images of films obtained on highly oriented pyrolytic graphite (HOPG) after spontaneous grafting of (**A**) 4-nitrobenzenediazonium and (**B**) 3,5-bis-*tert*-butylbenzenediazonium salts. (**C**) AFM (atomic force microscopy) image of a monolayer obtained from 4-nitrobiphenyldiazonium salt (l = 1.21 nm) on PPF, a near-atomically flat carbon; in the black square, the layer was erased by AFM scratching. From References [57,58] with permission of the American Chemical Society.

Thick films up to ~100 nm were obtained by electrografting 4-nitrobenzenediazonium at the reduction potential of the nitrophenyl group [59]. At this potential, the radical anion of the nitrophenyl group was formed, and electrons could transfer through the film, reach the surface, and reduce a diazonium cation; this process permitted the thickening of the film.

Conversely, many efforts were devoted to the formation of monolayers. Indeed, such monolayers would be very useful for the preparation of biosensors as they should provide faster and uniform electron transfer to a bioreceptor. Figure 10B shows the STM image of a monolayer (thickness ~0.6–0.8 nm) of 3,5-bis-*tert*-butylbenzenediazonium [57]. The steric hindrance of the two bulky *tert*-butyl groups prevented the aryl radicals from reacting of the aromatic ring and, consequently, the growth of the film [60]. However, with this method, post-modification was not possible; later on, it was modified to permit further reactions on the film [61].

A more general method [62] involves electrografting in the presence of redox mediators. Monolayers (0.6–0.9 nm) of 4-bromo, 4-iodo, 4-methoxy, and 4-diethylamino phenyl groups were obtained in the presence of three redox mediators: 2,2-diphenyl-1-picrylhydrazyl, chloranil, and dichlone. The efficiency of the method rests on a fast redox cross-reaction in the diffusion layer between the diazonium compound and the reduced form of the selected inhibitor. This method should permit preparing, in a repetitive manner, reactive monolayers that would be useful for biosensors due to their fast electron transfer.

Disordered oligomeric films without regular patterns on the surface were obtained from diazonium salts [2]. Using high-quality graphene and a diazonium with a long alkyl chain ($C_{22}H_{45}$-O-$C_6H_4N_2^+$ BF_4^-) [63], it was possible to obtain patterns of adsorbed molecules (imaged by STM); however, when the diazonium salt was reduced, grafting occurred and a new pattern was obtained. In this way, the authors obtained a pattern of grafted molecules; it is, however, surprising that the thickness of the grafted pattern was only 2 nm. As the diazonium was *para*-substituted, the molecules should have been more or less vertically aligned on the surface sp^3 carbons (Figure 11).

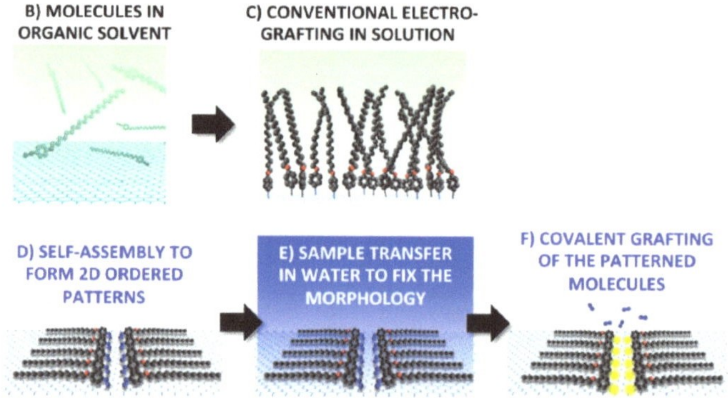

Figure 11. Adsorption and electrografting of long-chain benzenediazonium tetrafluoroborate on graphene. From Reference [63] with permission of the American Chemical Society.

2. Applications to Biosensors

A biosensor is composed of a bioreceptor (probe), which selectively binds the analyte of interest (target), and a transducer, allowing the transformation of the probe/target recognition event into a physical signal. Research and development of biosensors is extensively studied because they permit easy, rapid, low-cost, highly sensitive, and selective detection of analytes. They allow ultrasensitive point-of-care detection of markers for diseases and should lead to advances for next-generation medicinal applications such as personalized medicine. In addition to biomedical applications, biosensors are also capable of responding to the current needs for environmental monitoring. For both fields of applications, the ongoing trend is miniaturization, parallelization, and integration of sensors into everyday objects. However, despite the intense research and development activities around biosensors, very few of them actually reached the market because of their non-optimal performances [64].

The sensor selectivity, sensitivity, and robustness (stability, reproducibility, etc.) are mainly controlled by the intrinsic bioreceptor characteristics, such as its affinity toward the target, as well as its stability in the sensor operating conditions. In a biosensor, the bioreceptor is usually grafted onto a surface, i.e., in the close vicinity of the transducer. Hence, sensor sensitivity and robustness depend also on the methodology deployed for the bioreceptor immobilization onto the transducing surface. The accessibility of the target to the recognition site of the biomolecule, the grafting stability, and the distance between the receptor and the transducer (surface) are all parameters to be optimized in order to improve the device analytical performance. All these parameters depend on the method chosen to immobilize the biomolecule. Historically, conducting polymers (CP) were widely used as a conductive matrix to produce affinity (DNA, proteins, etc.) or enzymatic electrochemical sensors. The possibility of inserting the biomolecule inside the transducer material (CP) led to many ultrasensitive enzymatic sensors. However, CPs do not allow thin-film production, leading to sensors showing long response times (long delays to reach a stable signal). The other widely used approach is to immobilize the bioreceptor by self-assembly, including chemisorption of thiols. This approach led to the development of complex (controlled immobilization of several bioreceptors) and ultrathin (few nm between the binding site and the transducer) sensing systems presenting excellent analytical performance in terms of sensitivity. However, critical disadvantages remain, such as (i) the weak stability of the metal–S bond, (ii) the limited number of substrates (essentially noble metals), and (iii) the difficulty of localizing the film deposition. Indeed, in the development of chips comprising several sensors, it is necessary to have a functionalization method for addressing the deposition step. In this respect, electrodeposition is definitely advantageous as it offers the possibility of functionalizing specific (polarized) zones directly by the bioreceptor or an anchoring function thereof.

Electroreduction of diazonium salts is a rarely used functionalization method compared to CP and self-assembly, despite the achievement of extremely stable surface modifications (covalent bonding) that contribute to the stability of the biosensor. In addition, the high reactivity of the diazonium function allows a fast and extremely dense grafting on a wide range of substrates.

Indeed, the interest in the diazonium salt electroreduction approach is largely related to the remarkable reactivity of the diazonium function. However, its limited use in the field of biosensors is probably due to the same reason. Indeed, this method generally leads to "multilayered" structures that may increase the bioreceptor/surface (transducer) distance and, therefore, potentially provoke a loss of sensitivity (issue 1). Moreover, in the case when the biomolecule itself is modified by a diazonium salt, to ensure high affinity, it must be oriented with respect to the surface (accessibility to the target, structural reorganization associated with recognition, etc.) and not denatured upon grafting. The high reactivity of diazonium groups can lead to a random distribution and orientation of the receptors (issue 2), as well as to their degradation (issue 3); the diazonium group attached to, for example, a protein can react with a wide range of biological functional groups (phenols, amines, etc.) of the same protein or another molecule. These issues need to be completely addressed in order to take advantage of diazonium chemistry for the realization of biosensors.

In order to bind the receptor, an aryldiazonium group must be firstly equipped; this is generally achieved by peptidic coupling. Conversely, a platform can be created by reacting the surface with a diazonium, including a 4-substituent designed to react with the bioreceptor.

The many reviews (Table 1) published for the construction of biosensors using diazonium salts as anchoring molecules testified to the interest in this method [7,8,65–69]. The objective of this section is to carry out a review of biosensors using the electroreduction of diazonium salts structured by analyte type, each of them having specificities in terms of sensor typology and expected analytical performances. For each target, any progress made to address the aforementioned issues is highlighted.

Table 1. Reviews on biosensors built with the help of diazonium chemistry.

[1]	Some of advantages and disadvantages of alkanethiol and diazonium salts for the modification of electrode surfaces, and selected examples
[66]	The use of diazonium salts as surface modifiers and coupling agents, as well as some applications in biosensing
[67]	A general overview of biosensors including different methods that permit attaching sensing groups to a surface
[68]	A review of the author's work concerning biosensing of antibodies, oligonucleotides, and enzymes onto conductive supports
[69]	Advances in the use of aryldiazonium salts for modifying interfaces in sensors and biosensors
[70]	A review that outlines the potential of diazonium chemistry to prepare single or multianalyte electrochemical affinity biosensors on screen-printed electrodes (SPEs)
[71]	A review that evaluates the methods through which redox proteins can be attached to electrode surfaces in a "wired" configuration that facilitates direct electron transfer
[72]	Copper(I)-catalyzed click chemistry as a tool for the preparation of electrochemical (bio)sensors

2.1. Detection of Small Molecules of Biological Interest

Glucose. The detection of glucose is a challenge related to diabetes; a number of papers were published dealing with the detection and quantification of this molecule through diazonium chemistry. Note that the normal glucose content of blood is from 3.9 to 7.1 mmol/L (70 to 130 mg/dL). However, on the one hand, blood samples require the use of several membranes to avoid sensor biofouling, and, on the other hand, current glucose monitoring approaches target tears or sweat as glucose vectors in order to avoid body penetration. Therefore, accurate continuous glucose monitoring devices need a concentration linearity range in the micromolar range.

Glucose is electrochemically detected through its oxidation into gluconolactone; this reaction is catalyzed by glucose oxidase (GOx), which is reduced to its reduced form ($GOxH_2$), which is then reoxidized either directly by electron transfer from the electrode or more often through a mediator such as ferrocene (Fc in Figure 12). In agreement with the subject of this review, we describe the experiments where GOx is attached to the surface of the electrode by diazonium chemistry. Table 2 gives an overview of the different papers published on the subject. This table is divided into two parts: electron transfer through a mediator or direct electron transfer.

As an example of the methods involving a mediator, we present the first paper describing the use of a diazonium salt for detecting glucose. It involved the attachment of glucose oxidase to a glassy carbon surface modified with 4-phenylacetic acid diazonium fluoroborate through carbodiimide coupling [73]. Glucose was detected (Figure 12) through the electrochemical signal of the ferrocene/ferricinium–methanol couple (Fc/Fc^+) acting as an electron shuttle between the electrode and $GOx/GOxH_2$. Based on the catalytic regeneration of ferrocene, it was possible to determine the surface concentration of the active enzyme as $\Gamma \sim 1.8 \times 10^{-13}$ mol cm^{-2}, about one-tenth of the estimated value of a monolayer of GOx ($\Gamma \sim 1.7 \times 10^{-12}$ mol·cm^{-2}). A similar system was constructed from *trans*-cinnamic acid, showing good selectivity for the various possible compounds interfering in glucose analysis, namely, ascorbic acid and 4-acetamidophenol [74].

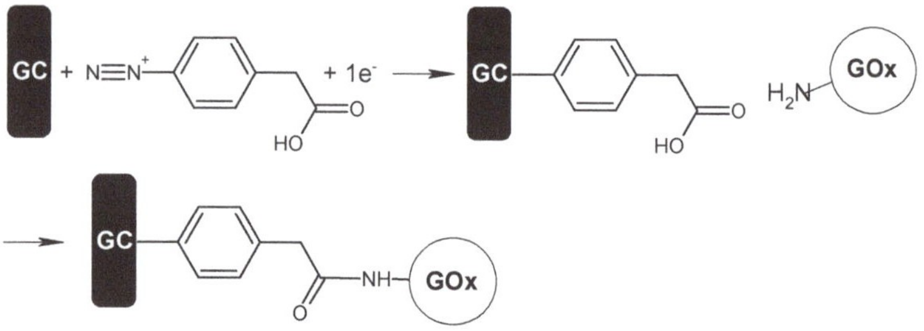

$$Fc - 1e^- \rightleftharpoons Fc^+$$
$$GOxH_2 + 2Fc^+ \longrightarrow GOx + 2Fc + 2H^+$$
$$GOx + Glucose \longrightarrow GOxH_2 + Gluconolactone$$

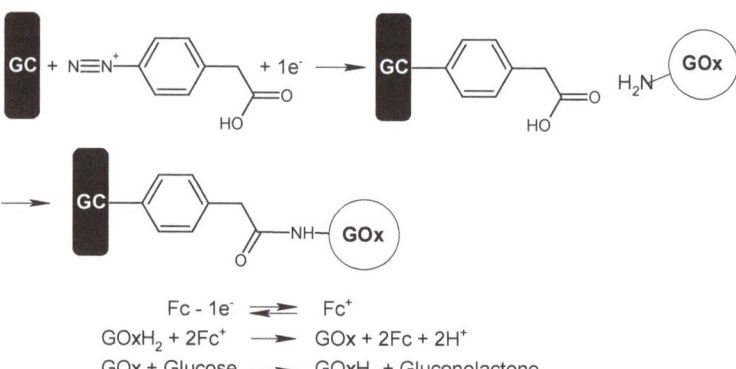

Figure 12. Detection of glucose by attaching glucose oxidase (GOx) to a glassy carbon surface. Catalytic scheme for the electro-catalyzed oxidation of its reduced form ($GOxH_2$).

Table 2. Glucose sensors based on diazonium grafting. GOx—glucose oxidase.

Surface [(a)]	Attached Aryl Group		Characteristics c = Concentration Range Γ (GOx) mol·cm^{-2}, k_{ET} T = Turnover S= Sensitivity	Reference
		Mediator		
GC	4-phenylacetic	$Fc\text{-}CH_2OH$ [(b)]	Γ ~1.8×10^{-13}	[73]
GC	4-phenylcinnamic	$Fc\text{-}CH_2OH$ [(b)]	-	[74]
GC	4-phenylacetic + layer by-layer [(c)]	$Fc\text{-}CH_2OH$ b)	Γ ~1.1×10^{-12}	[77]
Aligned CNT	4-aminophenyl [(d)]	Pt nanoparticles	c = 1×10^{-2}–7 mM Stability 20 days	[78]
CNT	4-(2-aminoethyl) phenyl	$Fc\text{-}CH_2\text{-}CH_2\text{-}COOH$ [(e)]	c = 5–50 mM s = 0.83 μA mM^{-1}	[79]
GC	Nile blue	O_2 f)	c = up to 2.5 mM	[80]
GC [(g)]	4-nitrophenyl	$Fe(CN)_6^{3-/4-}$	c= up to10 mM	[81]
Pt [(h)]	4-fluoro 4-methyl 4-methoxy	$Fc\text{-}CH_2OH$	c = 0.2–10 mM In the presence of ascorbic acid and uric acid	[82]
Porous HOPG [(i)]	To create the porosity	$Fc\text{-}CH_2OH$	5 μM–100 mM	[83]
		Direct Electron Transfer		
GC-Ar-GO-nP	4-carboxyphenyl	Direct electron transfer to GOx	C = 0.3–20 mM k_{ET} = 8.3 s^{-1} T = 112 s^{-1}.	[76]
GC	4-carboxyphenyl + oligo(phenylethynyl) (j)	Direct electron transfer to GOx	0–25 mM T = 1.1 s^{-1}	[75]
GC	3-phenylboronic	Direct complexation of GOx	–50 mM	[84]
GC + GO [(k)]	Thionine	Electrostatic adsorption of negatively charged GOx on positively charged functions of thionine	c = 0.5–6.0 mM s = 43.2 mA mM^{-1} cm^{-2}	[85]
Pt/GO-SO_3^-/PPy [(l)]	4-carboxyphenyl		c = 0.2 × 12 mM s = 0.56 μA mM^{-1} cm^{-2} In the presence of ascorbic acid (AA) and uric acid (UA)	[86]

[(a)] CNT: carbon nanotube, GC: glassy carbon, GO: graphene oxide, HOPG: highly oriented pyrolytic graphite. [(b)] Fc: ferrocene, [(c)] 4-phenylacetic group + precursor film (PF) composed of one layer of poly(styrenesulfonate) (PSS) sandwiched between two layers of poly(dimethyldiallylammonium) (PDDA); GOx is cast on this layer-by-layer assembly and finally Nafion™ is deposited. [(d)] By reduction of the attached 4-nitrophenyl groups. [(e)] $Fc\text{-}CH_2\text{-}CH_2\text{-}COOH$ attached to an amino function of the grafted aryl group. [(f)] Oxygen is used to reoxidize reduced GOx ($GOxH_2$). [(g)] Interdigitated array of 1:1 aspect ratio carbon nanoelectrodes. [(h)] A crosslinked chitosan–glutaraldehyde–GOx gel is deposited on the Pt electrode. [(i)] Graphene is made porous by grafting diazonium salts, and GOx bonded to pyrenebutyric acid is adsorbed in the pores of HOPG. [(j)] Assembled by simultaneous reduction of two diazonium salts; 20-Å-long molecular wire. [(k)] Thionine diazonium cation is covalently attached onto the glassy carbon electrode via graphene nanosheets. [(l)] Nanocomposite layers are deposited onto platinum electrodes through the electrochemical polymerization of pyrrole monomer in the presence of reduced graphene oxide bearing phenylsulfonyl groups and further modified with carboxyphenyl groups via electrochemical reduction of 4-carboxybenzenediazonium.

Direct electron transfer from the electrode to a GOx enzyme is also possible by diazonium chemistry but requires fine control of the overall structure. Indeed, to obtain an efficient enzyme wiring, a bottom-up approach needs to be implemented. The active center of GOx is a flavine adenine dinucleotide (FAD) buried deep inside the pocket of a proteinic structure of the enzyme. Gooding et al. designed a molecular wire (a 20-Å-long oligo(phenylethynyl)) able to reach the FAD active site and providing a fast electron transfer to GOx [75]. In addition, these bonded molecular wires were diluted in 4-carboxyphenyl groups (30/1) that served the twin purposes of being a spacer between molecular wires and an anchor to maintain the attached GOx on the surface via covalent peptidic coupling (Figure 13). The surface coverage of active GOx was calculated to be 2.41×10^{-12} mol·cm^2, and the rate of electron transfer was $k_{ET} = 78\ s^{-1}$.

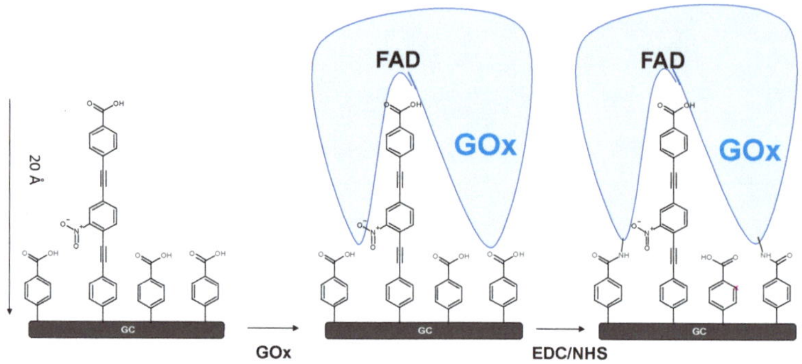

Figure 13. Biosensor of glucose based on the direct electron transfer to the of GOx flavine adenine dinucleotide (FAD) co-enzyme through a molecular wire.

Another complex assembly is presented in Figure 14, where grafting of diazonium salts was used for attaching (i) gold nanoparticles to graphene oxide (GO), (ii) modified GO to the glassy carbon (GC) electrode, and finally (iii) GOx to gold nanoparticles [76]. These two examples underline the ability of diazonium chemistry to form complex (nano)structures similar to those obtained by self-assembly.

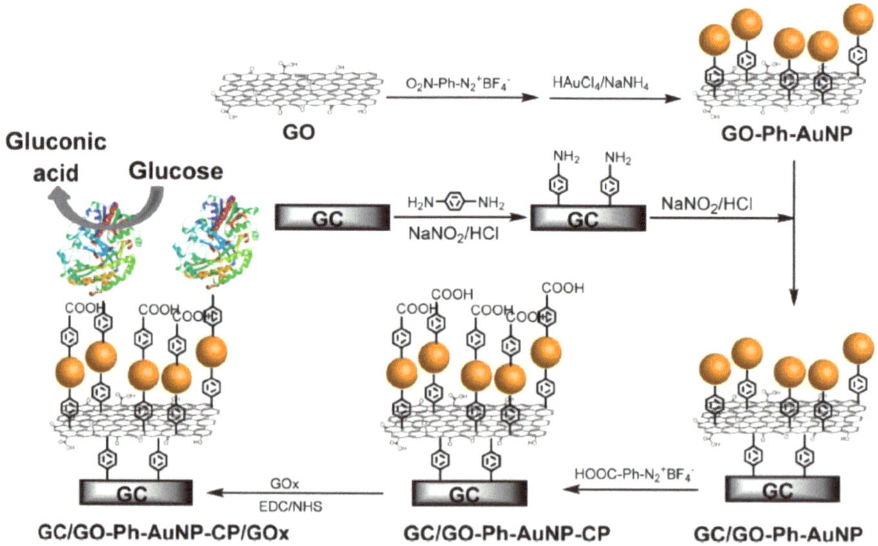

Figure 14. Attachment of glucose oxidase through graphene oxide and gold nanoparticles. From Reference [76] with permission of the Royal Chemical Society.

Table 2 gathers the different architectures for glucose biosensors obtained through diazonium salt chemistry.

NAD^+/NADH. The nicotinamide adenine dinucleotide redox co-factor, NAD^+, is the coenzyme of over 300 dehydrogenase enzymes (e.g., lactate dehydrogenase, alcohol dehydrogenase, glucose dehydrogenase). The quantitative detection of the reduced form, NADH, can be used as a measure of enzymatic activity. Amperometric biosensors based on this strategy were developed to assay of the corresponding enzymatic substrate molecules (e.g., lactate, malate, and ethanol).

NADH was detected [87–90] on an array of five electro-addressable electrodes. All the electrodes were grafted with 4-nitrophenyl groups; then, on two of these electrodes, the nitrophenyl group was reduced to aminophenyl to which pyrroloquinoline quinone was bonded by peptidic coupling. This quinone acts as a mediator for the oxidation of NADH [87]. In a similar way, anthraquinone, [89] toluidine blue [88], and azure A (for the detection of ethanol) [90] were attached to electrodes.

Other biomolecules. Some sensors were described that, based on diazonium salts, can detect different drugs; for example, ranitidine, a histamine H2 receptor antagonist, was detected on aminophenyl-modified gold nanoparticle films deposited on a GC electrode by differential pulse voltammetry (DPV) [91]. Calcitonin, a tumor marker, was detected on a GC electrode modified by 4-carboxy or 4-nitrobenzenediazonium and attachment of gold nanoparticles and graphene oxide; the increased surface area of the immunosensor translated into an enhanced sensitivity [92]. Uric and ascorbic acids were also detected by diazonium-based sensors [93–95]. For example, uric acid was detected with an Au gate field-effect transistor (FET)-based sensor where the gold surface was modified by a monolayer of 4-nitrobenzenediazonium in the presence of DPPH (2,2-diphenyl-1-picrylhydrazyl) [95]. Estradiol was detected by attachment of an aptamer (NH_2-APT) to a GC-reduced graphene oxide surface modified by reaction of 4-carboxybenzenediazonium [96].

Toxins. Biosensors were developed against dangerous toxins. Aflatoxins are- highly toxic mycotoxins produced by fungi. Among them, AFM1 can be found in commercially available milk; it is, therefore, considered as one of the most serious problems of food safety, and the level of aflatoxin in milk is subject to safety regulations. Okadaic acid is one of the most common marine biotoxins, which is ingested through filter feeding mechanisms by various species of shellfish such as mussels. Ochratoxin A is a carcinogenic mycotoxin that was identified as a contaminant in cereals, coffee, cocoa, dried fruits, and pork. Therefore, electrochemical detection of these toxins is important, and biosensors were constructed either via modifying a surface by attaching antibodies and aptamer or via modifying the toxin itself and attaching the modified species (Table 3).

Table 3. Detection of toxins. SWCNT—single-walled carbon nanotube.

Surface	Diazonium Salt + Attached Recognizing Group	Analyte	Detectable Label and Detection Limit	Reference
Attachment of toxins to surfaces modified by diazonium chemistry				
SPE [(a)]	4-carboxybenzenediazonium + hexaethyleneglycol-modified 21-mer oligonucleotide	Aflatoxin M1 in milk 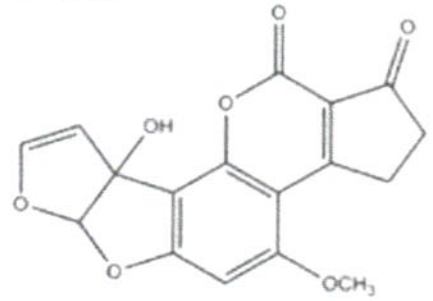	Direct detection by EIS [(b)] and CV [(c)] c = 20–1000 ng/kg.	[97]
SPE	4-carboxybenzenediazonium+ anti-okadaic acid monoclonal antibody (anti-OA-MAb)	Okadaic acid 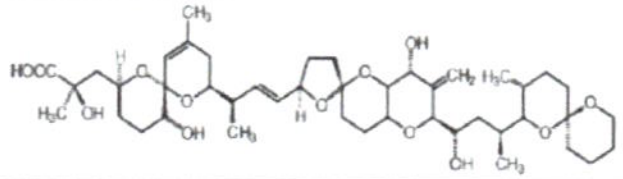	Direct detection by EIS	[98]
Graphene-modified SPE			Direct detection by SWV [(d)]	[99]
SPE	4-carboxybenzenediazonium+ streptavidine + biotin-anti-TGFβ1	TGFβ1 protein	CV of viologen-modified SWCNT	[100]
SPR chip with Cu/graphene layer	3,5-bis-fluoro 4-carboxybenzenediazonium	HT-2 toxin 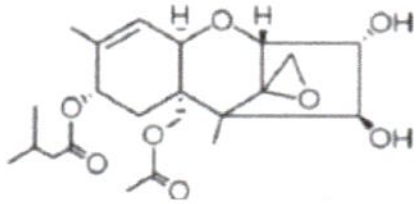	Label-free, surface plasmon resonance	[26]
Toxins modified with an aryl diazonium group				
Boron-doped diamond electrochemical microcell	Anti-ochratoxin polyclonal antibodies modified by a diazonium salt after coupling with 4-carboxymethylaniline	Ochratoxin A 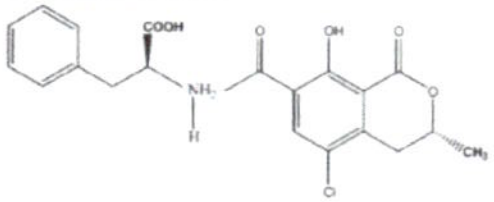		[101]

[(a)] Carbon screen-printed electrode, [(b)] electrochemical impedance spectroscopy, [(c)] cyclic voltammetry, [(d)] square wave voltammetry.

For example, okadaic acid was recognized by its attached antibody and detected by Electron Impedance Spectroscopy (EIS), as presented in Figure 15.

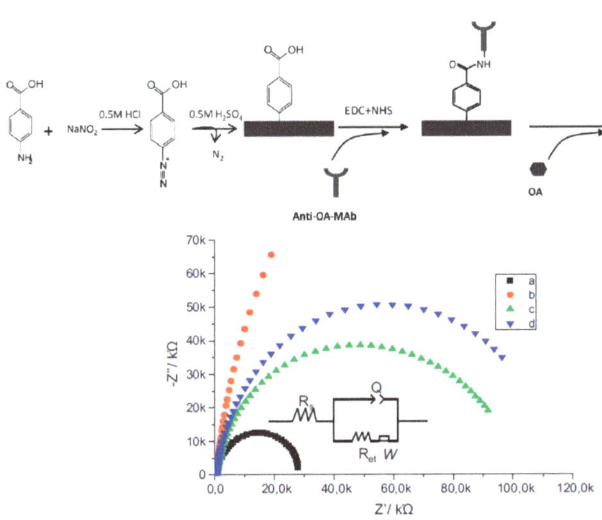

Figure 15. A schematic presentation of the surface chemistry used for immobilization anti-okadaic acid (OA) monoclonal antibody on an SPE electrode, and Nyquist plots of 1 mM $[Fe(CN)_6]^{4-/3-}$ for (**a**) bare SPE, (**b**) SPE modified by 4-carboxybenzenediazonium, (**c**) and an antibody, (**d**) in the presence of 10 μg/L ochratoxin. The insert is the equivalent electric circuit applied to fit the impedance spectroscopy. From Reference [98] with permission of Elsevier.

Biogenic amines. Biogenic amines are synthesized and degraded during normal metabolism of animals, plants, and microorganisms. Histamine, putrescine, cadaverine, tyramine, tryptamine, spermine, and spermidine are considered to be some of the most important biogenic amines in food. They were detected by attaching monoamine oxidase to the surface of SPE detection in the presence of ferrocene methanol as a detector [102].

2.2. Detection of Polypeptides and Proteins

Many polypeptides, proteins, and enzymes are important in biological or medical processes. Their detection and quantification in very minute quantities in body fluids is necessary; this can be achieved by using diazonium salts that provide an anchor to attach proteins to surfaces where they are detected. Two methods were used: (i) the protein is modified with an aminophenyl group (mainly by peptidic coupling), and this aminophenyl modified protein is then transformed to a diazonium salt that is attached to the surface; (ii) a diazonium salt with an appropriate 4- substituent (mainly carboxylic and amino groups) is attached to the surface and further reacts with the protein. These two methods (Table 4) permit creating a diazonium-based sandwich immunoassay; an analyte (most often a protein) is detected through the use of a diazonium-anchored antibody (also termed immunoglobulin IgG) (or antigen), and the assay is completed by attaching a detectable group to this construct (for example, by luminescence). Table 4 gathers examples of such biosensors.

Table 4. Detection of proteins. IgG—immunoglobulin G.

Surface	Diazonium Salt + Attached Recognizing Group	Analyte	Detectable Label and Detection Limit	Reference
Proteins modified with an aryl diazonium group				
SPE array	Immunoglobulin modified by a diazonium salt after coupling with 4-carboxymethylaniline	Anti-rabbit IgG antibodies	Peroxidase-labeled anti-rabbit or anti-human IgG antibodies. LOD: 50 fmol [a]	[103]
SPE array	Anti-human IgG	Human IgG	Horseradish peroxidase (HRP)-modified secondary antibody. Detection limit: 60 nm human IgG	[104–106] [b]
SPE array	Rabbit IgG modified by a diazonium salt after coupling with 4-carboxymethylaniline	Rheumatoid factor (RF)	Horseradish peroxidase (HRP)-modified secondary antibody. Detection range: 5.3–485 IU·mL^{-1}	
SPE array	HRP modified by a diazonium salt after coupling with 4-carboxymethylaniline	HRP		
Gold NPs [c] on SPE	Human prostate-specific antigen (PSA) modified by a diazonium salt after coupling with 4-carboxymethylaniline	Prostate-specific antigen (PSA) monoclonal antibody labeled with biotin	Horseradish peroxidase-labeled streptavidin Detection range of 5–80 ng/mL	[107]
(SPRi) [e] gold chip	Anti-ovalbumin IgG modified by a diazonium salt after coupling with 4-carboxymethylaniline	Ovalbumin	Direct reflectivity change. LOD: 100 ng/mL (2 nM).	[108] [d]
GC, Au	Biotinylated anti-TNF-antibody (Tumor Necrosis Factor) modified by a diazonium salt after coupling with 4-carboxyaniline	Anti-TNF-antibody	Avidin modified gold NPs or avidin-HRP	[109] [e]
Au	Carboxybetaine aryldiazonium derivative + lectin	Prostate-specific antigen (PSA)	EIS detection of anti-PSA antibody-modified Au nanoshells with a magnetic core	[110] [f]
Attachment of proteins to surfaces modified by diazonium chemistry				
	Modified Surface	**Analyte**	**Detectable Label and Detection Limit**	
SPRi gold chip	Gold surface modified	Protein A	Direct reflectivity change upon coupling the protein to the attached carboxylic group	[111] [g]
CNT	CNT modified by reaction of 4-carboxymethylbenzenediazonium + attachment of mouse olfactory receptor proteins	Various odoriferant organic compounds in vapor phase	Detection of odors: e-nose For example, 7 ppb of toluene	[112] [h]
Graphene-modified GC	Graphene surface modified by reaction of 4-aminobenzenediazonium + coupling with A-Fetoprotein antibody	A-Fetoprein	HRP-labeled anti-AFP antibody LOD: 0.03 ng mL^{-1}	[113] [i]

Table 4. *Cont.*

Surface	Diazonium Salt + Attached Recognizing Group	Analyte	Detectable Label and Detection Limit	Reference
Attachment of proteins to surfaces modified by diazonium chemistry				
	Modified Surface	**Analyte**	**Detectable Label and Detection Limit**	
SPE	Gold surface modified by reaction of 4-carboxybenzenediazonium	Lysozyme	Biotinylated antibody avidin–alkaline phosphatase LOD: 4.3 fM	[114] [j]
Gold	Gold NPs modified with 4-mercaptoaniline, attached to gold surface by diazonium coupling	Five-amino-acid polypeptide with a biotin group	Peroxidase-labeled streptavidin	[115]
Nanocomposite: gold NPs loaded on reduced graphene oxide	Gold NPs modified with 4-carbxyphenyl and 4-aminophenylphosphorylcholine + coupling with anti-TNF-α capture antibody	Cytokine tumor necrosis factor-alpha (TNF-α)	Coupling anti-TNF-α detection antibody (Ab2) attached to graphene oxides modified with ferrocenyl groups. Electrochemical detection of ferrocene. LOD: 0.1 $pg{\cdot}mL^{-1}$	[116] [k]
Indium tin oxide (ITO)	Gold surface modified by reaction of the diazonium salt of 4-aminobutyric acid + peptidic coupling of antibody (Ab1)	Tumor necrosis factor	HRP-conjugated detection antibody (Ab2) LOD: 10 pg/mL	[117] [l]
SPRi biochip	4-Carboxybenzenediazonium + anti-ovalbumin antibody	Ovalbumine	Direct detection by SPR	[118]
SPE	SPE surface modified by reaction of 4-methoxybenzenediazonium + oxidation of the grafted methoxy group+ attachment of anti-growth hormone antibodies	Growth hormone	EIS detection of 100 pg/mL growth hormone in undiluted whole blood LOD: 5 $pg{\cdot}mL^{-1}$	[119] [m]
Al–Mo nanoparticle membrane on Si	4-formylbenzenediazonium	Monoclonal antibodies specific for bovine herpes virus 1 (BHV-1)	Bovine herpes virus 1 Measurement of the resonance frequency shift of the membrane	[120]
Biomembrane-like films				
GC	Gold NPs capped with myoglobin on 4-carboxaldehyde diazonium salt + docecyl thiol	Myoglobin	Electrochemical detection of H_2O_2 0.3 μM	[121]

[a] LOD: Limit of Detection. Sequential modification of the electrodes is possible without interference; [b] the binding of non-RF human antibodies was found to be negligible; [c] NPs: nanoparticles; [c] SPRi: surface plasmon resonance imaging surface; [d] the integrity of the immobilized carboxymethylaniline-modified antibodies; the detection of ovalbumin by immobilized IgG against ovalbumin was evaluated; [e] response time 5 s, selective detection, selectivity of a given cytokine; [f] LOD: 34 $fg{\cdot}mL^{-1}$ and 108% recovery when spiking in a female serum; [g] non-specific absorption of the rabbit IgG remained negligible on the inter-spot areas; [h] remained active with stable normalized responses for ~5 days when stored in a humid environment; device responses to odorants show broad agreement with heterologous technique; [i] relative standard deviation was 4.5%, selectivity vs. bovine serum albumin and carcinoembryonic antigen, stability: 10 days under proper conditions; [j] detection range: 5 fM to 5 nM, selectivity vs. bovine serum albumin, cytochrome c, recovery 95–102% in wine samples; [k] selectivity vs. bovine serum albumin, prostate-specific antigen, cancer antigen-125, and mouse IgG, stability 30 days under proper conditions; [l] selectivity vs. hemoglobin, human serum albumin, comparison with ELISA test; [m] relative standard deviation less than 6% in whole-blood and plasma samples, mean recoveries ranging from 94% ± 3% to 103% ± 2%.

The first examples of such diazonium-based immunoassays were published by Marquette [103,104]; an array of individually addressable screen-printed electrodes was modified with an IgG. This was achieved by (i) coupling the carboxylic group of 4-carboxyaniline to an amino group of IgG by peptidic coupling (DCC/NHS, N-hydroxysuccinimide N,N′-dicyclohexylcarbodiimide), (ii) diazotizing the pending amino group in acidic water (20 mM HCl and 20 mM $NaNO_2$), and (iii) electrografting the diazonium-labeled IgG to a connected electrode of the array. This electrografting was characterized, as for other diazonium salts, by a decrease of the drawn-out wave of the diazonium salt upon repetitive scanning. This reaction sequence is presented in Figure 16.

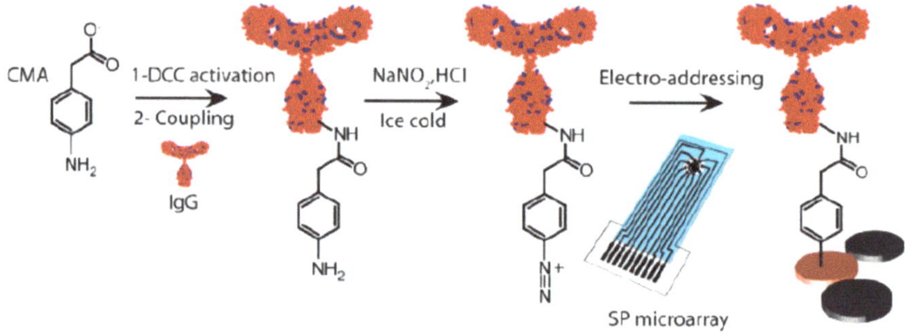

Figure 16. Electrografting of an immunoglobulin G (IgG) on an array of screen-printed electrodes. From Reference [104] with permission of Wiley.

The final step of the assembly is presented in Figure 17 in the case of a rheumatoid factor (a family of human antibodies largely involved in rheumatoid diseases) that binds to the surface-attached IgG. The detection was achieved by binding a secondary antibody labeled by a horseradish peroxidase; this final enzyme catalyzed the oxidation of luminol with light emission at 428 nm. Therefore, detection of this emission permitted quantifying the presence of rheumatoid factor in human serum in the range 5.3–485 IU·mL^{-1}. Detection was also achieved on SPRi (surface plasmon resonance imaging) surfaces by direct reflectivity change.

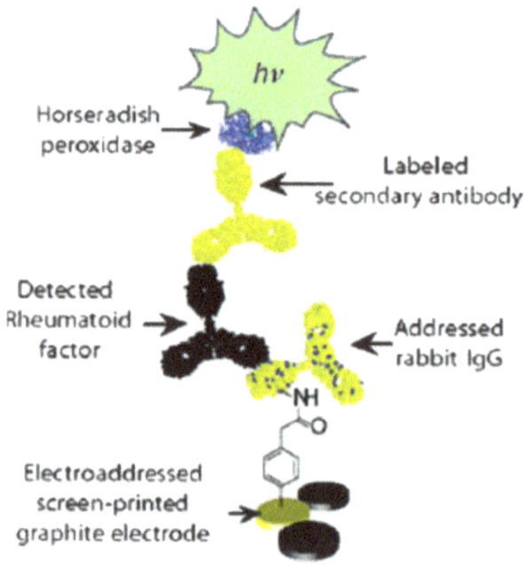

Figure 17. A sandwich immunoassay for the detection of rheumatoid factor. From Reference [104] with permission of Wiley.

The second strategy involves the fabrication by diazonium chemistry of platforms that can be connected to proteins. An excellent example was provided by a paper of Gooding that described the detection of tumor necrosis factor α (TNF-α) in whole blood [117]. TNF-α is a typical early-stage indicator of an inflammatory reaction, in response to infection or cancer. Affinity biosensors are difficult to operate in whole blood because biofouling of electrode surfaces compromises the performance of the final device. To prevent this phenomenon, a platform was prepared on ITO by reduction of two diazonium salts derived from 4-aminophenyl phosphorylcholine (PPC) and 4-(4-aminophenyl) butyric acid (PBA). Therefore, this mixed surface comprised phosphorylcholine groups that prevented biofouling of the electrode and phenylbutyic acid groups that permitted the attachment of antibodies as biorecognition elements. Other anti-biofouling molecules could be attached to electrode surfaces such as polyethyleneglycol molecules, but these types of long-chain molecules give rise to passivated surfaces with high impedance. Upon electrochemical reduction, the diazonium salts of PPC (E_p = −0.55 V/(Ag/AgCl) and PBA (E_p = −0.58 V/(Ag/AgCl), as well as their mixture, gave rise to the typical pattern of diazonium salts, where the current decreases upon repetitive scanning, as shown in Figure 18.

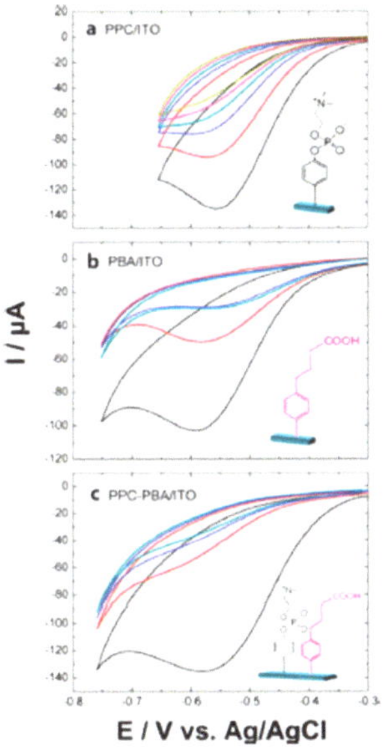

Figure 18. Cyclic voltammetry of the diazonium salts of 4-aminophenyl phosphorylcholine (PPC), 4-(4-aminophenyl) butyric acid (PBA), and their mixture on an indium tin oxide (ITO) electrode. From Reference [117] with permission of the American Chemical Society.

These modified ITO electrodes were characterized by XPS (X-ray Photoelectron Spectroscopy), cyclic voltammetry of redox probes ($Fe(CN)_6^{3-/4-}$), and EIS. The immunosensor was on this platform as presented in Figure 19. The capture antibody (Ab1) was immobilized onto the PPC–PBA/ITO surface via the classical EDC/NHS (EDC: 1-éthyl-3-(3-diméthylaminopropyl)carbodiimide) conjugation reactions between COOH groups on the mixed layer surface and residual amino groups of the Ab1. The final steps involved the binding of the analyte TNF-α and finally of the HRP-conjugated detection antibody. The consumption of H_2O_2 by HRP was detected by amperometry, and TNF-α concentrations in the range of 0.01–500 ng/mL were detected. The interference of human serum albumin or hemoglobin was limited by the presence of phosphorylcholine on the surface [117]. This example illustrates the possibility of realizing mixed layers with controlled structures. This new approach enhances the analytical performance by implementing both a bioreceptor and a non-specific adsorption reducer. Prior to this publication, only thiol-type self-assembly and co-polymerization approaches were used to obtain such structures.

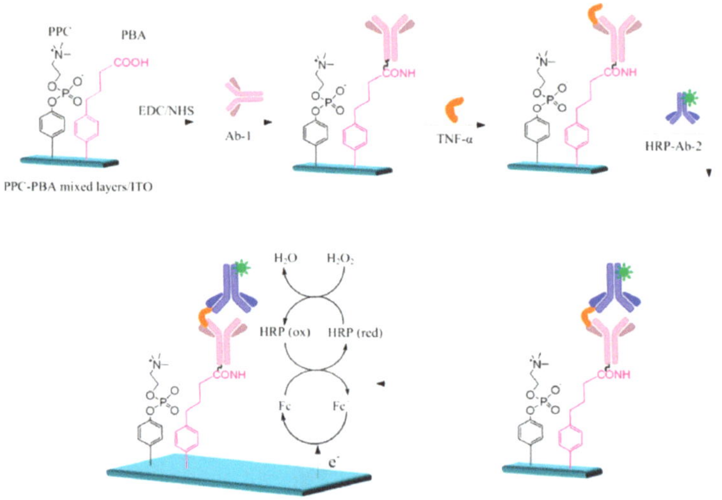

Figure 19. Immunoassay of tumor necrosis factor α (TNF-α) factor based on a mixed antifouling platform. From Reference [117] with permission of the American Chemical Society.

An "e-nose" capable of detecting different odors (eugenol, *n*-amylacetate, etc.) was constructed as shown in Figure 20 by coupling mouse olfactory receptor proteins (ORs) with carbon nanotube transistors. The CNT was modified with 4-carboxybenzenediazonium, and the ORs were attached by peptidic coupling. The resulting devices transduced signals associated with odorant binding to ORs in the gas phase under ambient conditions [112].

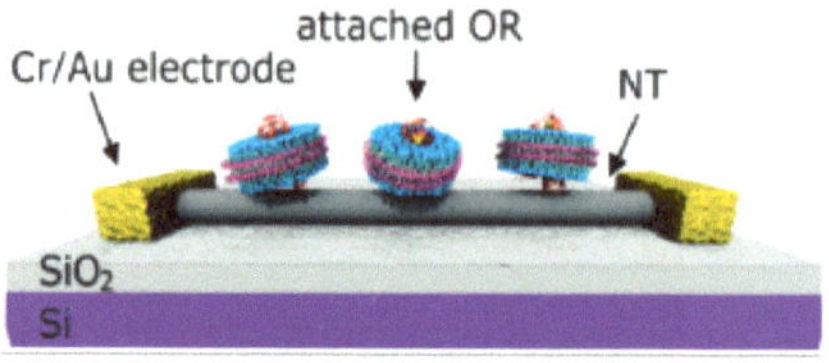

Figure 20. Transistor for the detection of odors ("e-nose"). From Reference [112] with permission of the American Chemical Society.

2.3. Detection of DNA

Affinity sensors using nucleic acids as bioreceptors are probably the most common biosensors. Indeed, nucleic acids are used to bind another nucleotide sequence (DNA, RNA, microRNA (miRNA), etc.) or, in the case of aptamers, a different type of target (e.g., small organic molecules, metal cations, proteins, etc.). Biosensors including a diazonium electroreduction step in their fabrication process already proved their relevance for DNA sensing. Methods for recognition of DNA sequences were as follows: (i) the DNA sequence to be recognized was equipped with a diazonium salt, electrografted and recognized in different ways [103]; (ii) the target DNA sequence was bonded to the surface through avidin–biotin recognition [122] or another method [123]; (iii) the target DNA sequence was linked to the surface and recognized by its complementary sequence [124,125]; (iv) a symmetric reaction where the complementary sequence was bonded to the surface was also possible [126]. The detection was achieved either by fluorescence or by electrochemistry (DPV, EIS) (Table 5).

Table 5. Detection of DNA

Surface	Diazonium Salt + Attached Recognizing Group	Analyte	Detectable Label and Detection Limit	Reference
		Modification of the Surface		
Carbon nanofibers	4-Nitrobenzenediazonium and reduction of 4-nitrophenyl to 4-aminophenyl groups after electrografting + reaction of a maleimide linker [(a)]	Thiol-terminated DNA attaches to the linker	Fluorescently labeled, perfect complement to the grafted oligonucleotides	[123]
Vertically aligned diamond nano-wires	4-Nitrobenzenediazonium and reduction of 4-nitrophenyl to 4-aminophenyl groups after electrografting + maleimide linker [(a)] + thiol-terminated DNA, 23-mer cancer marker cytokeratin	The complementary DNA sequence	The complementary sequence detected by Differential Pulsed Voltammetry (DPV) LOD ~2 pM	[124] [(b)]
Array of gold electrodes	4-Carboxybenzenediazonium and a bis-diazonium salt with a COOH terminal group + amino-terminated DNA	Human papillomavirus sequences terminated by tetramethylbenzidine	Electrochemical detection of tetramethylbenzidine	[127]
Vertically aligned carbon nanotubes	Diazonium salt of an aminophenyl group substituted + β-cyclodextrin (β-CD)	DNA probe substituted on one end by a dabcyl group and a CdS nanoparticle at the other end	In the presence of the complementary sequence, the probe could be captured by the β-CD-modified CNT electrode LOD by DPV: 5.0×10^{-13} M	[128] [(c)]
ITO, gold, GC	4-Carboxybenzenediazonium + attachment of avidin	Biotinylated DNA from influenza virus (type A)	Avidin–biotin recognition. Detection through the CV of ferro/ferricyanide LOD: 8.51×10^{-14} M	[122]
SPE	4-Nitrobenzenediazonium and reduction of 4-nitrophenyl to 4-aminophenyl groups after electrografting	Amine-modified (polyA)25 DNA probe.	Reaction between the diazonium group and NH_2-DNA and recognition of the hybridization by EIS and DPV LOD: 4.65 nm	[129] [(d)]
GC	4-Carboxybenzenediazonium and a naphthoquinone [(e)] to give a mixed layer + DNA probe attached to the surface	Fluorescent complementary DNA strand	Hybridization was detected by fluorescence and Alternative Current (AC) voltammetry Detection limit ca. 10 pM	[125]
Au	Mixed layer obtained from the diazonium salts bearing (i) a sulfobetaine group, (ii) a phenylmaleimido group + DNA probe attached to the surface	New Delhi metallo-β-lactamase (NDM)-Coding Gene	CV and chronoamperometric detection of the charges on the phosphate groups of DNA Detection limit: 54 pM	[130] [(f)]
	The diazonium salts of 4-aminophenylacetic acid + covalent immobilization of streptavidin and incubation of a biotinylated DNA capture probe	Biotinylated DNA target sequence associated with the human papillomavirus	HRP-DNA probe + electrochemical detection of tetramethyl benzidine Detection limit: 0.50 nM	[126] [(g)]
ITO	1-Naphthalenesulfonate diazonium salt	MicroRNA	Discrimination ability over single-mismatch, high sensitivity in the aM range thanks to the use of isothermal amplification strategy Detection limit: 25 aM	[131] [(h)]
SPE	4-Carboxybenzenediazonium + peptidic attachment of a DNA probe		Determination of DNA damage by various reagents using EIS	[132] [(i)]

[(a)] Sulfosuccinimidyl-4-(*N*-maleimidomethyl) cyclohexane-1-carboxylate; [(b)] concentration range: from 2 to 10×10^{-12} M, no degradation over 30 cycles of DNA hybridization/denaturation; [(c)] linear concentration range 1.0×10^{-7} to 1.0×10^{-12} M; [(d)] reproducibility: 7%, stability: retained 60% of its initial response after one month; [(e)] 5-hydroxy-1,4-naphthoquinone (juglone); [(f)] stability: >5 days; [(g)] linear range: 0–20 nM; [(h)] selectivity vs. concomitant miRNA, family members with sequence homology, recovery: 100% in 10% diluted human serum samples; [(i)] retained about 83% of its initial response after three weeks under proper conditions.

A mixed layer obtained from two diazonium salts (4-carboxybenzenediazonium + the diazonium salt of an amino derivative of 5-hydroxy-1,4-naphthoquinone, juglone) was prepared. The carboxylic group was activated by EDC/NHS, and a primary amine-functionalized DNA strand (NH_2-DNA probe) was coupled to the carboxylic acid. This attached DNA sequence recognized its complementary chain equipped with a fluorescent label, but hybridization was also detected by 3AC) voltammetry of the naphthoquinone group [128].

Human papillomavirus is a DNA virus responsible for cervical cancer. It can be detected based on the sequence of reactions presented in Figure 21. Carbon nano-onions are multilayered fullerenes concentrically arranged one inside the other; they were deposited onto a GC surface to form a stable micrometric film. This carbonaceous film was modified by the diazonium salt of phenylacetic acid. The COOH groups of the surface were activated using carbodiimide chemistry, followed by covalent immobilization of streptavidin and incubation of a biotinylated DNA capture probe (Figure 21a–c). Hybridization was accomplished with the target DNA sequence. Finally, an HRP-labeled reporter probe was introduced, and amperometric detection of the oxidation of tetramethylbenzidine permitted quantifying the analyte down to 0.11 $\mu A \cdot nM^{-1}$. Incorporation of carbon nano-onions on the surface resulted in better sensitivities and lower limits of detection than unmodified GC due to the increased surface area [130].

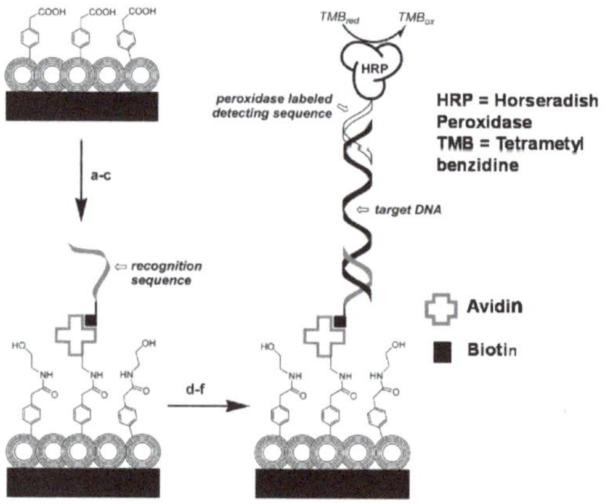

Figure 21. DNA sensors for human papillomavirus oncogene detection. From Reference [130] with permission of the American Chemical Society.

2.4. Cells

Gold SERS (surface-enhanced Raman spectroscopy) active substrates were modified with 4-carboxyphenyl groups, and the carboxylic groups were coupled to amino groups of folic acid. With such a surface, it was possible to test the presence of normal and melanoma-associated cells in a cultivation medium. The SERS spectra of the folic acid modified surface indicated the presence of differences arising from the interaction of the bio-liquid with the functional surface [133] (Figure 22).

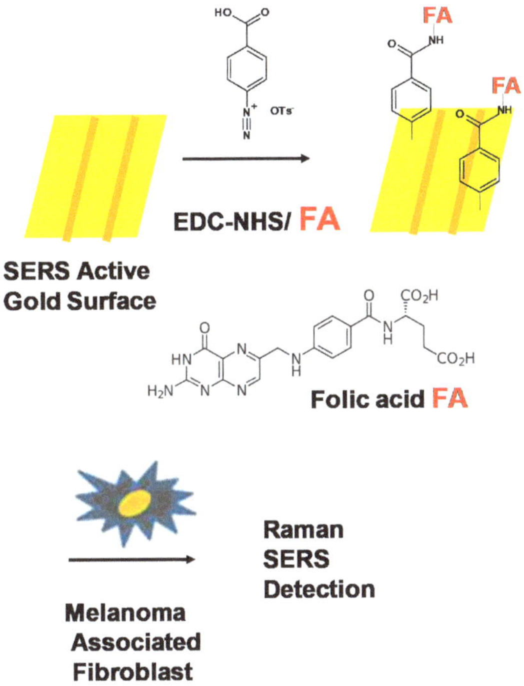

Figure 22. Detection of normal and melanoma-associated cell secretome by SERS (surface-enhanced Raman spectroscopy).

3. Concluding Remarks

The large and growing demand for robust and reliable detection devices continues to motivate much work around biosensors. Given the advantages of the electroreduction of diazonium salts in terms of stability of (bio)molecular buildings, it is very likely that this method will contribute to improving the reliability and stability of detection systems. However, before being able to take full advantage of diazonium chemistry, attention must be paid to reach a finer control of film thickness and composition, which is the current main challenge. Recent advances in controlling the thickness and composition of mixed layers obtained by reduction of diazonium salts are the first steps toward the realization of covalent buildings of controlled architecture. Beyond the classical approaches of biodetection, there is a tendency to integrate biosensors into everyday objects through printing methods. The reactivity of the diazonium function makes it possible to consider, as a future prospect, the development of a new generation of functional inks able to bind covalently to the surface of the substrate or to an underlying layer. This capability would allow the integration of a variety of functions, including biosensing ones, under ambient conditions over a wide range of surface areas.

Funding: This research was funded by Université de Paris, CNRS.

Conflicts of Interest: The authors declare no conflict of interest.

References

1. Allongue, P.; Delamar, M.; Desbat, B.; Fagebaume, O.; Hitmi, R.; Pinson, J.; Savéant, J.-M. Covalent Modification of Carbon Surfaces by Aryl Radicals Generated from the Electrochemical Reduction of Diazonium Salts. *J. Am. Chem. Soc.* **1997**, *119*, 201–207. [CrossRef]
2. Berisha, A.; Chehimi, M.M.; Pinson, J.; Podvorica, F.I. Electrode surface modification using diazonium salts. *Electroanal. Chem.* **2016**, *26*, 115–225.

3. Aryl Diazonium Salts. *New Coupling Agents in Polymer and Surface Science*; Chehimi, M.M., Ed.; Wiley-VCH: Weinheim, Germany, 2012.
4. Jiang, C.; Moraes Silva, S.; Fan, S.; Wu, Y.; Alam, M.T.; Liu, G.; Gooding, J.J. Aryldiazonium salt derived mixed organic layers: From surface chemistry to their applications. *J. Electroanal. Chem.* **2017**, *785*, 265–278. [CrossRef]
5. Assresahegn, B.D.; Brousse, T.; Belanger, D. Advances on the use of diazonium chemistry for functionalization of materials used in energy storage systems. *Carbon* **2015**, *92*, 362–381. [CrossRef]
6. Delaporte, N.; Belanger, R.L.; Lajoie, G.; Trudeau, M.; Zaghib, K. Multi-carbonyl molecules immobilized on high surface area carbon by diazonium chemistry for energy storage applications. *Electrochim. Acta* **2019**, *308*, 99–114. [CrossRef]
7. Bondarev, A.A.; Naumov, E.V.; Kassanova, A.Z.; Krasnokutskaya, E.A.; Stankevich, K.S.; Filimonov, V.D. First Study of the Thermal and Storage Stability of Arenediazonium Triflates Comparing to 4-Nitrobenzenediazonium Tosylate and Tetrafluoroborate by Calorimetric Methods. *Org. Process Res. Dev.* **2019**, *23*, 2405–2415. [CrossRef]
8. Pazo-Llorente, R.; Bravo-Diaz, C.; Gonzalez-Romero, E. pH Effects on Ethanolysis of Some Arenediazonium Ions: Evidence for Homolytic Dediazoniation Proceeding through Formation of Transient Diazo Ethers. *Eur. J. Org. Chem.* **2004**, 221–3226. [CrossRef]
9. Pazo-Llorente, R.; Maskill, H.; Bravo-Diaz, C.; Gonzalez-Romero, E. Dediazoniation of 4-Nitrobenzenediazonium Ions in Acidic MeOH/H_2O Mixtures: Role of Acidity and MeOH Concentration on the Formation of Transient Diazo Ethers that Initiate Homolytic Dediazoniation. *Eur. J. Org. Chem.* **2006**, 201–2209. [CrossRef]
10. Berisha, A.; Combellas, C.; Kanoufi, F.; Decorse, P.; Oturan, N.; Médard, J.; Seydou, M.; Maurel, M.; Pinson, J. Some Theoretical and Experimental Insights on the Mechanistic Routes Leading to the Spontaneous Grafting of Gold Surfaces by Diazonium Salts. *Langmuir* **2017**, *33*, 8730–8738. [CrossRef]
11. Bui-Thi-Tuyet, V.; Cannizzo, C.; Legros, C.; Andrieux, M.; Chaussé, A. Modification of fluorine-doped tin oxide surface: Optimization of the electrochemical grafting of diazonium salt. *Surf. Interfaces* **2019**, *15*, 110–116. [CrossRef]
12. Griffete, N.; Ahmad, R.; Benmehdi, H.; Lamouri, A.; Decorse, P.; Mangeney, C. Elaboration of hybrid silica particles using a diazonium salt chemistry approach. *Colloids Surf. A* **2013**, *439*, 145–150. [CrossRef]
13. Schuurman, J.C.; McNeill, A.R.; Martinez-Gazoni, R.F.; Scott, J.I.; Reeves, R.J.; Allen, M.W.; Downard, A.J. The effect of covalently bonded aryl layers on the band bending and electron density of SnO_2 surfaces probed by synchrotron X-ray photoelectron spectroscopy. *Phys. Chem. Chem. Phys.* **2019**, *21*, 17913–17922. [CrossRef] [PubMed]
14. Gam-Derouich, S.; Pinson, J.; Decorse, P.; Luo, Y.; Herbaut, R.; Royon, L.; Mangeney, C. Diazonium salt chemistry for the design of nano-textured anti-icing surfaces. *Chem. Commun.* **2018**, *54*, 8983–8986. [CrossRef] [PubMed]
15. Gam-Derouich, S.; Pinson, J.; Lamouri, A.; Decorse, P.; Bellynck, S.; Herbaut, R.; Royon, L.; Mangeney, C. Micro-patterned anti-icing coatings with dual hydrophobic/hydrophilic properties. *J. Mater. Chem. A* **2018**, *6*, 19353–19357. [CrossRef]
16. Joselevich, M.; Williams, F.J. Synthesis and Characterization of Diazonium Functionalized Nanoparticles for Deposition on Metal Surfaces. *Langmuir* **2008**, *24*, 11711–11717. [CrossRef]
17. Gorska, B.; Ratajczak, P.; Béguin, F. Faradaic processes on quinone-grafted carbons in protic ionic liquid electrolyte. *Electrochim. Acta* **2019**, *328*, 135090. [CrossRef]
18. Hossain, M.Z.; Shimizu, N. In Situ Functionalization of Graphene with Reactive End Group through Amine Diazotization. *J. Phys. Chem. C* **2017**, *121*, 25223–25228. [CrossRef]
19. Thaçi, B.S.; Gashi, S.T.; Podvorica, F.I. Preparation of heterogeneous reverse osmosis membranes undergoing modification process. *Desalin. Water Treat.* **2018**, *118*, 96–102. [CrossRef]
20. Zeb, G.; Viel, P.; Palacin, S.; Le, X.T. On the chemical grafting of titanium nitride by diazonium chemistry. *RSC Adv.* **2015**, *5*, 50298–50305. [CrossRef]
21. Bouriga, M.; Chehimi, M.M.; Combellas, C.; Decorse, P.; Kanoufi, F.; Deronzier, A.; Pinson, J. Sensitized Photografting of Diazonium Salts by Visible Light. *Chem. Mater.* **2013**, *25*, 90–97. [CrossRef]
22. Busson, M.; Berisha, A.; Combellas, C.; Kanoufi, F.; Pinson, J. Photochemical grafting of diazonium salts on metals. *Chem. Commun.* **2011**, *47*, 12631–12633. [CrossRef] [PubMed]

23. Nguyen, V.Q.; Ai, Y.; Martin, P.; Lacroix, J.-C. Plasmon-Induced Nanolocalized Reduction of Diazonium Salts. *ACS Omega* **2017**, *2*, 1947–1955. [CrossRef] [PubMed]
24. Tijunelyte, I.; Kherbouche, I.; Gam-Derouich, S.; Nguyen, M.; Lidgi-Guigui, N.; Lamy de la Chapelle, M.; Lamouri, A.; Lévi, G.; Aubard, J.; Chevillot-Biraud, A.; et al. Multi-functionalization of lithographically designed gold nanodisks by plasmon-mediated reduction of aryl diazonium salts. *Nanoscale Horiz.* **2018**, *3*, 53–57. [CrossRef]
25. Nguyen, M.; Kherbouche, I.; Gam-Derouich, S.; Ragheb, I.; Lau-Truong, S.; Lamouri, A.; Lévi, G.; Aubard, J.; Decorse, P.; Felidj, N.; et al. Regioselective surface functionalization of lithographically designed gold nanorods by plasmon-mediated reduction of aryl diazonium salts. *Chem. Commun.* **2017**, *53*, 11364–11367. [CrossRef]
26. Wu, F.; Thomas, P.A.; Kravets, V.G.; Arola, H.O.; Soikkeli, M.; Iljin, K.; Kim, G.; Kim, M.; Shin, H.S.; Andreeva, D.V.; et al. Layered material platform for surface plasmon resonance biosensing. *Sci. Rep.* **2019**, *9*, 1–10. [CrossRef]
27. Chehimi, M.M.; Pinson, J. (Eds.) *Applied Surface Chemistry of Nanomaterials*; Nova Publishers: New York, NY, USA, 2013.
28. Unwin, P.R.; Guell, A.G.; Zhang, G. Nanoscale Electrochemistry of sp2 Carbon Materials: From Graphite and Graphene to Carbon Nanotubes. *Acc. Chem. Res.* **2016**, *49*, 2041–2048. [CrossRef]
29. James, D.K.; Tour, J.M. Graphene: Powder, Flakes, Ribbons, and Sheets. *Acc. Chem. Res.* **2013**, *46*, 2307–2318. [CrossRef]
30. Schirowski, M.; Hauke, F.; Hirsch, A. Controlling the Degree of Functionalization: In-Depth Quantification and Side-Product Analysis of Diazonium Chemistry on SWCNTs. *Chem. Eur. J.* **2019**, *25*, 12761–12768. [CrossRef]
31. Romero-Ben, E.; Cid, J.J.; Assali, M.; Fernández-García, E.; Wellinger, R.E.; Khiar, N. Surface modulation of single-walled carbon nanotubes for selective bacterial cell agglutination. *Int. J. Nanomed.* **2019**, *14*, 3245–3263. [CrossRef]
32. Clément, P.; Trinchera, P.; Cervantes-Salguero, K.; Ye, Q.; Jones, C.R.; Palma, M. A One-Step Chemical Strategy for the Formation of Carbon Nanotube Junctions in Aqueous Solution: Reaction of DNA Wrapped Carbon Nanotubes with Diazonium Salts. *ChemPlusChem* **2019**, *84*, 1235–1238. [CrossRef]
33. Ejigu, A.; Kinloch, I.A.; Dryfe, R.A.W. Single Stage Simultaneous Electrochemical Exfoliation and Functionalization of Graphene. *ACS Appl. Mater. Interfaces* **2017**, *9*, 710–721. [CrossRef] [PubMed]
34. Ossonon, B.D.; Bélanger, D. Synthesis and characterization of sulfophenyl-functionalized reduced graphene oxide sheets. *RSC Adv.* **2017**, *7*, 27224–27234. [CrossRef]
35. Kirkman, P.M.; Güell, A.G.; Cuharuc, A.S.; Unwin, P.R. Spatial and Temporal Control of the Diazonium Modification of sp2 Carbon Surfaces. *J. Am. Chem. Soc.* **2014**, *136*, 36–39. [CrossRef]
36. Bellunato, A.; Schneider, G.F. Electrophilic radical coupling at the edge of Graphene. *Nanoscale* **2018**, *10*, 12011–12017. [CrossRef] [PubMed]
37. Bernal, M.M.; Di Pierro, A.; Novara, C.; Giorgis, F.; Mortazavi, B.; Saracco, G.; Fina, A. Edge-Grafted Molecular Junctions between Graphene Nanoplatelets: Applied Chemistry to Enhance Heat Transfer in Nanomaterials. *Adv. Funct. Mater.* **2018**, *28*, 1706954. [CrossRef]
38. van Nguyen, Q.; Martin, P.; Frath, D.; Della Rocca, M.L.; Lafolet, F.; Barraud, C.; Lafarge, P.; Mukundan, V.; James, D.; McCreery, R.L.; et al. Control of rectification in molecular junctions: Contact effects and molecular signature. *J. Am. Chem. Soc.* **2017**, *139*, 11913–11922. [CrossRef]
39. McCreery, R.L.; Bergren, A.J. Progress with Molecular Electronic Junctions: Meeting Experimental Challenges in Design and Fabrication. *Adv. Mater.* **2009**, *21*, 4303–4322. [CrossRef]
40. Supur, M.; Van Dyck, C.; Bergren, A.J.; McCreery, R.L. Bottom-up, Robust Graphene Ribbon Electronics in All-Carbon Molecular Junctions. *ACS Appl. Mater. Interfaces* **2018**, *10*, 6090–6095. [CrossRef]
41. Liu, Z.; Gan, F.; Dong, N.; Zhang, B.; Wang, J.; Chen, Y. Fabrication and nonlinear optical characterization of fluorinated zinc phthalocyanine covalently modified black phosphorus/PMMA films using the nanosecond Z-scan technique. *J. Mater. Chem. C* **2019**, *7*, 10789–10794. [CrossRef]
42. Li, D.O.; Gilliam, M.S.; Chu, X.S.; Yousaf, A.; Guo, Y.; Green, A.A.; Wang, Q.H. Covalent chemical functionalization of semiconducting layered chalcogenide nanosheets. *Mol. Syst. Des. Eng.* **2019**, *4*, 962–973. [CrossRef]

43. Wang, H.; Zhang, J.; Wu, Y.; Huang, H.; Jiang, Q. Chemically functionalized two-dimensional titanium carbide MXene by in situ grafting-intercalating with diazonium ions to enhance supercapacitive performance. *J. Phys. Chem. Solids* **2018**, *115*, 172–179. [CrossRef]
44. Kainz, Q.M.; Reiser, M. Polymer- and Dendrimer-Coated Magnetic Nanoparticles as Versatile Supports for Catalysts, Scavengers, and Reagents. *Acc. Chem. Res.* **2014**, *47*, 667–677. [CrossRef] [PubMed]
45. Griffete, N.; Herbst, F.; Pinson, J.; Ammar, S.; Mangeney, C. Preparation of Water-Soluble Magnetic Nanocrystals Using Aryl Diazonium Salt Chemistry. *J. Am. Chem. Soc.* **2011**, *133*, 1646–1649. [CrossRef]
46. Charlotte Boitard, C.; Lamouri, M.C.; Griffete, N. Whole Protein Imprinting over Magnetic Nanoparticles Using Photo-Polymerization. *ACS Appl. Polym. Mater.* **2019**, *1*, 928–932. [CrossRef]
47. Sanders, S.; Golden, T.D. Functionalization of Cerium Oxide Nanoparticles to Influence Hydrophobic Properties. *Langmuir* **2019**, *35*, 5841–5847. [CrossRef]
48. Mousli, F.; Chaouchi, A.; Hocine, S.; Lamouri, A.; Rei Vilar, M.; Kadri, A.; Chehimi, M.M. Diazonium-modified TiO_2/polyaniline core/shell nanoparticles. Structural characterization, interfacial aspects and photocatalytic performances. *Appl. Surf. Sci.* **2019**, *465*, 1078–1095. [CrossRef]
49. Jasmin, J.-P.; Cannizzo, C.; Dumas, E.; Chaussé, A. Fabrication and characterization of all-covalent nanocomposite functionalized screen-printed voltammetric sensors. *Electrochim. Acta* **2014**, *133*, 467–474. [CrossRef]
50. Mooste, M.; Kibena-Põldsepp, E.; Diby Ossonon, B.; Bélanger, D.; Tammeveski, K. Oxygen reduction on graphene sheets functionalised by anthraquinone diazonium compound during electrochemical exfoliation of graphite. *Electrochim. Acta* **2018**, *267*, 246–254. [CrossRef]
51. Laurentius, L.; Stoyanov, S.R.; Gusarov, S.; Kovalenko, A.; Du, R.; Lopinski, G.P.; McDermott, M.T. Diazonium-derived aryl films on gold nanoparticles: Evidence for a carbon gold covalent bond. *ACS Nano* **2011**, *5*, 4219–4227. [CrossRef]
52. Sampathkumar, K.; Diez-Cabanes, V.; Kovaricek, P.; del Corro, E.; Bouša, M.; Hošek, J.; Kalbac, M.; Frank, O. On the Suitability of Raman Spectroscopy to Monitor the Degree of Graphene Functionalization by Diazonium Salts. *J. Phys. Chem. C* **2019**, *123*, 22397–22402. [CrossRef]
53. Berger, F.J.; Luttgens, J.; Nowack, T.; Kutsch, T.; Lindenthal, S.; Kistner, L.; Muller, C.C.; Bongartz, L.M.; Lumsargis, V.A.; Zakharko, Y.; et al. Brightening of Long, Polymer—Wrapped Carbon Nanotubes by sp^3 Functionalization in Organic Solvents. *ACS Nano* **2019**, *13*, 9259–9269. [CrossRef] [PubMed]
54. Wepasnick, K.A.; Smith, B.A.; Bitter, J.L.; Fairbrother, D.H. Chemical and structural characterization of carbon nanotube surfaces. *Anal. Bioanal. Chem.* **2010**, *396*, 1003–1014. [CrossRef] [PubMed]
55. Van Dyck, C.; Bergren, A.J.; Mukundan, V.; Fereiro, J.A.; DiLabio, G.A. Extent of conjugation in diazonium-derived layers in molecular junction devices determined by experiment and modelling. *Phys. Chem. Chem. Phys.* **2019**, *21*, 16762–16770. [CrossRef]
56. Descroix, S.; Hallais, G.; Corinne Lagrost, C.; Pinson, J. Regular poly (para-phenylene) films bound to gold surfaces through the electrochemical reduction of diazonium salts followed by electropolymerization in an ionic liquid. *Electrochim. Acta* **2013**, *106*, 172–180. [CrossRef]
57. Greenwood, J.; Phan, T.H.; Fujita, Y.; Li, Z.; Ivasenko, O.; Vanderlinden, W.; Van Gorp, H.; Frederickx, W.; Lu, G.; Tahara, K.; et al. Covalent Modification of Graphene and Graphite Using Diazonium Chemistry: Tunable Grafting and Nanomanipulation. *ACS Nano* **2015**, *9*, 5520–5535. [CrossRef]
58. Anariba, F.; DuVall, S.H.; McCreery, R.L. Mono- and Multilayer Formation by Diazonium Reduction on Carbon Surfaces Monitored with Atomic Force Microscopy "Scratching". *Anal. Chem.* **2003**, *75*, 3837–3844. [CrossRef]
59. Ceccato, M.; Bousquet, A.; Hinge, M.; Pedersen, S.U.; Daasbjerg, K. Using a Mediating Effect in the Electroreduction of Aryldiazonium Salts To Prepare Conducting Organic Films of High Thickness. *Chem. Mater.* **2011**, *23*, 1551–1557. [CrossRef]
60. Combellas, C.; Kanoufi, F.; Pinson, J.; Podvorica, F.I. Sterically Hindered Diazonium Salts for the Grafting of a Monolayer on Metals. *J. Am. Chem. Soc.* **2008**, *130*, 8576–8577. [CrossRef]
61. Leroux, Y.R.; Fei, H.; Noël, J.-M.; Roux, C.; Hapiot, P. Efficient covalent modification of a carbon surface: Use of a silyl protecting group to form an active monolayer. *J. Am. Chem. Soc.* **2010**, *132*, 14039–14041. [CrossRef]
62. Pichereau, L.; López, I.; Cesbron, M.; Dabos-Seignon, S.; Gautier, C.; Breton, T. Controlled diazonium electrografting driven by overpotential reduction: A general strategy to prepare ultrathin layers. *Chem. Commun.* **2019**, *55*, 455–457. [CrossRef]

63. Xia, Z.; Leonardi, F.; Gobbi, M.; Liu, Y.; Bellani, V.; Liscio, A.; Kovtun, A.; Li, R.; Feng, X.; Orgiu, E.; et al. Electrochemical Functionalization of Graphene at the Nanoscale with Self-Assembling Diazonium Salts. *ACS Nano* **2016**, *10*, 7125–7134. [CrossRef] [PubMed]
64. Dincer, C.; Bruch, R.; Costa-Rama, E.; Fernández-Abedul, M.T.; Merkoçi, A.; Manz, A.; Urban, G.A.; Güder, F. Disposable Sensors in Diagnostics, Food, and Environmental Monitoring. *Adv. Mater.* **2019**, *31*, 1806739. [CrossRef] [PubMed]
65. Gooding, J.J. Advances in interfacial design sensors: Aryl diazonium salts for electrochemical biosensors and for modifying carbon and metal electrodes. *Electroanalysis* **2008**, *2*, 573–582. [CrossRef]
66. Mahouche-Chergui, S.; Gam-Derouich, S.; Mangeney, C.; Chehimi, M.M. Aryl diazonium salts: A new class of coupling agents for bonding polymers, biomacromolecules and nanoparticles to surfaces. *Chem. Soc. Rev.* **2011**, *40*, 4143–4166. [CrossRef] [PubMed]
67. Zhou, Y.; Chiu, C.-W.; Liang, H. Interfacial Structures and Properties of Organic Materials for Biosensors: An Overview. *Sensors* **2012**, *12*, 15036–15062. [CrossRef] [PubMed]
68. Abdellaoui, S.; Corgier, B.C.; Mandon, C.A.; Doumèche, B.; Marquette, C.A.; Blum, L.J. Biomolecules Immobilization Using the Aryl Diazonium Electrografting. *Electroanalysis* **2013**, *25*, 671–684. [CrossRef]
69. Cao, C.; Zhang, Y.; Jiang, C.; Qi, M.; Liu, G. Advances on Aryldiazonium Salt Chemistry Based Interfacial Fabrication for Sensing Applications. *ACS Appl. Mater. Interfaces* **2017**, *9*, 5031–5049. [CrossRef]
70. Yáñez-Sedeño, P.; Campuzano, S.; Pingarrón, J.M. Integrated Affinity Biosensing Platforms on Screen-Printed Electrodes Electrografted with Diazonium Salts. *Sensors* **2018**, *18*, 675. [CrossRef]
71. Yates, N.D.J.; Fascione, M.A.; Parkin, A. Methodologies for "Wiring" Redox Proteins/Enzymes to Electrode Surfaces. *Chem. A Eur. J.* **2018**, *24*, 12164–12182. [CrossRef]
72. Yáñez-Sedeño, P.; González-Cortés, A.; Campuzano, S.; Pingarrón, J.M. Copper(I)-Catalyzed Click Chemistry as a Tool for the Functionalization of Nanomaterials and the Preparation of Electrochemical (Bio)Sensors. *Sensors* **2019**, *19*, 2379. [CrossRef]
73. Bourdillon, C.; Delamar, M.; Demaille, C.; Hitmi, R.; Moiroux, J.; Pinson, J. Immobilization of glucose oxidase on a carbon surface derivatized by electrochemical reduction of diazonium salts. *J. Electroanal. Chem.* **1992**, *336*, 113–123. [CrossRef]
74. Yang, X.; Hall, S.B.; Tan, S.N. Electrochemical Reduction of a Conjugated Cinnamic Acid Diazonium Salt as an Immobilization Matrix for Glucose Biosensor. *Electroanalysis* **2003**, *15*, 885–891. [CrossRef]
75. Liu, G.; Paddon-Row, M.N.; Gooding, J.J. A molecular wire modified glassy carbon electrode for achieving direct electron transfer to native glucose oxidase. *Electrochem. Commun.* **2007**, *9*, 92218–92223. [CrossRef]
76. Qi, M.; Zhang, Y.; Cao, C.; Luc, Y.; Liu, G. Increased sensitivity of extracellular glucose monitoring based on AuNP decorated GO nanocomposites. *RSC Adv.* **2016**, *6*, 39180–39186. [CrossRef]
77. Ferreyra, N.; Coche-Guérente, L.; Labbé, P. Construction of layer-by-layer self-assemblies of glucose oxidase and cationic polyelectrolyte onto glassy carbon electrodes and electrochemical study of the redox-mediated enzymatic activity. *Electrochim. Acta* **2004**, *49*, 477–484. [CrossRef]
78. Zhao, K.; Zhuang, S.; Chang, Z.; Songm, H.; Dai, L.; He, P.; Fanga, Y. Amperometric Glucose Biosensor Based on Platinum Nanoparticles Combined Aligned Carbon Nanotubes Electrode. *Electroanalysis* **2007**, *19*, 1069–1074. [CrossRef]
79. Le Goff, A.; Moggia, F.; Debou, N.; Jegou, P.; Artero, V.; Fontecave, M.; Jousselme, B.; Palacin, S. Facile and tunable functionalization of carbon nanotube electrodes with ferrocene by covalent coupling and p-stacking interactions and their relevance to glucose bio-sensing. *J. Electroanal. Chem.* **2010**, *641*, 57–63. [CrossRef]
80. Nasri, A.; Shams, E.; Ahmadi, M. A glucose biosensor based on direct attachment of in situ generated Nile blue diazonium cations to the electrode surface. *J. Electroanal. Chem.* **2013**, *703*, 146–152. [CrossRef]
81. Sharma, D.; Lim, Y.; Lee, Y.; Shin, H. Glucose sensor based on redox-cycling between selectively modified and unmodified combs of carbon interdigitated array nanoelectrodes. *Anal. Chim. Acta* **2015**, *889*, 194–202. [CrossRef]
82. Raicopol, M.D.; Andronescu, C.; Atasiei, R.; Hanganu, A.; Vasile, E.; Brezoiu, A.M.; Pilan, L. Organic layers via aryl diazonium electrochemistry: Towards modifying platinum electrodes for interference free glucose biosensors. *Electrochim. Acta* **2016**, *206*, 226–237. [CrossRef]
83. Li, M.; Cui, L.; Niu, F.; Ji, X.; Xu, Y.; Liu, J. Efficient and Facile Fabrication of Glucose Biosensor Based on Electrochemically Etched Porous HOPG platform. *Electroanalysis* **2017**, *29*, 944–949. [CrossRef]

84. Morita, K.; Hirayama, N.; Imura, H.; Yamaguchi, A.; Teramae, N. Grafting of phenylboronic acid on a glassy carbon electrode and its application as a reagentless glucose sensor. *J. Electroanal. Chem.* **2011**, *656*, 192–197. [CrossRef]
85. Shervedani, R.K.; Amini, A.; Sadeghi, N. Electrografting of thionine diazonium cationon to the graphene edges and decorating with Au nano-dendrites or glucoseoxidase: Characterization and electrocatalytic applications. *Biosens. Bioelectron.* **2016**, *77*, 478–485. [CrossRef] [PubMed]
86. Ott, C.; Raicopol, M.D.; Andronescu, C.; Vasile, E.; Hanganue, A.; Prunaf, A.; Pilan, L. Functionalized polypyrrole/sulfonated graphene nanocomposites: Improved biosensing platforms through aryl diazonium electrochemistry. *Synth. Met.* **2018**, *235*, 20–28. [CrossRef]
87. Harper, J.C.; Polsky, R.; Dirk, S.M.; Wheeler, D.R.; Brozik, S.M. Electroaddressable Selective Functionalization of Electrode Arrays: Catalytic NADH Detection Using Aryl Diazonium Modified Gold Electrodes. *Electroanalysis* **2007**, *19*, 1268–1274. [CrossRef]
88. Nasri, Z.; Shams, E.; Ahmadi, M. Direct Modification of a Glassy Carbon Electrode with ToluidineBlue Diazonium Salt: Application to NADH Determination and Biosensing of Ethanol. *Electroanalysis* **2007**, *19*, 1268–1274.
89. Venarussoa, L.B.; Tammeveski, K.; Maia, G. Versatile charge transfer through anthraquinone films for electrochemical sensing applications. *Electrochim. Acta* **2011**, *56*, 8926–8933. [CrossRef]
90. Revenga-Parra, M.; Gomez-Anquela, C.; Garcia-Mendiola, T.; Gonzalez, E.; Parientea, F.; Lorenzo, E. Grafted Azure A modified electrodes as disposable β-nicotinamide adenine dinucleotide sensors. *Anal. Chim. Acta* **2012**, *747*, 84–91. [CrossRef]
91. Kesavan, S.; Revin, S.B.; John, S.A. Potentiodynamic formation of gold nanoparticles film on glassy carbon electrode using aminophenyl diazonium cations grafted gold nanoparticles: Determination of histamine H2 receptor antagonist. *Electrochim. Acta* **2014**, *119*, 214–224. [CrossRef]
92. Alarfaj, N.A.; El-Tohamy, M.F. A label-free electrochemical immunosensor based on gold nanoparticles and graphene oxide for the detection of tumor marker calcitonin. *New J. Chem.* **2017**, *41*, 11029–11035. [CrossRef]
93. William Richard, W.; Evrard, D.; Gros, P. A Novel Electrochemical Sensor Based on a Mixed Diazonium/PEDOT Surface Functionalization for the Simultaneous Assay of Ascorbic and Uric Acids. Towards an Improvement in Amperometric Response Stability. *Electroanalysis* **2014**, *26*, 1390–1399. [CrossRef]
94. Assaud, L.; Massonnet, N.; Evrard, D.; Vergnes, H.; Salvagnac, L.; Conédéra, V.; Noé, L.; Monthioux, M.; Gros, P.; Temple-Boyer, P.; et al. A new route for the integration of a graphene/diazonium/PEDOT electrode towards antioxidant biomarker detection. *J. Electroanal. Chem.* **2016**, *771*, 73–79. [CrossRef]
95. Himori, S.; Nishitani, S.; Sakata, T. Control of Potential Response to Small Biomolecules with Electrochemically Grafted Aryl-Based Monolayer in Field-Effect Transistor-Based Sensors. *Langmuir* **2019**, *35*, 3701–3709. [CrossRef] [PubMed]
96. Rather, J.A.; Khudaish, E.A.; Kannan, P. Graphene-amplified femtosensitive aptasensing of estradiol, an endocrine disruptor. *Analyst* **2018**, *143*, 1835–1845. [CrossRef]
97. Istamboulié, G.; Paniel, N.; Zara, L.; Reguillo Granados, L.; Barthelmebs, L.; Noguer, T. Development of an impedimetric aptasensor for the determination of aflatoxin M1 in milk. *Talanta* **2016**, *146*, 464–469. [CrossRef]
98. Hayat, A.; Barthelmebs, L.; Marty, J.-L. Electrochemical impedimetric immunosensor for the detection of okadaic acid in mussel sample. *Sens. Actuators B* **2012**, *171*, 810–815. [CrossRef]
99. Eissa, S.; Zourob, M. A graphene-based electrochemical competitive immunosensor for the sensitive detection of okadaic acid in shellfish. *Nanoscale* **2012**, *4*, 7593–7599. [CrossRef]
100. Sánchez-Tirado, E.; Arellano, L.M.; González-Cortés, A.; Yáñez-Sedeño, P.; Langa, F.; Pingarrón, J.M. Viologen-functionalized single-walled carbon nanotubes as carrier nanotags for electrochemical immunosensing. Application to TGF-β1 cytokine. *Biosens. Bioelectron.* **2017**, *98*, 240–247. [CrossRef]
101. Chrouda, A.; Sbartai, A.; Bessueille, F.; Renaud, L.; Maaref, A.; Jaffrezic-Renault, N. Electrically addressable deposition of diazonium functionalized antibodies on boron-doped diamond microcells for the detection of ochratoxin A. *Anal. Methods* **2015**, *7*, 2444–2451. [CrossRef]
102. Alonso-Lomillo, M.A.; Domínguez-Renedo, O.; Matos, P.; Arcos-Martín, M.J. Disposable biosensors for determination of biogenic amines. *Anal. Chim. Acta* **2010**, *665*, 26–31. [CrossRef]
103. Corgier, B.P.; Marquette, C.A.; Blum, L.J. Diazonium-Protein Adducts for Graphite Electrode Microarrays Modification: Direct and Addressed Electrochemical Immobilization. *J. Am. Chem. Soc.* **2005**, *127*, 18328–18332. [CrossRef] [PubMed]

104. Corgier, B.P.; Laurent, A.; Perriat, P.; Blum, L.J.; Marquette, C.A. A Versatile Method for Direct and Covalent Immobilization of DNA and Proteins on Biochips. *Angew. Chem. Int. Ed.* **2007**, *46*, 4108–4110. [CrossRef]
105. Corgier, B.P.; Li, F.; Blum, L.J.; Marquette, C.A. Direct electrochemical addressing of immunoglobulins: Immuno-chip on screen-printed microarray. *Biosens. Bioelectron.* **2007**, *22*, 1522–1526. [CrossRef] [PubMed]
106. Marquette, C.A.; Bouteille, F.; Corgier, B.P.; Degiuli, A.; Blum, L.J. Disposable screen-printed chemiluminescent biochips for the simultaneous determination of four point-of-care relevant proteins. *Anal. Bioanal. Chem.* **2009**, *393*, 1191–1198. [CrossRef] [PubMed]
107. Corgier, B.P.; Li, F.; Blum, L.J.; Marquette, C.A. On-Chip Chemiluminescent Signal Enhancement Using Nanostructured Gold-Modified Carbon Microarrays. *Langmuir* **2007**, *23*, 8619–8623. [CrossRef] [PubMed]
108. Corgier, B.P.; Bellon, S.; Anger-Leroy, M.; Blum, L.J.; Marquette, C.A. Protein-Diazonium Adduct Direct Electrografting onto SPRi-Biochip. *Langmuir* **2009**, *25*, 9619–9623. [CrossRef]
109. Polsky, R.; Harper, J.C.; Wheeler, D.R.; Dirk, S.M.; Arango, D.C.; Brozik, S.M. Electrically addressable diazonium-functionalized antibodies for multianalyte electrochemical sensor applications. *Biosens. Bioelectron.* **2008**, *23*, 757–764. [CrossRef]
110. Bertok, T.; Lorencova, L.; Hroncekova, S.; Gajdosova, V.; Jane, E.; Hires, M.; Kasak, P.; Kaman, O.; Sokol, R.; Bella, V.; et al. Advanced impedimetric biosensor configuration and assay protocol for glycoprofiling of a prostate oncomarker using Au nanoshells with a magnetic core. *Biosens. Bioelectron.* **2019**, *131*, 24–29. [CrossRef]
111. Mandon, C.A.; Blum, L.J.; Marquette, C.A. Aryl Diazonium for Biomolecules Immobilization onto SPRi Chips. *ChemPhysChem* **2009**, *10*, 3273–3277. [CrossRef]
112. Goldsmith, B.R.; Mitala, J.J., Jr.; Josue, J.; Castro, A.; Lerner, M.B.; Bayburt, T.H.; Khamis, S.M.; Jones, R.A.; Brand, J.G.; Sligar, S.G.; et al. Biomimetic Chemical Sensors Using Nanoelectronic Readout of Olfactory Receptor Proteins. *ACS Nano* **2011**, *5*, 5408–5416. [CrossRef]
113. Wang, R.; Xue, C. A sensitive electrochemical immunosensor for alpha-fetoprotein based on covalently incorporating a biorecognition element onto a graphene modified electrode via diazonium chemistry. *Anal. Methods* **2013**, *5*, 5195–5200. [CrossRef]
114. Ocaña, C.; Hayat, A.; Mishra, R.; Vasilescu, A.; del Vallea, M.; Marty, J.-L. A novel electrochemical aptamer–antibody sandwich assay for lysozyme detection. *Analyst* **2015**, *140*, 4148–4153. [CrossRef] [PubMed]
115. Adabo, A.H.; Zeggari, R.; Saïd, N.M.; Bazzi, R.; Elie-Caille, C.; Marquette, C.; Martini, M.; Tillement, O.; Perriat, P.; Chaix, C.; et al. Enhanced chemiluminescence-based detection on gold substrate after electrografting of diazonium precursor-coated gold nanoparticles. *J. Coll. Interface Sci.* **2016**, *467*, 271–279. [CrossRef] [PubMed]
116. Qi, M.; Zhang, Y.; Cao, C.; Zhang, M.; Liu, S.; Liu, G. Decoration of Reduced Graphene Oxide Nanosheets with Aryldiazonium Salts and Gold Nanoparticles toward a Label-Free Amperometric Immunosensor for Detecting Cytokine Tumor Necrosis Factor-α in Live Cells. *Anal. Chem.* **2016**, *88*, 9614–9621. [CrossRef]
117. Jiang, C.; Alam, M.T.; Moraes Silva, S.; Taufik, S.; Fan, S.; Gooding, J.J. Unique Sensing Interface That Allows the Development of an Electrochemical Immunosensor for the Detection of Tumor Necrosis Factor α in Whole Blood. *ACS Sens.* **2016**, *1*, 1432–1438. [CrossRef]
118. Fioresi, F.; Rouleau, A.; Maximova, K.; Vieillard, J.; Boireau, W.; Elie Caille, C.; Soulignac, C.; Zeggarib, R.; Clamens, T.; Lesouhaitier, O.; et al. Electrografting of diazonium salt for SPR application. *Mater. Today Proc.* **2019**, *6*, 340–344. [CrossRef]
119. Li, N.; Chow, A.M.; Ganesh, H.V.S.; Ratnam, M.; Brown, I.R.; Kerman, K. Diazonium-Modified Screen-Printed Electrodes for Immunosensing Growth Hormone in Blood Samples. *Biosensors* **2019**, *9*, 88. [CrossRef]
120. van den Hurk, R.; Baghelani, M.; Chen, J.; Daneshmand, M.; Evoy, S. Al-Mo nanocomposite functionalization for membrane-based resonance detection of bovine Herpesvirus-1. *Sens. Actuators A* **2019**, *296*, 186–191. [CrossRef]
121. Anjum, S.; Qi, W.; Gao, W.; Zhao, J.; Hanif, S.; Rehman, A.-U.; Xu, G. Fabrication of biomembrane-like films on carbon electrodes using alkanethiol and diazonium salt and their application for direct electrochemistry of myoglobin. *Biosens. Bioelectron.* **2015**, *65*, 159–165. [CrossRef]
122. Chung, D.-J.; Oh, S.-H.; Komathi, S.; Gopalan, A.I.; Lee, K.-P.; Choi, S.-H. One-step modification of various electrode surfaces using diazonium salt compounds and the application of this technology to electrochemical DNA (E-DNA) sensors. *Electrochim. Acta* **2012**, *76*, 394–403. [CrossRef]

123. Baker, S.E.; Tse, K.-Y.; Hindin, E.; Nichols, B.M.; Clare, T.L.; Robert, J.; Hamers, R.J. Covalent Functionalization for Biomolecular Recognition on Vertically Aligned Carbon Nanofibers. *Chem. Mater.* **2005**, *17*, 4971–4978. [CrossRef]
124. Nebel, C.E.; Yang, N.; Uetsuka, H.; Osawa, E.; Tokuda, N.; Williams, O. Diamond nano-wires, a new approach towards next generation electrochemical gene sensor platforms. *Diam. Relat. Mater.* **2009**, *18*, 910–917. [CrossRef]
125. Hai, L.V.; Reisberg, S.; Chevillot-Biraud, A.; Noël, V.; Pham, M.C.; Piro, B. Simultaneous Electroreduction of Different Diazonium Salts for Direct Electrochemical DNA Biosensor Development. *Electrochim. Acta* **2014**, *140*, 49–58. [CrossRef]
126. Bartolome, J.P.; Echegoyen, L.; Fragoso, A. Reactive Carbon Nano-Onion Modified Glassy Carbon Surfaces as DNA Sensors for Human Papillomavirus Oncogene Detection with Enhanced Sensitivity. *Anal. Chem.* **2015**, *87*, 6744–6751. [CrossRef]
127. Civit, L.; Fragoso, A.; O'Sullivan, C.K. Thermal stability of diazonium derived and thiol-derived layers on gold for application in genosensors. *Electrochem. Commun.* **2010**, *12*, 1045–1048. [CrossRef]
128. Yang, L.; Xu, Y.; Wang, X.; Zhu, J.; Zhang, R.; He, P.; Fang, Y. The application of cyclodextrin derivative functionalized aligned carbon nanotubes for electrochemically DNA sensing via host–guest recognition. *Anal. Chim. Acta* **2011**, *689*, 39–46. [CrossRef]
129. Revenga-Parra, M.; Garcia-Mendiola, T.; Gonzalez-Costas, J.; Gonzalez-Romero, E.; Garcia Marin, A.; Pau, J.L.; Pariente, F.; Lorenzo, E. Simple diazonium chemistry to develop specific gene sensing platforms. *Anal. Chim. Acta* **2014**, *813*, 41–47. [CrossRef]
130. Kuo, T.M.; Shen, M.Y.; Huang, S.Y.; Li, Y.K.; Chuang, M.C. Facile Fabrication of a Sensor with a Bifunctional Interface for Logic Analysis of the New Delhi Metallo-β-Lactamase (NDM)-Coding Gene. *ACS Sens.* **2016**, *1*, 124–130. [CrossRef]
131. Ge, L.; Wang, W.; Li, F. Electro-Grafted Electrode with Graphene-Oxide-Like DNA Affinity for Ratiometric Homogeneous Electrochemical Biosensing of MicroRNA. *Anal. Chem.* **2017**, *89*, 11560–11567. [CrossRef]
132. Mousavisania, S.Z.; Raoof, J.-B.; Turner, A.P.F.; Ojani, R.; Mak, W.C. Label-free DNA sensor based on diazonium immobilisation for detection of DNA damage in breast cancer 1 gene. *Sens. Actuators B* **2018**, *264*, 59–66. [CrossRef]
133. Guselnikova, O.; Dvorankova, B.; Kakisheva, K.; Kalachyova, Y.; Postnikov, P.; Svorcik, V.; Lyutakov, O. Rapid SERS-based recognition of cell secretome on the folic acid-functionalized gold gratings. *Anal. Bioanal. Chem.* **2019**, *411*, 3309–3319. [CrossRef] [PubMed]

biosensors

MDPI

Article

Diazonium-Modified Screen-Printed Electrodes for Immunosensing Growth Hormone in Blood Samples

Nan Li [1], Ari M. Chow [1,2], Hashwin V. S. Ganesh [1], Melanie Ratnam [2], Ian R. Brown [2] and Kagan Kerman [1,*]

1 Department of Physical and Environmental Sciences, University of Toronto Scarborough, Toronto, ON M1C 1A4, Canada
2 Centre for the Neurobiology of Stress, Department of Biological Sciences, University of Toronto Scarborough, Toronto, ON M1C 1A4, Canada
* Correspondence: kagan.kerman@utoronto.ca

Received: 1 June 2019; Accepted: 15 July 2019; Published: 17 July 2019

check for updates

Abstract: Altered growth hormone (GH) levels represent a major global health challenge that would benefit from advances in screening methods that are rapid and low cost. Here, we present a miniaturized immunosensor using disposable screen-printed carbon electrodes (SPCEs) for the detection of GH with high sensitivity. The diazonium-based linker layer was electrochemically deposited onto SPCE surfaces, and subsequently activated using covalent agents to immobilize monoclonal anti-GH antibodies as the sensing layer. The surface modifications were monitored using contact angle measurements and X-ray photoelectron spectroscopy (XPS). The dissociation constant, K_d, of the anti-GH antibodies was also determined as 1.44 (±0.15) using surface plasmon resonance (SPR). The immunosensor was able to detect GH in the picomolar range using a 20 μL sample volume in connection with electrochemical impedance spectroscopy (EIS). The selectivity of the SPCE-based immunosensors was also challenged with whole blood and serum samples collected at various development stages of rats, demonstrating the potential applicability for detection in biological samples. Our results demonstrated that SPCEs provided the development of low-cost and single-use electrochemical immunosensors in comparison with glassy carbon electrode (GCE)-based ones.

Keywords: immunosensor; electrochemical impedance spectroscopy; growth hormone; real samples; diazonium grafting

1. Introduction

In an immunosensor, the integration of the recognition element with the signal transducer is usually achieved by a chemical layer modification that enables the immobilization of antibodies. Aryl diazonium salts have become increasingly popular given their ease of use for modifying a wide variety of surfaces and their stability as a chemical linker [1–4]. Eissa et al. [5] reported the electrochemical modification of graphene-modified screen-printed carbon electrodes (SPCEs) with 4-nitrophenyl diazonium salt, which enabled the covalent attachment of antibodies for the detection of a milk allergen, β-lactoglobulin. Eissa and Zourob [6] have also reported the development of an electrochemical competitive immunosensor for the detection of okadaic acid in shellfish. Graphene modified SPCEs were functionalized by the electrochemical reduction of in situ generated 4-carboxyphenyl diazonium salt in acidic aqueous solution [6]. The sensitive detection of egg allergen ovalbumin was also achieved with a detection limit of 0.83 pg/mL in phosphate buffer solution (PBS) using graphene modified SPCEs with a carboxyphenyl film on the graphene surface [7]. SPCEs can be mass-produced at low-cost, and each experiment can be performed on a fresh and analogous surface to prevent possible cross-contamination issues [8]. Each SPCE can be disposed after use and requires small volumes of reagents for measurement. Furthermore, current advances in instrumentation have allowed SPCEs to become compatible with cellphone-size portable devices for convenient on-site measurements [9].

GH is a peptide hormone with two isoforms (22 kDa and 20 kDa) released from the somatotroph cells in the anterior pituitary gland, serving multiple functions in many tissue targets [10]. GH plays a critical role in the regulation of blood glucose, longitudinal bone growth, and enhancement of muscle mass [11]. Deficiencies in GH can have severe developmental consequences, such as hypoglycemia in newborns, stunted growth in childhood, and physical/psychological symptoms well into adulthood [12,13]. On the other hand, over expression of GH can result in gigantism, acromegaly, and impaired glucose tolerance [14–16]. Beyond a clinical setting, GH doping is observed extensively in professional athletes to enhance overall tissue maintenance, repair, and muscle growth [17]. The use of synthetic GH as an ergogenic drug has become increasingly prevalent, which resulted in its ban from professional athletes by the World Anti-Doping Agency (WADA) [18]. To face these major global health challenges, advances in GH screening are required to facilitate early detection and treatment [19]. At present, reported GH screening methods include enzyme-linked immunosorbent assays (ELISA) [20], radioimmunoassays (RIA) [21], mass spectrometry [22], and surface plasmon resonance (SPR) [23]. Current issues concerning these methods include sensitivity, specificity, cost and time. There is an urgent need for an ultrasensitive screening procedure for the detection of GH that is rapid and low cost [24–30]. Our immunosensor provides an economical approach to a miniaturized electrochemical system utilizing SPCEs. As a model system, we modified SPCEs with anti-GH antibodies using diazonium grafting and compared the analytical characteristics of SPCE-based immunosensors with similarly prepared GCE (glassy carbon electrode)-based ones. Then, we demonstrated the detection of GH in blood and serum samples using SPCE-based immunosensors in comparison with a commercially available kit.

2. Materials and Methods

2.1. Chemicals and Reagents

Rat growth hormone (GH), full-length protein (ab68388) and monoclonal mouse anti-growth hormone (GH) antibody (ab9821) were obtained from Sigma-Aldrich (Oakville, ON, Canada). Potassium dihydrogen orthophosphate (KH_2PO_4), dipotassium orthophosphate (K_2HPO_4), 1-ethyl-3-(3-dimethylaminopropyl) carbodiimide hydrochloride (EDC), N-hydroxysuccinamide (NHS), potassium ferricyanide ($K_4Fe(CN)_6$), potassium ferrocyanide ($K_3Fe(CN)_6$), 4-methoxybenzenediazonium tetrafluroborate, acetonitrile (HPLC grade), and bovine serum albumin (BSA) were obtained from Sigma-Aldrich (Oakville, ON, Canada). Tetraethylammonium tetrafluoroborate was purchased from Alfa Aesar (Mississauga, ON, Canada). All solutions were prepared using ultra-pure water from a Cascada LS water purification system (Pall Co., Port Washington, NY, USA) at 18.2 MΩ. Male Sprague Dawley rats (Charles River, MA, USA) were anesthetized with isoflurane and then maintained under isofluorane by a vaporizer with a nose cone attachment. Blood was drawn by cardiac puncture in the right ventricle. The blood samples were subsequently analyzed using ELISA and our electrochemical immunosensor for growth hormone levels. For control studies, the GH sample concentration used for the blood and plasma was 100 pg/mL. An undiluted sample (also coined as blood stock or plasma stock) indicates that the blood or plasma sample was used as received without any further dilutions using buffer solutions. All procedures using animals were approved by the Animal Care Committee of the University of Toronto and were in accordance with the guidelines and codes established by the Canadian Council on Animal Care.

2.2. Electrode Preparation and Modification

A glassy carbon electrode (GCE, CH Instruments, Austin, TX, USA) was polished for 2 min using 1.0, 0.3, and 0.05 μm alumina. The electrode was rinsed with ultrapure water and sonicated for 5 min between each polishing step to remove any alumina present on the surface. Screen-printed carbon electrodes (SPCEs, DEP-Chip EP-N) were purchased from BioDevice Technology Ltd. (Ishikawa, Japan). As shown in Figure 1B, SPCEs had a carbon ink-based working electrode surface that enabled working with small volumes of reagents and samples (1–20 μL). The counter and reference electrodes were printed using carbon and silver ink, respectively. The electrical contacts were protected from sample solutions with a hydrophobic coating as shown in the inset of Figure 1B.

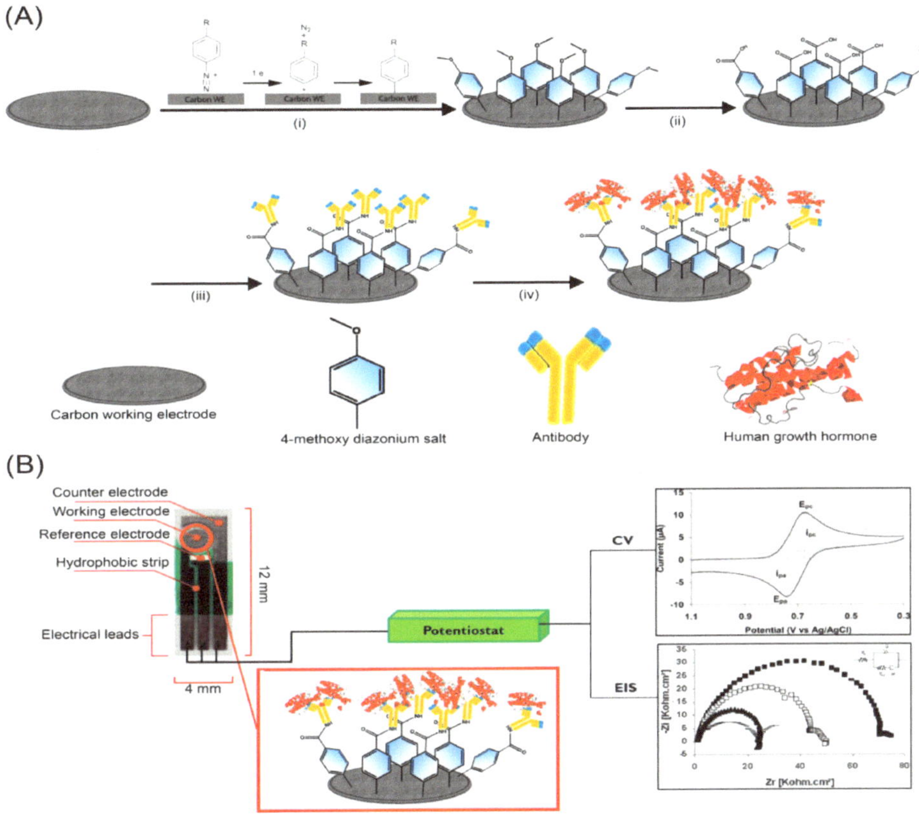

Figure 1. Conceptual illustration of the impedimetric detection of rat growth hormone (GH). (**A**) A film of 4-methoxybenzenediazonium tetrafluroborate (4-MBD) was immobilized on an electrode surface by electrodeposition (i); surface-confined 4-MBD molecules were electrochemically oxidized to carboxylic acid groups (ii) that were activated with 1-ethyl-3-(3-dimethyl-aminopropyl) carbodiimide (EDC) and N-hydroxysuccinimide (NHS) to allow subsequent covalent immobilization of antibodies via lysine residues (iii); GH is captured with antibodies on the surface (iv). (**B**) screen-printed carbon electrodes (SPCE) of dimension 4 × 12 mm in length with carbon ink-based working and counter electrodes in connection with a silver ink-based reference electrode. Electrochemical impedance spectroscopy (EIS) and cyclic voltammetry (CV) were employed to characterize SPCE and glassy carbon electrode (GCE) surfaces.

The surface modification of GCEs and SPCEs was performed following the same procedures of electrodeposition. Briefly, aryl diazonium salt (1 mM) in acetonitrile was prepared and de-aerated for 15 min. Electrodeposition was performed using tetraethylammonium tetrafluoroborate as the electrolyte with CV. A scan rate of 100 mV/s was applied for two cycles between +1.0 V and −1.0 V, and then rinsed with copious amounts of electrolyte and water [1,2]. Next, the electrode was exposed to a biasing potential of +1.0 V for 1 min in 50 mM KH_2PO_4/K_2HPO_4 (PBS) at pH 7.4. The activation of exposed carboxylic acid groups on the surface was accomplished using 10 mM EDC and 40 mM NHS for 1 h and rinsed with PBS [4]. The antibodies were incubated on the electrode surface, to allow covalent attachment between the free amine groups of lysine residues in the Fc region and the carboxyl groups on the diazonium grafted surfaces for 18 h at 4 °C, then rinsed with PBS. GH sample solutions were prepared at various concentrations in PBS and used as the target analyte. An aliquot (10 μL) of each sample solution (real samples or PBS spiked with the desired amount of GH) was added to the electrode surface and incubated for 30 min at room temperature with moderate shaking, to ensure contact of the GH with the surface-confined antibodies.

2.3. Electrochemistry

Cyclic voltammetry (CV) and electrochemical impedance spectroscopy (EIS) were performed using a μAutolabIII Electrochemical Analyzer (Metrohm, Utrecht, The Netherlands) in conjunction with its general-purpose electrochemistry software and a frequency response analyzer (Metrohm, Utrecht, The Netherlands). CV and EIS were measured with 10 mM $[Fe(CN)_6]^{3-/4-}$ in PBS with 100 mM KCl at room temperature using a three-electrode system with GCE as the working electrode, a Pt wire as the counter electrode, an Ag/AgCl reference electrode and also SPCE as depicted in the inset of Figure 2. CV measurements were performed before and after aryl diazonium salt modification at a scan rate of 100 mV/s between −0.5 V and +0.5 V. EIS was performed with a frequency range from 100 kHz to 100 mHz at a biasing voltage of 0.20 V.

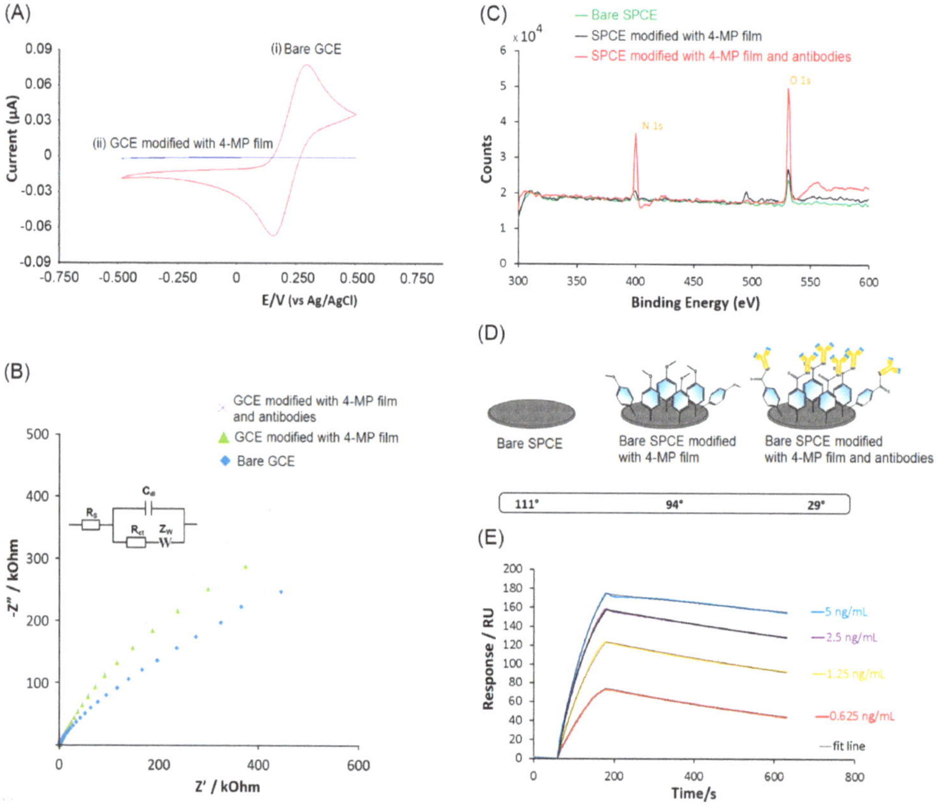

Figure 2. Surface characterization of electrode surface modifications. (**A**) Cyclic voltammogram of (i) bare glassy carbon electrode (GCE) and (ii) 4-methoxyphenyl (4-MP) film modified GCE using 10 mM $[Fe(CN)_6]^{3-/4-}$ in PBS with 100 mM KCl at 100 mV/s as described in the Materials and Methods section.

(**B**) EIS measurements demonstrated with Nyquist plot (−Z″ vs Z′) and fitted with Randles equivalent circuit for the characterization of bare GCE, 4-MP film-modified electrodes and antibody-modified electrodes. (**C**) XPS-based characterization of bare screen-printed carbon electrode (SPCE), 4-MP film modified electrode, and antibody modified electrode. (**D**) Contact angle measurements of bare SPCE, 4-MP film modified electrode, and antibody-modified electrode. (**E**) SPR-based immunosensor measurements to determine the K_d of antibody with targeted GH as the analyte. SPR responses increased as the concentration of GH increased from 0.625 ng/mL (red line), 1.25 ng/mL (yellow line), 2.5 ng/mL (purple line) to 5 ng/mL (blue line). SPR sensorgrams were modeled with a fit line (black line) to determine the binding parameters. The sensorgram displays the average data of three consecutive measurements for each concentration of GH on renewed sensorchip surfaces (n = 3).

2.4. *Surface Plasmon Resonance (SPR)*

All SPR experiments were performed using a Biacore X100 system (GE Healthcare, Mississauga, ON, Canada) with a CM5 sensorchip. Experiments were conducted at 25 °C and the SPR running buffer (SPR running buffer, 0.01 M HEPES, 0.15 M NaCl, 0.05 mM EDTA, 0.05% surfactant P20, pH 7.4) was sterile filtered (0.2 µm). Two flow cells of the sensorchip were used, one (reference flow cell, FC-1) to detect the non-specific adsorption for background subtraction, and the other one (detection flow cell, FC-2) was used to detect the specific binding of GH. CM5 sensorchips contain carboxymethylated dextran covalently attached to Au surfaces. The anti-GH antibodies were covalently coupled onto the CM5 sensorchip surfaces using the Amine Coupling Kit (GE Healthcare, Mississauga, ON, Canada), following the standard covalent attachment protocol using EDC and NHS in a similar fashion as described in the preparation of electrochemical biosensors in Section 2.2. The target GH samples were injected into both flow cells using four different concentrations at 30 µL/min for 2 min. The setup was fully automated using the Biacore X100 software (GE Healthcare, Mississauga, ON, Canada). All concentrations of the target GH were performed in triplicates, with zero concentration blanks before and after each injection of the sample. The binding affinity of the interaction was determined using Biacore Evaluation Software (GE Healthcare, Mississauga, ON, Canada).

2.5. *X-Ray Photoelectron Spectroscopy (XPS)*

XPS spectra were recorded using a Physical Electronics (PHI) Quantera II spectrometer (Laval, QC, Canada) equipped with an Al anode source for X-ray generation and a quartz crystal monochromator for focusing the generated X-rays. A monochromatic Al K-α X-ray (1486.7 eV) source was operated at 50 W and 15 kV, and a pass energy of 280 eV was used to obtain all collected survey data. All spectra were obtained at 45° take off angles, and a dual beam charge compensation system was used to neutralize all samples. The system base pressure was no higher than 1.0×10^{-9} Torr, with an operating pressure that did not exceed 2.0×10^{-8} Torr. The instrument was calibrated using a sputter-cleaned piece of Ag, where the Ag 3d5/2 peak had a binding energy of 368.3 ± 0.1 eV and full width at half maximum for the Ag 3d5/2 peak was at least 0.52 eV. Data manipulation was performed using PHI MultiPak Version 9.5.1.0 software (Laval, QC, Canada).

2.6. *Contact Angle Goniometry*

Contact angles of modified surfaces were measured using a Future Digital Scientific OCA35 system (Westbury, NY, USA). Milli-Q water (18.2 Ohms) was used to determine the contact angles. The static sessile drop method was used and a small droplet (~0.2 µL) of Milli-Q water was placed on the sample surface. A picture was captured for each droplet on modified surfaces, and the contact angle was calculated using Young's equation.

2.7. *ELISA-Based Detection*

GH detection kit was purchased from EMD Millipore (Etobicoke, ON, Canada). In the sandwich-based assay, the GH samples were captured by the pre-tittered anti-GH polyclonal antibodies on the 96-well plate, and then, the binding of a second biotinylated anti-GH polyclonal antibody would form a "sandwich" by capturing the target protein on the surface. Streptavidin-conjugated horseradish peroxidase (HRP) was then exposed to the biotinylated antibodies. Upon addition of TMB (3,3′,5,5′-tetramethylbenzidine) substrate, the quantitative detection was achieved under acidic conditions with the formation of yellow-colored product at 450 nm. The concentration of GH in blood and plasma samples was derived by interpolation from a calibration curve generated in the assay, with reference standards of known concentrations of rat GH at 0.07, 0.21, 0.62, 1.9, 5.6, 16.7, and 50 ng/mL. We were able to determine the GH concentration in blood samples using the calibration plot that was constructed using the standard additions of known concentrations of GH in blood samples. We have also used this ELISA-based commercial kit to confirm our EIS data obtained with the standard addition method.

3. Results and Discussion

Following the chemical reactions as illustrated in Figure 1A, 4-methoxyphenyl (4-MP) film was formed through electrodeposition (i). Upon electrochemical grafting, liberation of nitrogen gas occurred, forming an aryl radical intermediate. The radical then formed a covalent bond with the carbon surface with high stability over time [31–33]. A conditioning potential of +1.0 V (vs. Ag/AgCl) was applied to form carboxylic acid on the film (ii). Subsequently, the carboxylic acid groups were activated with 1-ethyl-3-(3-dimethylaminopropyl) carbodiimide (EDC) and N-hydroxysuccinamide (NHS) (iii) to immobilize the antibodies by forming amide bonds with the lysine residues of the Fc region. Various concentrations of GH could be detected using the antibodies immobilized on the surface (iv) through antibody–antigen interactions. As shown in Figure 1B, both cyclic voltammetry (CV) and electrochemical impedance spectroscopy (EIS) [34–37] were used to characterize the modifications on carbon surfaces.

As shown in Figure 2A-i, the bare GCE displayed prominent oxidation and reduction peaks for the $[Fe(CN)_6]^{3-/4-}$ redox couple at 0.28 V and 0.16 V, respectively. The electrodeposition of 4-MP film resulted in a significant decrease in cyclic voltammograms, which suggested that the 4-MP film created an insulating layer that suppressed the reaction of the $[Fe(CN)_6]^{3/4-}$ redox couple on the surface (Figure 2A-ii).

EIS was used to detect the interfacial properties for bare GCE, 4-MP film-modified GCE, and antibody-modified GCE. As shown in Figure 2B, the impedimetric measurements were demonstrated with Nyquist plots as the sum of the Z′ and Z″. The Z′ (real Z) accounts for double-layer resistance and Z″ (imaginary Z) accounts for capacitance. The Randles equivalent circuit shown in the inset of Figure 2B was selected to reflect the electrochemical process. The equivalent circuit includes the ohmic resistance of the electrolyte solution, R_s, and the Warburg impedance, Z_w, resulting from the diffusion of ions from the bulk electrolyte to the interface. The electrode double-layer capacitance, C_{dl}, and the charge transfer resistance, R_{ct}, depends on the dielectric and insulating properties at the interface. R_{ct} values continued to increase with the step-wise modifications with 4-MP film and then antibodies on the electrode surfaces.

The increase in R_{ct} was consistent with the previous CV data suggesting the aryl diazonium salt film had an insulating property. A further increase in R_{ct} suggested that antibodies were immobilized onto the GCE surface. The EIS results were confirmed using XP. As shown in Figure 2C, the increases in O 1s′ counts—shown in green lines and red lines—were contributed to by the oxygen atoms in the 4-MP film and antibodies, and the significant increase in N 1s′ counts was attributed to the primary amine groups of the antibodies, suggesting the immobilization of the antibodies on the SPCE surface. As shown in Figure 2D, the contact angle measurement had a small decrease after the immobilization of 4-MP film, and then, a significant decrease was observed after the immobilization of antibodies. The contact angle measurements were in agreement with the XPS results (Figure 2C), where the small increase in O 1s after immobilization of 4-MP film, followed by a significant increase in O 1s after the immobilization of antibodies, would contribute to the decreasing hydrophobicity of the electrode surface. In order to determine the binding dissociation constant, K_d, value of the anti-GH antibody, SPR (Figure 2E) was performed using anti-GH-modified CM5 sensorchips. The anti-GH antibodies that were covalently immobilized on CM5 sensorchips displayed a strong binding affinity towards GH at 1.44 (±0.15) nM (n = 3).

The anti-GH antibody-modified GCE enabled the detection of GH with a dynamic range between 100 pg/mL and 1000 pg/mL, as shown in Figure 3A. The initial (R_i) and final (R_f) R_{ct} values were measured before and after the GH binding to the antibody-modified electrodes. These R_{ct} ratios, between blank and a range of GH concentrations, are summarized in Figure 3B,D. The immunosensor was significantly improved using SPCEs with miniaturized and portable detection capabilities. As shown in Figure 3C, the R_{ct} values corresponding to each modification of the SPCEs continued to increase. The Nyquist plots increased with increasing concentrations of GH, and the R_{ct} ratios are displayed in Figure 3D. The linear range of GH detection using GCE was from 100 pg/mL to 1000 pg/mL, with a regression line formula y = 0.0021x + 1.1805, R^2 = 0.982. Even though the dynamic range using SPCE was similar to that of the GCE surface, the linear range of GH detection was smaller, from 10 pg/mL to 100 pg/mL, with a regression line formula of y = 0.0035x + 0.2431, R^2 = 0.9944. Using the formula for the limit of detection = 3 × σ_{blank}/m, where σ_{blank} is the standard deviation of blank measurements and m is the slope of the calibration curve, we have determined the limit of detection as 5 pg/mL, which is highly comparable with literature values. Ozhikandathil et al. [38] reported a detection limit of 25 ng/mL using an evanescent-cascaded waveguide coupler design. Sadabadi et al. [39] reported the detection of GH using the LSPR response of gold nanoparticles grown in microfluidics with a detection limit of 3.7 ng/mL. In order to demonstrate the applicability of our GH detection method to biological samples containing high concentrations of potentially interfering proteins, a SPCE-based immunosensor was challenged with blood and plasma samples from male rats and the results are shown in Figure 4.

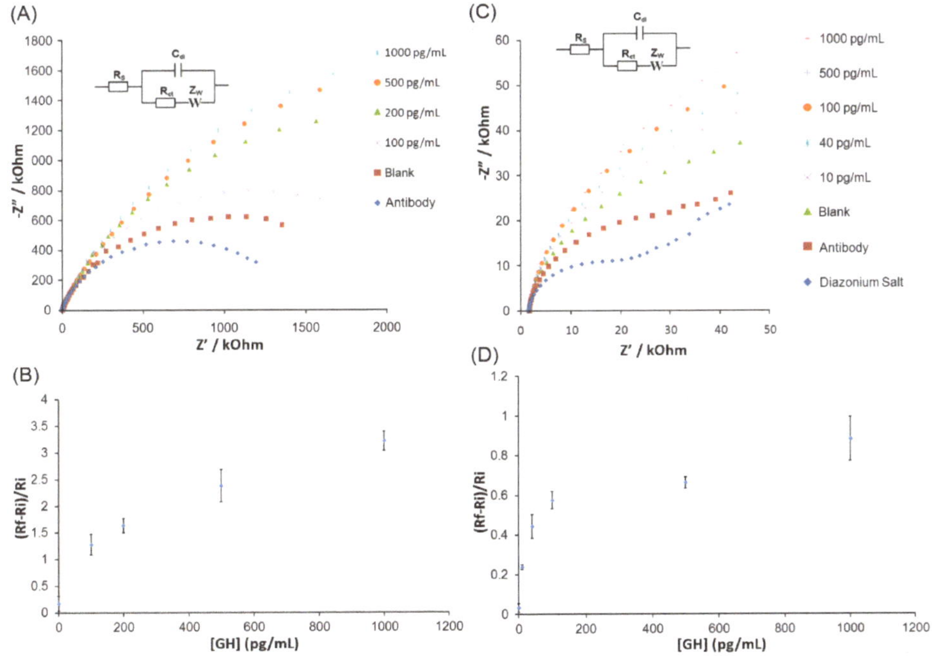

Figure 3. Impedimetric detection of growth hormone (GH) using antibody-modified carbon electrodes; (**A**) Nyquist plots for the detection of various concentrations of GH on glassy carbon electrodes (GCE);

(**B**) Dependence of GH concentration on the R_{ct} values obtained from the Randles equivalent circuit at GCE. (**C**) Nyquist plots for the detection of various GH concentrations on screen-printed carbon electrodes (SPCE). (**D**) Dependence of GH concentration on the R_{ct} values obtained from the Randles equivalent circuit at the SPCE. Error bars indicate three consecutive measurements (n = 3) of each GH concentration using renewed GCE surfaces and using a new single-use SPCE.

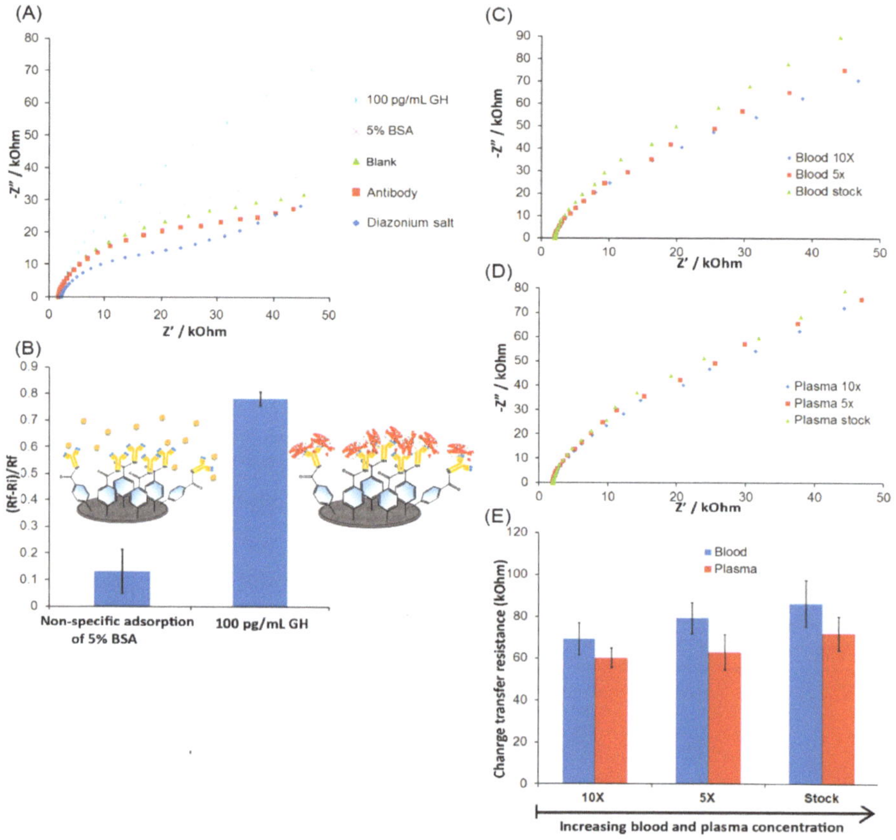

Figure 4. (**A**) Nyquist plots for the detection of growth hormone (GH) and bovine serum albumin (BSA) in diluted blood using screen-printed carbon electrodes (SPCE); (**B**) R_{ct} ratio comparison between attachment of BSA and GH in diluted blood (n = 3). (**C**) Nyquist plots for the detection of 100 pg/mL GH in various concentrations of blood. (**D**) Nyquist plots for the detection of 100 pg/mL GH in various concentrations plasma; (**E**) R_{ct} ratios for the detection of GH in various concentrations of blood and plasma (n = 3). Error bars indicate the standard deviation of three consecutive measurements using SPCEs (n = 3).

As shown in Figure 4A, the Nyquist plots displayed an increase with 4-MP film modification and the immobilization of antibodies. As shown in Figure 4B, BSA (5%, w/v) and GH (100 pg/mL) were spiked into 10-fold diluted blood samples. The Ri and Rf were measured before and after the binding of BSA or GH at the antibody-modified SPCEs. A p-value of 1.9×10^{-5} was calculated, by conventional criteria, the difference between R_{ct} ratios from the binding of BSA and the binding of GH was considered to be statistically highly significant. The R_{ct} ratio was significantly higher for GH-spiked blood samples, suggesting that the antibody-modified SPCEs were effective in facilitating the ultrasensitive detection of GH in the presence of complex protein mixtures present in blood. Hence, our immunosensor has promising potential for applications using clinical samples.

As shown in Figure 4C,D, GH (100 pg/mL) was detected in 10-fold and 5-fold diluted blood samples, as well as in the undiluted whole blood (blood stock) and plasma (plasma stock) samples. To investigate the reproducibility of our SPCEs, EIS responses of all 18 biosensors were recorded in these real samples that were spiked with 100 pg/mL GH, as shown in Figure 4E. The relative standard deviation was found to be less than 6% (n = 3 for each 10-fold and 5-fold diluted blood sample, as well as undiluted whole blood and plasma samples) with mean recoveries ranging from 94% ± 3% to 103% ± 2%. In Figure 4E, the average R_{ct} increased by 10 kOhm from 10-fold to 5-fold diluted blood samples, and 7 kOhm from the 5-fold diluted blood sample to the whole blood sample. This increase in R_{ct} was attributed to the possible non-specific adsorption of proteins in the whole blood matrix. The *p*-value was 4.6×10^{-2} between 10-fold and 5-fold diluted blood samples, and it was considered to be statistically significant by conventional criteria. However, the p-value was 2.4×10 between 5-fold diluted and whole blood samples, and it was not considered to be statistically significant by conventional criteria. In addition, the repeatability of the immunosensor was evaluated by performing three experiments with the similarly prepared three SPCEs for three consecutive days, and after each measurement, the SPCEs were disposed of. Prior to use, the immunosensors were soaked in 50 mM PBS (pH 7.4) and stored at 4 °C. SPCE-based immunosensors revealed a good repeatability performance with a relative standard deviation of less than 5% (n = 3 using 10-fold diluted plasma samples, data not shown).

Blood plasma samples with a simpler protein composition resulted in lower R_{ct} than whole blood samples in 10-fold diluted, 5-fold diluted and undiluted samples. The p-value was 4.6×10^2 between 10-fold and 5-fold diluted plasma samples, and 8.6×10^2 between 5-fold diluted and undiluted plasma samples. Both *p*-values were not considered to be statistically significant by conventional criteria and suggested increasing plasma concentrations did not result in a significant increase with non-specific adsorption and surface fouling.

For real sample studies, Vance et al. [40] have reported a comparative analysis of human growth hormone in serum using SPRi, nano-SPRi and ELISA assays. An ultrasensitive bimodal waveguide biosensor enabled the direct detection of human GH in undiluted urine samples in the 10 pg/mL range [41]. A nano-integrated suspended polymeric microfluidics platform provided a detection limit of 2 ng/mL for the detection of bovine growth hormones [42]. A summary of articles [43–52] reporting the detection of GH using various surface modifications and detection techniques with limits of detection is presented in Table 1. Rezaei et al. [45] modified gold electrodes with gold nanoparticles using 1,6-hexanethiol and managed to develop an ultrasensitive EIS-based immunosensor, which detected GH with a limit-of-detection of 0.64 pg/mL.

Our immunosensor was applied to monitor GH in blood samples collected from male rats aged from 2 weeks to 19 months. EIS results of SPCEs were compared with those obtained using a commercial GH ELISA kit. As shown in Figure 5, GH levels decreased with advancing age rats using both detection systems. In general, the EIS data indicated that the rats had higher GH compared to the data obtained using the commercial kit. For the data that were obtained in months 1, 2, 18 and 19 using EIS-based immunosensors and commercial kit, the *p*-values ranged between 2.7×10^2 and 4.3×10^2. These *p*-values were not considered to be statistically significant by conventional criteria and suggested that the data collected in those months were similar. In months 6, 14 and 16, the fluctuation of GH created data with *p*-values ranging between 3.5×10^{-4} and 1.8×10^{-3}, indicating that the data were significantly different from each other. This was attributed to the possible fluctuations of GH in adult rats during these months. The lower GH concentration determined by the commercial kit may be attributed to the complicated sample preparation procedures, as well as the short lifetime of the light-sensitive TMB (3,3′,5,5′-tetramethylbenzidine) dye, which was observed to decrease in absorbance rapidly over time.

Table 1. A summary of published literature on the detection of GH using biosensors with optical and electrochemical detection techniques.

Biosensor Surface	Detection Technique	Limit of Detection	Reference
Silica-on-silicon (SOS) with a cascaded waveguide coupler	Evanescent wave-based fluoroimmunoassay	25 ng/mL	[38]
Gold nanoparticles synthesized in a poly(dimethylsiloxane) (PDMS) microfluidic chip	LSPR-based immunoassay	3.7 ng/mL	[39]
Anti-hGH coated with near-infrared quantum dots	SPRi (SPR imaging)- & Nano-SPRi-based immunoassay	0.03 ng/mL–100 ng/mL	[40]
Anti-hGH-modified interferometer	Bimodal waveguide interferometry-based immunoassay	10 pg/mL	[41]
Nano-integrated suspended polymeric microfluidics (SPMF) platform	Microcantilever-based immunoassay	2 ng/mL	[42]
Anti-hGH modified gold surfaces	SPR-based immunoassay	6 ng/mL	[43]
Anti-hGH modified gold surfaces	SPR-based immunoassay	1-6 ng/mL	[44]
Gold nanoparticles immobilized on gold electrodes using 1,6-hexanedithiol	EIS-based immunoassay	0.64 pg/mL	[45]
Sandwich-based immunoassay using horseradish peroxidase (HRP)-labeled secondary antibody	SPR, pulsed amperometry (PA), electrochemically-assisted chemiluminescence (ECL), CV	0.051 nM by SPR 0.027 nM by PA 0.061 nM by ECL 0.056 nM by CV	[46]
Tosyl-activated magnetic microparticles on screen-printed gold electrodes	Square-wave voltammetry (SWV) of 4-aminophenyl phosphate as the substrate of alkaline phosphatase	0.005 ng/mL	[47]
Protein A-gold binding domain fusion protein	SPR-based immunoassay	90 ng/mL	[48]
Anti-hGH immobilized on gold surfaces	PA & CV	75 nM by PA 108 nM by CV	[49]
Oriented anti-hGH-modified gold surfaces using biotin-streptavidin	SPR-based immunoassay	0.9 ng/mL for 22K and 20K hGH isoforms	[50]
Plasmonic gold decorated multi-walled carbon nanotube nanocomposite	LSPR-based immunoassay	1 ng/mL	[51]
Carbon fiber microelectrode	Differential pulse voltammetry for in vivo and ex vivo measurements using rats	2 µg/µL	[52]
Anti-GH modified SPCE	EIS-based immunoassay	5 pg/mL	This work

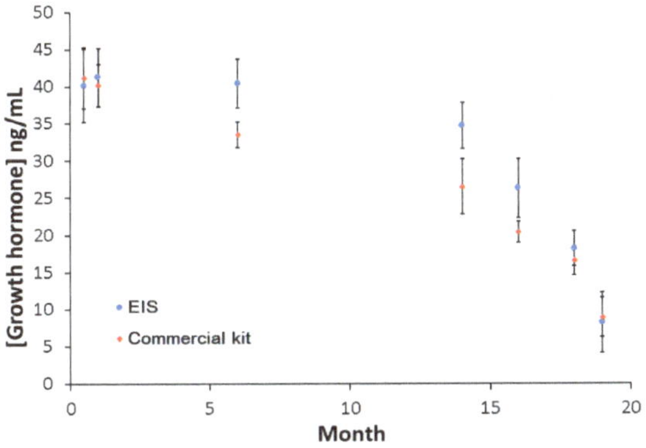

Figure 5. Dependence of growth hormone concentration on the age of rats. The results of EIS-based immunosensor using screen-printed carbon electrodes (SPCE) was compared with those obtained using the commercial detection kit as described in the Section 2.7. Error bars indicate the standard deviation of three consecutive measurements (n = 3) performed using the samples collected on the same day of each month.

4. Conclusions

An immunosensor was developed using aryl diazonium salt modification for the covalent immobilization of antibodies on screen-printed carbon electrodes (SPCE) and glassy carbon electrode (GCE) surfaces. SPCE-based immunosensors provided the development of low-cost and single-use detection systems, avoiding possible cross-contamination issues that can be observed using GCE surfaces. The impedimetric detection of GH exhibited a high analytical performance, demonstrating good sensitivity with good reproducibility and repeatability. The results demonstrated that the immunosensor could detect 5 pg/mL GH with a linear range between 10 pg/mL and 100 pg/mL GH. The detection of GH in the presence of increasing concentrations of complex protein mixtures present in blood demonstrated the applicability of the technique to real samples.

Author Contributions: Conceptualization, N.L., A.M.C., I.R.B., and K.K.; methodology, N.L., A.M.C., I.R.B. and K.K.; formal analysis, N.L., A.M.C., H.V.S.G. and M.R.; investigation, N.L., A.M.C., H.V.S.G. and M.R.; writing—original draft preparation, N.L., A.M.C., H.V.S.G. and M.R.; writing—review and editing, I.R.B. and K.K.; supervision, I.R.B. and K.K.

Funding: This work was supported by the Canada Research Chair Tier-2 award to K. Kerman for "Bioelectrochemistry of Proteins" (Project No. 950-231116), the Ontario Ministry of Research and Innovation (Project No. 35272), Discovery Grant (Project No. 3655) from the Natural Sciences and Engineering Research Council of Canada (NSERC), and the Canada Foundation for Innovation (Project No. 35272).

Acknowledgments: The authors gratefully acknowledge NSERC Discovery Grants to K.K. and I.R.B., Canada Research Chair (Tier I) in Neuroscience to I.R.B and Canada Research Chair (Tier II) in Bioelectrochemistry of Proteins to K.K. Authors also thank Rana N.S. Sodhi for his technical assistance with XPS and contact angle measurements in the Ontario Centre for the Characterization of Advanced Materials in University of Toronto.

Conflicts of Interest: The authors declare no conflict of interest. The funders had no role in the design of the study; in the collection, analyses, or interpretation of data; in the writing of the manuscript, or in the decision to publish the results.

References

1. Gooding, J.J. Advances in interfacial design for electrochemical biosensors and sensors: Aryl diazonium salts for modifying carbon and metal electrodes. *Electroanalysis* **2007**, *20*, 573–582. [CrossRef]
2. Gooding, J.J.; Ciampi, S. The molecular level modification of surfaces: From self-assembled monolayers to complex molecular assemblies. *Chem. Soc. Rev.* **2011**, *40*, 2704–2718. [CrossRef]

3. Liu, G.; Luais, E.; Gooding, J.J. The fabrication of stable gold nanoparticle-modified interfaces for electrochemistry. *Langmuir* **2011**, *27*, 4176–4183. [CrossRef]
4. Liu, G.; Gooding, J.J. An interface comprising molecular wires and poly(ethylene glycol) spacer units self-assembled on carbon electrodes for studies of protein electrochemistry. *Langmuir* **2006**, *22*, 7421–7430. [CrossRef]
5. Eissa, S.; Tlili, C.; L'Hocine, L.; Zourob, M. Electrochemical immunosensor for the milk allergen β-lactoglobulin based on electrografting of organic film on graphene modified screen-printed carbon electrodes. *Biosens. Bioelectron.* **2012**, *38*, 308–313. [CrossRef]
6. Eissa, S.; Zourob, M. A Graphene-based electrochemical competitive immunosensor for the sensitive detection of okadaic acid in shellfish. *Nanoscale* **2012**, *4*, 7593–7599. [CrossRef]
7. Eissa, S.; L-Hocine, L.; Siaj, M.; Zourob, M. A graphene-based label-free voltammetric immunosensor for sensitive detection of the egg allergen ovalbumin. *Analyst* **2013**, *138*, 4378–4384. [CrossRef]
8. Won, Y.H.; Jang, H.S.; Kim, S.M.; Stach, E.; Ganesana, J.; Andreescu, S.; Stanciu, L.A. Biomagnetic glasses: Preparation, characterization, and biosensor applications. *Langmuir* **2010**, *26*, 4320–4326. [CrossRef]
9. Kolliopoulos, A.V.; Metters, J.P.; Banks, C.E. Electroanalytical sensing of selenium(IV) utilising screen printed graphite macro electrodes. *Anal. Methods* **2013**, *5*, 851–856. [CrossRef]
10. Wei, Y.; Wang, Y.; Kang, A.; Wang, W.; Ho, S.V.; Gao, J.; Ma, G.; Su, Z. A novel sustained-release formulation of recombinant human growth hormone and its pharmacokinetic, pharmacodynamic and safety profiles. *Mol. Pharm.* **2012**, *9*, 2039–2048. [CrossRef]
11. Kwon, O.S.; Ahn, S.R.; Park, S.J.; Song, H.S.; Lee, S.H.; Lee, J.S.; Hong, J.Y.; Lee, J.S.; You, S.A.; Yoon, H.; et al. Ultrasensitive and selective recognition of peptide hormone using close-packed arrays of hPTHR-conjugated polymer nanoparticles. *ACS Nano* **2012**, *6*, 5549–5558. [CrossRef]
12. Bains, R.K.; Wells, S.E.; Flavell, D.M.; Fairhall, K.M.; Storm, M.; Le Tissier, P.; Robinson, I.C. Visceral obesity without insulin resistance in late-onset obesity rats. *Endocrinology* **2004**, *145*, 2666–2679. [CrossRef]
13. Higham, C.E.; Trainer, P.J. Growth hormone excess and the development of growth hormone receptor antagonists. *Exp. Physiol.* **2008**, *93*, 1157–1169. [CrossRef]
14. Hecht, M.L.; Tsai, Y.H.; Liu, X.; Wolfrum, C.; Seeberger, P.H. Synthetic inositol phosphoglycans related to GPI lack insulin-mimetic activity. *ACS Chem. Biol.* **2010**, *5*, 1075–1086. [CrossRef]
15. Wei, H.; Li, H.; Mao, S.; Lin, J.M. Cell signaling analysis by mass spectrometry under coculture conditions on an integrated microfluidic device. *Anal. Chem.* **2011**, *83*, 9306–9313. [CrossRef]
16. Wu, X.; Schultz, P.G. Synthesis at the interface of chemistry and biology. *J. Am. Chem. Soc.* **2009**, *131*, 12497–12515. [CrossRef]
17. Mukhopadhyay, R. Deciphering cellular conversations. *Anal. Chem.* **2007**, *79*, 3963–3965. [CrossRef]
18. Thomas, A.; Delahaut, P.; Krug, O.; Schänzer, W.; Thevis, M. Metabolism of growth hormone releasing peptides. *Anal. Chem.* **2012**, *84*, 10252–10259. [CrossRef]
19. Holt, R.I.G.; Soenksen, P.H. Growth hormone, IGF-I and insulin and their abuse in sport. *J. Pharmacol.* **2008**, *154*, 542–556. [CrossRef]
20. Nelson, A.E.; Ho, K.K. A robust test for growth hormone doping-present status and future prospects. *Asian J. Androl.* **2008**, *10*, 416–425. [CrossRef]
21. Powrie, J.K.; Bassett, E.E.; Rosen, T.; Jørgensen, J.O.; Napoli, R.; Sacca, L.; Christiansen, J.S.; Bengtsson, B.A.; Sönksen, P.H. Detection of growth hormone abuse in sport. *Growth Horm. IGF Res.* **2007**, *17*, 220–226. [CrossRef]
22. Cung, L.; Baxter, R.C. Detection of growth hormone responsive proteins using SELDI-TOF mass spectrometry. *Growth Horm. IGF Res.* **2009**, *19*, 383–387.
23. Yang, S.Y.; Yang, J.A.; Kim, E.S.; Jeon, G.; Oh, J.E.; Choi, W.Y.; Hahn, S.K.; Kim, J.K. Single-file diffusion of protein drugs through cylindrical nanochannels. *ACS Nano* **2010**, *4*, 3817–3822. [CrossRef]
24. Dong, W.; Shang, S.; Li, J.; Tan, Z.; Dean, T.; Maeda, A.; Gardella, T.J.; Danishefsky, S.J. Engineering of therapeutic polypeptides through chemical synthesis: Early lessons from human parathyroid hormone and analogues. *J. Am. Chem. Soc.* **2012**, *134*, 15122–15129. [CrossRef]
25. Kausaite-Minkstimiene, A.; Ramanaviciene, A.; Kirlyte, J.; Ramanavicius, A. Comparative study of random and oriented antibody immobilization techniques on the binding capacity of immunosensor. *Anal. Chem.* **2010**, *82*, 6401–6408. [CrossRef]

26. Seth, J.; Ellis, A.; Al-Sadie, R. Serum growth hormone measurements in clinical practice: An audit of performance from the UK national external quality assessment scheme. *Horm. Res.* **1993**, *51*, 13–19. [CrossRef]
27. Bauman, G.; Stolar, M.W.; Buchanan, T.A. The metabolic clearance, distribution, and degradation of dimeric and monomeric growth hormone (GH): Implications for the pattern of circulating GH forms. *Endocrinology* **1986**, *119*, 1497–1501. [CrossRef]
28. Chatelain, P.; Bouillat, B.; Coehn, R.; Sassolas, G.; Souberbielle, J.C.; Ruitton, A.; Joly, M.O.; Job, J.C. Assay of growth hormone levels in human plasma using commercial kits: Analysis of some factors influencing the results. *Acta Paediatr.* **1990**, *370*, 56–61. [CrossRef]
29. Reiter, E.O.; Morris, A.H.; Macgillivray, M.H.; Weber, D. Variable estimates of serum growth hormone concentrations by different radioassay systems. *J. Clin. Endocrinol. Metab.* **1988**, *66*, 68–71. [CrossRef]
30. Jan, T.; Shaw, M.A.; Baumann, G. Effects of growth hormone-binding proteins on serum growth hormone measurements. *J. Clin. Endocrinol. Metab.* **1991**, *72*, 387–391. [CrossRef]
31. Tasca, F.; Harreither, W.; Ludwig, R.; Gooding, J.J.; Gorton, L. Cellobiose dehydrogenase aryl diazonium modified single walled carbon nanotubes: Enhanced direct electron transfer through a positively charged surface. *Anal. Chem.* **2011**, *83*, 3042–3049. [CrossRef]
32. Liu, J.; Wang, R.; Cui, L.; Tang, J.; Liu, Z.; Kong, Q.; Yang, W.; Gooding, J.J. Using molecular level modification to tune the conductivity of praphene papers. *J. Phys. Chem. C* **2012**, *116*, 17939–17946. [CrossRef]
33. Boland, S.; Barriere, F.; Leech, D. Designing stable redox-active surfaces: Chemical attachment of an osmium complex to glassy carbon electrodes prefunctionalized by electrochemical reduction of an in situ-generated aryl diazonium cation. *Langmuir* **2008**, *24*, 6351–6358. [CrossRef]
34. Li, N.; Brahmendra, A.; Veloso, A.J.; Prashar, A.; Cheng, X.R.; Hung, V.W.S.; Guyard, C.; Terebiznik, M.; Kerman, K. Disposable immunochips for the detection of legionella pneumophila using electrochemical impedance spectroscopy. *Anal. Chem.* **2012**, *84*, 3485–3488. [CrossRef]
35. Li, Q.; Li, N.; Le Tisser, P.; Grattan, D.; Kerman, K. Miniaturized electrochemical immunosensors for the detection of growth hormone. *Electroanalysis* **2012**, *24*, 1272–1276. [CrossRef]
36. Pei, Y.; Travas-Sejdic, J.; Williams, D.E. Water structure change-induced expansion and collapse of zwitterionic polymers surface-grafted onto carbon black. *Langmuir* **2012**, *28*, 8072–8083. [CrossRef]
37. Liu, G.Z.; Liu, J.Q.; Bocking, T.; Eggers, P.K.; Gooding, J.J. The modification of glassy carbon and gold electrodes with aryl diazonium salt: The impact of the electrode substrate on the rate of heterogeneous electron transfer. *Chem. Phys.* **2005**, *319*, 136–146. [CrossRef]
38. Ozhikandathil, J.; Packirisamy, M. Detection of recombinant growth hormone by evanescent cascaded waveguide coupler on silica-on-silicon. *J. Biophotonics* **2013**, *6*, 457–467. [CrossRef]
39. Sadabadi, H.; Badilescu, S.; Packirisamy, M.; Wüthrich, R. Integration of gold nanoparticles in PDMS microfluidics for lab-on-a-chip plasmonic biosensing of growth hormones. *Biosens. Bioelectron.* **2013**, *44*, 77–84. [CrossRef]
40. Vance, S.; Zeidan, E.; Henrich, V.C.; Sandros, M.G. Comparative analysis of human growth hormone in serum using SPRi, nano-SPRi and ELISA assays. *J. Vis. Exp.* **2016**, *7*, e53508. [CrossRef]
41. Gonzalez-Guerrero, A.B.; Maldonado, J.; Dante, S.; Grajales, D.; Lechuga, L.M. Direct and label-free detection of the human growth hormone in urine by an ultrasensitive bimodal waveguide biosensor. *J. Biophotonics* **2017**, *10*, 61–67. [CrossRef]
42. Sadabadi, H.; Packirisamy, M. Nano-integrated suspended polymeric microfluidics (SPMF) platform for ultra-sensitive biomolecular recognition of bovine growth hormones. *Sci. Rep.* **2017**, *7*, 10969. [CrossRef]
43. Trevino, J.; Calle, A.; Rodriguez-Frade, J.M.; Mellado, M.; Lechuga, L.M. Determination of human growth hormone in human serum samples by surface plasmon resonance immunoassay. *Talanta* **2009**, *78*, 1011–1016. [CrossRef]
44. Trevino, J.; Calle, A.; Rodriguez-Frade, J.M.; Mellado, M.; Lechuga, L.M. Surface plasmon resonance immunoassay analysis of pituatry hormones in urine and serum samples. *Clin. Chim. Acta* **2009**, *403*, 56–62. [CrossRef]
45. Rezaei, B.; Khayamian, T.; Majidi, N.; Rahmani, H. Immobilization of specific monoclonal antibody on Au nanoparticles for hGH detection by electrochemical impedance spectroscopy. *Biosens. Bioelectron.* **2009**, *25*, 395–399. [CrossRef]

46. Ramanaviciene, A.; German, N.; Kausaite-Minkstimiene, A.; Voronovic, J.; Kirlyte, J.; Ramanavicius, A. Comparative study of surface plasmon resonance, electrochemical and electroassisted chemiluminescence methods based immunosensor for the determination of antibodies against human growth hormone. *Biosens. Bioelectron.* **2012**, *36*, 48–55. [CrossRef]
47. Serafin, V.; Ubeda, N.; Agui, L.; Yanez-Sedeno, P.; Pingarron, J.M. Ultrasensitive determination of human groth hormone (hGH) with a disposable electrochemical magneto-immunosensor. *Anal. Bioanal. Chem.* **2012**, *403*, 939–946. [CrossRef]
48. de Juan-Franco, E.; Caruz, A.; Pedrajas, J.R.; Lechuga, L.M. Site-directed antibody immobilization using a protein A-gold binding domain fusion protein for enhanced SPR immunosensing. *Analyst* **2013**, *138*, 2023–2031. [CrossRef]
49. German, N.; Kausaite-Minkstimiene, A.; Kirlyte, J.; Makaraviciute, A.; Ramanavicius, A.; Mikoliunaite, L.; Ramanaviciene, A. Determination of antibodies against human growth hormone using a direct immunoassay format and different electrochemical methods. *Analyst* **2013**, *138*, 1427–1433. [CrossRef]
50. de Juan-Franco, E.; Radriguez-Frade, J.M.; Mellado, M.; Lechuga, L.M. Implementation of a SPR immunosensor for the simultaneous detection of the 22K and 20K hGH isoforms in human serum samples. *Talanta* **2013**, *114*, 268–275. [CrossRef]
51. Ozhikandathil, J.; Badilescu, S.; Packirisamy, M. Plasmonic gold decorated MWCNT nanocomposite for localized plasmon resonance sensing. *Sci. Rep.* **2015**, *5*, 13181. [CrossRef]
52. Crespi, F. Central [CNS] and peripheral [Gastric tissue] selective monitoring of somatostatin (SRIF) with micro-sensor and voltammetry in rats: Influence of growth factors (GH, EGF). *Biosensors* **2017**, *7*, 53. [CrossRef]

biosensors

MDPI

Article

Biochar from Brewers' Spent Grain: A Green and Low-Cost Smart Material to Modify Screen-Printed Electrodes

Rocco Cancelliere [1], Katya Carbone [2], Mauro Pagano [3], Ilaria Cacciotti [4] and Laura Micheli [1,*]

1 Department of Chemical Sciences and Technologies, University of Rome "Tor Vergata", Via della Ricerca Scientifica, 00133 Rome, Italy; rocco.cancelliere@uniroma2.it
2 CREA, Research Centre for Olive, Citrus and Tree Fruit, Via di Fioranello 52, 00134 Rome, Italy; katya.carbone@crea.gov.it
3 CREA Research Centre for Engineering and Agro-Food Processing, Via Della Pascolare 16, Monterotondo, 00015 Rome, Italy; mauro.pagano@crea.gov.it
4 Engineering Department, University of Rome "Niccolò Cusano", Via Don Carlo Gnocchi 3, 00166 Rome, Italy; ilaria.cacciotti@unicusano.it
* Correspondence: laura.micheli@uniroma2.it

Received: 4 October 2019; Accepted: 27 November 2019; Published: 3 December 2019

check for updates

Abstract: In the present study, biochar from brewers' spent grain was used, for the first time, to develop screen-printed electrodes. After having investigated the dispersion behaviour of biochar in different organic solvents, a biochar-based screen-printed electrode was prepared with the drop-casting technique. In order to understand the electrochemical potentiality and performances of the biochar/sensor tool, different electroactive species, i.e., ferricyanide, benzoquinone, epinephrine, ascorbic, and uric acids, were used. The results were compared with those of the same electrodes that were modified with commercial graphene, confirming that the proposed electrode showed improved electrochemical behaviour in terms of resolution, peak-to-peak separation, current intensity, and resistance to charge transfer. Furthermore, a tyrosinase biosensor was developed by direct immobilisation of this enzyme on the biochar/screen printed electrode, as an example of the potential of biochar for disposable biosensor development. The efficiently occurred immobilisation of the biochar on the screen printed electrode's (SPE's) surface was demonstrated by the observation of the working electrode with a scanning electron microscope. The detection was performed by measuring the current due to the reduction of the corresponding quinone at low potential, equal to −0.310 V for epinephrine. The experimental conditions for the tyrosinase immobilization and the analytical parameters, such as applied potential and pH of buffer, were studied and optimized. Under these conditions, the electrochemical biosensors were characterized. A linear working range of epinephrine was obtained from 0.05 up to 0.5 mM. The detection limit was 2×10^{-4} mM for the biosensor.

Keywords: brewers' spent grain; biochar; screen-printed electrode; biosensor; tyrosinase

1. Introduction

The beer brewing process is one of the most polluting industrial processes, generating a huge amount of wastewater effluent and solid wastes (i.e., spent grain and yeast), which must be disposed or treated in the least costly way to meet strict discharge regulations set by government entities [1]. Particularly, spent grain, consisting of grain husks and other residual compounds not converted into fermentable sugars in the mashing process, can constitute as much as 85% of a brewery's total by-products [2,3]. One of the main challenges of the brewery sector is, therefore, the recovery and valorisation of these wastes through the application of a circular economy model. Brewers' spent grain

(BSG) is available at low or no cost throughout the year and it is produced in large quantities by small and big breweries. Several attempts have been made to recover and valorise BSG in animal feeding; in biotechnological processes, such as cultivation of mushrooms and actinobacteria; and as a source of value-added products, such as phenolic acids or sugar alcohols [4,5]. However, a diffuse employment of BSG as an industrial feedstock is hampered by its chemical deterioration and its susceptibility to microbial attacks due to its high water content (about 70%–80%). The wet and unstable nature of BSG limits the transportation, and the high moisture content prevents efficient or direct energetic utilization. Therefore, an effective treatment of BSG needs to be developed in order to produce a solid, stable carbon source or high-value materials that are beneficial in terms of waste valorisation. In fact, nowadays, the rational use of food waste represents a major challenge in terms of environmental protection, but also from an economic perspective. In this regard, new technologies of pyrolysis and gasification have been developed, especially for thermal processing of biomass [6]. Thermal methods are promising technologies that allow transforming certain types of waste to quality fuel or valuable chemical raw materials. Recently, Sperandio et al. [1] proposed a pyro gasification process for the conversion of BSG into syngas and biochar, which can offer a sustainable approach to by-product disposal to benefit both the environment and craft breweries' economic outputs. Indeed, this approach allows one to provide energy in terms of syngas for the satisfaction of farms' energy demands, and at the same time, to replace commercial fertilizer with the use of biochar, obtained from waste biomasses, as a soil improver. Biochar has gained great attention since its production, in combination with its storage in soils, has been suggested as one possible means of reducing the atmospheric CO_2 concentration. From an agronomic point of view, biochar can improve agricultural productivity, particularly in low-fertility and degraded soils, reducing losses of nutrients, and improve the water-holding capacity of soils [7]. On the other hand, since biochar is a highly porous carbonaceous material, which consists of an inert internal structure and a highly functionalized surface (condensed or residual aliphatic compounds, condensed aromatics) with the ability to interact with different compounds, it is also gaining the attention of analytical chemists [8–10].

These structural characteristics of biochar are very similar to those of nanomaterials widely used in electrochemistry (i.e., graphene, nanotubes, and nanofibers), making it a potential alternative for the manufacture of screen printed electrodes (SPEs) based on renewable and biocompatible sources. Nowadays, in fact, there is a growing interest in the use of eco-friendly materials for electronics, giving rise to an innovative generation of high-performance green modifiers [10].

In the literature, several applications of the carbon paste electrode (BCPE), modified with commercial biochar, are reported for the determination of organic and inorganic pollutants in the environment [11–13], based on the direct interaction between biochar and pollutants. An example of the application of biochar from tea waste was reported by Bal Altuntas et al. [14] for the development of the glucose biosensor. In that study, the biochar was mixed with graphite and mineral oil to develop the biochar-BCPE and to demonstrate the possibility to use this waste for the production of novel composite electrode. Recently, a review about the disposable electrodes produced by waste materials has been published, where the use of waste paperboard for the support of these electrodes or of vegetables waste for the production of nanoparticles for electrochemical application are reported [10]. Anyway, to the best of our knowledge, no study covering the electroanalytical use of biochar coupled with screen printed electrodes (biochar/SPEs) has been reported in the literature until now. The application of other carbon based materials, such as single-walled carbon nanotubes (SWCNTs), multi-walled carbon nanotubes (MWCNTs) and graphene, has been carried out for the modification of the surface of screen-printed electrodes (SPEs) in order to improve their electrochemical performances. For example, Gomez et al. [15] studied the electrochemical behaviour of SPEs modified with SWCNTs, MWCNTs and graphene, respectively, to detect melatonin (MT) and serotonin (5-HT), with a remarkable improvement in terms of selectivity, reproducibility, and detection limit. Pérez-Ràfols et al. [16] and Apetrei et al. [17], who used SPEs modified with carbon nanofibers, obtained the same improvements in the electrochemical performance. In the present study, biochar from spent grain was used for the first time, in order to modify the graphite-working electrode of SPEs. The modification was carried out by drop casting, using a stable dispersion of biochar (biochar/SPE), demonstrating the possibility of recycling this waste material. At the same time, the SPEs were modified with commercial graphene in order to compare the electrochemical performances (in terms of sensibility, working range, and detection limit) of both electrodes. Moreover, these modified sensors were used as support for the immobilization of tyrosinase, an enzyme selective for catecholamines [18].

In light of these considerations, in the present study, for the first time, biochar from agricultural food waste was used to modify a cheap printed electrode for tyrosinase biosensor development. In this work, the biochar from BSG was studied as an electrochemical enhancer, comparing its performance to those of printed electrodes modified with graphene. This study demonstrated that the use of biochar, instead of graphene, for the fabrication of screen-printed electrode, is very promising, reducing the cost of these devices while increasing the sustainability of beer production.

2. Materials and Methods

2.1. Chemicals

All reagents were of high purity and they were used without further purification. Graphene, epinephrine (EP), and tyrosinase (Ty) from mushrooms were purchased from Sigma (Merk Life Science S.r.l., Milan, Italy). A 5 mg/mL solution of Ty in 50 mM phosphate buffer (PB), pH 7.4, was used for the enzyme immobilization. The working buffer was 50 mM PB + 10 mM KCl, pH 7.0. Ultrapure water was used for the preparation of all aqueous solutions.

2.2. Preparation and Characterization of Biochar From Brewers' Spent Grain

Pellets from brewers' spent grain were used as a feedstock for biochar production, as previously reported by Sperandio et al. [1]. Pellets were subjected to a pyrolytic micro-gasification process (T = 400 °C) in an Elsa D17 micro pyrolytic reactor (Bluecomb Ltd., Udine, Italy). After the pyrolysis, biochar samples were manually ground to improve their homogeneity, and then used without further modifications for the sensors' fabrication [19].

2.3. FTIR Analysis and SEM Investigation of Biochar Samples

FTIR spectra of biochar samples were acquired using a Thermo-Scientific instrument (mod. iS 50 Nicolet, Thermo Scientific Inc., Madison, WI, USA), equipped with a single-reflectance horizontal ATR cell with a diamond crystal. Ground and homogenized biochar samples were placed at the surface of the diamond crystal and pressed with a system press tip flap. Samples were scanned at wavenumbers ranging from 4000 to 600 cm^{-1} (scans: 32; scan speed: 0.20 cm/s; resolution: 4 cm^{-1}) and corrected against the background spectrum of air. To obtain an averaged spectrum, three replicates of each sample were scanned and averaged. OMNIC™ software (Thermo Fisher Scientific Inc., Waltham, MA, USA) was used for processing the acquired spectra. In order to investigate the morphology of the biochar particles, in terms of shape, average size, and surface porosity, scanning electron microscopy (SEM) micrographs were acquired by means of field emission gun scanning electron microscopy (FEG-SEM) (Cambridge Leo Supra 35, Carl Zeiss, Jena, Germany), after sputter-coating with gold under argon atmosphere (25 mA, 120 s). The average particles diameter and pore sizes were determined considering randomly selected particles from the acquired SEM micrographs (ImageJ, NIH, Bethesda, MA, USA).

2.4. Electrochemical Characterization of Biochar Samples

All the electrochemical characterizations were performed by cyclic voltammetry (CV) and amperometry (CA), using screen printed electrodes (SPEs), home produced by the Laboratory of Analytical Chemistry of the University of Rome "Tor Vergata." The diameter of the working electrode was 0.3 cm, resulting in a geometric area of 0.07 cm^2. The measurements were carried out using an

Autolab electrochemical system (Eco Chemie, Utrecht, The Netherlands), equipped with PGSTAT-12 and GPES software (Eco Chemie, Utrecht, The Netherlands)) [20,21].

2.5. Biochar Modified SPEs (Biochar/SPE) Preparation and Characterization

The biochar dispersions were prepared in ethanolic medium at a concentration of 1 mg/mL, and then treated in an ultrasonic bath for 60 min. The biochar/SPEs were assembled via drop casting with 6 μL of biochar dispersions on bare graphitic SPE; then, the solvent was allowed to volatilize (37 °C). Electrochemical measurements were performed in drop (70 μL) detection mode. To evaluate the biochar distribution on the SPE, biochar/SPE was observed at FEG-SEM, after sputter-coating with gold under argon atmosphere (25 mA, 120 s), and compared with bare SPE.

2.6. Graphene-Modified SPEs (Graphene/SPE)

Graphene dispersions were prepared following the procedure described by Cinti et al. [21]: 10 mg of commercial reduced graphene oxide powder (rGO) was added to 10 mL of solvent (a mixture dimethylformamide (DMF): water (1:1, v/v)) and sonicated for 60 min at 59 kHz. Using this solution, the SPEs were modified via drop casting, adding 6 μL of graphene dispersion (1 mg/mL) on a working electrode, as described in detail in Section 2.5.

2.7. Fabrication and Characterization of Biosensor

The biosensor (Ty/biochar/SPE) was prepared immobilizing Ty on biochar/SPE by drop casting technique, followed by cross-linking with glutaraldehyde. A quantity of 50 μL of 50 mM PB containing 5 mg/mL of Ty was added onto the graphite biochar modified working electrode. After drying, the Ty/biochar/SPE was exposed to a 2.5% (v/v) glutaraldehyde solution (in 50 mM PB + 10 mM KCl, pH 7.0) for 20 min at room temperature. The enzyme-immobilized film was dried at room temperature and rinsed with PB to remove any unbound enzyme from the electrode surface. Finally, the biosensors were stored at 4 °C.

To study the repeatability, the stability, and the storage of the Ty/biochar/SPE biosensor, amperometric measurements were carried out with different EP solutions, in 50 mM PB at pH 7.0, using the same biosensor device. To demonstrate that enzymatic immobilization had occurred on the modified biochar/SPE, SEM micrographs were acquired by means of FEG-SEM, after sputter-coating with gold under argon atmosphere (25 mA, 120 s).

3. Results and Discussions

3.1. FTIR Analysis of Biochar Samples

In the present study, FTIR spectroscopy was used to analyse the functional groups on the surface of the biochar particles employed for sensor fabrication. In fact, besides the porosity, the adsorption behaviour of a biochar is influenced by the chemical reactivity of its surface, especially in the form of chemisorbed oxygen in various functional groups. Figure 1 shows a typical, averaged FT-IR spectrum (media of 16 spectra) of the biochar material investigated. The absence of the bands due to aromatic C–H stretch at 3050 cm^{-1} and aliphatic C–H stretch at 2900 cm^{-1} suggests that BSG biochar is comprised of two main structural fractions: graphite-like and randomly ordered, amorphous aromatic structures [22]. The IR spectrum of BSG biochar was characterized essentially by a restricted group of frequencies linked to the presence of aryl ring and phenol features. The peaks at around 2360 and 2340 cm^{-1} indicate carboxyl and carbonyl groups. The aromatic ring vibrations in the wavenumber range around 1600–1450 cm^{-1} confirm the presence of the aforementioned aromatic structures (C=C–C absorption bands) [23,24]. The peaks between 1350 and 1050 cm^{-1} can be ascribed to the presence of primary, secondary, and tertiary alcohols; phenols; ethers; and esters showing C–O stretching and O–H deformation vibrations.

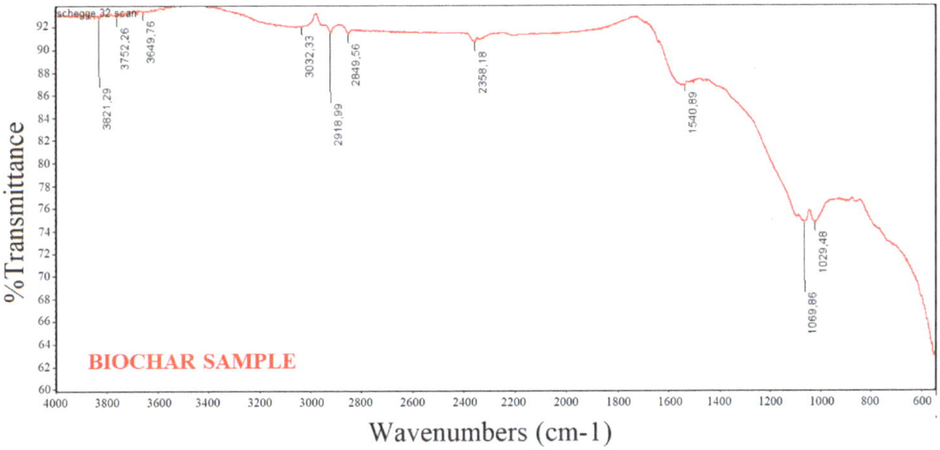

Figure 1. FTIR spectrum of the biochar sample.

Finally, the peaks between 850 and 630 cm^{-1} correspond to aromatic C–H stretching vibrations that indicate the presence of adjacent aromatic hydrogens in biochar samples [25].

The morphologies of the biochar samples were analysed at SEM. Figure 2a–d shows low and high magnification SEM micrographs of the biochar from BSG.

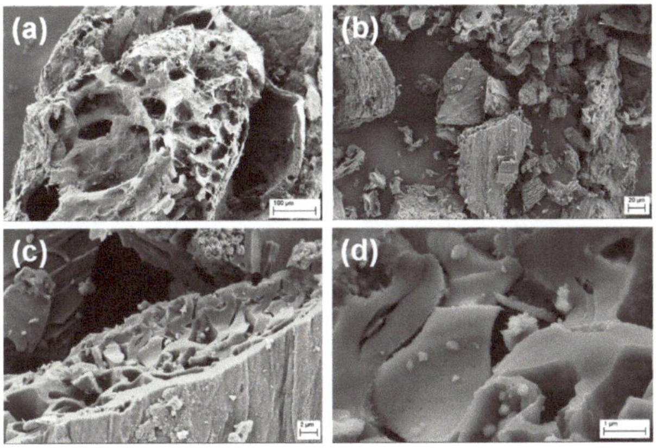

Figure 2. SEM micrographs of the biochar sample at different magnifications: (**a**) 500×, (**b**) 1000×, (**c**) 10,000×, (**d**) 50,000×.

Figure 2b underlines that biochar powder was composed of irregularly shaped, significantly amorphous and heterogeneous macro- and micro-particles with sponge-like structures, characterized by the presence of several pores of different sizes and many hollow channels (average diameters around 10–20 μm), even if little defined, uneven, and not uniform (Figure 2a). It is possible to observe very coarse, heterogeneous and plane cleavage surfaces, due to the pyrolysis process that is able to stabilize the volatile hydrocarbons, smoothening the biochar surface, and broken edges with tarry deposits on the surface. The vesicles on the biochar surface resulted from the gradual release of different volatile compounds formed in the softened biomass matrix during the pyrolysis process through a melt phase of cellular components (Figure 2c,d) [25–28].

It is important to take into account that the lignin and high volatile matter content in the starting biomass waste significantly affects the formation of porosities in the resulting biochar sample [29]. Moreover, the heat transfer during the gasification process strongly depends on the bulk density of the starting biomass waste material because lower ratios of air/fuel are achieved for lower bulk densities (i.e., lesser amount of biomass in the same volume of reactor) [30]. Generally, the number of pores and their size increase with the pyrolysis temperature (up to 550 °C), with consequent increment of the specific surface area as a function of temperature [31]. However, the process tends to combustion, increasing both temperature and reaction velocity, and higher carbonization degrees correspond to an increase in the carbon amount and a decrease in the oxygen content of the biochar, with a resultant increment of its surface hydrophobicity [30]. For these reasons, the pyrolysis was performed at 400 °C. Indeed, it has been reported that at pyrolysis temperatures higher than 550 °C, the biochar has a lower specific surface area due to the shrinkage of chars at post-softening and swelling temperatures, resulting in narrowing or closing pores, and a distortion of the pore structures occurs starting from 700 °C [32]. Indeed, it is very important to obtain and preserve these porous structures, since they provide a high internal surface area, and thus, high adsorption ability.

3.2. Biochar-Modified SPEs (Biochar/SPE)

With the aim of expanding its processability and future practical applications for SPE modification, we previously investigated the dispersion behaviour of biochar in different organic solvents. In this study, three different solvents were used for the dispersion of biochar: *N,N*-dimethylformamide (DMF)/H_2O (1:1, v/v), 1-Methyl-2-pyrrolidinone (NMP), and ethanol.

These dispersions (1 mg/mL), after sonication (60 min, 59 kHz), were used to modify the surface of the working electrode (WE) via drop casting; then the electrochemical response was tested and compared using cyclic voltammetry (CV) (Figure 3). Two test solutions were used: 50 mM PB + 10 mM KCl, pH 7.0, and 1 mM ferricyanide ($K_3Fe(CN)_6$), respectively (70 μL/SPE).

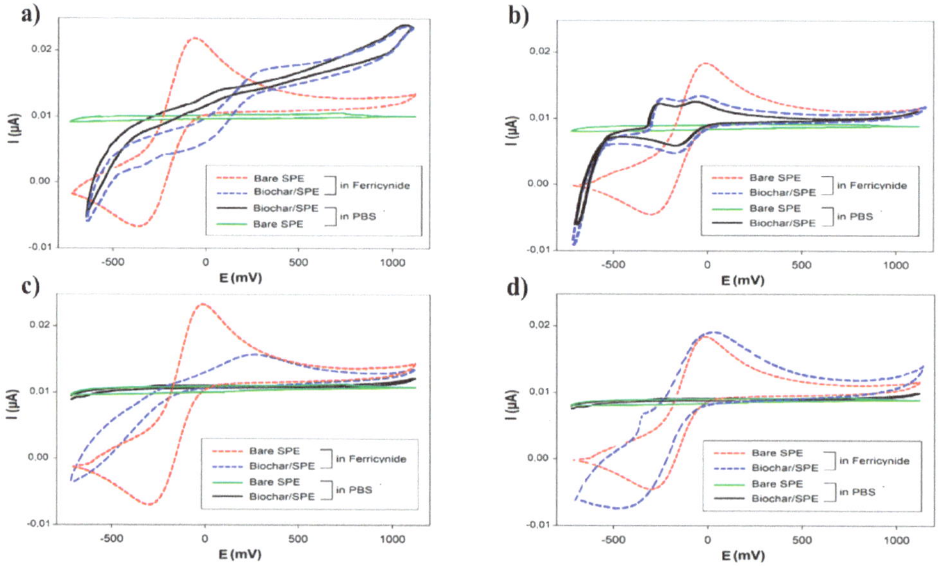

Figure 3. Cyclic voltammograms of bare screen printed electrodes (SPEs) and biochar/SPEs sensors obtained with dispersion of (**a**) DMF:H_2O (1:1 v/v), (**b**) DMF, (**c**) NMP, or (**d**) ethanol. Measurement conditions: scan rate 30 mVs^{-1}.

The results, shown in Figure 3, indicated that the solvent to be used for the modification of SPEs had to be the ethanol, because it guaranteed the most homogenous dispersion of the biochar, in relation to the recorded electrochemical signal. Using this solvent, the voltammogram of ferricyanide was closer to the ideal reversible behaviour of this compound; for this reason, it was chosen as the working solvent.

The electrochemical characterization of biochar/SPEs ($n = 6$) was carried out using five different electroactive compounds (Figure 4, Table 1): ferricyanide, epinephrine (EP), benzoquinone, uric acid (UA), and ascorbic acid (AA).

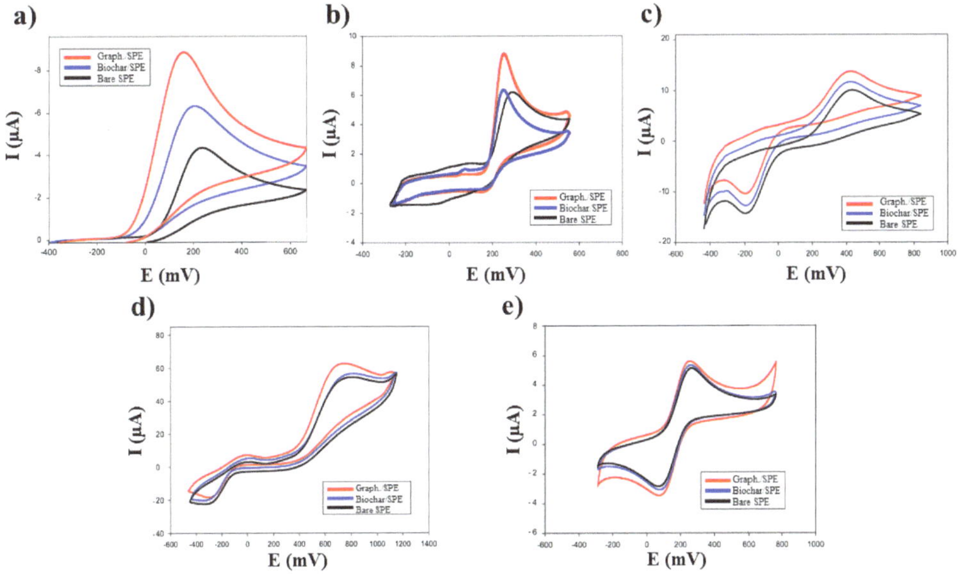

Figure 4. Cyclic Voltammograms of biochar based, graphene based, and bare SPE sensors obtained with: (**a**) 20 mM ascorbic acid, (**b**) 20 mM uric acid, (**c**) 20 mM benzoquinone, (**d**) 20 mM epinephrine, and (**e**) 20 mM ferricyanide in 50 mM phosphate buffer + 10 mM KCl, pH 7.4; scan rate 30 mVs^{-1}.

Table 1. Comparison of anodic peak current repeatability for biochar/SPE and graphene/SPE sensors for different electrochemical substrates (RSD% = standard deviation/μA, $n = 5$).

	Biochar/SPE RSD%	**Graphene/SPE RSD%**
Ascorbic acid	13	2
Uric acid	13	6
Benzoquinone	12	2
Epinephrine	12	1
Ferrcyanide	12	4

The results obtained with biochar/SPEs were compared with those of commercial graphene/SPE sensors, using the same substrates (Table 1).

The performances of biochar/sensors were comparable to those based on commercial graphene, widely used for this type of modification [17]. This experimental evidence shows that biochar can actually be used as an electrochemical enhancer with electrochemical performance similar to that of commercial graphene.

3.3. Electrochemical Behavior of Biochar/SPE

A further investigation was carried out in order to understand the electrochemical behaviour of the biochar/SPE interface (Figure 5, Table 2).

The voltammetric peak heights (I_p), obtained for biochar and graphene modified-SPEs in the scan rate study (Figure 5), were plotted against the square root of the scan rate. The following slopes ($\mu Acm^{-1}s^{-1/2}$) were obtained: 0.86 (R^2 = 0.997), 0.82 (R^2 = 0.995), and 0.47 (R^2 = 0.992), corresponding to graphene/SPE, biochar/SPE, and bare SPE, respectively.

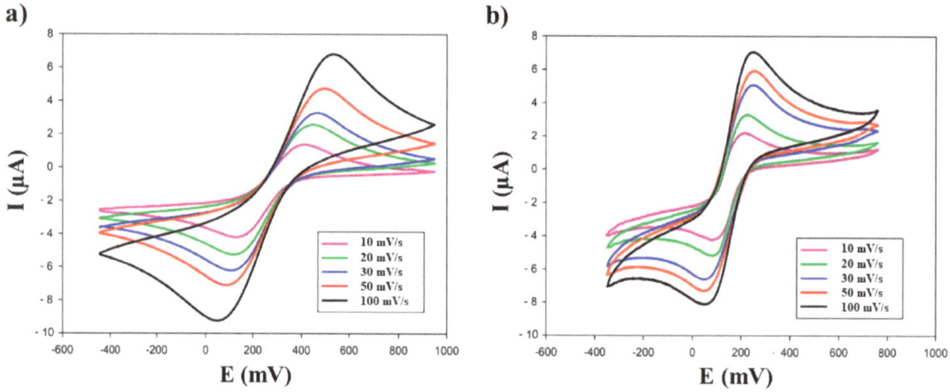

Figure 5. Scan rate study: comparison of cyclic voltammograms obtained with (**a**) biochar/SPEs and (**b**) graphene/SPEs sensors using 10 mM ferricyanide in 50 mM phosphate buffer + 10 mM KCl, pH 7.4.

The effective electrode area was calculated (Table 2) by Randless–Sevcik equation (Equation (1)) [33, 34] using the diffusion coefficient D_O and D_R, described by Konopka et al. (i.e., $D_O = 7.26 \times 10^{-6}$ cm^2/s, $D_R = 6.67 \times 10^{6}$ cm^{2}s^{-1}), and the I_p values, obtained for scan rate 30 mVs^{-1}:

$$I_p = (0.4463)nFAC\sqrt{\frac{nFvD_O}{RT}}, \quad (1)$$

where F is the constant of Faraday (mol^{-1}), R the universal constant of gas (JK^{-1}mol^{-1}), n the number of electrons exchanged, A the electrodic surface (cm^2), C the analyte concentration (molcm^{-3}), D_O the diffusion coefficient (cm^{2}s^{-1}), and v the scan rate (mVs^{-1}), respectively.

Table 2. Comparison of the effective active area for two different biochar/SPE and graphene/SPE sensors.

	A (cm^2)		M ± σ
	Anodic	**Cathodic**	
biochar/SPE	0.041	0.045	(4.3 ± 0.5) × 10^{-2}
graphene/SPE	0.053	0.049	(5.1 ± 0.6) × 10^{-2}

Furthermore, the heterogeneous rate constants (k^0) for the redox process $[Fe(CN)_6]^{3-} + 1e^- \leftrightarrows [Fe(CN)_6]^{4-}$ were calculated, using the Equation (2):

$$k^0 = \frac{\left[D_O \pi v \left(\frac{nF}{RT}\right)\right]^{\frac{1}{2}}}{\left(\frac{D_O}{D_R}\right)^{\frac{\alpha}{2}}}\varphi \quad (2)$$

where D_0 and D_R are the diffusion coefficients for the ferricyanide (D_0) and ferrocyanide (D_R), v is the scan rate (Vs^{-1}), n is the number of electrons involved in the process, F is the Faraday constant (mol^{-1}), T is the temperature (K), R is the universal gas constant ($JK^{-1}mol^{-1}$), and α the dimensional transfer coefficient [34].

For the D_0 and D_R, the relative values described by Konopka et al. (i.e., $D_0 = 7.26 \times 10^{-6}$ cm^2/s, $D_R = 6.67 \times 10^6$ cm^{2}s^{-1}) were used [34]. The α parameter was chosen to be equal to 0.5, assuming the ratio of the anodic and cathodic peak equal approximately to 1 ($I_{pa}/I_{pc} = 1$).

The parameter φ can be obtained using the Nickolson method [35], where for each ΔE there is a correspondence with a Ψ value. For a better evaluation of this parameter, the equation based on the Nickolson theory was used [36]:

$$\Psi = \frac{(-0.6288 + 0.0021 \cdot \Delta E)}{(1 - 0.017 \cdot \Delta E)} \quad (3)$$

The results obtained show a variation in the heterogeneous rate constants, demonstrating that biochar modified SPEs have a slower process of electronic transfer than the graphene modified platforms (Table 3).

Table 3. Comparison of k^0 standard electron rate constant for biochar/SPE and graphene/SPE sensors.

	k^0 (cms^{-1})
biochar/SPE	$(2.3 \pm 0.1) \times 10^{-3}$
graphene/SPE	$(1.6 \pm 0.1) \times 10^{-3}$

3.4. Tyrosinase Biosensor: An Example of Biochar/SPE Application

Liu at al. [37] reported an interesting work about the development of a tyrosinase/Chitosan/GOx SPE with a good sensitivity (22 nM) and a broad linear range (0.1–500 μM) compared with existing electrochemical sensors. Taking inspiration from the literature, biochar was used, for the first time, for the fabrication of biosensors by using biochar/SPE as the support for tyrosinase (Ty) immobilization, in order to demonstrate the possibility of using it like the widely commercial nanomaterials. This Ty/Biochar/SPE was designed for the development of biosensors with an enhanced electrochemical active area and enhanced electronic transfer properties. The enzymatic substrate chosen was epinephrine (one of the catecholamines).

In order to investigate the enzyme immobilization on the surface of the biochar modified SPEs, the bare SPE, biochar/SPE, and Ty/biochar/SPE samples were observed by means of SEM. Figure 6 compares their low and high magnification SEM micrographs. In all cases, low magnification micrographs allowed us to demonstrate the uniform and homogeneous distribution of the used ink on the surface of the working electrode, characterized by graphite micrometre-sized flakes bound together with an inert polymeric binder and covered of small cross-linking particles, present in the ink we employed. Additionally, biochar/SPE presented structures ascribable to the biochar addition, even if smaller than the starting particles, due to the ultrasonic treatment applied in order to obtain a good dispersion in the ethanol. In the case of Ty/biochar/SPE, the presence of a uniform and homogeneous layer on the biochar/SPE is evident, testifying that the enzyme immobilization allows for a homogenous and well-anchored enzymatic membrane.

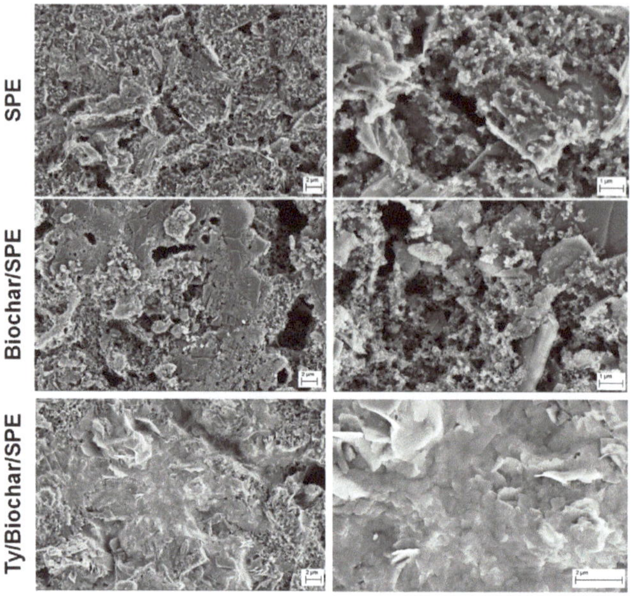

Figure 6. SEM micrographs of bare SPE, biochar/SPE and Ty/biochar/SPE.

The cyclic voltammograms are shown in Figure 7 using 100 μM epinephrine in 50 mM PB + 10 mM KCl, pH 7.0, where a reversible electrochemical behaviour was observed for epinephrine.

The oxidation peak is due to the oxidation of catecholamine to o-quinone, as shown in the following Reaction (1) [17,18]:

$$\text{Epinephrine} + \text{Tyrosinase } (O_2) \rightarrow \text{o-Epinephrinequinone} + H_2O. \quad (1)$$

The o-epinephrinquinone is electrochemically reduced to epinephrine on biosensor surface (2):

$$\text{o-Epinephrinequinone} + 2H^+ + 2e^- \rightarrow \text{Epinephrine}. \quad (2)$$

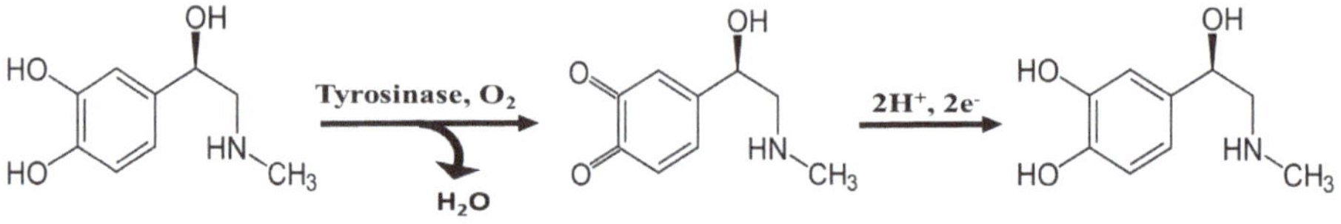

In EP solution (supporting electrolyte 50 mM PB, pH 7.0), the CV gave two well-defined peaks: cathodic peak at −0.030 V and anodic peak at +0.25 V.

These experimental data demonstrate that the tyrosinase enzyme retains its bioactivity when immobilized on biochar thick film. Tyrosinase, immobilized on biochar/SPE, efficiently catalyses the oxidation of EP. The sharp and intense oxidation and reduction peaks reveal a fast electron transfer at tyrosinase immobilized on biochar/SPE [17,18]. These results confirm that this method allows the determination of compounds belonging to the catecholamine family with satisfactory results, as also explained by Arduini et al. [38] and Maciejewska et al. [39].

Additionally, the effect of potential scan rate on the current peak of this catecholamine was studied.

In Figure 7, it can also be seen that the oxidation peak shifts to a more positive value, while the reduction peak to more negative values proportional to the increasing scan rates with the increment of the current for this catecholamine.

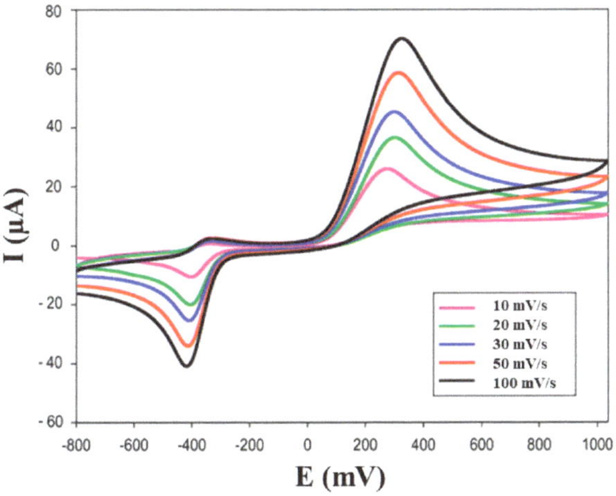

Figure 7. Cyclic Voltammograms of Ty-biochar/SPE biosensor of scan rate study: comparison of cyclic voltammograms obtained with Ty/biochar/SPEs biosensors using 2 mM epinephrine solution (in 50 mM phosphate buffer + 10 mM KCl, pH 7.4).

BSG biochar presents, not only, similarities with graphene, such as the ability to improve electrochemical performance and to be easily dispersed in stable and homogeneous suspensions, but has benefits such as cost-effectiveness—being configured as a cheap and easy-to-use material—for the development of electrochemical sensors.

The characterization of biochar as a supporting material to modify the SPE was carried out studying several parameters, such as the pH of the working buffer used for the electrochemical measurements, working range, stability, and reproducibility of the biosensor, and the storage conditions of the biochar-based biosensor. The influence of the pH (in the range of 4.0–9.0) on the electrochemical behaviour of EP in several buffer solutions was investigated using cyclic voltammetry. The pH of working buffer showed a significant effect on the electrochemical behaviour of EP (100 µM) at the surface of the modified biosensor.

It was observed that the peak potentials decreased (with a shift towards negative potential values) with the increase of the pH. In Figure 8, the graph of the cathodic peak current (Ipc) versus pH is reported. The better electrochemical performances were obtained when the pH was 7.0.

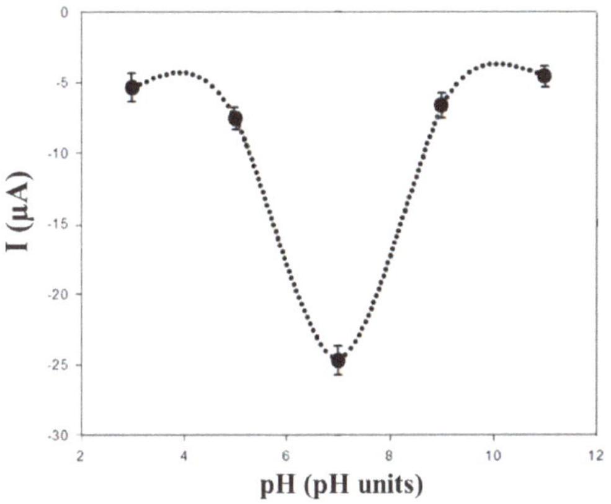

Figure 8. Current-pH dependence in 10 mM EP solution (in 50 mM phosphate buffer + 10 mM KCl, at different pH values).

The dependence of the biosensor's response on the electrode potential is shown in Figure 9 using 100 μM EP in 50 mM PB + 10 mM KCl, pH 7.0. The maximum of the electrochemical signal was obtained at −0.32 V; this potential was applied to EP detection after the immobilization of a specific enzyme for this molecule.

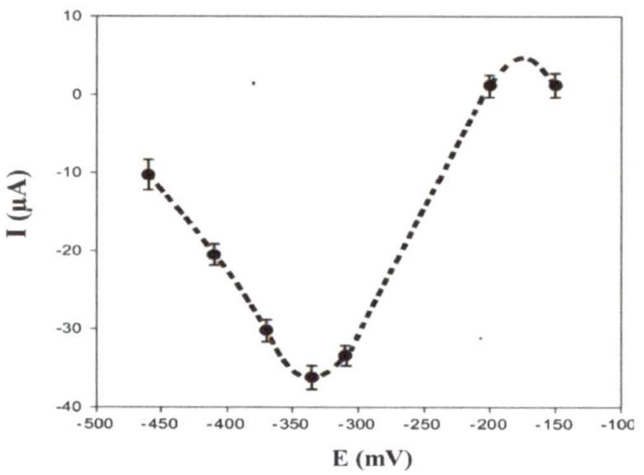

Figure 9. Current-potential dependence in 10 mM EP solution (in 50 mM phosphate buffer + 10 mM KCl, pH 7.0).

3.5. Amperometric Response of the Biosensor

Figure 10 illustrates a typical amperometric response for the Ty/biochar/SPE biosensor after the addition of successive aliquots of EP using 50 mM PB + 10 mM KCl (pH 7.0) and applying −0.32 V as the working potential. The reduction of current, proportional to the concentration of catecholamine, was due to the electrochemical reduction of o-epinephrinequinone, the enzymatic product of tyrosine when EP is present.

The measurements of EP showed a linear current response in a concentration range of 0.05–0.5 mM (Figure 10). The detection limit, calculated as *3sb/m* criterion (*m* is the slope of the calibration graph; *sd* is the standard deviation (n = 5) of the current signals of the substrate at the concentration level corresponding to the lowest concentration of the calibration plot) [40]) resulted in 2.4×10^{-4} mM. This value was within the range of the detection limit values found for tyrosinase biosensors present in literature [18].

Using these results, the Hill coefficient (h) can be calculated as the $\log[I/(I_{max} - I)]$ versus log [EP] (the logarithm of the substrate concentration), obtaining 0.89 for biochar/SPE biosensor ($r^2 = 0.990$) and 1.01 for graphene/SPE biosensor ($r^2 = 0.999$), respectively. The h parameter, calculated from the corresponding Hill's plot, was close to 1, demonstrated that the kinetics of the enzymatic reaction fitted into a Michaelis–Menten kinetics model [41]. The enzymatic kinetic parameters were obtained using the linearization of Lineweaver–Burk by Equation (4) [40]:

$$\frac{1}{I} = \frac{1}{I_{max}} + \frac{K_M^{app}}{I_{max}[S]}, \quad (4)$$

where $[S]$ is the concentration of the substrate, I the cathodic current at −0.32V, and K_M^{app} the apparent Michaelis–Menten constant for the enzymatic reaction and I_{max} the steady-state current. Using the Lineweaver–Burk equation and representing $1/I$ versus 1/[EP], the apparent Michaelis-Menten constant was calculated from the slope and the I_{max} from the intercept.

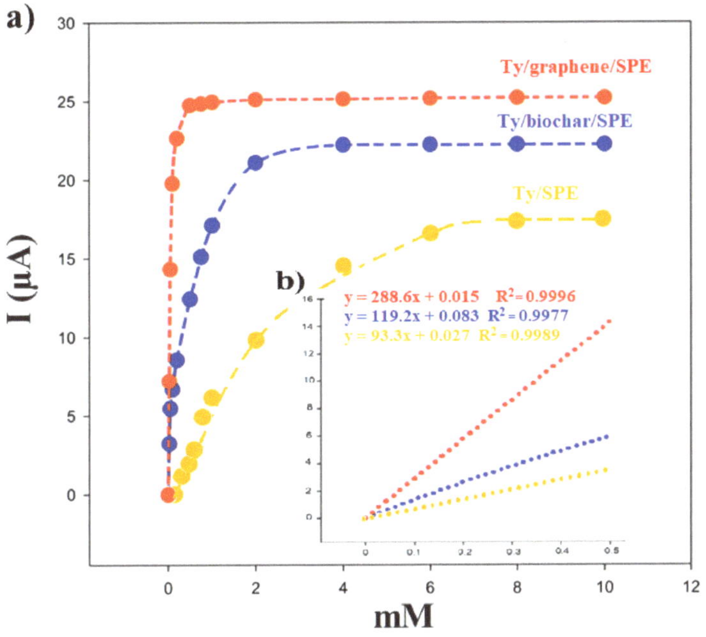

Figure 10. (**a**) Calibration curve between the cathodic current and the concentration of EP in 50 mM phosphate buffer + 10 mM KCl, pH 7.4: (blue) Ty/biochar/SPE, (red) Ty/GPH/SPE, and (yellow) Ty/bare SPE biosensor. (**b**) Linear range.

From the data shown in the Table 4 we can see a greater affinity of the enzyme when it is immobilized on graphene; however, as for all the experimental results obtained in this work, in this case, the biosensor performance based on the biochar is comparable with that of Ty/graphene/SPE.

Table 4. Comparison of analytical parameters for different biosensors: Ty/bare SPE, Ty/Biochar/SPE, and Ty/Graphene/SPE biosensors.

	Linear Range (mM)	LOD (mM)	K_M^{app} (mM)
Ty/biochar/SPE	0.02–0.50	2.4×10^{-4}	0.15
Ty/graphene/SPE	0.02–0.25	1.0×10^{-4}	0.09
Ty/bare SPE	0.02–2.0	9.2×10^{-4}	0.25

3.6. Repeatability and Stability of the Biosensor

To study the repeatability of Ty/biochar and graphene based SPEs, amperometric measurements for epinephrine detection were carried out with a 0.25 mM EP solution in PB, using the same biosensors devices for 10 times. The relative standard deviation (%RSD) for the Ty/biochar/SPEs was equal to 6%, while for Ty/graphene platform it was 4%.

The stability of the biosensor was studied by monitoring the amperometric response of the same biosensor at regular intervals of 24 hours for 20 days. Between measurements, the biosensor was stored in a refrigerator at 4 °C in dry condition. The amperometric responses of the biosensor with and without biochar remained quite constant in the first 7 days of storage. In particular, biochar-based biosensor maintained 98.2% of its initial current response with a decrease to 87.6% after two weeks, while the Ty/graphene-SPE showed an initial 98.5%, decreasing to 92.2% after two weeks. After this time, the biosensor's amperometric response constantly decreased, until 55.7% (biochar) and 75.6%

(graphene) of the initial signal in 20 days. Therefore, both graphene and biochar-based platforms can be considered "ready to use" and stored at 4 °C within the first 7–10 days, obtaining good results for EP determination.

4. Conclusions

In the present study, brewers' spent grain biochar was investigated as an innovative, eco-friendly graphene-alternative for the production of screen-printed electrodes.

From SEM investigation, it is evident that the biochar powder we produced was made of irregularly shaped macro- and micro-particles with a sponge-like structure, characterized by the presence of several pores of different sizes and many hollow channels, suitable for enzymes' immobilization and electrochemical applications. Following this approach, the biochar obtained from spent grain was used to improve the performances of screen-printed electrodes through the modification of their surfaces with the proposed material by drop casting. Moreover, the good and uniform distribution of biochar on the SPE surface was demonstrated by observation at SEM, as well as the tyrosinase immobilization in the case of Ty/biochar/SPE. The Ty/biochar/SPEs exhibited similar characteristics to graphene/SPEs, widely used for biosensor development, in terms of the electron transfer kinetics for electroactive compounds. Large peak potential separation and high peak current could be obtained using CV on the developed electrode. Good sensitivity and detection limit for epinephrine promote the Ty/biochar/SPE to be an effective sensor for direct determination of this molecular target in real samples.

Author Contributions: Conceptualization, L.M. and K.C.; methodology, L.M. and K.C.; validation, R.C., K.C., M.P. and I.C.; formal analysis, R.C. and L.M.; investigation, R.C., K.C. and I.C.; data curation, R.C. and L.M.; writing—original draft preparation, R.C., K.C., I.C. and L.M.; writing—review and editing R.C., K.C., I.C. and L.M.; supervision, K.C. and L.M.; project administration, K.C.

Funding: Not specific funding was used for this research.

Conflicts of Interest: The authors declare no conflict of interest.

References

1. Sperandio, G.; Amoriello, T.; Carbone, K.; Fedrizzi, M.; Monteleone, A.; Tarangioli, S.; Pagano, M. Increasing the value of spent grain from craft microbreweries for energy purposes. *Chem. Eng. Trans.* **2017**, *58*, 487–492.
2. Mussatto, S.I.; Dragone, G.; Roberto, I.C. Brewers' spent grain: Generation, characteristics and potential applications. *J. Cereal Sci.* **2006**, *43*, 1–14. [CrossRef]
3. Townsley, P.M. Preparation of commercial products from brewer's waste grain and trub [Protein flours]. *Tech. Q. Master Brew. Assoc. Am.* **1979**, *4*, 130–134.
4. Mussatto, S.I. Brewer's spent grain: A valuable feedstock for industrial applications. *J. Sci. Food Agric.* **2014**, *94*, 1264–1275. [CrossRef] [PubMed]
5. Faulds, C.B.; Mandalari, G.; LoCurto, R.; Bisignano, G.; Waldron, K.W. Arabinoxylan and mono-and dimeric ferulic acid release from brewer's grain and wheat bran by feruloyl esterases and glycosyl hydrolases from *Humicola insolens*. *Appl. Microbiol. Biotechnol.* **2004**, *64*, 644–650. [CrossRef] [PubMed]
6. Pham, T.P.T.; Kaushik, R.; Parshetti, G.K.; Mahmood, R.; Balasubramanian, R. Food waste-to-energy conversion technologies: Current status and future directions. *Waste Manag.* **2015**, *38*, 399–408. [CrossRef]
7. Woolf, D.; Amonette, J.E.; Street-Perrott, F.A.; Lehmann, J.; Joseph, S. Sustainable biochar to mitigate global climate change. *Nat. Commun.* **2010**, *1*, 56. [CrossRef]
8. Krull, E.S.; Baldock, J.A.; Skjemstad, J.O.; Smernik, R.J. Characteristics of biochar: Organo-chemical properties. In *Biochar for Environmental Management*; Lehmann, J., Joseph, S., Eds.; Routledge: London, UK, 2009; pp. 53–66.
9. Zhao, L.; Cao, X.; Zheng, W.; Wang, Q.; Yang, F. Endogenous minerals have influences on surface electrochemistry and ion exchange properties of biochar. *Chemosphere* **2015**, *136*, 133–139. [CrossRef]
10. Moro, G.; Bottari, F.; Van Loon, J.; Du Bois, E.; De Wael, K.; Moretto, L.M. Disposable electrodes from waste materials and renewable sources for (bio)electroanalytical applications. *Biosens. Bioelectron.* **2019**, *146*, 111758. [CrossRef]

11. De Oliveira, P.R.; Kalinke, C.; Gogola, J.L.; Mangrich, A.S.; Junior, L.H.M.; Bergamini, M.F. The use of activated biochar for development of a sensitive electrochemical sensor for determination of methyl parathion. *J. Electroanal. Chem.* **2017**, *799*, 602–608. [CrossRef]
12. Gevaerd, A.; de Oliveira, P.R.; Mangrich, A.S.; Bergamini, M.F.; Marcolino-Junior, L.H. Evaluation of antimony microparticles supported on biochar for application in the voltammetric determination of paraquat. *Mater. Sci. Eng. C* **2016**, *62*, 123–129. [CrossRef] [PubMed]
13. Agustini, D.; Mangrich, A.S.; Bergamini, M.F.; Marcolino-Junior, L.H. Sensitive voltammetric determination of lead released from ceramic dishes by using of bismuth nanostructures anchored on biochar. *Talanta* **2015**, *142*, 221–227. [CrossRef] [PubMed]
14. Bal Altuntas, D.; Akgül, G.; Yanik, J.; Anik, U. A biochar-modified carbon paste electrode. *Turk. J. Chem.* **2017**, *41*, 455–465. [CrossRef]
15. Gomez, F.J.V.; Aída, M.; Silva, M.F.; Escarpa, A.A. Screen-printed electrodes modified with carbon nanotubes or graphene for simultaneous determination of melatonin and serotonin. *Microchim. Acta* **2015**, *182*, 1925–1931. [CrossRef]
16. Pérez-Ràfols, C.; Serrano, N.; Díaz-Cruz, J.M.; Ariño, C.; Esteban, M. Glutathione modified screen-printed carbon nanofiber electrode for the voltammetric determination of metal ions in natural samples. *Talanta* **2016**, *155*, 8–13. [CrossRef] [PubMed]
17. Apetrei, I.M.; Apetrei, C. Study of different carbonaceous materials as modifiers of screen-printed electrodes for detection of catecholamines. *IEEE Sens. J.* **2015**, *15*, 3094–3101. [CrossRef]
18. Apetrei, I.M.; Popa, C.V.; Apetrei, C.; Tutunaru, D. Biosensors based on graphene modified screen-printed electrodes for the detection of catecholamines. *Rom. Biotechnol. Lett.* **2014**, *19*, 9801–9809.
19. Abenavoli, L.M.; Longo, L.; Proto, A.R.; Gallucci, F.; Ghignoli, A.; Zimbalatti, G.; Colantoni, A. Characterization of biochar obtained from olive and hazelnut prunings and comparison with the standards of European Biochar Certificate (EBC). *Procedia Behav. Sci.* **2016**, *223*, 698–705. [CrossRef]
20. Idegami, K.; Chikae, M.; Nagatani, N.; Tamiya, E.; Takamura, Y. Fabrication and Characterization of Planar Screen-Printed Ag/AgCl Reference Electrode for Disposable Sensor Strip. *Jpn. J. Appl. Phys.* **2010**, *49*, 097003. [CrossRef]
21. Cinti, S.; Arduini, F.; Carbone, M.; Sansone, L.; Cacciotti, I.; Moscone, D.; Palleschi, G. Screen-Printed Electrodes Modified with Carbon Nanomaterials: A Comparison among Carbon Black, Carbon Nanotubes and Graphene. *Electroanalysis* **2015**, *27*, 2230–2238. [CrossRef]
22. Reeves, J.B., III. Mid-infrared spectroscopy of biochars and spectral similarities to coal and kerogens: What are the implications? *Appl. Spectrosc.* **2012**, *66*, 689–695. [CrossRef] [PubMed]
23. Coates, J. Interpretation of infrared spectra, a practical approach. *Encycl. Anal. Chem.* **2000**, *12*, 10815–10837.
24. Angın, D. Effect of pyrolysis temperature and heating rate on biochar obtained from pyrolysis of safflower seed press cake. *Bioresour. Technol.* **2013**, *128*, 593–597. [CrossRef] [PubMed]
25. Inyang, M.; Gao, B.; Pullammanappallil, P.; Ding, W.; Zimmerman, A.R. Biochar from anaerobically digested sugarcane bagasse. *Bioresour. Technol.* **2010**, *101*, 8868–8872. [CrossRef]
26. Angın, D.; Şensöz, S. Effect of pyrolysis temperature on chemical and surface properties of biochar of rapeseed (*Brassica napus* L.). *Int. J. Phytoremediation* **2014**, *16*, 684–693. [CrossRef]
27. Liu, Z.; Zhang, F.S.; Wu, J. Characterization and application of chars produced from pinewood pyrolysis and hydrothermal treatment. *Fuel* **2010**, *89*, 510–514. [CrossRef]
28. Guerrero, M.; Ruiz, M.P.; Alzueta, M.U.; Bilbao, R.; Millera, A. Pyrolysis of eucalyptus at different heating rates: Studies of char characterization and oxidative reactivity. *J. Anal. Appl. Pyrolysis* **2005**, *74*, 307–314. [CrossRef]
29. Lehmann, J.; Rillig, M.C.; Thies, J.; Masiello, C.A.; Hockaday, W.C.; Crowley, D. Biochar effects on soil biota—A review. *Soil Biol. Biochem.* **2011**, *43*, 1812–1836. [CrossRef]
30. Shaaban, A.; Se, S.M.; Dimin, M.F.; Juoi, J.M.; Husin, M.H.M.; Mitan, N.M.M. Influence of heating temperature and holding time on biochars derived from rubber wood sawdust via slow pyrolysis. *J. Anal. Appl. Pyrolysis* **2014**, *107*, 31–39. [CrossRef]
31. Jindarom, C.; Meeyoo, V.; Kitiyanan, B.; Rirksomboon, T.; Rangsunvigit, P. Surface characterization and dye adsorptive capacities of char obtained from pyrolysis/gasification of sewage sludge. *Chem. Eng. J.* **2007**, *133*, 239–246. [CrossRef]

32. Guo, J.; Chong Lua, A. Characterization of chars pyrolyzed from oil palm stones for the preparation of activated carbons. *J. Anal. Appl. Pyrolysis* **1998**, *46*, 113–125. [CrossRef]
33. Wang, J. *Analytical Electrochemistry*; John Wiley and Sons: Hoboken, NJ, USA, 2000; pp. 31–33.
34. Konopka, S.J.; McDuffie, B. Diffusion coefficients of ferri-and ferrocyanide ions in aqueous media, using twin-electrode thin-layer electrochemistry. *Anal. Chem.* **1970**, *42*, 1741–1746. [CrossRef]
35. Nicholson, R.S. Theory and Application of Cyclic Voltammetry for Measurement of Electrode Reaction Kinetics. *Anal. Chem.* **1965**, *37*, 1351–1355. [CrossRef]
36. Vicentini, F.C.; Ravanini, A.E.; Figueiredo-Filho, L.C.; Iniesta, J.; Banks, C.E.; Fatibello-Filho, O. Imparting improvements in electrochemical sensors: Evaluation of different carbon blacks that give rise to significant improvement in the performance of electroanalytical sensing platforms. *Electrochim. Acta* **2015**, *157*, 125–133. [CrossRef]
37. Liu, C.Y.; Chou, Y.C.; Tsai, J.H.; Huang, T.M.; Chen, J.Z.; Yeh, Y.C. Tyrosinase/Chitosan/Reduced Graphene Oxide Modified Screen-Printed Carbon Electrode for Sensitive and Interference-Free Detection of Dopamine. *Appl. Sci.* **2019**, *9*, 622. [CrossRef]
38. Arduini, F.; Giorgio, F.D.; Amine, A.; Cataldo, F.; Moscone, D.; Palleschi, G. Electroanalytical Characterization of Carbon Black Nanomaterial Paste Electrode: Development of Highly Sensitive Tyrosinase Biosensor for Catechol Detection. *Anal. Lett.* **2010**, *43*, 1688–1702. [CrossRef]
39. Maciejewska, J.; Pisarek, K.; Bartosiewicz, I.; Krysiński, P.; Jackowska, K.; Biegunski, A.T. Selective detection of dopamine on poly(indole-5-carboxylic acid)/tyrosinase electrode. *Electrochim. Acta* **2011**, *56*, 3700–3706. [CrossRef]
40. Lavín, Á.; Vicente, J.D.; Holgado, M.; Laguna, M.F.; Casquel, R.; Santamaría, B.; Maigler, M.V.; Hernández, A.L.; Ramírez, Y. On the Determination of Uncertainty and Limit of Detection in Label-Free Biosensors. *Sensors* **2018**, *18*, 2038. [CrossRef]
41. Kurganov, B.; Lobanov, A.; Borisov, I.; Reshetilov, A. Criterion for Hill equation validity for description of biosensor calibration curves. *Anal. Chim. Acta* **2001**, *427*, 11–19. [CrossRef]

biosensors

MDPI

Article

Exploiting Laser-Ablation ICP-MS for the Characterization of Salt-Derived Bismuth Films on Screen-Printed Electrodes: A Preliminary Investigation

Carlo Dossi [1,*], Gilberto Binda [1], Damiano Monticelli [2], Andrea Pozzi [2], Sandro Recchia [2] and Davide Spanu [2]

[1] Dipartimento di Scienze Teoriche ed Applicate, Università degli Studi dell'Insubria, Via Dunant, 3, 22100 Varese, Italy; g.binda2@uninsubria.it

[2] Dipartimento di Scienza ed Alta Tecnologia, Università degli Studi dell'Insubria, Via Valleggio, 11, 21100 Como, Italy; Damiano.Monticelli@uninsubria.it (D.M.); andrea.pozzi@uninsubria.it (A.P.); sandro.recchia@uninsubria.it (S.R.); d.spanu1@uninsubria.it (D.S.)

* Correspondence: carlo.dossi@uninsubria.it; Tel.: +39-031-2386235

Received: 30 July 2020; Accepted: 8 September 2020; Published: 9 September 2020

check for updates

Abstract: The use of insoluble bismuth salts, typically $BiPO_4$, is known to be a viable alternative to classical Bi^{3+} ion electrochemical reduction for the preparation of bismuth film electrodes (BiFE) on screen-printed electrodes. The freshly prepared electrodes are indefinitely stable, and the active bismuth film is simply formed by in situ reduction. Two aspects are still to be investigated, namely the bismuth distribution on the working electrode and the possible residual presence of the counteranion, namely phosphate. High-vacuum techniques such as electron microscopy or spectroscopy, which are commonly employed for this purpose, cannot be safely used: the bismuth surface is well-known to reconstruct and recrystallize under the electron beam in vacuum. Here, we demonstrate the suitability and the effectiveness of laser ablation ICP-MS (LA-ICP-MS, a technique that vaporizes and analyzes the surface material under flowing helium at atmospheric pressure) for the characterization of BiFE. Fast and stable measurements of bismuth and phosphorous distribution are achieved with the advantage of a minimum alteration of the sample surface, avoiding possible interferences. This investigation evidenced how, upon reductive activation, the bismuth film is distributed with a radial symmetry and the phosphate counteranion is completely absent on the working electrode surface.

Keywords: laser ablation ICP-MS; surface characterization; bismuth film electrodes; screen-printed electrodes; trace electroanalysis

1. Introduction

The pioneering works by Wang and Hocevar on bismuth [1,2] had the merit to draw considerable interest in this "green" element to prepare electroactive surface films as a low-toxicity alternative to conventional mercury films. In fact, bismuth shows evident analogies with mercury in forming alloys and adsorptive complexes with many metals [3]. Although many years have passed since the introduction of voltammetry and stripping for the detection of trace heavy metals, research on these electroanalytical techniques is still a hot topic, since many advantages may be seen compared to classical optical techniques, not simply their low cost. Interestingly, increased interest to "green" metal modifiers other than bismuth is also observed [4–7]. It is finally noteworthy that bismuth, thanks to its very low toxicity, will play a definite role for the electroanalytical studies of organic as well as biological molecules [8–10].

The preparation of an active bismuth film on a screen-printed electrode (BiFE-SPE) is generally carried out under the conventional electrochemical reduction of Bi^{3+} ion in aqueous electrolytes. This process is not a facile one, since it is reported to be sensitive to the nature of the bismuth precursor, the possible occurrence of kinetics limitations, and the sensitivity to the cleanliness of the carbon surface of the working electrode [11,12]. This wide range of experimental conditions leading to the optimal and active bismuth film may be rationalized on the basis of the rich chemistry of bismuth in aqueous solutions [13]. In this context, a decisive advantage was gained applying surface modification with polymers: this treatment permits to mitigate interferences and to improve the mechanical stability of the bismuth film [14,15]. As an additional benefit, some polymeric layers are reported to have a synergistic sensing effect (e.g., using carboxymethyl cellulose which is capable of chelating cadmium ions) [16].

An alternative and innovative approach was proposed by Brainina et al. [17]. This strategy involves the in situ formation of an insoluble bismuth salt (e.g., $BiPO_4$) during the electrode preparation with the polymeric film. The bismuth film is then formed by reductive activation. However, the deposition of the insoluble salt precursor may be critical, and the following activation to the active bismuth surface may be complicated by liquid/solid phase kinetics. These conditions require a precise microscopical characterization of the bismuth film and the electrode surface, both before and after the electrochemical activation to ensure electrode performances. Two aspects are of main interest for their characterization: the bismuth distribution and the possible residual presence of the counteranion of the insoluble salt, namely phosphate.

A number of groups have used scanning and transmission electron microscopy [4–6,10,17–19] for the investigation of these features on bare and modified metal films on SPE, showing very different shapes and morphology. However, bismuth (or mercury) is a soft metal, and possible changes of the phase state of the thin film have been evidenced, due to the combined effect of high-vacuum and high-energy beam conditions [20]. Therefore, it should be taken into consideration the possibility that the observed shape and morphology of the metal surface may be profoundly dependent upon the experimental conditions of the high-vacuum experiment. Surface spectroscopies, primarily X-ray photoelectron spectroscopy (XPS), are also reported to be used to characterize metal films, as well as nanomodifiers, on screen-printed electrodes [21–23]. XPS is, in fact, very effective to probe the chemical composition of the outermost few nanometers of materials and the electrochemical activity strictly depends on the first atomic layers in contact with the electrolytic solution. However, no imaging information is likely to be obtained.

In this paper, the laser ablation ICP-MS technique has been proposed as a potential substitute (or complementary) technique of electron microscopy or surface spectroscopy for the investigation of the bismuth surface. Lasers have been used for the modification and functionalization of electrodes, and laser-induced forward transfer (LIFT) has emerged as a powerful physisorption method for the deposition of various materials on the electrode surface [24].

Here, we focused only on the analytical use of a scanning laser beam that vaporizes the surface material under flowing helium at atmospheric pressure. Such conditions should be fit for purpose without compromising bismuth stability. Although this technique possesses a low resolution, it has been already used for many applications for solid material characterization, giving information on the surface topology and on the qualitative and quantitative elemental composition [25,26]. We demonstrate here that the application of this technique is also suitable in the case of the analysis of bismuth films on screen-printed electrodes for electroanalysis. Laser ablation ICP-MS can permit a fast and stable measurement of the distribution of this metal on the electrode, with the advantage of a reduced alteration of the sample surface, avoiding possible interferences.

2. Materials and Methods

2.1. Apparatus and Reagents

Screen-printed electrodes (SPEs) have been a kind gift by Prof. Giuseppe Palleschi and his group at the University of Roma "Tor Vergata". The printing procedures of the electrodes on a flexible polyester film have been described in the literature [27–29].

An optical microscopy view of the electrode is shown in Figure 1A, where its three-electrode structure is evidenced. The carbon working electrode is in the center, with a diameter of 0.3 cm, along with the silver pseudo-reference electrode on the top right, and the surrounding carbon counter-electrode on the left.

Hydrochloric and nitric acid were TraceSelect from Fluka, and they were purified by sub-boiling distillation in a quartz-Teflon apparatus [30]. Chemical reagents (disodium hydrogen phosphate, potassium nitrate, sodium sulphate) were purchased from Aldrich. Poly (sodium-4-styrene sulfonate) (PSS, Aldrich, St. Louis, MO, USA) was used as such.

Standard stock solutions of Cd and Pb were purchased from Aldrich, and bismuth stock solution was prepared from bismuth nitrate pentahydrate (CarloErba). These 1000-mg/L solutions were diluted as required for the preparation of standards.

For standards and solutions, as well as glassware cleaning, high-purity MilliQ water from MilliPore was used.

2.2. Preparation of Bismuth Films Screen-Printed Electrodes from Insoluble Bismuth Phosphate

Pristine screen-printed electrodes have been used as received after cleaning in HCl solution. Polystyrene sulfonate (PSS)-modified Bi/PO_4 electrodes were prepared via deposition of the polymer layer and of the insoluble bismuth phosphate, following the 2-step recipe in Reference [13]:

Step 1: 2 µL of a 25 mM solution of PSS is drop-casted onto the working electrode with a microsyringe and left for drying.

Step 2: equal volumes of Na_2HPO_4 (4 mM) and $Bi(NO_3)_3$ (2 mM) were mixed, and 5 µL of this solution is then drop-casted onto the working electrode and left for drying. These electrodes can be stored in air at room temperature.

Prior of their use for the electrochemical analysis, the active bismuth film is formed by cycling eight times in 0.01 M HCl (pH = 2) under the electroanalytical conditions for the differential pulse stripping analysis described below in Section 2.3. The activated electrodes must be stored in a diluted acid solution without contact to air.

For the nitrate or sulphate interference study, sodium sulphate or potassium nitrate were added equimolar to 0.01 M HCl during repetitive scan experiments.

2.3. Measurement Procedures

PalmSens (Palm Instrument BV, Houten, The Netherlands) and Amel4330 (Amel srl, Milano, Italy) instruments were used. All instrument control as well as the full data analysis were handled by the Windows-based VApeak software.

A special cell was designed in collaboration with Amel, in order to optimize mass transport to the electrodes. To avoid the drawbacks of the internal pseudo-reference electrode, an external 3 M KCl Ag/AgCl reference electrode was used in order to reference all potentials.

The analysis of cadmium and lead was performed in 0.01 M HCl, or 0.01 M HNO_3, using differential pulse stripping voltammetry (DPV). High-purity nitrogen for degassing and blanketing was used throughout the entire stripping experiment. Both the activation of the electroactive bismuth film and the stripping analysis were done under the sane conditions. Potential was initially set at −1.05 V for 60 s at a stirring rate of 300 rpm; then, stirring was stopped, and the DPV analysis started, scanning the potential from −0.95 to −0.25 V, with a step potential of 5 mV, pulse time of 100 ms, and pulse amplitude of 50 mV.

Peak areas measured as (μA × mV) were used for quantitation. The best-fit linear calibration line was calculated as y = mx + q. Analytical sensitivity was defined as the slope (m) of the best-fit calibration line. Limits of detection (LOD) were calculated as 3σ/m and reported as μg/L. σ was determined from the standard deviation of 10 replicate analyses on a standard Pb and Cd sample with concentrations near the limit of detection.

2.4. LA-ICP-MS Analysis

An inductively coupled plasma-mass spectrometry (ICP-MS), model Thermo Scientific ICAP Q, coupled with a fourth harmonic Nd: YAG 266 nm laser system, model New Wave UP Series 266, was used for the laser ablation-ICP-MS (LA-ICP-MS) measurements. The sample chamber for laser scanning (a viscous film sample chamber was employed in this work, see reference [31] for details concerning the design and analytical performances) was flushed with high-purity helium, that was then interfaced with the ICP torch. A careful setting of the instrumental parameters has been done in order to optimize the mass signal, in terms both of speed and signal-to-noise ratio. The optimized parameters of the laser ablation ICP-MS instrument are reported in Table 1.

Table 1. List of instrumental parameters for the laser ablation (LA)-ICP-MS experiments.

Laser Ablation Parameters	
Scan Speed	20 μm/s
% output	20%
Step rate	20 Hz
Laser spot size	80 μm
Distance between each laser scan	260 μm
Mass Spectrometer Parameters	
Dwell Time	40 ms
PC Detector	3300

Optical microscopy views of the screen-printed electrode prior and after a full laser ablation scan are reported in Figure 1A,B, respectively. The six laser scans are clearly visible as thin dark lines. During the laser scan, the ablated material is transported by the helium flow into the ICP torch, and mass signals at ^{13}C, ^{31}P, and ^{209}Bi are continuously recorded using a dwell time of 40 ms. The ^{13}C signal is used as background to reference phosphorous and bismuth *m*/*z* signals in order to improve the signal stability and signal-to-noise ratios.

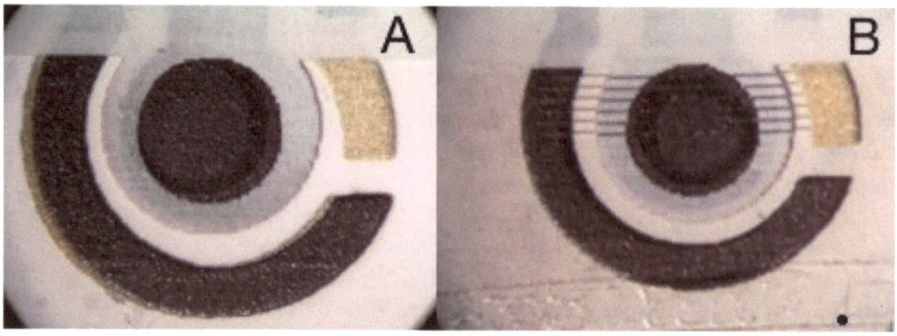

Figure 1. Optical micrographs of a freshly prepared, untreated Bi/PO_4/PSS electrode before (**A**) and after (**B**) the laser ablation ICP-MS experiment.

3. Results

3.1. Electrode Preparation

The concentration of bismuth in the pre-deposited $BiPO_4$ phase as a precursor of the active Bi^0 metal phase in screen-printed electrodes is known to have a profound effect on analytical performances. In both References [13,17], the best performances are shown by the lowest Bi concentration, with an ideal value around 1 mM in terms of peak shape and increased linearity at low analyte concentrations. For laser ablation studies, instead, the need to increase bismuth loadings was immediately evident, with respect to the best working low-loading electrode, in order to increase signal-to-noise ratios. Therefore, in this paper, bismuth film electrodes have been prepared and studied using an intermediate concentration of bismuth, namely 2 mM, trying to find a good compromise between analytical performance and S/N ratios in ICP-MS.

These bismuth film electrodes (BiFE) have been tested in the trace analysis of Cd and Pb under differential pulse stripping and the results are reported in Figure 2.

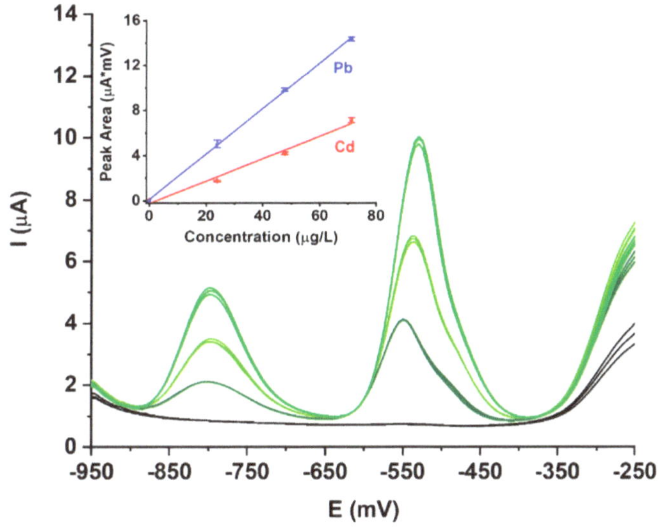

Figure 2. Differential pulse, anodic stripping voltammetry analyses of Cd and Pb solutions, and corresponding calibration curves (inset) at 0, 25, 50 and 75 μg/L on a working $Bi/PO_4/PSS$ electrode.

Baselines are very reproducible and free of noise; peaks are found at the expected potential, i.e., around −800 mV for Cd and around −530 mV for Pb. They show only a minor broadening observed as a shoulder, which is slightly more evident for Pb. For this reason, calibration curves are done on the integrated signals.

Calibration curves are very straight, and limits of linearity were not reached at the maximum concentration of 75 μg/L used in the calibration. Low-concentration nonlinearity is negligible and it can be seen by the small deviation of the X-intercept from zero (see inset in Figure 2). The non-linearity of bismuth film electrodes is well-known but does not significantly affect the limits of detection (LODs) for Cd and Pb which are in the μg/L range around 1.2 μg/L for Cd and 0.9 μg/L for Pb. These results compare well with those reported for $BiPO_4$-derived BiFEs.

The stability of the bismuth phase has been checked by running a replicate analytical scan of Pb and Cd at 15 μg/L and checking baselines after each repetition. Baselines are always stable and free of noise, and the integrated analytical signals are not significantly decreased after 20 replicates. This is in full agreement with those previously reported, and, therefore, this is not reported here.

The effect of anions different from chloride in the matrix has been studied, in order to understand whether residual phosphate ions after activation may affect the analytical performance. Adding nitrate ions during a repetitive scan indicated a small shift in the peak maxima of about +20 mV and a small, but significant, loss of signal of about 10% (Figure 3).

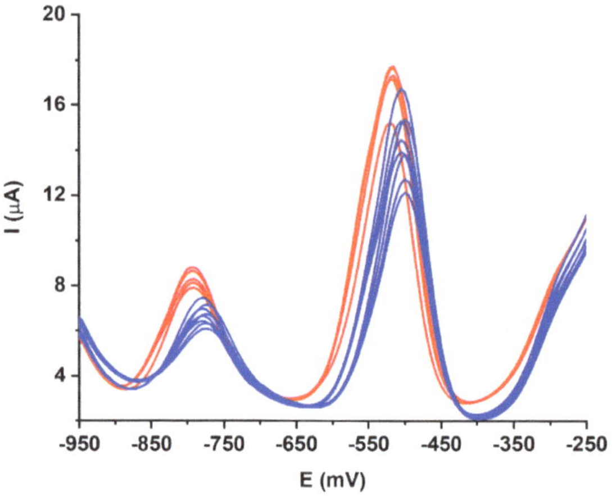

Figure 3. Effect of nitrate addition (blue trace) on Pb and Cd analysis in HCl at pH = 2 (red trace) on a working Bi/PO_4/PSS electrode.

This effect is then linked to a decrease in the analytical sensitivity, as evidenced in Figure 4 by comparing the calibration lines for Pb analysis in HCl or HNO_3 matrix solutions at pH = 2. A similar effect was found with sulphate ion (data not shown here).

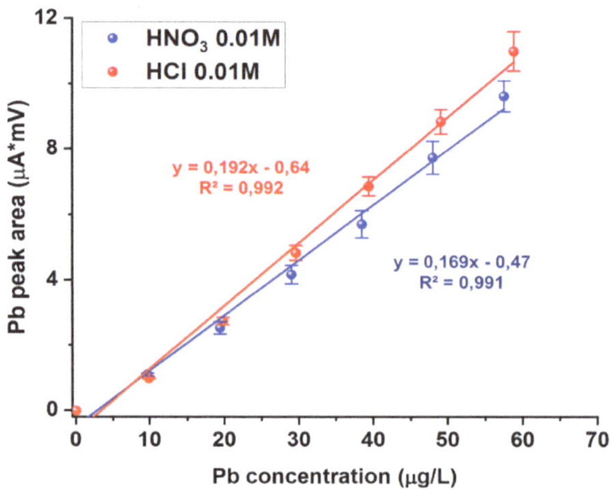

Figure 4. Comparison between calibration lines for Pb analysis in 0.01 M HCl (red line) and 0.01 M HNO_3 (blue line) on a working Bi/PO_4/PSS electrode.

Therefore, a detailed characterization was thought to be interesting, both for investigating the structure of the active bismuth layer as well as for fully understanding the chemistry occurring during the reductive activation of the bismuth salt layer upon activation and electrochemical analysis, to understand whether phosphate ions are fully removed prior the analysis, or are still playing a role in the electrochemical analysis.

3.2. Laser Ablation-ICP-MS Characterization

The characterization of bismuth film deposited onto the screen-printed working electrode was carried out by scanning the high-power UV laser beam through the *X*-axis of the dry electrode, while the ablated material is analyzed in real time by ICP-MS.

The analysis is first performed on an untreated screen-printed electrode right after the preparation of the $BiPO_4$ layer via drop-casting deposition onto the carbon working electrode. The first result on a single scan line is reported in Figure 5, where the traces corresponding at the two isotopes of ^{209}Bi and ^{31}P are compared. The signal is, however, a little bit noisy, and a 10-point moving average smoothing is routinely applied throughout the paper.

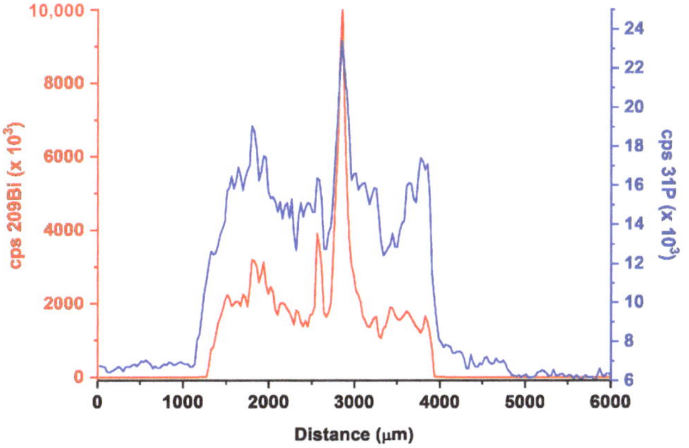

Figure 5. Comparison of the 209 Bi (red line) and 31P (blue line) laser ablation profiles located at 260 μm on a freshly prepared, untreated Bi/PO_4/PSS electrode.

It is immediately evident that the background signal for ^{209}Bi is very low, since no bismuth is expected to be present on the pristine (untreated) screen-printed electrode. The ^{209}Bi signal is almost perfectly confined at the center of the scan, which corresponds exactly to the borders of the carbon working electrode, with minimal spreading outside (please note in Figure 5 that the ^{209}Bi signal is appreciable for a length of 0.3 cm, i.e., the diameter of the working electrode). This may demonstrate that the careful choice of the drop-casting procedure may lead to the preferential formation of the $BiPO_4$ phase on the carbon surface. Obviously, this may be helped by the hydrophobic character of the plastic surface of the screen-printed electrode (SPE), preventing unwanted spreading during evaporation of the two subsequent layers of PSS and Na_2HPO_4 + $Bi(NO_3)_3$ solution.

Looking in more detail at the concentration profile of ^{209}Bi, the presence of peaks and valleys reveals the inhomogeneous nature of the Bi distribution over the working electrode. Since the intensity signal is essentially correlated with the thickness of the bismuth layer onto the surface, the irregular shape of the bismuth signal may suggest an uneven Bi distribution with the presence of much thicker aggregates giving the most intense signals.

It is interesting to compare the traces of ^{209}Bi and ^{31}P on the same scanning line. The phosphorous signal is much less intense than the bismuth one (please note that different scales were used for ^{209}Bi and ^{31}P signals in Figure 5): this is a result of the much lower sensitivity of m/z = 31 of phosphorous with respect to the m/z = 209 signal of bismuth, which is characterized by a very high sensitivity. Secondly, some phosphorous has to be pristinely present in the polymeric base of the electrode, leading to high background values. Both facts have the consequence that the ^{31}P signal shows low S/N ratios, and its plots are intrinsically less diagnostic than ^{209}Bi. Thirdly, it is noteworthy to evidence that the concentration profile of ^{31}P closely resembles that of ^{209}Bi, although it is less defined as said before. This is a clear indication that $BiPO_4$ is the prevailing chemical form present on the electrode surface, with only a residual presence of Bi-rich or P-rich phases which could be formed upon evaporating the aqueous impregnation solution of Na_2HPO_4 and $Bi(NO_3)_3$.

In Figure 6, the contouring plots of ^{209}Bi (Figure 6A) and ^{31}P (Figure 6B) are shown.

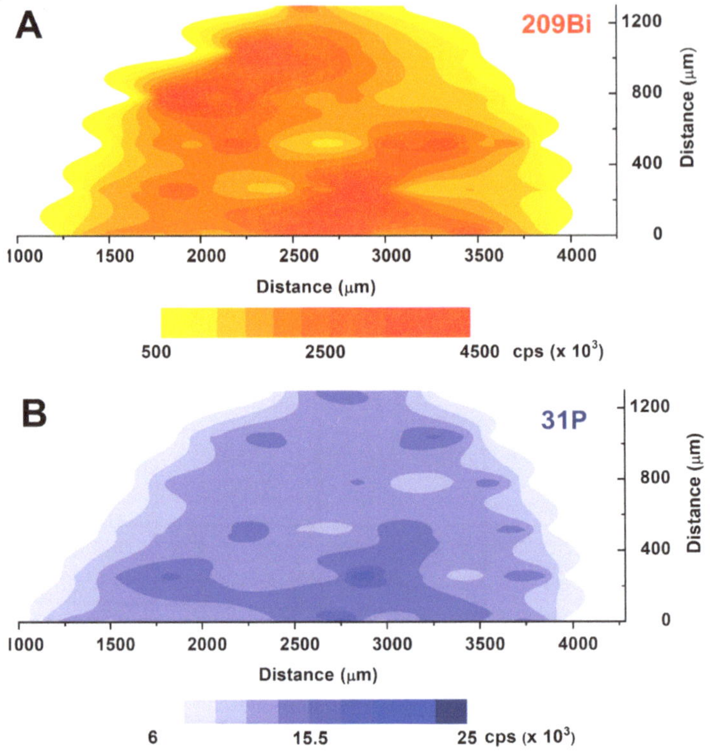

Figure 6. LA-ICP-MS contouring plots for 209Bi ((**A**), **top**) and 31P ((**B**), **bottom**) on a freshly prepared, untreated Bi/PO_4/PSS electrode.

The spatial resolution is, unfortunately, quite poor due to the spot size of the laser beam which is necessary for having a reasonable signal for phosphorous. In any case, it can be immediately evidenced the presence of large aggregates of a microcrystal, where bismuth and phosphorous seem to be simultaneously present in the form of $BiPO_4$ crystals.

The laser ablation ICP-MS results on the used electrodes after electrochemical activation are reported in Figure 7 and show a totally different chemical and structural situation.

In fact, the plot at $m/z = 209$ (Figure 7A) shows a pretty regular, spherical distribution of bismuth, with the maximum concentration on the center of the graphite surface of the working electrode. Instead, the phosphorous signal at $m/z = 31$ of Figure 7B is completely lost on the electrode surface, with some residual signals being only left at the borders, where the potential density during the electrochemical measurement is minimal. These results are very important, in that they confirm our previous hypothesis that no phosphorous should be left upon electrochemical activation of the $BiPO_4$ surface. The possible effect of anions different from chloride, that was shown to slightly decrease the analytical sensitivity and peak intensity, is to be independent from the residual presence of phosphate, or phosphorous species, after electrode activation, as it was previously supposed. This is to be related to two possible causes: (i) to the different diffusivity of the anions within the PSS layer, or (ii) to the complexing properties of chloride ions, with respect to nitrate or sulphate. This second hypothesis is thought to be the prevailing factor, as it may be evidenced by the positive potential shift in the peak position of Pb and Cd. In fact, bismuth surface activation is due to a reductive elimination reaction when negative potential is applied, leading to the reduction of Bi(III) to Bi(0) and the parallel elimination of phosphate ions for the redox balance. The experimental conditions that have been chosen then allow the zerovalent bismuth phase to slowly grow, reaching its maximum thickness at the center of the working electrode. In fact, the kinetics of this reductive elimination process is mainly governed by the solid–liquid electrolyte interphase. This regular shape of the bismuth surface may also explain the good reproducibility and stability of the stripping analyses.

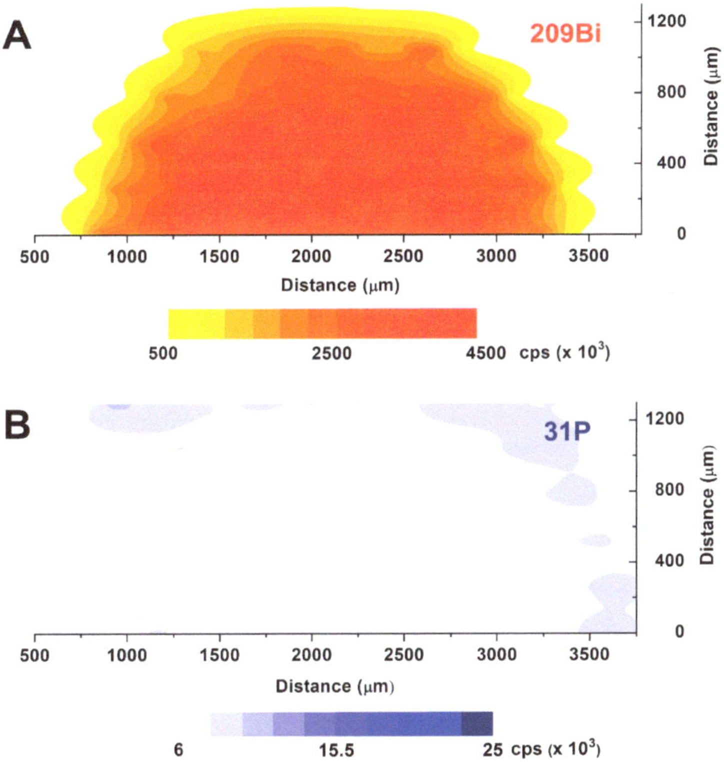

Figure 7. LA-ICP-MS contouring plots for 209Bi ((**A**), **top**) and 31P ((**B**), **bottom**) on a working Bi/PO_4/PSS electrode after electrochemical activation and real analysis.

4. Conclusions

Our preliminary investigation has shown how laser ablation ICP-MS may play an important role in the characterization of electrochemical surfaces. In fact, the electrode surface is simply dried in a nitrogen atmosphere before placing it under the laser compartment under helium at atmospheric pressure. In this way, no contamination from air is possible, and no restructuring of the mobile bismuth (or mercury) surface can occur, as would be taking place for conventional techniques, such as scanning electron microscopy, using high-energy electron or photon beams under high-vacuum conditions.

This investigation clearly evidenced how the phosphate counteranion is completely lost upon reductive activation in the electrochemical cell, leading to a very clean, "ball-type" bismuth active surface. The formation of such a bismuth surface may account for any anion-type interference, as well as of the non-linearities at low analyte concentrations with increasing bismuth loadings.

Finally, these screen-printed electrodes, based on Brainina's original ideas, are demonstrated to be viable alternatives to the in situ-prepared bismuth film electrodes. They are indefinitely stable in air, easily activated under the reductive conditions of the stripping electroanalytical experiment, and their final properties may be tuned by varying the bismuth concentration and deposition conditions.

Author Contributions: Conceptualization, C.D., D.M., A.P. and S.R.; methodology, C.D., D.M., A.P. and S.R.; formal analysis, C.D.; data curation, C.D., G.B. and D.S; writing—original draft preparation, C.D., G.B. and D.S.; All authors have read and agreed to the published version of the manuscript.

Funding: Financial support from Fondo di Ateneo per la Ricerca (FAR)—Università degli Studi dell'Insubria is greatly acknowledged.

Acknowledgments: The authors wish to thank Giuseppe Palleschi and his group at the University of Rome "Tor Vergata" for kindly providing the screen-printed electrodes and for scientific discussions. Research collaboration with Amel srl is greatly acknowledged.

Conflicts of Interest: The authors declare no conflict of interest.

References

1. Wang, J.; Lu, J.; Hocevar, S.B.; Farias, P.A.; Ogorevc, B. Bismuth-Coated Carbon Electrodes for Anodic Stripping Voltammetry. *Anal. Chem.* **2000**, *72*, 3218–3222. [CrossRef] [PubMed]
2. Hoĉevar, S.B.; Ogorevc, B.; Wang, J.; Pihlar, B. A Study on Operational Parameters for Advanced Use of Bismuth Film Electrode in Anodic Stripping Voltammetry. *Electroanalysis* **2002**, *14*, 1707–1712. [CrossRef]
3. Grabarczyk, M.; Adamczyk, M. Bismuth film electrode and chloranilic acid as a new alternative for simple, fast and sensitive Ge(IV) quantification by adsorptive stripping voltammetry. *RSC Adv.* **2018**, *8*, 15215. [CrossRef]
4. Lu, Y.; Liang, X.; Niyungeko, C.; Zhou, J.; Xu, J.; Tian, G. A review of the identification and detection of heavy metal ions in the environment by voltammetry. *Talanta* **2018**, *178*, 324–338. [CrossRef]
5. Yao, Y.; Wu, H.; Ping, J. Simultaneous determination of Cd(II) and Pb(II) ions in honey and milk samples using a single-wallet carbon nanohorns modified screen-printed electrochemical sensor. *Food Chem.* **2019**, *274*, 8–15. [CrossRef]
6. Liu, X.; Yao, Y.; Ying, Y.; Ping, J. Recent advances in nanomaterial- enabled screen-printed electrochemical sensors for heavy metal detection. *TrAC* **2019**, *115*, 187–202. [CrossRef]
7. Economou, A. Screen-printed electrodes modified with "Green" metals for electrochemical stripping analysis of Toxic Elements. *Sensors* **2018**, *18*, 1032. [CrossRef]
8. Lezi, N.; Vyskočil, V.; Economou, A.; Barek, J. Electroanalysis of Organic Compounds at Bismuth Electrodes: A Short Review. In *Sensing in Electroanalysis*; Kalcher, K., Metelka, R., Svancara, I., Vytras, K., Eds.; University Press Centre: Pardubice, Czech Republic, 2012; Volume 7, pp. 71–78.
9. Tyszczuk-Rotko, K.; Surowiec, K.; Szwagierek, A. Application of Eco-friendly Bismuth Film Electrode for the sensitive determination of Rutin. *Curr. Pharm. Anal.* **2018**, *14*, 571–577. [CrossRef]
10. Vladislavic, N.; Buzuk, M.; Brinic, S.; Byljac, M.; Bralic, M. Morphological characterization of ex-situ prepared bismuth film electrodes and their application in the electroanalytical determination of the biomolecules. *J. Sol. State Electr.* **2016**, *20*, 2241–2250. [CrossRef]
11. Mandil, A.; Amine, A. Screen-Printed Electrodes Modified by Bismuth Film for the Determination of Released Lead in Moroccan Ceramics. *Anal. Lett.* **2009**, *42*, 1245–1257. [CrossRef]
12. Calvo Quintana, J.; Arduini, F.; Amine, A.; Van Velzen, K.; Palleschi, G.; Moscone, D. Part two: Analytical optimisation of a procedure for lead detection in milk by means of bismuth-modified screen-printed electrodes. *Anal. Chim. Acta* **2012**, *736*, 92–99. [CrossRef] [PubMed]
13. Dossi, C.; Monticelli, D.; Pozzi, A.; Recchia, S. Exploiting chemistry to improve performance of screen-printed, bismuth film electrodes (SP-BiFE). *Biosensors* **2016**, *6*, 38. [CrossRef] [PubMed]
14. Bedin, K.C.; Mitsuyasu, E.Y.; Ronix, A.; Cazetta, A.L.; Pezoti, O.; Almeida, V.C. Inexpensive bismuth-film electrode supported on pencil-lead graphite for determination of Pb(II) and Cd(II) ions by anodic stripping voltammetry. *Int. J. Anal. Chem.* **2018**, 1473706. [CrossRef] [PubMed]

15. Kefala, A.; Economou, A.; Voulgaropoulos, A. A study of Nafion-coated bismuth-film electrodes for the determination of trace metals by anodic stripping voltammetry. *Analyst* **2004**, *129*, 1082–1090. [CrossRef] [PubMed]
16. Ning, J.; Luo, X.; Wang, F.; Huang, S.; Wang, J.; Liu, D.; Liu, D.; Chen, D.; Wei, J.; Liu, Y. Synergetic sensing effect of sodium carboxymethyl cellulose and bismuth on cadmium detection by differential pulse anodic stripping voltammetry. *Sensors* **2019**, *19*, 5482. [CrossRef] [PubMed]
17. Malakhova, N.A.; Stojko, N.Y.; Brainina, K.Z. Novel approach to bismuth modifying procedure for voltammetric thick film carbon containing electrodes. *Electrochem. Comm.* **2007**, *9*, 221–227. [CrossRef]
18. Vladislavic, N.; Buzuk, M.; Buljac, M.; Kozuk, S.; Bralic, M.; Brinic, S. Sensitive electrochemical determination of folic acid using ex-situ prepared bismuth film electrodes. *Croat. Chem. Acta* **2017**, *90*, 231–239. [CrossRef]
19. Bernardelly, J.K.B.; Lapolli, F.R.; Gomes da Silva Cruz, C.M.; Floriano, J.B. Determination of zinc and cadmium with characterized electrodes of carbon and polyurethane modified by a bismuth film. *Mat. Res.* **2011**, *14*, 366–371. [CrossRef]
20. Yu Kolosov, V.; Yushkov, A.A.; Veretennikov, L.M. Recrystallization and investigation of bismuth thin film IOP by means of electron beams in transmission electron microscopy. *J. Phys. Conf. Ser.* **2018**, *1115*, 032087. [CrossRef]
21. Tyszczuk-Rotko, K.; Metelka, R.; Vytřas, K.; Barczak, M.; Sadok, I.; Mirosław, B. A simple and easy way to enhance sensitivity of Sn(IV) on bismuth film electrodes with the use of a mediator. *Mon. Chem. Chem. Mon.* **2016**, *147*, 61–68. [CrossRef]
22. Xie, R.; Zhou, L.; Lan, C.; Fan, F.; Xie, R.; Tan, H.; Xie, T.; Zhao, L. Nanostructured carbon black for simultaneous electrochemical determination of trace lead and cadmium by differential pulse stripping voltammetry. *R. Soc. Open Sci.* **2018**, *5*, 180282. [CrossRef] [PubMed]
23. Mazzaracchio, V.; Tomei, M.R.; Cacciotti, I.; Chiodoni, A.; Novara, C.; Castellino, M.; Scordo, G.; Amine, A.; Moscone, D.; Arduini, F. Inside the different types of carbon black as nanomodifiers for screen-printed electrodes. *Electrochim. Acta* **2019**, *317*, 673–683. [CrossRef]
24. Milano, F.; Giotta, L.; Chirizzi, D.; Papazoglou, S.; Kryou, C.; De Bartolomeo, A.; De Leo, V.; Guascito, M.R.; Zergioti, I. Phosphate Modified Screen Printed Electrodes by LIFT Treatment for Glucose Detection. *Biosensors* **2018**, *8*, 91. [CrossRef]
25. Monticelli, D.; Di Iorio, A.; Ciceri, E.; Castelletti, A.; Dossi, C. Tree ring microanalysis by LA–ICP–MS for environmental monitoring: Validation or refutation? Two case histories. *Microchim. Acta* **2009**, *164*, 139–148. [CrossRef]
26. Amman, A.A. Inductively coupled plasma mass spectrometry (ICP MS): A versatile tool. *J. Mass Spectrom.* **2007**, *42*, 419–427. [CrossRef] [PubMed]
27. Arduini, F.; Amine, A.; Moscone, D.; Ricci, F.; Palleschi, G. Fast, sensitive and cost-effective detection of nerve agents in the gas phase using a portable instrument and an electrochemical biosensor. *Anal. Bioanal. Chem.* **2007**, *388*, 1049–1057. [CrossRef]
28. Arduini, F.; Cassisi, A.; Amine, A.; Ricci, F.; Moscone, D.; Palleschi, G. Electrocatalytic oxidation of thiocholine at chemically modified cobalt hexacyanoferrate screen-printed electrodes. *J. Electroanal. Chem.* **2009**, *626*, 66–74. [CrossRef]
29. Palchetti, I.; Laschi, S.; Mascini, M. Miniaturised stripping-based carbon modified sensor for in field analysis of heavy metals. *Anal. Chim. Acta* **2005**, *530*, 61–67. [CrossRef]
30. Monticelli, D.; Castelletti, A.; Civati, D.; Recchia, S.; Dossi, C. How to Efficiently Produce Ultrapure Acids. *Int. J. Anal. Chem.* **2019**, *2019*, 1–5. [CrossRef]
31. Monticelli, D.; Civati, D.; Giussani, B.; Dossi, C.; Spanu, D.; Recchia, S. A viscous film sample chamber for Laser Ablation Inductively Coupled Plasma—Mass Spectrometry. *Talanta* **2018**, *179*, 100–106. [CrossRef]

MDPI
St. Alban-Anlage 66
4052 Basel
Switzerland
Tel. +41 61 683 77 34
Fax +41 61 302 89 18
www.mdpi.com

Biosensors Editorial Office
E-mail: biosensors@mdpi.com
www.mdpi.com/journal/biosensors

www.ingramcontent.com/pod-product-compliance
Lightning Source LLC
LaVergne TN
LVHW072314090526
838202LV00019B/2278
* 9 7 8 3 0 3 6 5 1 3 2 2 5 *